Foreword

Cyanobacteria are ubiquitous organisms represented in almost all diversified habitats. The toxic metabolites of cyanobacteria, known as cyanotoxins, can cause severe adverse impacts on the health of humans and animals. Risks caused by Cyanotoxins are a serious concern, especially by anatoxins (ATX). To address these important issues, this book illustrates and provides user-friendly protocols for the most commonly experienced cyanotoxins, namely microcystins, cylindrospermopsins, and saxitoxins.

This book contains sections dedicated to protocols for studying cyanobacteria including: Part I—Covering sampling, isolation and cultivation methods of cyanobacteria; Part II—Focusing on toxicity analysis of cyanotoxins; Part III—Describing cyanotoxin extraction, detection, and quantification methods; and Part IV—Introducing and outlining some advanced methods for cyanotoxin detection and analysis. Collectively, 66 protocols are written by reputed researchers from the USA, the UK, China, Canada, Spain, Korea, India, Italy, Mexico, Hungary, Finland, Kenya, Sri Lanka, and Malaysia. The protocols are written in easily understandable language. The book summarizes basic principles, procedures from isolation and cultivation methods to advanced analytical methods of cyanobacterial sampling, and extraction, detection, and toxin analyses. As such, this manual will be an extremely valuable guide for young (undergraduate) and more advanced (postgraduate) researchers interested in acquiring knowledge about basic and cutting-edge methodologies in the field of cyanobacteria. Of particular importance, is the inclusion of protocols like Molecular Imprinting Polymer (MIP) techniques, molecular detection of *cyr* C genes, nanosensors in the detection of cyanotoxins and CRISPR techniques involved in finding genes in cyanobacterial genomes. These protocols will provide a unique resource to researchers in the advanced arena of cyanobacteria.

I congratulate all of the contributors for their painstaking efforts that provide simplified versions of these important protocols. Also, I congratulate the Editors, Prof. N. Thajuddin, Dr. A. Sankara narayanan, and Dr. D. Dhanasekaran for compiling this book that makes available these important techniques and approaches in such a wonderful way. Finally, a special thanks to Springer Nature Publishers for their tireless work that has made it possible to bring volumes of this type to the scientific community which have a tremendous value.

Rensselaer Polytechnic Institute
New York, USA

Sandra A. Nierzwicki-Bauer

Department of Biological Sciences
Darrin Fresh Water Institute
New York, USA

Preface

The *Protocols for Cyanobacteria Sampling and Detection of Cyanotoxin* provides detailed illustration on collection of cyanobacteria, extraction, quantification, and detection of cyanotoxins from *Anabaena, Anabaenopsis*, *Cylindrospermopsis*, *Microcystis*, *Nodularia, Nostoc, Schizothrix*, *Lyngbya, Raphidiopsis*, *Oscillatoria*, and *Planktothrix* in aquatic resources and the upcoming manual also sprawls its wings on narrating details about the procedures are pertaining to basic, advanced, and emerging techniques as well as in vitro, in vivo, and in silico approaches that are employed in the field of cyanotoxins.

The emerging trend of cyanotoxin due to cyanobacterial blooms affect the quality and safety of water, food, and also accumulated in marine animals. A steady increasing trend of cyanotoxins from small ponds to ocean is a serious concern and threat to humanity and animals in marine and freshwater bodies throughout the world. Reports reveal that there are five different types of toxins: (a) hepatotoxins (cylidrospermopsins, microcystins, and nodularins), (b) neurotoxins (anatoxins and saxitoxins), (c) dermatotoxins, (d) cytotoxins, and (e) irritant toxins (lipopolysaccharide). Further the research reports reiterated that a single genus of cyanobacteria can produce both hepato- and neurotoxins, for example, *Anabaena* can produce microcystins (hepatotoxin), anatoxin (neurotoxin), and β-*N*-methylamino L-alanine (BMAA). Detection and toxicity assay of cyanotoxins are the major concerns in the book. Further the researchers are highly concerned that when compared to bacterial and fungal toxins, cyanotoxins are neglected and less studied.

The target audiences include B.Sc., M.Sc., or Ph.D. in Microbiology, Food Technology, Aquaculture and Technology, Botany and Zoology, Microbial Biotechnology, Pharmacology, or Agriculture. The secondary audiences are faculty members, researchers, and food industrial peoples can be use the protocol for standardizing and quantification of cyanotoxins in food, water samples on commercial basis are the targeted readers of this manual. The book consists of four broad sections under which different titles are prepared based on the total content of 66 protocols. It includes the main division of collection, identification of toxic cyanobacteria from water, food samples, toxicity analysis, extraction, detection and quantification of cyanotoxin, and advanced methods in cyanotoxins. This edited protocol is contributed by reputed researchers from the USA, the UK, China,

Canada, Spain, Korea, Italy, Mexico, Hungary, Finland, Kenya, Sri Lanka, Malaysia, and India. This manual helps all the researchers who are interested in working on identification of toxic blooms of cyanobacteria in Mangrove Forest, estuarine, and salt pan. Detection of cyanotoxins using Bacterial, plants bioassays methods, detection of cylindrospermopsins, anatoxins, saxitoxins, guanitoxin from aquatic samples by HPLC and MS/MS, LC-MS, UHPLC-MS/MS, Rapid detection method of microcystin in water, qPCR assay in *sxtA* gene in saxitoxin producing cyanobacteria, molecular imprinting polymer in the detection of microcystin, cyclic peptide hepatotoxins in cyanobacteria using PCR, nodularin producing cyanolichen by Polymerase Chain Reaction (PCR), identification of microcystin, nodularin synthetase gene clusters in toxic cyanobacteria using anti SMASH pipeline, nanosensor devices on the detection of cyanotoxin.

We are extremely thankful to all the authors who contributed chapters and for their prompt and timely responses. We extend our earnest appreciation to Ms. Dr. Bhavik Sawhney, Senior Editor-Biomedicine, Springer Nature, and their team for their constant encouragement and help in bringing out the volume in the present form. I am also indebted to Springer Nature, the Authorities of Bharathidasan University, Tiruchirappalli, Tamil Nadu, India. We gratefully acknowledge the financial support from the Department of Biotechnology (DBT) for the establishment of the National Repository for Microalgae and Cyanobacteria - Freshwater (Phase II) (BT/PR29901/PBD26/694/2018/ dated 11.03.2023) and Sri Sathya Sai University for Human Excellence in Kalaburagi, Karnataka, India, for their support in the task of publishing this book.

Chennai, India — N. Thajuddin
Kalaburagi, India — A. Sankara narayanan
Tiruchirappalli, India — D. Dhanasekaran

Protocols for Cyanobacteria Sampling and Detection of Cyanotoxin

Editors
N. Thajuddin
B.S. Abdur Rahman Crescent Institute of Science & Technology (Deemed to be University)
Chennai, Tamil Nadu, India

Department of Microbiology, National Repository for Microalgae and Cyanobacteria - Freshwater (NRMC-F)
School of Life Sciences, Bharathidasan University
Tiruchirappalli, Tamil Nadu, India

A. Sankara narayanan
Department of Life Sciences
Sri Sathya Sai University for Human Excellence
Kalaburagi, Karnataka, India

D. Dhanasekaran
Department of Microbiology, National Repository for Microalgae and Cyanobacteria - Freshwater (NRMC-F)
Bharathidasan University
Tiruchirappalli, Tamil Nadu, India

ISBN 978-981-99-4513-9 ISBN 978-981-99-4514-6 (eBook)
https://doi.org/10.1007/978-981-99-4514-6

This Springer imprint is published by the registered company Springer Nature Singapore Pte Ltd.
The registered company address is: 152 Beach Road, #21-01/04 Gateway East, Singapore 189721, Singapore

N. Thajuddin •
A. Sankara narayanan • D. Dhanasekaran
Editors

Protocols for Cyanobacteria Sampling and Detection of Cyanotoxin

Springer

Contents

Part I Isolation and Cultivation Methods for Cyanobacteria

1 **Collection of Toxic Cynobacteria from Fish Samples** 3
Priyanka Jayam Rajendran and D. Dhanasekaran

2 **Identification of Toxic Blooms of Cyanobacteria in Freshwater (Lake, Ponds) Habitat** 7
Zati Sharip, Mohd Hafiz Zulkifli, and Mohd Nur Farhan Abdul Wahab

3 **Identification of Toxic Blooms of Cyanobacteria in Marine Water Habitat** 17
Simon Okomo Aloo, Denish Obiero Aloo, Selvakumar Vijayalakshmi, and Deog-Hwan Oh

4 **Identification of Toxic Blooms of Cyanobacteria in Mangrove Forest** 27
Silambarasan Tamil Selvan, Sanjivkumar Muthusamy, and Bhuvaneswari Santharam

5 **Identification of Toxic Blooms of Cyanobacteria in Estuarine Habitat** 35
Shree Kumari G R, Ranjitha Dhevi V. Sundar, Ajith Kumar K, and Lokesh Ravi

6 **Identification of Toxic Blooms of Cyanobacteria in Saltpan Habitat** 47
Ramasamy Thangaraj, Mohamed Arief Mohideen Pitchai, and N. Thajuddin

7 **Sampling and Identification of Toxic Cyanobacteria in the Landfill Leachate** 53
Vishnu Chitthan, Nisha Banu, N. Thajuddin, and D. Dhanasekaran

8 **Handling and Processing of Cyanobacterial Toxin Containing Pond and Lake Water Samples** 61
Tamalika Chakraborty, Sutripto Ghosh, Sumana Roy, and Dipanjan Mandal

9 **Isolation of Toxin-Producing Cyanobacteria from Aquatic Samples with *Anabaena*** 71
Ramasamy Thangaraj, SakthiThangavel Ramnath, and N. Thajuddin

10 **Isolation of Toxin Producing Cyanobacteria from Aquatic Samples—*Anabaenopsis* sp.** 77
M. Veerapagu, K. R. Jeya, and A. Sankara narayanan

11 **Isolation of Toxin Producing Cyanobacteria from Aquatic Samples with *Cylindrospermopsis* sp.** 85
O. Archana and Lokesh Ravi

12 **Isolation of Toxin Producing Cyanobacteria from Aquatic Sample with *Nodularia* sp.** 95
Gomathi Selvam, Ambikapathy Varatharaju, and Panneerselvam Annamalai

13 **Isolation of Toxin Producing Cyanobacteria from Aquatic Samples with *Nostoc* sp.** 99
Rajasekar Thirunavukkarasu, P. Gayathri, and V. Jeyanthi

14 **Isolation of Toxins Producing Cyanobacteria from Aquatic Sample with Schizothrix** 107
Vajiha Banu Habeeb Mohamed, Sumithra Pasumalaiarasu, and Kavitha Parangusadoss

15 **Isolation of Toxin Producing Cyanobacteria from Aquatic Samples with *Lyngbya*** 113
S. Nisha Nandhini, Kallingil Gopi Divya, and Nithaniyal Stalin

16 **Isolation of Toxin Producing Cyanobacteria from Aquatic Samples with *Raphidiopsis*** 119
Nithaniyal Stalin and S. Nisha Nandhini

17 **Isolation of Toxin Producing Cyanobacteria from Aquatic Samples: *Oscillatoria*** 123
S. Nisha Nandhini and Nithaniyal Stalin

18 **Isolation of Toxin Producing Cyanobacteria from Aquatic Samples: *Planktothrix*** 127
S. Nisha Nandhini and Nithaniyal Stalin

19 **Analysis of Cyanobacterial Species Composition and Variation Analysis Among Toxic Blooms** 131
Anu Sharon and Lokesh Ravi

20 **Confocal Laser Scanning Microscope (CLSM) Identification of Toxic Cyanobacteria from Algal Bloom** 141
Shakena Fathima Thajuddin, Fayaazuddin Thajuddin, N. Thajuddin, and D. Dhanasekaran

21 **Determination of Cyanotoxin Patterns in Strains and Species of Cyanobacteria** 149
Thangaraj Ramasamy, Nizhanthini Chellapandian, Ashok Muthumari, Ponmanickam Ponnirul, and N. Thajuddin

22 **Chemical and Toxicological Studies of the Toxic Freshwater Cyanobacterium: *Microcystis aeruginosa*** 155
K. R. V. Bandara, P. M. Samudrani, H. A. S. N. Abeysiri, G. Y. Liyanage, P. M. Sandupama, D. M. P. I. Dissanayake, M. M. Pathmalal, and S. A. Kulasooriya

23 **Chemical and Toxicological Studies of the Toxic Freshwater Cyanobacterium: *Anabaena Flos-aquae*** 161
Thangaraj Ramasamy, Nivetha Marimuthu, Naveena Kamalahasan, and N. Thajuddin

24 **Risk Assessment of Microcystin in Water Resources** 165
P. A. B. P. Abeyrathne, K. P. A. Imanthi, H. A. S. N. Abeysiri, G. Y. Liyanage, M. M. Pathmalal, and S. A. Kulasooriya

25 **Extraction and Quantification of BMAA from Water Samples** 173
Milena Bruno and Valentina Messineo

26 **Extraction and Quantification of BMAA from Fish Tissue** 181
Milena Bruno and Valentina Messineo

27 **Postmortem Sampling of *Cyanobacteria* in the Case of Suspected Drowning** 189
Viktor Soma Poór, Vivien Fejes, Dominika Szűcs, and Gábor Simon

Part II Toxicity Analysis of Cyanotoxin

28 **Detection of Cyanotoxins Using Vertebrate Bioassays in Mice** 197
K. T. Dilrukshi, H. A. S. N. Abeysiri, K. P. A. Imanthi, P. A. K. C. Wijerathna, P. M. Manage, and S. A. Kulasooriya

29 **Detection of Cyanotoxin Toxicity in Fish** 205
Ramakrishnan Kumar and D. Dhanasekaran

30 **Detection of Cyanotoxins Using Invertebrate Bioassays in Brine Shrimp** 211
V. Jeyanthi, Rajasekar Thirunavukkarasu, and Kumaran Subramanian

31 **Identification of Cyanobacterial Toxins from Lichen Thalli** 217
Ulla Kaasalainen

32 **Detection of Cyanotoxins Using Bacterial Bioassays** 227
Thenmozhli Geetha Saravanan, Kishore Kumar Annamalai, Abirami Baskaran, Ranjani Singaraj, Radhakrishnan Manikkam, and Manigundan Kaari

33 **Detection of Cyanotoxins: Bioassay Using Plants** 233
Neelam Mishra and Sheetal Gupta

34 **In Vitro Assay for Determining Cyanotoxin Using Serological Methods** 243
Sakshi R. Varshney, V. Z. Misbah Rehman, and Lokesh Ravi

35 **In Vitro Assay for Determining Cyanotoxin Using Cell Line Method: Hepatotoxicity (Cell Lines—HepG2, Caco-2, and V79)** . . . 253
H. A. S. N. Abeysiri, K. T. Dilrukshi, S. A. Kulasooriya, and P. M. Manage

36 **In Vitro Assay for Determining Cyanotoxin Using Cell Line Method: Neurotoxicity (Neuro-2a Neuroblastoma Cell)** 259
M. Haripriyaa and K. Suthindhiran

37 **Genotoxicity of Cyanotoxins** 265
H. D. D. Sadeepa, K. R. V. Bandara, G. Y. Liyanage, P. M. Manage, and S. A. Kulasooriya

38 **In Silico Analysis of Cyanotoxin Using Computational Tools** 273
Santhiya Panchalingam, Govindaraju Kasivelu, and Raja Muthuramalingam Thangavelu

Part III Cyanobacterial Toxin Extraction, Detection and Quantification

39 **Extraction of Cyanotoxins Using Filtration and Other Methods** . . . 285
K. P. A. Imanthi, P. A. B. P. Abeyrathne, P. A. K. C. Wijerathna, G. Y. Liyanage, P. M. Manage, and S. A. Kulasooriya

40 **Extraction of Cyanotoxins by Filtration and Other Methods from Biological Materials (Animal Tissue: Fish and Mussels)** 295
Milena Bruno, Valentina Messineo, Ilaria Di Marco Pisciottano, and Pasquale Gallo

41 **Extraction and Quantification of Nodularins from Fish Samples by LC-MS/MS** 303
Guhanraj Radhamanalan and D. Dhanasekaran

42 **Extraction and Quantification of Nodularins from Shrimp Samples by ELISA** 309
Guhanraj Radhamanalan and D. Dhanasekaran

43 **Extraction and Quantification of Cylindrospermopsins from Aquatic Samples by ELISA** 315
Milena Bruno and Valentina Messineo

44 **Extraction and Quantification of Anatoxins from Aquatic Samples by HPLC with Fluorescence Detection** 325
Gayathri Ganesan, Gomathi Selvam, Ambikapathy Varatharaju, and Panneerselvam Annamalai

45 **Extraction and Quantification of Saxitoxins from Aquatic Samples by LC-MS/MS** 329
Sharon Beaula, Manigundan Kaari, Ramesh Kumar, Kishore Kumar Annamalai, Radhakrishnan Manikkam, Gopikrishnan Venugopal, and Sakthivel Krishnan

46 **Extraction and Quantification of Guanitoxin from Aquatic Samples by HPLC-MS** 335
Angel Alisha, Manigundan Kaari, Masilamani Selvam, Kishore Kumar Annamalai, Radhakrishnan Manikkam, Gopikrishnan Venugopal, and Sakthivel Krishnan

47 **Detection of Anatoxins from Aquatic Samples by LC-MS** 339
A. Swedha, G. Kanimozhi, and A. Panneerselvam

48 **Detection of Guanitoxin from Aquatic Samples by HPLC/MS** 343
Stany B, Shatakshi Mishra, and Lokesh Ravi

49 **Detection of Cyanotoxin in Algal Food and Feed Supplements Using UHPLC-MS/MS** 353
Ajith Kumar K, Shree Kumari G R, and Lokesh Ravi

50 **Rapid Detection Method of Microcystin in Water** 363
G. Y. Liyanage, S. P. Manage, H. A. S. N. Abeysiri, P. M. Manage, and S. A. Kulasooriya

Part IV Advanced Methods in Cyanotoxins

51 **Screening of Polyketide Synthase Genes in the Anatoxin-a-Producing Cyanobacteria** 375
G. Y. Liyanage, S. A. Kulasooriya, S. P. Manage, D. M. P. I. Dissanayake, K. R. V. Bandara, Budhdhika Prabhath, and P. M. Manage

52 **NRPS and PKS Gene Analysis in Cyanobacteria Using PCR** 381
Nazar Reehana, Mohamed Yousuff Mohamed Imran, Mohamed Iqbal Mohamed Mahroop Raja, N. Thajuddin, and D. Dhanasekaran

53 **Identification of Microcystin, Nodularin Synthatase Gene Clusters in Toxic Cyanobacteria Using AntiSMASH Pipeline** 387
Fayaazuddin Thajuddin, Asraf Sithikka Rasheed, N. Thajuddin, and D. Dhanasekaran

54 qPCR Assay in *sxtA* Gene in Saxitoxin-Producing Cyanobacteria 397
Zhongkun Zhou, Yunhao Ma, Yuqing Niu, Juan Lu, Lixue Tu, and Peng Chen

55 qPCR Assay for the Detection of Hepatotoxigenic Cyanobacteria 403
Megha J K and Lokesh Ravi

56 Molecular Detection of *Mcy* Genes in Toxic Cyanobacteria: *Anabaena, Microcystis, Planktothrix* 421
H. A. S. N. Abeysiri, G. Y. Liyanage, K. Sirisena, S. A. Kulasooriya, and M. M. Pathmalal

57 Molecular Imprinting Polymer (MIP) in the Detection of Microcystin-LR 427
Yashaswini DM, Sathvik G, and Lokesh Ravi

58 Molecular Identification of the Cyclic Peptide Hepatotoxins in Cyanobacteria 435
F. S. Idroos, P. M. Manage, and S. A. Kulasooriya

59 Molecular Identification of Microcystin Synthetase Genes *mcyE* in Cyanobacteria Using PCR 443
G. Y. Liyanage, S. A. Kulasooriya, S. P. Manage, D. M. P. I. Dissanayake, and P. M. Manage

60 Molecular Identification of Cyr C Gene in Toxic Cyanobacteria in Photobiont 451
Lekhavani Ramesh, Karthiyayini Balakrishnan, Vishnu Raja Vijayakumar, and D. Dhanasekaran

61 Detection of Nodularin-Producing Cyanolichen by Polymerase Chain Reaction (PCR) 459
Vishnu Raja Vijayakumar, Lekhavani Ramesh, Karthiyayini Balakrishnan, and D. Dhanasekaran

62 Discrimination of Nodularin Toxin-Producing Strains from Non-producing Strains Using PCR 469
Janet Jan-Roblero, Juan A. Cruz-Maya, and Juan C. Cancino-Diaz

63 Detection of Cyanobacterial Cell Wall Components by Cellular Signaling Biosensors 475
K. R. V. Bandara, L. D. C. Peiris, D. M. P. I. Dissanayake, P. M. Manage, and S. A. Kulasooriya

64 Nanosensor Devices on the Detection of Cyanotoxin 483
Charu Gupta, Mahendra K. Gupta, Mir Sajad Rabani, and Madhu Gupta

65 In Silico Insights into the Cyanobacterial Genomes to Reveal Their Metabolic Interaction 489
Gargi Sen, Sandipan Ghosh, Indrani Sarkar, and Arnab Sen

66 CRISPR Gene Finding in the Genome of *Oscillatoria* sp. and *Lyngbya* sp. 497
Fayaazuddin Thajuddin, Peer Muhamed Noorani, Upaasna Sunilkumar, and N. Thajuddin

Index 503

Editors and Contributors

About the Editors

N. Thajuddin was born on August 6, 1962, at Thirumanilaiyur, the suburbs of Karur, Tamil Nadu, India.

Dr. Thajuddin received his PhD from Bharathidasan University, Tiruchirappalli, and DSc from Periyar University, Salem, Tamil Nadu. He had 1-year post-doctoral training at the Department of Biology, Rensselaer Polytechnic Institute, Troy, New York, USA, through the Department of Biotechnology (Govt. of India) long-term overseas award and presently assumed charge of Pro Vice-Chancellor, B.S. Abdur Rahman Crescent Institute of Science & Technology, (Deemed to be University), Vandalur, Chennai - 600 048, Tamil Nadu, India.

He has 32 years of teaching and 35 years of research experience in microbiology. Joining as a lecturer in 1991, Dr. Thajuddin rose to the level of Professor in the Department of Microbiology, Bharathidasan University. He held several administrative positions such as Senate Member; Syndicate Member; Department Head; Dean, Faculty of Science, Engineering and Technology; Chair and Coordinator, School of Life Sciences; Convener, World Class Curriculum Development Cell; Coordinator, DST–PURSE Program; Member Secretary, Institutional Bio-Safety Committee and Institutional Ethical Committee and several top-level committees of Government Funding agencies and several Universities.

On his research credit, he has published 383 research and review papers, including 3 papers in Nature Group Journals and 70 international collaborative publications with 52 universities and research institutions from 21 countries, 6 edited books, 3 lab manuals, and 1 textbook in Tamil under under different international publishers.

His publications received 10,223 citations with total impact factor—48, h index—47, and i10 index—169. He deposited 630 gene sequences in GenBank and developed barcodes for 6 fungi. Four Indian patents and two US patents were filed and published in the patent office Journal, of which one US patent was granted in April 2022. He signed five MoUs with the universities and institutions at Republic of Korea, P.R. China, and Taiwan and two companies in Tamil Nadu on phycoremediation of municipal waste leachate and textile dye effluents.

Dr. Thajuddin received major research grants to the tune of around 6.24 crores and received Rs. 18.39 crores (other than personal research grants) from UGC-Non-SAP, DST–PURSE, and FIST. The establishment of National Repository for Microalgae and Cyanobacteria–Freshwater (NRMC-F) at Bharathidasan University funded by the Department of Biotechnology, Govt. of India, is one of his key contributions to the nation.

Dr. Thajuddin guided 10 post-doctoral, 35 PhDs (including 10 candidates as co-guide), 31 MPhil scholars, and 131 MSc project students. Presently guiding 8 doctoral scholars. He organized several national and international level symposia, workshops, refresher courses, DST-PURSE programs; participated in several international and national level conferences with his research team and presented around 450 papers of which 52 papers received best paper awards. As a resource person, delivered Lead and Plenary lectures, Keynote, Inaugural, Valedictory and Graduation Day addresses in 230 conferences/symposia/seminars/workshops and chaired 38 technical sessions, besides 4 talks in All India Radio.

Dr. Thajuddin received several awards including Tamil Nadu Scientist Award (TANSA); Dr. G. S. Venkataraman Memorial Best Scientist Award and M.S. Swaminathan Leadership Award of National Academy of Biological Sciences; Elected Fellow of Royal Society in Biology—FRSB (London), Fellow of Mycological Society of India, Fellow of The Linnean Society— FLS (London), Fellow of Microbiologists

Society, India (FMBSI), and Fellow of National Academy of Biological Sciences (FNABS). His name is listed under Top 2% Scientist of the World in Microbiology, published by Stanford University, USA (2020 & 2021).

He visited the USA, the UK, KSA, Honk Kong, Singapore, Malaysia, Germany, South Korea, Sweden, Denmark, Finland, and Italy to disseminate his expertise and to keep himself abreast of the advanced techniques in the field of microbiology. For more details about him, visit: https://www.bdu.ac.in/schools/life-sciences/microbiology/docs/faculty/dr_n_thajuddin.pdf

A. Sankara narayanan is an Associate Professor in the Department of Life Sciences, Sri Sathya Sai University for Human Excellence, Kalaburagi, Karnataka, India, from June 2021 onwards. His current research focus is on fermented food products. He has published 15 books under different international publications and 45 chapters, 80 research articles in International and National journals of repute, guided 5 PhDs, 16 MPhil, scholars, and operated 5 minor funded projects in Microbiology.

From 2002 to 2015, he worked as an Assistant Professor and Head, Department of Microbiology, K.S.R. College of Arts and Science, Tiruchengode, Tamil Nadu and in August 2015–May 2021 he was associated with Uka Tarsadia University, Surat, Gujarat, India. He has been awarded with the Indian Academy of Sciences (IASc), National Academy of Sciences (NAS), and The National Academy of Sciences (TNAS) sponsored summer research fellowship for young teachers consecutively for 3 years and his name is included as a Mentor in DST-Mentors/Resource persons for summer/winter camps and other INSPIRE initiatives, Department of Science and Technology, Govt. of India, New Delhi. He is a grant reviewer in the British Society for Antimicrobial Chemotherapy (BSAC), UK. He is an active member in Editorial boards of various National and International Journals. He is Doctoral Committee member, Board of Study member in Microbiology in various colleges and Universities in India, and Reviewer in many International reputed Journals. One among of his edited book received Outstanding Academic Title (OAT) Award from one of the famous international publications in the year 2020. This OAT Award represents the highest caliber of scholarly titles, overall excellence in presentation and importance relative to other literature in the field. His name is included in the World Scientists and University Rankings for the Year 2023; ranking list is released by AD Scientific Index.

D. Dhanasekaran is working as an Professor, Department of Microbiology, School of Life Sciences, Bharathidasan University, Tiruchirappalli, India. He has experience in fields of Actinobacteriology and Phycology. His current research focus is on actinobacteria, microalgae, fungi, and mushroom for animal and human health improvement. He has been awarded UGC-Raman Post-Doctoral Fellowship and worked in the Department of Molecular, Cellular and Biomedical Sciences, University of New Hampshire, Durham, USA. He has been awarded elected Fellow of Linnean Society, London, UK, and INSA Visiting Scientist Fellow by Indian National Science Academy, New Delhi, Government of India. He is serving as State President of Microbiologists Society, India, for Tamil Nadu (2022–2023). He is serving as Extraordinary Professor in the Food Security and Safety at the Mafikeng Campus of the North-West University, South Africa. He has qualified the Tamil Nadu State Eligibility Test (SET) for Lectureship in Life Science. He is subject expert in the Board of Study Chairman and member in Microbiology for seven institutions and designed the structure of MSc, MPhil Microbiology Curriculum. He is External member in Research Advisory Committee (RAC), University College of Engineering, Bharathidasan Institute of Technology (BIT) campus, Anna University, Tiruchirappalli, Food Safety Officers' examiner, Tamil Nadu, Food Safety and Drug Administration Department. He is currently working on microbiome profiling using metagenomics approaches and genome and comparative analysis of actinobacterial biosynthetic gene clusters. He has filed 2 Indian patents. He has deposited 113 nucleotide sequences 16Sr RNA, 17 draft genome, 5 metagenomes in GenBank, 5 bioactive compounds in PubChem, published 116 research and review articles including 1 paper in Nature Group Journal Scientific Report and 61 book chapters. He has H-index of 35 with total citation of 3802 as per Google scholar. He has edited 14 books international publisher. He has guided 12 PhD candidates, 1 Post-Doctoral Fellow and organized 10 national and international level symposia, conference, workshop, and training programs. He received major research projects from the Department of Biotechnology, University Grant Commission, Indian Council for Medical research, and International Foundation for Science, Sweden. He is member in editorial boards in National, International Journals, Guest editor in Frontiers in Cellular and Infection Microbiology, Genes, Doctoral Committee member, Board of Study member in Microbiology and Reviewer in the scientific Journals and research grants. As per the reports of Indian Journal of Experimental Biology, 51, 2013, Dr. D. Dhanasekaran is rated in the second position among the top five institutions in the field of actinobacteria research in India.

Contributors

P. A. B. P. Abeyrathne Centre for Water Quality and Algae Research, Department of Zoology, University of Sri Jayewardenepura, Nugegoda, Sri Lanka
Department of Zoology, Centre for Water Quality and Algae Research, University of Sri Jayewardenepura, Nugegoda, Sri Lanka

H. A. S. N. Abeysiri Centre for Water Quality and Algae Research, Department of Zoology, University of Sri Jayewardenepura, Nugegoda, Sri Lanka
Faculty of Graduate Studies, University of Sri Jayewardenepura, Nugegoda, Sri Lanka

Angel Alisha Department of Biotechnology, Sathyabama Institute of Science and Technology, Chennai, Tamil Nadu, India

Denish Obiero Aloo Department of Microbiology, CSI International Ltd., Nairobi, Kenya

Simon Okomo Aloo Department of Food Science and Biotechnology, College of Agriculture and Life Sciences, Kangwon National University, Chuncheon, South Korea

Kishore Kumar Annamalai Natural Product Discovery Laboratory, Centre for Drug Discovery and Development, Col. Dr. Jeppiaar Research Park, Sathyabama Institute of Science and Technology, Chennai, Tamil Nadu, India

Panneerselvam Annamalai PG and Research Department of Botany, A.V.V.M Sri Pushpam College (Autonomous), Poondi, (Affiliated to Bharathidasan University, Tiruchirappalli), Thanjavur, Tamil Nadu, India

O. Archana Department of Botany, St Joseph's University, Bengaluru, Karnataka, India

Karthiyayini Balakrishnan Department of Microbiology, School of Life Sciences, Bharathidasan University, Tiruchirappalli, Tamil Nadu, India

K. R. V. Bandara Centre for Water Quality and Algae Research, Department of Zoology, University of Sri Jayewardenepura, Nugegoda, Sri Lanka
Faculty of Graduate Studies, University of Sri Jayewardenepura, Nugegoda, Sri Lanka

Nisha Banu Department of Microbiology, Bharathidasan University, Tiruchirappalli, Tamil Nadu, India

Abirami Baskaran Centre for Drug Discovery and Development, Sathyabama Institute of Science and Technology, Chennai, Tamil Nadu, India

Sharon Beaula Department of Biotechnology, Sathyabama Institute of Science and Technology, Chennai, Tamil Nadu, India

Milena Bruno FAST, Core Facilities, National Institute of Health, Rome, Italy

Juan C. Cancino-Diaz Departamento de Microbiología, Escuela Nacional de Ciencias Biológicas, Instituto Politécnico Nacional, Mexico City, Mexico

Tamalika Chakraborty Division of Life Science, Guru Nanak Institute of Pharmaceutical Science and Technology, Kolkata, West Bengal, India

Nizhanthini Chellapandian Department of Microbiology, Ayya Nadar Janaki Ammal College, Sivakasi, Tamil Nadu, India

Peng Chen School of Pharmacy, Lanzhou University, Lanzhou, People's Republic of China

Vishnu Chitthan Department of Microbiology, Bharathidasan University, Tiruchirappalli, Tamil Nadu, India

Juan A. Cruz-Maya Unidad Profesional Interdisciplinaria en Ingeniería y Tecnologías Avanzadas, Instituto Politécnico Nacional, Mexico City, Mexico

D. Dhanasekaran Department of Microbiology, School of Life Sciences, Bharathidasan University, Tiruchirappalli, Tamil Nadu, India

K. T. Dilrukshi Centre for Water Quality and Algae Research, Department of Zoology, University of Sri Jayewardenepura, Nugegoda, Sri Lanka
Faculty of Graduate Studies, University of Sri Jayewardenepura, Nugegoda, Sri Lanka

Kallingil Gopi Divya Department of Pharmacognosy, Siddha Central Research Institute, Chennai, Tamil Nadu, India

Aakash Welgamage Don School of Pharmacy and Life Sciences, Robert Gordon University, Aberdeen, Scotland

Vivien Fejes Department of Forensic Medicine, Medical School, University of Pécs, Pécs, Hungary

Pasquale Gallo Coordination of Chemistry Department, Experimental Zoo prophylactic Institute of Southern Italy, Naples, Italy

Gayathri Ganesan PG and Research Department of Botany, A.V.V.M Sri Pushpam College (Autonomous), (Affiliated to Bharathidasan University, Thanjavur, Tamil Nadu, India

P. Gayathri Department of Medical Laboratory Technology (MLT), School of Allied Health Sciences (AHS), Chennai, Tamil Nadu, India

Sandipan Ghosh Department of Botany, University of North Bengal, Siliguri, West Bengal, India

Sutripto Ghosh Division of Life Science, Guru Nanak Institute of Pharmaceutical Science and Technology, Kolkata, West Bengal, India

K. Govindaraju Centre for Ocean Research (DST-FIST Sponsored Centre), Sathyabama Institute of Science and Technology, Chennai, India

Charu Gupta School of Studies in Microbiology, Jiwaji University, Gwalior, Madhya Pradesh, India

Madhu Gupta Microbiology Research Lab, School of Studies in Botany, Jiwaji University, Gwalior, Madhya Pradesh, India

Mahendra K. Gupta Microbiology Research Lab, School of Studies in Botany, Jiwaji University, Gwalior, Madhya Pradesh, India

Sheetal Gupta Department of Botany, St Joseph's University, Bengaluru, Karnataka, India

M. Haripriyaa Marine Biotechnology and Bioproducts lab, Department of Biomedical Sciences, School of Biosciences and Technology, Vellore Institute of Technology, Vellore, India

F. S. Idroos Centre for Water Quality and Algae Research, Department of Zoology, University of Sri Jayewardenepura, Nugegoda, Sri Lanka

K. P. A. Imanthi Centre for Water Quality and Algae Research, Department of Zoology, University of Sri Jayewardenepura, Nugegoda, Sri Lanka
Faculty of Graduate Studies, University of Sri Jayewardenepura, Nugegoda, Sri Lanka

Mohamed Yousuff Mohamed Imran National Repository for Microalgae and Cyanobacteria-Freshwater (NRMC-F) (Sponsored by the DBT, Government of India), Department of Microbiology, School of Life Sciences, Bharathidasan University, Tiruchirappalli, Tamil Nadu, India

Janet Jan-Roblero Departamento de Microbiología, Escuela Nacional de Ciencias Biológicas, Instituto Politécnico Nacional, Mexico City, Mexico

K. R. Jeya Department of Biotechnology, Government Arts and Science College (Women), Perambalur, Tamil Nadu, India

V. Jeyanthi Department of Biotechnology, SRM Arts and Science College, Chennai, Tamil Nadu, India

Manigundan Kaari Centre for Drug Discovery and Development, Sathyabama Institute of Science and Technology, Chennai, Tamil Nadu, India
Natural Product Discovery Laboratory, Centre for Drug Discovery and Development, Col. Dr. Jeppiaar Research Park, Sathyabama Institute of Science and Technology, Chennai, Tamil Nadu, India

Ulla Kaasalainen Finnish Museum of Natural History, University of Helsinki, Helsinki, Finland

Naveena Kamalahasan Department of Microbiology, Ayya Nadar Janaki Ammal College, Sivakasi, Tamil Nadu, India

G. Kanimozhi PG and Research Department of Microbiology, A.V.V.M. Sri Pushpam College (Autonomous) (Affiliated to Bharathidasan University), Thanjavur, Tamil Nadu, India

Sakthivel Krishnan Crop Protection Section, ICAR-Indian Institute of Oilseed Research, Hyderabad, Telangana, India

S. A. Kulasooriya National Institute of Fundamental Studies, Kandy, Sri Lanka

S. Kumaran Department of Biotechnology, SRM Arts and Science College, Chennai, Tamil Nadu, India

Ajith Kumar K Department of Life Sciences, Kristu Jayanti College (Autonomous), Bengaluru, Karnataka, India

Ramakrishnan Kumar Department of Microbiology, School of Life Sciences, Bharathidasan University, Tiruchirappalli, Tamil Nadu, India

Ramesh Kumar Department of Biotechnology, Sathyabama Institute of Science and Technology, Chennai, Tamil Nadu, India

G. Y. Liyanage Department of Aquatic Bio resources, Faculty of Urban and Aquatic Bioresources, University of Sri Jayewardenepura, Nugegoda, Sri Lanka
Centre for Water Quality and Algae Research, Department of Zoology, University of Sri Jayewardenepura, Nugegoda, Sri Lanka

Juan Lu School of Pharmacy, Lanzhou University, Lanzhou, People's Republic of China

P. M. Manage Department of Zoology, Centre for Water Quality and Algae Research, University of Sri Jayewardenepura, Nugegoda, Sri Lanka

Faculty of Graduate Studies, University of Sri Jayewardenepura, Nugegoda, Sri Lanka

S. P. Manage Centre for Water Quality and Algae Research, Department of Zoology, University of Sri Jayewardenepura, Nugegoda, Sri Lanka

Dipanjan Mandal Division of Pharmacology, Guru Nanak Institute of Pharmaceutical Science and Technology, Kolkata, West Bengal, India

Radhakrishnan Manikkam Centre for Drug Discovery and Development, Sathyabama Institute of Science and Technology, Chennai, Tamil Nadu, India
Natural Product Discovery Laboratory, Centre for Drug Discovery and Development, Col. Dr. Jeppiaar Research Park, Sathyabama Institute of Science and Technology, Chennai, Tamil Nadu, India

Nivetha Marimuthu Department of Microbiology, Ayya Nadar Janaki Ammal College, Sivakasi, Tamil Nadu, India

Yunhao Ma School of Pharmacy, Lanzhou University, Lanzhou, People's Republic of China

J. K. Megha Department of Botany, School of Life Science, St Joseph's University, Bengaluru, Karnataka, India

Valentina Messineo FAST, Core Facilities, National Institute of Health, Rome, Italy

Neelam Mishra Department of Botany, St Joseph's University, Bengaluru, Karnataka, India

Shatakshi Mishra Department of Biotechnology, School of Bio Sciences and Technology, Vellore Institute of Technology, Vellore, India

Vajiha Banu Habeeb Mohamed PG and Research Department of Microbiology, Jamal Mohamed College (Autonomous), Tiruchirappalli, Tamil Nadu, India

Ashok Muthumari Department of Microbiology, Ayya Nadar Janaki Ammal College, Sivakasi, Tamil Nadu, India

Sanjivkumar Muthusamy Department of Microbiology, K.R. College of Arts and Science, Kovilpatti, Tamil Nadu, India

S. Nisha Nandhini Department of Botany, St. Joseph's College (Autonomous) (affiliated to Bharathidasan University), Tiruchirappalli, Tamil Nadu, India

Yuqing Niu School of Pharmacy, Lanzhou University, Lanzhou, People's Republic of China

Peer Muhamed Noorani Department of Microbiology, School of Life Sciences, Bharathidasan University, Tiruchirappalli, Tamil Nadu, India

Deog-Hwan Oh Department of Food Science and Biotechnology, College of Agriculture and Life Sciences, Kangwon National University, Chuncheon, South Korea

A. Panneerselvam PG and Research Department of Microbiology, A.V.V.M. Sri Pushpam College (Autonomous) (Affiliated to Bharathidasan University), Thanjavur, Tamil Nadu, India

Kavitha Parangusadoss PG and Research Department of Microbiology, Srimad Andavan Arts and Science College (Autonomous), Tiruchirappalli, Tamil Nadu, India

D. Pasindu Centre for Water Quality and Algae Research, Department of Zoology, University of Sri Jayewardenepura, Nugegoda, Sri Lanka

Sumithra Pasumalaiarasu PG and Research Department of Microbiology, Srimad Andavan Arts and Science College (Autonomous), Tiruchirappalli, Tamil Nadu, India

M. M. Pathmalal Centre for Water Quality and Algae Research, Department of Zoology, University of Sri Jayewardenepura, Nugegoda, Sri Lanka
Faculty of Graduate Studies, University of Sri Jayewardenepura, Nugegoda, Sri Lanka

Dinithi Peiris Department of Zoology, Faculty of Applied Sciences, University of Sri Jayewardenepura, Nugegoda, Sri Lanka

Ilaria Di Marco Pisciottano Coordination of Chemistry Department, Experimental Zoo prophylactic Institute of Southern Italy, Naples, Italy

Mohamed Arief Mohideen Pitchai Department of Microbiology, Ayya Nadar Janaki Ammal College, Sivakasi, Tamil Nadu, India

Ponmanickam Ponnirul Department of Zoology, Ayya Nadar Janaki Ammal College, Sivakasi, Tamil Nadu, India

Viktor Soma Poór Department of Forensic Medicine, Medical School, University of Pécs, Pécs, Hungary

Budhdhika Prabhath Centre for Water Quality and Algae Research, Department of Zoology, University of Sri Jayewardenepura, Nugegoda, Sri Lanka
Faculty of Graduate Studies, University of Sri Jayewardenepura, Nugegoda, Sri Lanka

Mir Sajad Rabani Microbiology Research Lab, School of Studies in Botany, Jiwaji University, Gwalior, Madhya Pradesh, India

Guhanraj Radhamanalan Department of Microbiology, School of Life Sciences, Bharathidasan University, Tiruchirappalli, Tamil Nadu, India

Mohamed Iqbal Mohamed Mahroop Raja PG and Research Department of Microbiology, Jamal Mohamed College (Autonomous) [Affiliated to Bharathidasan University], Tiruchirappalli, Tamil Nadu, India

T. Rajasekar Centre for Drug Discovery and Development, Sathyabama Institute of Science and Technology, Chennai, India

Priyanka Jayam Rajendran Department of Microbiology, School of Life Sciences, Bharathidasan University, Tiruchirappalli, India

Thangaraj Ramasamy Department of Microbiology, Ayya Nadar Janaki Ammal College, Sivakasi, Tamil Nadu, India

Lekhavani Ramesh Department of Microbiology, School of Life Sciences, Bharathidasan University, Tiruchirappalli, Tamil Nadu, India

SakthiThangavel Ramnath Department of Microbiology, Ayya Nadar Janaki Ammal College, Sivakasi, Tamil Nadu, India

Asraf Sithikka Rasheed School of Life Sciences, Bharathidasan University, Tiruchirappalli, Tamil Nadu, India

Lokesh Ravi Department of Food Technology, Faculty of Life and Allied Health Sciences, MS Ramaiah University of Applied Sciences, Bengaluru, India

Nazar Reehana PG and Research Department of Microbiology, Jamal Mohamed College (Autonomous) [Affiliated to Bharathidasan University], Tiruchirappalli, Tamil Nadu, India

V. Z. Misbah Rehman Department of Chemical Engineering, CPDMED, TBI, Indian Institute of Science, Bengaluru, Karnataka, India
Hoyasan Labs Pvt. Ltd., Bengaluru, Karnataka, India

Sumana Roy Division of Pharmaceutics, Guru Nanak Institute of Pharmaceutical Science and Technology, Kolkata, West Bengal, India

H. D. D. Sadeepa Centre for Water Quality and Algae Research, Department of Zoology, University of Sri Jayewardenepura, Nugegoda, Sri Lanka

P. M. Samudrani Centre for Water Quality and Algae Research, Department of Zoology, University of Sri Jayewardenepura, Nugegoda, Sri Lanka

P. M. Sandupama Centre for Water Quality and Algae Research, Department of Zoology, University of Sri Jayewardenepura, Nugegoda, Sri Lanka

A. Sankara narayanan Department of Life Sciences, Sri Sathya Sai University for Human Excellence, Kalaburagi, Karnataka, India

Bhuvaneswari Santharam Department of Biochemistry, K.R. College of Arts and Science, Kovilpatti, Tamil Nadu, India

P. Santhiya Centre for Ocean Research (DST-FIST Sponsored Centre), Sathyabama Institute of Science and Technology, Chennai, India

Thenmozhli Geetha Saravanan Centre for Drug Discovery and Development, Sathyabama Institute of Science and Technology, Chennai, Tamil Nadu, India

Indrani Sarkar Bioinformatics Facility, University of North Bengal, Siliguri, West Bengal, India

G. Sathvik Department of Botany, School of Life Science, St Joseph's University, Bengaluru, Karnataka, India

Gomathi Selvam PG and Research Department of Botany, A.V.V.M Sri Pushpam College (Autonomous) (Affiliated to Bharathidasan University, Thanjavur, Tamil Nadu, India

Masilamani Selvam Department of Biotechnology, Sathyabama Institute of Science and Technology, Chennai, Tamil Nadu, India

Arnab Sen Bioinformatics Facility, University of North Bengal, Siliguri, West Bengal, India
Department of Botany, University of North Bengal, Siliguri, West Bengal, India
Biswa Bangla Genome Centre, University of North Bengal, Siliguri, West Bengal, India

Gargi Sen Bioinformatics Facility, University of North Bengal, Siliguri, West Bengal, India

Thajuddin Shakena Fathima Department of Microbiology, School of Life Sciences, Bharathidasan University, Tiruchirappalli, Tamil Nadu, India

Zati Sharip Research Centre for Water Quality and Environment, National Water Research Institute of Malaysia, Seri Kembangan, Malaysia

Anu Sharon Department of Botany, School of Life Sciences, St Joseph's University, Bengaluru, Karnataka, India

Shree Kumari G R Department of Biomedical Sciences, School of Bio Sciences and Technology, Vellore Institute of Technology, Vellore, India

Gábor Simon Department of Forensic Medicine, Medical School, University of Pécs, Pécs, Hungary

Ranjani Singaraj Centre for Drug Discovery and Development, Sathyabama Institute of Science and Technology, Chennai, Tamil Nadu, India

K. Sirisena Department of Environmental Technology, Faculty of Technology, University of Colombo, Colombo, Sri Lanka

Nithaniyal Stalin Department of Botany, St. Joseph's University, Bengaluru, Karnataka, India

Stany B Department of Biotechnology, School of Bio Sciences and Technology, Vellore Institute of Technology, Vellore, India

Ranjitha Dhevi V. Sundar Department of Biotechnology, School of Bio Sciences and Technology, Vellore Institute of Technology, Vellore, India

Upaasna Sunilkumar School of Life Sciences, Bharathidasan University, Tiruchirappalli, Tamil Nadu, India

K. Suthindhiran Marine Biotechnology and Bioproducts lab, Department of Biomedical Sciences, School of Biosciences and Technology, Vellore Institute of Technology, Vellore, India

A. Swedha PG and Research Department of Microbiology, A.V.V.M. Sri Pushpam College (Autonomous), (Affiliated to Bharathidasan University), Thanjavur, Tamil Nadu, India

Dominika Szűcs Department of Forensic Medicine, Medical School, University of Pécs, Pécs, Hungary

Silambarasan Tamil Selvan Department of Microbiology, School of Allied Health Sciences, VIMS Hospital Campus, Vinayaka Missions Research Foundation (DU), Salem, Tamil Nadu, India

Fayaazuddin Thajuddin School of Life Sciences, Bharathidasan University, Tiruchirappalli, Tamil Nadu, India

N. Thajuddin B.S. Abdur Rahman Crescent Institute of Science & Technology (Deemed to be University), Chennai, Tamil Nadu, India
Department of Microbiology, National Repository for Microalgae and Cyanobacteria - Freshwater (NRMC-F), School of Life Sciences, Bharathidasan University, Tiruchirappalli, Tamil Nadu, India

Ramasamy Thangaraj Department of Microbiology, Ayya Nadar Janaki Ammal College, Sivakasi, Tamil Nadu, India

Raja Muthuramalingam Thangavelu Department of Plant Pathology and Ecology, The Connecticut Agricultural Experiment Station, New Haven, CT, USA

Rajasekar Thirunavukkarasu Centre for Drug Discovery and Development, Sathyabama Institute of Science and Technology, Chennai, Tamil Nadu, India

Lixue Tu School of Pharmacy, Lanzhou University, Lanzhou, People's Republic of China

Ambikapathy Varatharaju PG and Research Department of Botany, A.V.V.M Sri Pushpam College (Autonomous) (Affiliated to Bharathidasan University), Thanjavur, Tamil Nadu, India

Sakshi R. Varshney Department of Chemical Engineering, CPDMED, TBI, Indian Institute of Science, Bengaluru, Karnataka, India
Hoyasan Labs Pvt. Ltd., Bengaluru, Karnataka, India

M. Veerapagu Department of Biotechnology and Microbiology, National College (Autonomous), Tiruchirappalli, Tamil Nadu, India

Gopikrishnan Venugopal Natural Product Discovery Laboratory, Centre for Drug Discovery and Development, Col. Dr. Jeppiaar Research Park, Sathyabama Institute of Science and Technology, Chennai, Tamil Nadu, India

Vishnu Raja Vijayakumar Department of Microbiology, School of Life Sciences, Bharathidasan University, Tiruchirappalli, Tamil Nadu, India

Selvakumar Vijayalakshmi Department of Food Science and Biotechnology, College of Agriculture and Life Sciences, Kangwon National University, Chuncheon, South Korea

Mohd Nur Farhan Abd Wahab Research Centre for Water Quality and Environment, National Water Research Institute of Malaysia, Seri Kembangan, Malaysia

P. A. K. C. Wijerathna Department of Zoology, Centre for Water Quality and Algae Research, University of Sri Jayewardenepura, Nugegoda, Sri Lanka
Faculty of Graduate Studies, University of Sri Jayewardenepura, Nugegoda, Sri Lanka

D. M. Yashaswini Department of Botany, School of Life Science, St Joseph's University, Bengaluru, Karnataka, India

Zhongkun Zhou School of Pharmacy, Lanzhou University, Lanzhou, People's Republic of China

Mohd Hafiz Zulkifli Research Centre for Water Quality and Environment, National Water Research Institute of Malaysia, Seri Kembangan, Malaysia

Part I

Isolation and Cultivation Methods for Cyanobacteria

Collection of Toxic Cynobacteria from Fish Samples

1

Priyanka Jayam Rajendran and D. Dhanasekaran

Abstract

Blue-green algae, sometimes referred to as cyanobacteria, are capable of producing cyanotoxins. The majority of cyanobacteria may be found in lakes and oceans, where they are extensively scattered. There are prerequisites for their exponential reproduction, which results in flowering. Cyanotoxins produced by flowering cyanobacteria have been shown to harm and even kill humans and animals at high enough quantities. The strong growth of cyanobateria blooms is favoured by eutrophic freshwater habitats. The rising eutrophication of several lakes, lagoons, and reservoirs has made possible for cyanobacteria to frequently predominate in phytoplankton. In addition, cyanotoxins can develop in several animal species, including fish, shellfish, and crustaceans, and can be hazardous when eaten. Many fish species, however, are unable to avoid ingesting these poisonous organisms in habitats like small lakes and aquaculture ponds. The main objective of this protocol is outlining the collection process of toxic cyanobacteria from fish.

Keywords

Cyanobacteria blooms · Lakes · Lagoons · Fish ponds · Fish tissue · Fish net · Microcystins and nodularins

P. J. Rajendran (✉)
Department of Microbiology, School of Life Sciences, Bharathidasan University, Tiruchirappalli, India

D. Dhanasekaran
Department of Microbiology, National Repository for Microalgae and Cyanobacteria - Freshwater (NRMC-F), Bharathidasan University, Tiruchirappalli, Tamil Nadu, India

N. Thajuddin et al. (eds.), *Protocols for Cyanobacteria Sampling and Detection of Cyanotoxin*, https://doi.org/10.1007/978-981-99-4514-6_1

1.1 Introduction

Ancient photosynthetic prokaryotic microbes known as cyanobacteria (blue-green algae) need light, water, carbon, and inorganic materials to grow. Since the Precambrian era, they have existed for more than 3.5 billion years. They contributed to its creation by releasing oxygen into the atmosphere. The main sources of nitrogen and phosphate enrichment in freshwaters and coastal marine waterbodies are agricultural fertilisers, farm animal waste, everyday detergents, and industrial or urban waste. Moreover, sunlight and warm temperatures encourage the bloom of microorganisms that produce oxygen. Cyanotoxins are secondary metabolites made by poisonous organisms. Cyanotoxins, which are secondary metabolites, are produced by toxic cyanobacteria. Toxic cyanotoxins are secondary metabolites made by cyanobacteria. Cyclic peptides, alkaloids, and lipopolysaccharides are the three categories used to classify cyanotoxins, whereas hepatotoxins, neurotoxins, and cytotoxins are used to classify them according to their toxicity. Cyclic peptides, alkaloids, and lipopolysaccharides are the three main categories of cyanotoxins, whereas hepatotoxins, neurotoxins, and cytotoxins are used to categorise them according to their toxicity. The two most common cyanotoxins are microcystins (MCYST) and nodularins (NODLN). A peptide and polyketide synthetases enzyme complex makes up their non-ribosomal structure in the ecosystems frequently experience eutrophication during warm months, particularly in spring and early autumn (Kalaitzidou et al. 2016). It has long been recognised that cyanobacterial hepatotoxins and neurotoxins can kill animals and sicken livestock quickly. Since the beginning of the twentieth century, mass intoxications have been recorded. These occurrences have contributed to the realisation that cyanobacterial water blooms may be dangerous. A special amino acid called Adda is what gives cyanotoxins called microcystins, which are cyclic heptapeptides, their poisonous activity. It has been reported that there are more than 70 microcystin analogues. The literature on microcystins in marine environments is sparse, despite the fact that they have been extensively investigated in fresh and brackish waters. Seawater from the Atlantic Ocean, Caribbean Sea, Pacific Ocean, Indian Ocean, Arabian Sea, Marmara Sea, and Mediterranean Sea has been found to contain microcystins, which have been found to be present in the marine environment. Swimming in sea waters where a cyanobacteria bloom had taken place has been linked to numerous incidents of skin and eye irritation, diarrhoea, asthma attacks, and allergic responses. Moreover, poisonous marine cyanobacteria pose a risk to divers, water sports instructors, cleaners and maintainers of coastlines, fish and shellfish farms, and fisherman. Moreover, it has been noted that poisonous marine cyanobacteria can poison crabs as well as zooplankton by preventing them from feeding on them (Filliousis et al. 2015).

1.2 Materials

1. Fish
2. Pure methanol
3. Fishing net
4. Hexane
5. Neutral-buffered formalin
6. Toxic cyanobacteria
7. Centrifuge
8. Sonicator

1.3 Methods

1. Cyanobacteria accumulate on the surface layer of waterbodies during the "water bloom phenomenon" and produce scum, foams, or mats. Then toxic species would develop cyanotoxin and other secondary metabolites, which cause risks to people, animals, and the environment.
2. A complete description of the sampling site should be the initial step in isolating toxic cyanobacteria, together with the site's address and description,
3. It is important to record variables including light intensity, temperature, pH, and salinity.
4. Collect the fish using fish net from the pond which contains blooms.
5. Instantaneously use 10% neutral-buffered formalin to preserve all samples.
6. Collect freeze-dried muscle tissue to look for any cyanotoxin that might have accumulated.
7. Sonicate the samples are treated with pure methanol (3 mL), centrifuge at 3300 rpm for 10 min, and collect the supernatants in a fresh microfuge tube.
8. Add hexane (6 mL) to the supernatants to remove interfering lipids and discard it after phase separation.
9. Add 10% methanol (5 mL) and evaporate the samples, the liquid is sonicated, and then it is processed through reversed-phase cartridges to retain the cyanotoxin.
10. The next step is to elute the cartridges with 100% methanol (3 mL), after which the leftovers are evaporated and then dissolved in 75% aqueous methanol (200 mL) (Drobac et al. 2016).
11. LC-MS/MS is used to detect cyanotoxin in fish muscle samples. (Drobacbacković et al. 2021).

1.4 Conclusion

Oxygen in a body of water may be depleted during the decomposition of a harmful algal bloom brought on by cyanobacteria. Fish and other aquatic creatures may become oxygen-starved and die as a result, which is when this situation happened.

Additionally, animals with gills may struggle to breathe in the water due to high levels of cyanobacteria (Centers for Disease Control and Prevention 2017). Certain strains of the cyanobacteria *Microcystis, Anabaena, Oscillatoria, Nodularia, Nostoc, Cylindrospermopsin*, and *Umezakia* produce hepatotoxins that are harmful to the liver. *Cylindroapermopsis raciborskii* species cyanobacteria can also release harmful alkaloids that can harm humans by producing renal illness or gastrointestinal complaints. There may still be toxin-producing cyanobacteria of these species, as not all of them are known to be toxic. The main way that people are exposed to cyanobacteria toxins is through drinking or swimming in water. (Drobacbacković et al. 2021).

References

Centers for Disease Control and Prevention (2017) Harmful algal bloom (hab)-associated illness. https://www.cdc.gov/habs/general.html. 2018

Drobac D, Tokodi N, Lujić J, Marinović Z, Subakov-simić G, Dulić T, Važić T, Nybom S, Meriluoto J, Codd GA, Svirčev Z (2016) Cyanobacteria and cyanotoxins in fishponds and their effects on fish tissue. Harmful Algae 55:66–76

Drobacbacković D, Tokodi N, Marinović Z, Lujić J, Dulić T, Simić SB, Đorđević NB, Kitanović N, Šćekić I, Urbányi B, Meriluoto J (2021) Cyanobacteria, cyanotoxins, and their histopathological effects on fish tissues in Fehérvárcsurgó reservoir, Hungary. Environ Monit Assess 193(9):554

Filliousis M, Petridou G, Economou E, Theodoridis A, Aggelidis P (2015) Isolation of toxic marine cyanobacteria and detection of microcystins in thermaikos gulf in Central Macedonia in Greece. Haicta, 832–841

Kalaitzidou M, Petridou E, Economou V, Theodoridis A, Angelidis P (2016) Toxic cyanobacteria. A biological hazard for animal and public health: a review. Air Water Borne Dis 5:130

Identification of Toxic Blooms of Cyanobacteria in Freshwater (Lake, Ponds) Habitat

2

Zati Sharip, Mohd Hafiz Zulkifli,
and Mohd Nur Farhan Abdul Wahab

Abstract

Cyanobacteria are blue-green algae that can develop a massive population to form thick biomass at the surface and outcompete other algae. The massive population releasing harmful toxins becomes habitat toxic for other living beings. These harmful algal blooms (HAB) are a major concern as it reduces water quality, causing fish kills and deteriorates overall lake ecosystem health. However, various factors are causing cyanobacterial blooms and impacting water resources. Lack of identification techniques have restricted the monitoring of toxic blooms in freshwater habitats, especially in developing countries. As such their occurrences remain unnoticed until it becomes severe. This chapter summarizes the identification of toxic blooms of cyanobacteria in freshwater habitat in the field and laboratory, respectively.

Keywords

Ecosystem health, eutrophication · Harmful algal bloom (HAB) · Phytoplankton · Transparency · Waterbodies

2.1 Introduction

Cyanobacteria, also known as blue green algae, are planktonic bacteria within the group of Cyanophyta that are naturally occurring in freshwater habitats such as lakes, natural and artificial, and ponds. Cyanobacteria, depending on species, can

Z. Sharip (✉) · M. H. Zulkifli · M. N. F. Abdul Wahab
Research Centre for Water Quality and Environment, National Water Research Institute of Malaysia, Seri Kembangan, Malaysia
e-mail: zati@nahrim.gov.my

N. Thajuddin et al. (eds.), *Protocols for Cyanobacteria Sampling and Detection of Cyanotoxin*, https://doi.org/10.1007/978-981-99-4514-6_2

form water bloom when they reproduce rapidly under suitable environmental conditions (Graham et al. 2008; Graham et al. 2009). High nutrient especially phosphorus and nitrogen cause eutrophication and promote dense of cyanobacteria concentration or bloom (Lad et al. 2022). The widely used definition of a bloom is when the population of cyanobacteria are large enough with high cell densities greater than 20,000 cells per millilitre (Graham et al. 2008). Dissimilar to other phytoplankton blooms, Cyanobacterial blooms are toxic as they can produce toxins that are harmful to human, animals, and aquatic biota. These cyanotoxins toxins can cause skin rashes, nausea, abdominal pain, diarrhoea, gastrointestinal distress, numbness, fatigue, and paralysis besides can affect liver and brain function (Lad et al. 2022). There are four major group of cyanotoxins namely cytotoxins, hepatotoxins, neurotoxins, and dermatotoxins (Kaur et al. 2021). The most common toxic cyanobacteria genera in freshwater habitats are *Microcystis*, *Oscillatoria*, *Anabaena, Planktothrix*, and *Aphanocapsa* (Chorus and Welker 2021) while the most frequently detected cyanotoxins are microcystins, nodularins, cylindrospermopsins, and neurotoxins (Buratti et al. 2017; WHO 2003).

Noxious cyanobacterial blooms decrease water clarity, hinder recreational activities, and create offensive tastes and odours in drinking water (Lad et al. 2022).

Identification of cyanobacteria usually requires enumeration and microscopic analysis in the laboratory. However, few key water characteristics that can be used as physical and visual indicators for identification of harmful algal blooms in freshwater habitat (IEPA 2021; California Water Board 2022). Three main characteristics in identifying potentially toxic algal bloom in freshwater habitat are appearance, colour, and odour. Appearance are physical indictor where the cyanobacteria blooms can look like (1) coloured paint or soup spilled into the water, (2) thick puffy scum, bubbling or spit-like floating foams on the surface of the water, (3) a guise in the form of coloured crust along the shoreline or swirling colours beneath the surface of the water (IEPA 2021) (Fig. 2.1). Other physical indicators are colour where cyanobacteria blooms often look green, blue-green, green-brown, or red, and odour where bloom exhibits distinctive smell, sometimes described as rotting plant, gasoline, septic, or fishy (IEPA 2021; California Water Board 2022). Use the jar and stick tests in the field to differentiate cyanobacterial bloom from other green algae and macrophyte bloom (Interstate Technology Regulatory Council 2021). These visual methods are the easiest to perform and operative to be gathered by mass through citizen science activities.

In situ measurement for identification of cyanobacteria bloom usually involves lowering Secchi disk to measure water transparency and using probes to determine chlorophyll or photosynthetic blue-green pigment. These methods of physical measurement are not only easy to perform but also more effective than the other visualization methods. Low clarity, and high total chlorophyll and phycocyanin results, indicating the presence of water blooming. However, total chlorophyll measurement does not differentiate eukaryotic phytoplankton from cyanobacteria within phytoplankton assemblages. Phycocyanin (PC) measurement using fluorescence is becoming a suitable indicator for detecting the amount of cyanobacteria in water samples (Chorus and Welker 2021). The World Health Organization set

Fig. 2.1 Detection of toxic bloom in freshwater habitat (Image by Sharip)

moderate risk level of cyanobacterial bloom when the density reached 100,000 cyanobacteria cells/ml with the presence of cyanotoxins (Buratti et al. 2017). This level is equivalent to a concentration of 90 μg/L and 30 g/L of Chlorophyll *a* and PC, respectively (Buratti et al. 2017; Brient et al. 2008). Biomass analysis of water samples in the laboratory provides more accurate quantification of Chlorophyll *a* pigment (APHA 2005). Three standard methods that are widely used to determine photosynthetic pigments chlorophyll in the laboratory are the spectrophotometric, fluorometric and high performance liquid chromatographic (HPLC) techniques (Wetzel 2001).

Collection of water samples by bottles or Van Dorn samplers for microscopy analysis of phytoplankton species. Microscopic method is useful to determine the density of the plankton and identify cyanobacterial species in water samples (Chorus and Welker 2021). Quantitative analysis of cyanobacteria is usually determined by counting cells or Neubauer chambers using standard or inverted microscopes (Wetzel 2001).

2.2 Materials

- Camera
- Depth sounder
- Distilled water
- Global positioning system (GPS)
- Gloves
- Ice chests
- Jar bottle
- Microscope (standard or inverted compound)
- Opaque sample bottle or containers
- Lugol's solution (solution of 5% Iodine and 10% Potassium Iodide)
- Secchi disk, 8-inch (20 cm) circular disk with alternating black and white quadrants
- Sensor probe (Multi-parameter or single)
- Van Dorn sampler

2.3 Methods

2.3.1 Visual and Physical Method to Detect HAB (Chorus and Welker 2021; IEPA 2021; California Water Board 2022; Interstate Technology Regulatory Council 2021)

1. Observe the conditions of the water, specifically its appearance, and colour (Fig. 2.2). Any discoloration of water (such as blue, green, brown, or white) or the presence of surface scum, green streak, or crust indicate the presence of cyanobacteria bloom. If possible, photograph the water environment by using a camera for records.
2. Wear gloves and scoop up a handful of the bloom and let the water drain between your fingers. Cyanobacterial blooms are microscopic; they will disintegrate and cannot be lifted. Alternatively, use a stick to lift a strand or mat of algae and check whether the strand or mat is hanging or breaking (stick test). Collect water into a jar and test for a few minutes whether the algal floats or sinks into the bottom; Cyanobacteria have buoyancy to float to the top from gas vesicles (Jar test).
3. Observe the presence of distinctive smell or musty odour such as rotting plant, septic or fishy, or gasoline.

Fig. 2.2 Characteristics of cyanobacterial bloom (Image by Sharip)

2.3.2 Physical Measurement to Determine HAB (Chorus and Welker 2021; APHA 2005; Wetzel 2001)

2.3.2.1 Measuring Transparency (Chorus and Welker 2021; Wetzel 2001)

1. Prepare materials and devices as in Fig. 2.3. Go to the field and use the hand-held Global Positioning System (GPS) to determine sampling points. Measure water depth by using a depth sounder.
2. Lower a Secchi disk slowly into the water in the shade of a boat (or a jetty or landing ground) until the disk ceases to be visible from the surface and lift until it just reappears. Lift and lower the disk several times to determine accurately the depth at which the disk disappears and reappears.
3. Take the average of the two measurements and record the depth of transparency of water.
4. When immersing the Secchi disk, some of the patchy bloom will move away from the spot, wait a few seconds until they have redistributed before taking the measurement.

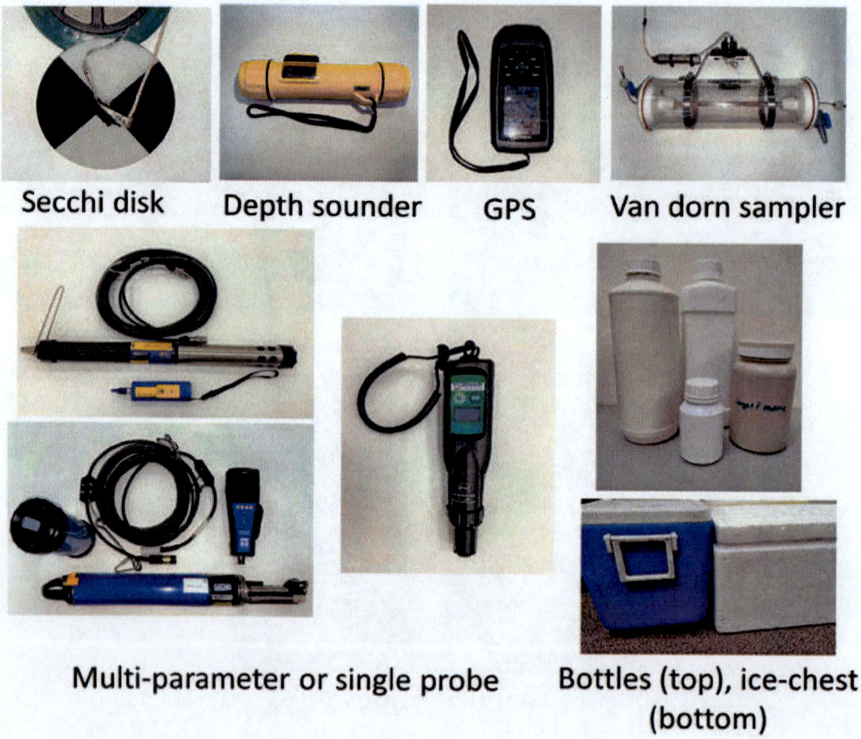

Fig. 2.3 Material and devices related to cyanobacterial toxic bloom

5. Do the procedure on clear skies or partly cloudy conditions (i.e. sun not covered by clouds).
6. Do not wear sunglasses while doing the procedures as that may distort the reading.

2.3.2.2 Measuring In Situ Chlorophyll and Phycocyanin (Chorus and Welker 2021)

1. Prepare materials and instruments as in Fig. 2.3. Calibrate all instruments prior to field measurement. Go to the field and use the GPS to determine sampling points.
2. Note sampling position, time, and meteorological conditions.
3. Deploy a single or multi-parameter probe (to measure Chlorophyll and Phycocyanin at surface or as a function of depth) and stabilize it for 5 min (Fig. 2.4).
4. If deploying a multi-parameter probe, pull the probe to the surface quickly and deploy it back to each station's maximum depth. Pull the probe from the depth to surface to record the value of temperature, pH, dissolved oxygen, and Chlorophyll *a* as a depth function. Report average of values collected from 0 to 10-m depth. If the probe has memory function, record the measured values at different intervals by lowering the sonde slowly from near surface until it reach the 0.5 m off bottom (depth profiling).
5. If using single fluoroprobe, lower the probe into water at surface and record the value of Chlorophyll or Phycocyanin.
6. Clean the devices with sterilized water before using at other sampling location (Fig. 2.5).

Fig. 2.4 Measuring water depth and transparency (Image by Hafiz and Farhan)

Fig. 2.5 *In situ* measurement and collection of water samples (Image by Farhan and Hafiz)

2.3.2.3 Collecting Samples for Chlorophyll *a* and Phytoplankton Enumeration (Chorus and Welker 2021; APHA 2005; Wetzel 2001)

1. Collect samples from the surface using bottles or sampling devices such as Van Dorn (Fig. 2.3) or Kemmerer-type sampler.
2. Clean bottles and sampling devices and rinse with distilled water prior to sampling.
3. In the presence of surface blooms or scums, grab samples by scooping water with a wide-mouthed bottle from or near the surface. A grab sample is useful for situational analysis such as determining maximum density of cyanobacteria. It is preferable to collect multiple samples to consider dispersed bloom or spatial heterogeneity of the freshwater habitat. Pool repetitive samples from one depth.
4. At deeper water where stratification is present, take integrated samples by lowering a sampling device in water at discrete depth of water column. Once lowered at the desired depth, release the messenger to trip the closing mechanism of the sampler. Take individual samples from several depths to collect integrated samples of the water column.
5. Pour water samples from the sampling device into labelled bottles. Use separate bottle samples for water quality analysis (such as Chlorophyll *a* and cyanobacterial toxin) and phytoplankton identification.
6. For phytoplankton enumeration, preserve the samples by using Lugol's solution; add 3 mL Lugol's solution to 1-L sample.
7. Store in ice chests or refrigerator (in the dark) and transport to the laboratory for enumeration.

2.3.3 Laboratory Analyses (APHA 2005; Wetzel 2001)

2.3.3.1 Chlorophyll *a* Analysis (APHA 2005; Wetzel 2001)

1. Filter water samples by using glass fibre filters to collect algae.
2. Extract Chlorophyll *a* from cells collected on filter using solvent (such as acetone, ethanol, or methanol) in a subdued light environment.
3. Extract absorbance is determined by using spectrophotometer, fluorometer or HPLC. Note: Please refer specific chapter for cyanobacterial toxin analysis.

2.3.3.2 Microscopic Analysis (APHA 2005)

1. Concentrate water samples in the laboratory such as using sedimentation or membrane filtration technique.
2. Use either a standard or an inverted compound microscope to examine/enumerate samples.
3. Place the samples in a cylindrical settling chamber with a thin, clear glass bottle. Allow a certain period for organisms to settle. Count organisms in the settling chamber and record the concentration or density results using similar reporting units (cells/units of organisms per mL).
4. Use standard references or current literature (possibly within the geographical region of the sampling area) to identify cyanobacterial species.
5. If possible, photograph organisms using a microscopic camera for records.

Acknowledgement The first authors acknowledge the National Water Research Institute of Malaysia for management support.

References

APHA (2005) Standard methods for the examination of water and waste water, 22nd edn. American Public Health Association (APHA), Washington, DC

Brient L, Lengronne M, Bertrand E, Rolland D, Sipel A, Steinmann D et al (2008) A phycocyanin probe as a tool for monitoring cyanobacteria in freshwater bodies. J Environ Monit 10(2): 248–255

Buratti FM, Manganelli M, Vichi S, Stefanelli M, Scardala S, Testai E, Funari E (2017) Cyanotoxins: producing organisms, occurrence, toxicity, mechanism of action and human health toxicological risk evaluation. Arch Toxicol 91(3):1049–1130

California Water Board (2022) https://mywaterquality.ca.gov.visualguide_fs.pdf. Accessed 15 Oct 2022

Chorus I, Welker M (2021) Toxic cyanobacteria in water: a guide to their public health consequences, monitoring and management. Taylor & Francis

Graham JL, Loftin KA, Ziegler AC, Meyer MT (2008) Guidelines for design and sampling for cyanobacterial toxin and taste-and-odor studies in lakes and reservoirs. U.S. Geological Survey Scientific Investigations Report 2008–5038

Graham JL, Loftin KA, Kamman N (2009) Monitoring recreational freshwaters. Lakelines 29:18–24

IEPA (2021) Identifying cyanobacteria blooms. www.illinois.gov/epa/topics/water-quality/monitoring/algal-bloom. Accessed 15 Oct 2022

Interstate Technology Regulatory Council (2021) Strategies for preventing and managing harmful cyanobacteria blooms (HCB-1). https://itrcweb.org/itrcwebsite/teams/training/hcb. Accessed 20 Oct 2022

Kaur S, Srivastava A, Ahluwalia AS, Mishra Y (2021) Cyanobacterial blooms and cyanotoxins: occurrence and detection. In: Algae. Springer, pp 339–352

Lad A, Breidenbach JD, Su RC, Murray J, Kuang R, Mascarenhas A et al (2022) As we drink and breathe: adverse health effects of microcystins and other harmful algal bloom toxins in the liver, gut, lungs and beyond. Life (Basel) 12(3):418

Wetzel RG (2001) Limnology: lake and river ecosystems, 3rd edn. Academic, San Diego

WHO (2003) Guidelines for safe recreational water environments, vol. 1. World Health Organization, Malta

Identification of Toxic Blooms of Cyanobacteria in Marine Water Habitat

3

Simon Okomo Aloo, Denish Obiero Aloo,
Selvakumar Vijayalakshmi, and Deog-Hwan Oh

Abstract

Cyanobacteria are a diverse group of photosynthetic bacteria found in various environments, including marine habitats. They play a vital role in marine ecosystems, and some species produce bioactive compounds with biotechnological applications. However, cyanobacteria can also produce toxins called cyanotoxins, which can be harmful to human health and the environment. Cyanobacterial blooms that produce these toxins have been responsible for numerous cases of illness and death in humans and animals. Therefore, it is important to monitor and manage cyanobacterial blooms to minimize risks associated with these toxins. This chapter outlines the methods used to identify toxic blooms of cyanobacteria in marine water habitats, including visual inspection, Polymerase chain reaction (PCR), and enzyme-linked immunosorbent assay (ELISA). The procedures for sample collection, examination under a microscope, DNA amplification using PCR, and detection of specific toxins using ELISA are described. A combination of these methods can provide a

S. O. Aloo · D.-H. Oh
Department of Food Science and Biotechnology, College of Agriculture and Life Sciences, Kangwon National University, Chuncheon, South Korea

D. O. Aloo
Department of Microbiology, CSI International Ltd., Nairobi, Kenya

S. Vijayalakshmi (✉)
Department of Food Science and Biotechnology, College of Agriculture and Life Sciences, Kangwon National University, Chuncheon, South Korea

Centre Of Molecular Medicine and Diagnostics (COMManD), Department of Biochemistry, Saveetha Dental College & Hospitals, Saveetha Institute of Medical and Technical Sciences, Saveetha University, Chennai, India

N. Thajuddin et al. (eds.), *Protocols for Cyanobacteria Sampling and Detection of Cyanotoxin*, https://doi.org/10.1007/978-981-99-4514-6_3

comprehensive understanding of the distribution, toxicity, and potential impacts of toxic cyanobacteria blooms on the ecosystem and public health.

Keywords

Toxic blooms · Cyanobacteria · Marine water habitat · Visual inspection · PCR · ELISA

3.1 Introduction

Cyanobacteria are a group of photosynthetic bacteria widely distributed in various environments, including freshwater, marine, and terrestrial habitats (Whitton and Potts 2000). They are commonly referred to as blue-green algae, even though they are not true algae (Sánchez-Baracaldo 2015). Marine cyanobacteria are diverse organisms that play an essential role in marine ecosystems. While the specific types of cyanobacteria present in a marine habitat can vary depending on temperature, salinity, and nutrient availability, several common genera are typically found in marine environments. One common genus of marine cyanobacteria is *Prochlorococcus*, which is believed to be one of the most abundant photosynthetic organisms on the planet (Scanlan et al. 2009). This genus is particularly prevalent in tropical and subtropical waters and is adapted to thrive in low-nutrient environments. Another common genus of marine cyanobacteria is *Synechococcus*, which is also widespread in marine habitats (Partensky et al. 1999). Like *Prochlorococcus*, *Synechococcus* are capable of photosynthesis and is typically found in nutrient-poor waters. Other genera of marine cyanobacteria commonly found in marine habitats include Oscillatoria, Trichodesmium, and Lyngbya (Paerl et al. 2001). Oscillatoria is a filamentous cyanobacterium that forms mats in intertidal zones and estuaries, while Trichodesmium is a free-living, nitrogen-fixing cyanobacterium that forms colonies and is particularly abundant in warm, oligotrophic oceans. Lyngbya, on the other hand, is a filamentous cyanobacterium that is found in tropical and subtropical waters.

Cyanobacteria are believed to have played a significant role in shaping the Earth's atmosphere by oxygenating it through the process of photosynthesis, which led to the evolution of aerobic organisms (Sánchez-Baracaldo 2015). Cyanobacteria are also known for their ability to fix atmospheric nitrogen, which is an essential nutrient for plant growth (Whitton and Potts 2000). They do this through the activity of specialized cells called heterocysts, which are involved in nitrogen fixation (Zehr et al. 2001). In addition to their ecological and evolutionary importance, cyanobacteria have attracted attention due to their potential for biotechnological applications. For example, some cyanobacteria produce bioactive compounds with pharmaceutical or industrial uses, such as cyanotoxins, pigments, and polysaccharides (Singh et al. 2021). However, cyanobacteria can also pose a threat to human health and the environment. For instance, even though industrially important, cyanotoxins produced by these organisms can cause harmful effects if they are

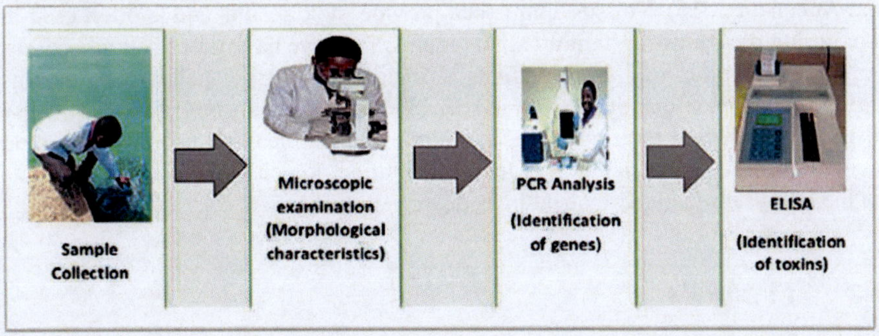

Fig. 3.1 Identification of toxic cyanobacteria blooms in marine water combining polymerase chain reaction (PCR) and enzyme-linked immunosorbent assay (ELISA)

ingested or when they come into contact with the skin (Carmichael et al. 2001). The two most common types of cyanotoxins are microcystins and cylindrospermopsins, which can cause liver damage and gastrointestinal illness in humans and animals (Falconer 2005). Other types of cyanotoxins include anatoxins, saxitoxins, and nodularins, which can cause neurological symptoms and respiratory paralysis. Cyanobacteria can also cause eutrophication in water bodies, which can lead to harmful algal blooms and hypoxic conditions. Most Cyanobacteria associated toxins, including cyanotoxins are produced under certain environmental conditions, especially with high nutrient levels, warm temperatures, and stagnant water (Paerl and Otten 2013). Thus, Cyanobacterial blooms that produce these toxins have been responsible for numerous cases of illness and death in humans and animals (Saker et al. 2005).

It is therefore important to monitor and manage cyanobacterial blooms to minimize the risks associated with toxins they produce. The identification of toxic blooms of cyanobacteria in marine water habitats involves several methods, including visual inspection, polymerase chain reaction (PCR), and enzyme-linked immunosorbent assay (ELISA). These methods use various materials to determine the type and abundance of cyanobacteria and their associated toxins. Water samples are collected from different locations in the water body where there is a suspected presence of toxic cyanobacteria. The samples are collected using sterile containers and stored in coolers with ice packs. Visual inspection of water samples is essential in identifying toxic cyanobacteria (Nienaber and Steinitz-Kannan 2018). The water samples should be examined under a microscope to determine the type and abundance of cyanobacteria. The presence of toxic cyanobacteria should be noted, and their abundance recorded (Paerl and Otten 2013). Polymerase chain reaction (PCR) is a molecular technique that amplifies the DNA of cyanobacteria in water samples (Soto-Liebe et al. 2010). This technique can help identify the specific species of cyanobacteria present in the water and whether they are toxic or not. Enzyme-linked immunosorbent assay (ELISA) is an immunological technique that detects the presence of specific toxins produced by cyanobacteria in water samples (Glibert and Anderson 2016). This technique can provide rapid results and is often used in monitoring programs for harmful algal blooms. Here we have described procedures for the most commonly used methods in identifying toxic cyanobacteria blooms (Microscopic examination, PCR, and ELISA). Overall, a combination of these methods can identify toxic cyanobacteria blooms in marine water habitats, providing a comprehensive understanding of their distribution, toxicity, and potential impacts on the ecosystem and public health, as described in Fig. 3.1.

3.2 Materials

- Map of the area, GPS device
- Sterile bottles
- Gloves
- Waterproof labels
- Waterproof pen
- Field notebook
- Cooler with ice packs
- Refrigerator or freezer
- Microscope
- Sterile water
- Clean glass slides
- Coverslips
- Microscope slides
- Blue-green algae (Cyanobacteria) identification guidebook
- 0.45-micron pore-size filter
- PCR machine
- Primers
- DNA extraction kit

For the ELISA test, the following materials are needed

- ELISA test kit
- Microtiter plate (96-well)
- Coating buffer (e.g., carbonate-bicarbonate buffer, pH 9.6)
- Antibody specific to the cyanotoxin of interest (e.g., anti-microcystin or anti-saxitoxin antibody)
- Blocking buffer (e.g., 1% BSA in PBS)
- Sample (marine water sample)
- A standard curve (purified cyanotoxin of known concentration)
- Detection antibody (e.g., anti-mouse IgG conjugated to HRP)
- Substrate solution (e.g., TMB substrate)
- Stop solution (e.g., 2 M sulfuric acid)
- ELISA reader

3.3 Methods

3.3.1 Sample Collection (Paerl and Otten 2013; Carmichael 1992; Department of Environment and Science 2018)

1. Identify the sampling sites based on the suspected presence of toxic cyanobacteria using maps of the area and/or GPS devices to locate the sampling sites accurately.
2. Collect water samples using sterile bottles. Wear gloves to avoid contamination. Store the bottles in a cooler with ice packs to maintain low temperatures and preserve the samples.
3. Label the bottles with the date, time, location, and relevant information about the sampling site. Record the information in a field notebook to avoid errors and confusion.
4. Transport the samples to the laboratory in a cooler with ice packs. Store the samples in a refrigerator or freezer until analysis. Keep the samples in the dark to prevent any growth of algae or other contaminants.

3.3.2 Visual Inspection/Microscopic Examination

3.3.2.1 Using a Binocular Microscope (Environmental Protection Agency 2018; Centers for Disease Control and Prevention 2017)

A binocular microscope is a type of microscope that allows for the observation of specimens through two eyepieces. Identification of cyanobacteria under the microscope typically involves observing their cellular morphology, pigmentation, and other features.

Procedure

1. Shake the sampling bottle well to mix the water sample before use.
2. Take a small amount of water sample and transfer it to a clean glass slide.
3. Add a drop of sterile water to the water sample on the slide.
4. Place a coverslip over the water sample and gently press it to flatten the sample.
5. Place the prepared slide under the binocular microscope.
6. Focus the microscope to get a clear view of the water sample.
7. Observe the water sample for the presence of Cyanobacteria (Use a blue-green algae (Cyanobacteria) identification guidebook to confirm the presence of toxic cyanobacteria species).
8. Record the observations made during the visual inspection, including the location, date, and observation time.
9. Document the species of toxic cyanobacteria identified if any.

3.3.2.2 Using a Compound Microscope (United States Environmental Protection Agency (EPA) 2019; World Health Organization (WHO) 1999)

A compound microscope uses a combination of lenses to achieve high magnification. The two primary lenses in a compound microscope are the objective lens and the eyepiece lens. Identification of cyanobacteria under the microscope typically involves observing their cellular morphology, pigmentation, and other features.

Procedure

1. Filter the water samples using a 0.45-micron pore-size filter. This will remove any larger particles that might interfere with the microscopic examination of the sample.
2. Examine the filtered water samples under a microscope. You can use a compound microscope with a magnification of 100× or 200×.
3. Identify the presence of Cyanobacteria in the water samples. Use morphological characteristics such as shape, size, and color to identify the Cyanobacteria species.

3.3.3 Polymerase Chain Reaction (PCR) (Saker and Griffiths 1999; Kurmayer and Kutzenberger 2003)

Polymerase Chain Reaction is a technique used to amplify a specific segment of DNA or RNA from a larger DNA or RNA sample. PCR is used to identify cyanobacteria based on the presence of specific DNA sequences. The PCR process involves amplifying a specific segment of DNA from a sample, which can then be analyzed to determine the presence or absence of the target sequence.

3.3.3.1 Procedure

1. Extract DNA from the water samples using a DNA extraction kit. Several commercial DNA extraction kits are available in the market, such as the DNeasy Blood & Tissue Kit (Qiagen) or the PowerWater DNA Isolation Kit (MoBio). Follow the manufacturer's instructions to extract DNA from the water samples.
2. Prepare the PCR reaction mix by combining the following reagents in a sterile PCR tube: Template DNA (extracted from the water sample), Taqpolymerase, Primers, Deoxynucleoside triphosphates (dNTPs), and Buffer.
   Note: Use primers specific to the target DNA sequence of cyanobacteria (For example, the primers used for detecting the cyanobacterial toxin genes microcystin-LR).
3. Amplify the target DNA sequence using a thermocycler or real-time PCR machine.
4. For the thermocycler, analyze the PCR products by running them on an agarose gel using gel electrophoresis. Use a DNA ladder to estimate the size of the PCR products.

5. For real-time PCR, analyze the data using specialized software provided by the real-time PCR machine manufacturer. The software will calculate the amount of fluorescence emitted in each cycle and generate a graph of the fluorescence intensity versus the cycle number. The amount of target DNA sequence can be quantified by comparing the fluorescence intensity to a standard curve generated using known concentrations of the target DNA sequence.

3.3.4 ELISA (Lawton and Edwards 2015; Meriluoto and Spoof 2017)

Enzyme-linked immunosorbent assay is a biomolecular technique that utilizes the specificity of an antibody, as well as the sensitivity of enzyme assays, to detect and quantify molecules. ELISA can detect and quantify specific cyanotoxins in a sample by using antibodies that are specific to the target toxin.

3.3.4.1 Steps

1. Coat the microtiter plate with the antibody specific to the cyanotoxin of interest. Dilute the antibody in the coating buffer and add it to the wells of the microtiter plate. Incubate the plate overnight at 4 °C or for 2–4 h at room temperature.
2. Wash the plate with a washing buffer (e.g., PBS-Tween) to remove any unbound antibodies.
3. Block the remaining binding sites on the plate with the blocking buffer. Add the blocking buffer to the wells of the plate and incubate for 1 h at room temperature.
4. Prepare a standard curve by diluting the purified cyanotoxin of known concentration in the blocking buffer. Add the standard curve to the plate in duplicate or triplicate and the sample to the plate in duplicate or triplicate.
5. Incubate the plate for 1–2 h at room temperature to allow the cyanotoxin to bind to the antibody.
6. Wash the plate with the washing buffer to remove any unbound cyanotoxin.
7. Add the detection antibody (e.g., anti-mouse IgG conjugated to HRP) to the plate wells. Dilute the detection antibody in the blocking buffer and incubate the plate for 1 h at room temperature.
8. Wash the plate with the washing buffer to remove any unbound detection antibody.
9. Add the substrate solution (e.g., TMB substrate) to the wells of the plate. Incubate the plate for 10–30 min at room temperature, depending on the rate of color development.
10. Stop the color development by adding the stop solution (e.g., 2 M sulfuric acid) to the wells of the plate.
11. Measure the absorbance of each well using an ELISA reader at a wavelength of 450 nm. Calculate the cyanotoxin concentration in the sample by comparing its absorbance to the standard curve.

References

Carmichael WW (1992) Cyanobacteria secondary metabolites—the cyanotoxins. J Appl Microbiol 72(6):445–459

Carmichael WW, Azevedo SM, An JS, Molica RJ, Jochimsen EM, Lau S et al (2001) Human fatalities from cyanobacteria: chemical and biological evidence for cyanotoxins. Environ Health Perspect 109(7):663–668

Centers for Disease Control and Prevention (2017) Harmful Algal Blooms (HABs). https://www.cdc.gov/habs/index.html

Department of Environment and Science (2018) Water sampling and analysis—field manual. https://www.des.qld.gov.au/__data/assets/pdf_file/0025/107404/water-sampling-analysis-field-manual.pdf

Environmental Protection Agency (2018) Cyanobacterial harmful algal blooms (CyanoHABs) in water. https://www.epa.gov/nutrientpollution/cyanobacterial-harmful-algal-blooms-cyanohabs-water

Falconer IR (2005) Cyanobacterial toxins of drinking water supplies: cylindrospermopsins and microcystins. CRC Press, Boca Raton

Glibert PM, Anderson DM (2016) Global ecology and oceanography of harmful algal blooms. In: Ecology of harmful algae. Springer, Cham, pp 119–133

Kurmayer R, Kutzenberger T (2003) Application of real-time PCR for quantification of microcystin genotypes in a population of the toxic cyanobacterium Microcystis sp. Appl Environ Microbiol 69(12):6723

Lawton LA, Edwards C (2015) Enzyme-linked immunosorbent assays (ELISAs) for detecting and quantifying cyanotoxins in water. In: Cyanobacterial harmful algal blooms: state of the science and research needs. Springer, pp 327–338

Meriluoto JA, Spoof L (2017) ELISA

Nienaber MA, Steinitz-Kannan M (2018) A guide to cyanobacteria: identification and impact. University Press of Kentucky

Paerl HW, Otten TG (2013) Harmful cyanobacterial blooms: causes, consequences, and controls. Microb Ecol 65(4):995–1010

Paerl HW, Fulton RS, Moisander PH, Dyble J (2001) Harmful freshwater and marine algal blooms. In: Wehr JD, Sheath RG (eds) Freshwater algae of North America: ecology and classification. Academic, San Diego, pp 1091–1138

Partensky F, Hess WR, Vaulot D (1999) Prochlorococcus, a marine photosynthetic prokaryote of global significance. Microbiol Mol Biol Rev 63(1):106–127

Saker ML, Griffiths DJ (1999) The effect of sample treatment method on PCR detection of microcystin-producing cyanobacteria in water. J Appl Microbiol 87(6):706–715. https://doi.org/10.1046/j.1365-2672.1999.00844.x

Saker ML, Griffiths DJ, Falconer IR (2005) Toxins produced by cyanobacteria. In: Water toxins. Springer, pp 1–332

Sánchez-Baracaldo P (2015) Origin of marine planktonic cyanobacteria. Science 350(6257): aac7375

Scanlan DJ, Ostrowski M, Mazard S, Dufresne A, Garczarek L, Hess WR et al (2009) Ecological genomics of marine picocyanobacteria. Microbiol Mol Biol Rev 73(2):249–299

Singh P, Gupta R, Kumari P (2021) Biotechnological potential of cyanobacteria. In: Cyanobacteria. Springer, Singapore, pp 257–282

Soto-Liebe K, Fuentes-Valdés JJ, Guerrero MA (2010) PCR detection methods for the identification of potentially toxic cyanobacteria in water and environmental samples. J Appl Microbiol 109(5):1575–1587

United States Environmental Protection Agency (EPA) (2019) Standard operating procedure for the collection and analysis of cyanobacterial samples from freshwater and marine environments. EPA/600/R-19/100. https://cfpub.epa.gov/si/si_public_file_download.cfm?p_download_id=540979&Lab=NRMRL

Whitton BA, Potts M (2000) The ecology of cyanobacteria: their diversity in time and space. Springer Science & Business Media

World Health Organization (WHO) (1999) Identification and quantification of microcystins (cyanobacterial toxins) in water samples. WHO/SDE/WSH/03.04/28. https://www.who.int/water_sanitation_health/dwq/iwachap7.pdf

Zehr JP, Mellon MT, Hiorns WD (2001) Phylogeny of cyanobacterial nifH genes: evolutionary implications and potential applications to natural assemblages. Microbiol Mol Biol Rev 65(1): 13–37

4 Identification of Toxic Blooms of Cyanobacteria in Mangrove Forest

Silambarasan Tamil Selvan, Sanjivkumar Muthusamy, and Bhuvaneswari Santharam

Abstract

Cyanobacteria or blue green algae are the most predominant group of photosynthetic and prokaryotes. The cyanobacteria are presented in the widespread of natural environment such as lakes, rivers, and ponds and have different pigments like chlorophylls, phycocyanin, and phycoerythrin. Hence, the cyanobacteria are involved in photosynthesis process through aerobic as well as anaerobic conditions. The association of cyanobacterial mass found on the reservoirs are called as cyanobacteria bloom. The blooms were highly accumulated on the surface of aquatic system, which are enriched with biological wastes or nutrients like nitrogen and phosphorus. The cyanobacterial blooms are typically denoted during in the season of spring (*Planktothrix* spp.) or in season of late summer (*Microcystis* spp.). These cyanobacterial blooms may produce metabolically active chemical components called as cyanotoxins against different aquatic predators. It may be characterized into dermato, cyto, hepato, and neurotoxins. These toxins are produced either dangerous effects to aquatic organisms as well as human beings or as any therapeutically importance. In this study, such metabolically active toxin producing cyanobacterial blooms were identified.

Keywords

Cyanobacteria · Cyanotoxins · Plankton net · Marine agar medium · PCR

S. Tamil Selvan
Department of Microbiology, K. R. College of Arts and Science, Kovilpatti, Tamil Nadu, India

S. Muthusamy (✉)
Department of Microbiology, School of Allied Health Sciences, VIMS Hospital Campus, Vinayaka Missions Research Foundation (DU), Salem, Tamil Nadu, India

B. Santharam
Department of Biochemistry, K.R. College of Arts and Science, Kovilpatti, Tamil Nadu, India

N. Thajuddin et al. (eds.), *Protocols for Cyanobacteria Sampling and Detection of Cyanotoxin*, https://doi.org/10.1007/978-981-99-4514-6_4

4.1 Introduction

Mangrove ecosystems are considered as the most unique environment and are recorded as one of the extremely productive natural ecosystems across the world. Dynamic mangrove ecosystem is enriched with enormous growth of microbes including phosphate solubilizing, nitrogen fixing, sulphate reducing and methanogenic bacteria, wood decaying fungi, together with photosynthetic microbes such as cyanobacteria and microalgae, which accomplish complex interactions for stabilizing nutrient cycle and ecological balances in the ecosystem (Rippka et al. 1979). Basic ecological factors are important for biodiversity of cyanobacteria. Mangrove forests constantly interface between of land and sea and fluctuate in various factors viz., salinity, light intensity, temperature, humidity, and other environmental factors (Rippka et al. 1979; Tanjuchi et al. 2012).

Blue-green algae are ubiquitous widespread ancient photosynthesis organisms. They contributed to the development of the atmosphere by producing oxygen through photosynthesis, and they play a vital role in the carbon, nitrogen biological fixation cycles (Tanjuchi et al. 2012). They are native occupants of a variety of habitats including littoral rocks, soils, fresh, brackish, and marine seas. In the mangrove environment, cyanobacteria are crucial as they are the main producers of organic carbon and nitrogen. Numerous cyanobacteria species a prominent group within the diazotrophic guild, contribute to the systems nitrogen supply by fixing nitrogen (Tanjuchi et al. 2012). Cyanobacteria have developed over billions of years to be able to adapt to and endure under many climatic conditions. This offers a significant environmental concern because cyanobacteria can now spread so quickly, creating harmful algae blooms for aquatic life. Additionally, a lot of cyanobacteria synthesize toxins, making them dangerous for both animals and human beings when they encounter (Roy-Lachapelle et al. 2021).

Cyanotoxins have significant impacts on aquatic species, such as lowering survival rates, preventing feeding, and paralyzing them. The most significant long-term effects of cyanotoxins on aquatic animals include decreased growth and fecundity, behavioural and biochemical changes in the activities by the key enzymes viz., phosphatases, glutathione-S-transferases, acetylcholinesterase, and proteases (Roy-Lachapelle et al. 2021). Cyanotoxins can be categorized according to two basic factors: mode of action on terrestrial vertebrates, particularly mammals, for example, hepatotoxins, neurotoxins, and dermatotoxins; and based on the chemical structure of the substance, such as cyclic peptides, alkaloids, or lipopolysaccharides (Du et al. 2019). The most frequent bloom-forming cyanobacteria found in freshwaters is *Microcystis*, and is implicated in several intoxication of human beings, domestic and wild animals. Several cyanobacterial genera can form hepatotoxins in freshwater bodies viz., alkaloid cylindrospermopsin, small peptide microcystins and neurotoxins such as anatoxins and saxitoxins. Many secondary metabolites that cyanobacteria can synthesize are hazardous to humans, animals, and plants (Zambrano-Intriago et al. 2021). Microcystins are more hepatotoxic in nature, and their hepatotoxicity is due to the particular inhibitory protein phosphatase 1 and 2A in liver cells, which causes acute liver failure by disrupting hepatocyte cytoskeletal elements (Roy-Lachapelle et al. 2021; Du et al. 2019; Zambrano-Intriago et al. 2021)

The toxicity nature of cyanotoxin is mostly manifested by the inhibition of enzymes involved in important metabolic processes, skin irritation, effects on the neurological system, cytotoxic effects, interference with receptor signalling, hormonal disruption, disorders of reproduction, tumour promotion, etc. Therefore, it is frequently not a simple process to sample, detect, and monitor cyanobacteria that may be harmful. The cyanobacterial toxins are identified using HPLC, LC-MS, MALDI, ESI, and FAB analytical methods.

4.2 Materials

Planktonic mesh, Erlenmeyer flasks, membrane filters, BG11 growth medium, centrifuge, PCR, etc.

4.2.1 Composition of BG11 Medium

The BG11 cyanobacteria growth medium consists of 1.5 $gL^{-1}$of $NaNO_3$,40 mgL^{-1} of K_2HPO_4, 75 mgL^{-1} of$MgSO_4.7H_2O$, 36 mgL^{-1} $CaCl_2.7H_2O$, 6 mgL^{-1} of citric acid, 1 mgL^{-1} of EDTA, 6 mgL^{-1} of ferric ammonium citrate, 2 mgL^{-1} of Na_2CO_3, 1 mL of trace salt, 10 g of Agar, and pH −7.1 (Rippka et al. 1979; Tanjuchi et al. 2012; Roy-Lachapelle et al. 2021).

4.2.2 Trace Salt Mixtures

The trace salt mixture contains 390 mgL^{-1} of $NaMoO_4.2H_2O$, 2.86 gL^{-1} of H_3BO_3, 220 mgL^{-1} of $ZnSO_4.7H_2O$, 1.81 gL^{-1} of $MnCl_2.4H_2O$, 50 mgL^{-1} of $Co(NO_3)_2.6H_2O$.

4.3 Methods

4.3.1 Sample Collection

1. Collections of water and sediment samples from fresh, marine water, and Mangrove Forest station.
2. Samples are collected from 3 m beneath depth from the surface using plankton mesh or portable hose sampler.
3. Collected samples are transfer into sterile plastic containers.
4. Evaluate the water qualities like pH, temperature, conductivity, salinity, and oxygen saturation.

5. Place the samples in an insulated ice chests and subsequently, filter using millipore membrane filter (0.45 μm pore size) connected with a pressurized vacuum pump.
6. Examine the membrane filter through microscopic analysis and keep the filters in −80 °C for further isolation process (Tanjuchi et al. 2012).

4.3.2 Isolation of Blooms of Cyanobacteria

1. Prepare the culture (BG11) medium in a 500 mL Erlenmeyer flask and autoclave at 121 °C, 15 psi for 15–30 min.
2. Transfer the membrane filter paper (from the filtration unit) into sterile BG11 medium supplemented with an antibiotic Kanamycin (50 μg/mL) and keep on a rotary shaker (120 rpm) at 28 ± 2 °C for 3 weeks.
3. Samples containing cyanobacterium will be serially diluted from 10^{-1} to 10^{-8} with sterile distilled water and each multiple will be inoculate onto sterile petri dishes with BG11 agar medium (Tanjuchi et al. 2012).
4. The cultured petri plates will seal using parafilm and incubated at ambient temperature 28 ± 2 °C for 3 weeks under 14:10 light dark cycles (illuminated by cool fluorescent light 700 lux) and the growth of the cyanobacteria will monitor at the 7th, 10th, 15th, and 21st day from the date of initial incubation.
5. Once a single colony will observe and transfer to BG11 medium. The culture will be subject to identification process.
6. The purity of the monoclonal cyanobacteria will check regularly using light microscope.
7. The morphological characterization of the isolate colonies will observe using a light microscope.

4.3.3 Identification of Cyanobacteria Toxins

The cyanobacteria toxins are identified using various analytical assay methods viz., chemical, biochemical, bioassays assays, and molecular analysis (Roy-Lachapelle et al. 2021).

4.3.3.1 Bioassays

- Bioassays are used and identified the cyanobacteria toxins presence in water samples.
- The determining cyanotoxins and toxin potency using mouse bioassay method and measured by the lethal dose concentration (LD_{50}) from infected animals' survival time.
- Plant bioassays are used to investigate the phytotoxic effects of cyanotoxins.
- Toxic effects of cyanotoxins are determined using plant tests, and the plants are sensitive to cyanotoxins. The physiological indicators are affected by toxicity.

4.3.3.2 Biochemical-Based Methods

- The detection of cyanotoxin interactions with biological macromolecules using biochemical assay methods viz., receptor bioassays, enzyme inhibition assays, and immunoassays.
- Biochemical methods are a simple and cost-effective assay to monitor the concentration of cyanotoxins such as protein phosphatase inhibitor assay (PPIA) and enzyme-linked immunosorbent assay (ELISA) are most common techniques.
- Their findings in cyanotoxin detection complements other approaches, such as molecular-based methods and provides confirmation of the presence of toxins in waterbodies.
- The detection of cyanotoxins will use antibody-based immunoassays (ELISA) and developed in recent years, including Fluorescence Immunoassay (FIA), immune-chromatography, as well as biosensor techniques.

4.3.3.3 Chemical Assays

- Chemical analysis is the most consistent technique for detecting and identifying cyanobacterial toxins because it is based on the physical and chemical properties of the toxins from samples.
- Cyanobacterial toxins have been detected and identified precisely and sensitively using NMR, HPLC, CE, LC-MS, GC, SPE-UHPLC-HRMS, and other advanced chemical analysis methods.
- The matrix-assisted laser desorption-ionization-time-of-flight mass spectrometry (MALDI-TOF-MS) method allows an individual colony or filament to be differentiated from non-toxic strains without the need to extract them by solvent.
- In recent years, MS-based methods are considered to be the most efficient approaches to assessing and identifying cyanotoxins.

4.3.3.4 AP_{tot} Enzyme-Linked Immunosorbent Assay (ELISA) Analysis

- The direct competitive ELISA based on antibody recognition of APs will be used to identify the cyanotoxins in cyanobacteria using AP_{tot}ELISA (Du et al. 2019; Zambrano-Intriago et al. 2021).
- Take 1:5 ratio (deionized water:buffer) and wash wells.
- Add 50 μL of standards (1.5 μg L^{-1}) and control (0.80 μg L^{-1}) in the AP_{tot}ELISA wells.
- Add HRP conjugate enzyme solution 50 μL add in the well and add 55 μL of antibody solution in the same well.
- Cover the wells using parafilm and incubate at 37 °C for 1 h. After incubation, discard the mixture and wash the wells three times with 250 μL of wash buffer.
- Invert microplate for 5 min to dry the wells and remove excess remaining buffers.
- Add 150 μL of substrate solution in the wells and cover with parafilm. Incubate the microplate at 37 °C in a dark place for 20 min.
- In the microplate, the wells will be filled with 100 mL of a stop solution and the plate will be placed on the ELISA reader.
- The wells will be read at 450 nm within 15 min and values will be recorded.

4.3.3.5 SPE-UHPLC-HRMS Analysis

- The quantitative of cyanotoxins will be analyzed using SPE-UHPLC-HRMS method (Du et al. 2019; Zambrano-Intriago et al. 2021).
- 1 mL of test sample will be injected in the sample loading port and sample flow rate will be set on 1 mL min^{-1}.
- Then, pass 0.5 mL solvent through column and wash the column. After washing, pass 0.5 mL of MeOH solvent through the column with backflush and transfer the elution directly through the analytical column using an analytical pump gradient.
- The chromatographic separation mobile phases viz., A and B will be used and the organic solvents will be used as a water and acetonitrile solvent, both solvents will be acidified with 0.1% of formic acid.
- The 525 μL min^{-1} flow rate of mobile phase solvents will be run at 8 min in chromatographic column.
- In the initial process, load 10% of acetonitrile (mobile phase B) and wash the online SPE column for 1.5 min and increase 100% solvent flow for 2 min.
- As a final step, the mobile phase is returned to the initial conditions and maintained at 1.5 min.
- In the capillary column temperature range of 350 °C, ionization will be performed using heated electrospray ionization interface (HESI-II) in the positive mode at +3500 V in ionization spray.
- The vaporizer will be set at 250 °C with a 60% gas flow and 15 arbitrary units of pressure.
- Full scan mode will be used to scan the chromatographic spectrum with parameters set to 400–1200 m/z and 1×10^6 ion capacity automated gain control (AGC).
- The ion time (IT) will be set at 100 ms filling and the parallel reaction monitoring (PRM) pattern will have an FWHM scan range of 17,500 ms.
- AGC and IT will be set at 50 ms for 50 ion capacity and 200 m/z for MC-LA resolving power.

4.4 Observation

The isolated toxin blooms of cyanobacteria are identified and characterized.

Acknowledgement We sincerely acknowledge the Management and Principal of K.R. College of Arts & Science for their moral support throughout this protocol.

References

Du X, Liu H, Yuan L, Wang Y, Ma Y, Wang R, Chen X, Losiewicz MD, Guo H, Zhang H (2019) The diversity of cyanobacterial toxins on structural characterization, distribution and identification: a systematic review. Toxins 11(9):530

Rippka R, Deruelles J, Waterbury JB, Herdman M, Stanier RY (1979) Generic assignments, strains, histories and properties of pure culture of cyanobacteria. J Gen Microbiol 111:1–61

Roy-Lachapelle A, Solliec M, Sauve S, Gagnon C (2021) Evaluation of ELISA-based method for total anabaeno-peptins determination and comparative analysis with on-line SPE-UHPLC-HRMS in freshwater cyanobacterial blooms. Talanta 223:121802

Tanjuchi Y, Chen YLL, Chen HY, Tsai ML, Ohki K (2012) Isolation and characterization of the unicellular diazotrophic cyanobacterium group C TW3 from the tropical western Pacific Ocean. Environ Microbiol 14:641–654

Zambrano-Intriago LA, Amorim CG, Rodríguez-Díaz JM, Montenegro MCB, Araújo AN (2021) Electrochemical biosensing of algal toxins. In: Advances in the domain of environmental biotechnology: microbiological developments in industries, wastewater treatment and agriculture, pp 227–252

Identification of Toxic Blooms of Cyanobacteria in Estuarine Habitat

5

Shree Kumari G R, Ranjitha Dhevi V. Sundar, Ajith Kumar K, and Lokesh Ravi

Abstract

Our entire environment is surrounded with millions of hidden microscopic worlds with diverse life forms. One such microscopic life forms are these photosynthetic bacteria, evolved over 2 billion years ago, known as cyanobacteria. They are found ubiquitously. We will be focusing on their presence particularly in an Estuarian Ecosystem. These photosynthetic cyanobacteria can also produce toxins when they grow excessively and can also cause harm to the environment, plants and animals living under water. They can also cause harmful effects to humans as well sometimes. These toxins producing cyanobacterial blooms can be identified by different studies.

Keywords

Super blooms of cyanobacteria · Nuisance phytoplankton blooms · Estuaries · Cyanotoxin(s) · Microcystin · Nodularin · Toxicogenic cyanobacteria

S. K. G. R.
Department of Biomedical Sciences, School of Bio Sciences and Technology, Vellore Institute of Technology, Vellore, Tamil Nadu, India

R. D. V. Sundar
Department of Biotechnology, School of Bio Sciences and Technology, Vellore Institute of Technology, Vellore, India

A. K. K.
Department of Life Sciences, Kristu Jayanti College (Autonomous), Bengaluru, India

L. Ravi (✉)
Department of Food Technology, Faculty of Life and Allied Health Sciences, MS Ramaiah University of Applied Sciences, Bengaluru, India

N. Thajuddin et al. (eds.), *Protocols for Cyanobacteria Sampling and Detection of Cyanotoxin*, https://doi.org/10.1007/978-981-99-4514-6_5

5.1 Introduction

We are surrounded by hidden microscopic worlds filled with fascinating life forms. Thousands of microbial organisms live within a single drop of water. A pond ecosystem is composed of microbial wildernesses that ages over 2 billion years, that are yet to be explored. These organisms were the first to live by photosynthesis. These organisms used sunlight to make sugars and then these sugars gave power to the cell. A by-product of photosynthesis is oxygen, and 2 billion years ago, these organisms became so abundant that they completely changed the Earth's atmosphere. It went from the one that was very oxygen poor, to the oxygen-rich atmosphere that we live in today. These microbes are called cyanobacteria (Albrecht et al. 2017). They are just as abundant today as they were 2 billion years ago. Cyanobacteria can be found all over the world. There are currently around 3000 described species with different morphologies and habitats from freshwater ponds to arctic oceans. They even live in soil. These cyanobacteria are also known as photosynthetic bacteria (United States Environmental Protection Agency 2020).

5.1.1 Estuaries

It is an ecosystem by itself. They are simple terms known as the transition area where the freshwater (river or stream) meets the ocean. Here the salinity of water increases. It is a semi-enclosed body of water with an opening to the ocean and feed by freshwater. They are also known as the "nursey of the sea" as they provide shelter for aquatic habitat as it provides a supportive environment for the wild to raise their young (Elliott and McLusky 2002). They have the producers which are the phytoplankton which is a part of the food chain. They are known for the brackish water which is the combination of salt and fresh water. They are important as they help in filtering the sediment and pollutants from the water before it flows into the ocean. In addition, they buffer the ocean and land which helps decrease the effect of flooding and storm surges (Elliott and McLusky 2002; Perillo 1995). These estuaries can be classified according to their geological features. The classification includes (Elliott and McLusky 2002; Perillo 1995; Lehman et al. 2022):

1. Coastal plain estuary: Also known as a lagoon, they resemble valleys with gentle slopping bottoms, their depth increases towards the river's mouth. Very commonly encountered throughout the world, for example, The Chesapeake Bay.
2. Tectonic estuary: Created when the tectonic eruption occurs or when the sea fills in the hole or a basin that is formed from sinking land, for example, San Francisco Bay
3. Bar-Built estuary: formed when sand bars build up along the cost line and they partially cut off the water behind them from the sea.
4. Fjord estuary: they are narrow with steep sides and are usually straight and long. Usually found in areas covered by glaciers.

5.1.2 Cyanotoxins

Cyanobacteria are considered the most problematic bacteria in the freshwater ecosystems. These photosynthetic bacteria are ubiquitous, naturally occurring but under certain conditions like presence of high phosphorous and nitrogen (nutrient) content or climate change (excess sunlight/dry or rain/wet conditions) they can grow out of control (overgrow) unchecked by natural cycles in the water. The uncontrolled growth of photosynthetic bacteria in a short period of time are called 'Blooms of Cyanobacteria' (Rastogi et al. 2015). Some the effects of these Cyanobacterial Blooms are they consume oxygen and block the sunlight by not letting it penetrate into the waterbody for other plants and animals living underwater. When these Cyanobacterial Blooms start decaying they consume excess oxygen creating hypoxic conditions which results in die-offs of many plants and animals living under water (Bormans et al. 2020). Under favorable conditions i.e., photo rich and nutrient rich environment, some species of these photosynthetic bacteria produce some toxic secondary metabolites or when there is release of gases from the ruptured dead cells of the cyanobacteria it is called a toxin/cyanotoxin (Fristachi et al. 2008).

5.1.3 Toxic Blooms of Cyanobacteria in Estuaries

Ideally toxigenic cyanobacteria occur mostly in freshwater habitats. Since an estuary is an area where freshwater meets ocean, it is ideally a mixture of freshwater as well as saline water (A semi–enclosed body of water were saltwater mixes with freshwater). Estuaries are usually referred to have Brackish water which shows growth of toxigenic cyanobacteria. The blooms of these toxigenic cyanobacteria are found in majority of the estuaries throughout the world. They cause harmful effects to the estuarine environment, to animals, humans as well as spoil recreational opportunities along the coast line. Not all blooms are harmful (Fristachi et al. 2008). Blooms known as "Nuisance blooms" can discolor water, smell bad, and cause the water or fish to taste bad. Nuisance blooms are not usually dangerous to people, pets, and livestock because they do not produce toxins. However, they can discourage people from visiting beaches, drinking tap water, or eating fish from water with an ongoing bloom (Paerl 1988). On the other hand, "Super Blooms" of cyanobacteria that produce biotoxins that are harmful and cause global health issues are caused by certain toxicogenic cyanobacteria (Miller et al. 2010). These toxic cyanobacterial blooms can affect the water in different ways. One of them is by taste and odor, they create water that may tase like mud or moldy water. The cyanotoxins are compounds that may cause health problems aby affecting the liver, kidney or nerves system effects in humans (Brown et al. 2018). There are three different classes of cyanobacteria: Neurotoxin (synthetically or naturally occurring toxins can directly damage/affect the nervous system), Hepatotoxin (synthetically or naturally occurring toxins that damage the liver) & Dermatoxins (synthetically or naturally occurring toxins that damages the skin cells).

Table 5.1 Toxins produced by cyanobacteria in Estuaries

Sl. no.	Cyano-bacteria	Toxin produced	Chemical structure	Toxin category	References
1.	*Nodularia spumigena*	Nodularin	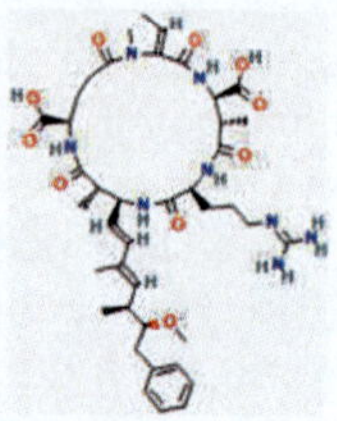	Hepatotoxin	Paerl (1988)
2.	*Oscillatoria* sp.	Anatoxin-a	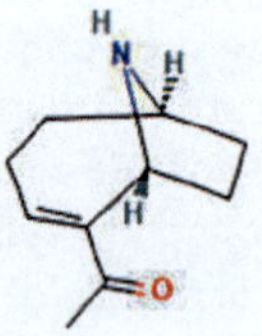	Neurotoxin	Paerl (1988)
3.	*Microcystis aeruginosa*	Microcystin	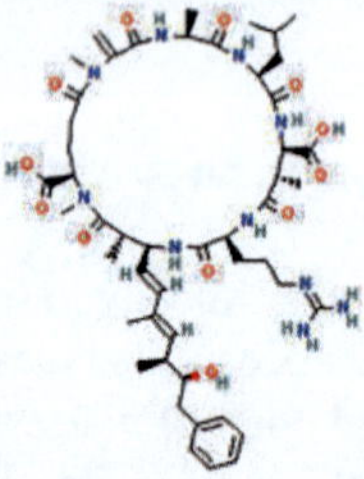	Hepatotoxin	Gibble and Kudela (2014)

Usually, toxic blooms of species of *Nodularia* and *Oscillatoria* were initially observed, identified and studied in estuarian ecosystems (Paerl 1988). Later, studies from the 2000s have confirmed presence of toxic blooms of species of Microcystis as well. This is because usually *Microcystis* are intolerant to saline conditions (Fristachi et al. 2008). But there were many studies that reported the presence of *Microcystis* sp. in estuarine habitat can be because of wet climate (Lehman et al. 2022). The wet climatic conditions decrease the salinity of water as there is continuous mixing of freshwater to the saline water with in turn reduces the salinity and creates a favorable environmental condition for the growth of blooms of *Microcystis* sp. as well. Some signs of toxin producing cyanobacteria are (1) they can make the estuary pigmented, (2) they make the environment that they are present very unpleasant by releasing gases/ toxins that give unpleasant odor, taste etc. (3) while collection of water sample there might be presence of tiny, floating, green particles. Though the above-mentioned signs may be considered as the preliminary characteristics to suspect if a particular estuary has toxin producing blooms of cyanobacteria or not (United States Environmental Protection Agency 2020; Rastogi et al. 2015). Because one cannot tell if a water body has a toxin producing blooms of cyanobacteria just by looking at it. To find out what species of toxin producing cyanobacterial bloom is living in that particular estuary a microscopic analysis is required. The Table 5.1

below discusses the toxin produced by the most prominent species of cyanobacteria that produce toxins in estuarian ecosystem. Frequency and Severity ranking of presence of toxin producing cyanobacteria is as follows Freshwater > Estuarine (Brackish) > Marine (Fristachi et al. 2008).

Some of the impacts of toxin producing cyanobacteria on waterbody health and ecosystem viability are because of the following. Though the factors responsible for bloom formation and toxin production still remains not well understood. Some of the researchers consider that Chemical factors, Hydrological factors, Light and Temperature may be involved factors in bloom formations and toxin production. Invasion of water milfoil with bacterial blooms are occurred in estuarine habitats. In brackish water cyanobacterial blooms producing toxins get accumulated in fish and leads to its death. Fish kills in estuaries can be associated with cyanobacterial blooms (Fristachi et al. 2008).

5.1.3.1 Microcystis

Microcystis is a cyanobacterium that is the most prevalent generator of freshwater blooms on all continents except Antarctica, and these blooms are typically linked with some degree of toxins (Sellner et al. 2003). Microcystis is classified as a toxic CHAB (cyanobacterial harmful algal bloom) because some species produce potent hepatotoxins known as microcystins, which induce cancer and tumor growth in the livers of mammals and humans. It also causes surface scum, which obstructs recreation, lessens dissolved oxygen content, reduces aesthetics and causes odor and taste issues in drinking water. Survival and growth rate of rotifers, cladocera and copepods are affected by microcystins found in zooplankton food or dissolved in the water column (Lehman et al. 2010). Microcystis, Phormidium, Anabaena genera, Planktothrix, Nostoc and Oscillatoria are the primary producers of Microcystins, with Microcystis being the first known producer (Lopes and Vasconcelos 2011). In fresh and brackish eutrophic environments such as lakes, rivers and estuaries, they can generate cyanobacterial hazardous algal blooms and dominate phytoplankton populations. Natural toxins produced by blooms can induce liver cancers, neurological toxicity and developmental toxicity. Neurotoxin β-methylamino-L-alanine, lipopolysaccharides and the metabolite microcystin are among the natural toxins (Acuña et al. 2020).

5.1.3.2 Nodularia

Nodularia spumigena is a brackish water heterocytous cyanobacterium that has been observed blooming in estuary lakes throughout Europe, South Africa, Canada, the Mediterranean, Australia, the United States and New Zealand (Kaur et al. 2015). Toxic *N. spumigena* blooms have regularly been documented in the Baltic Sea in Europe, and as a result, nodularin is regarded as one of the most numerous naturally occurring compounds in this brackish-water environment (Lopes and Vasconcelos 2011). Nodularin is the most well-known toxin generated by the brackish water cyanobacteria *Nodularia spumigena* (Lopes and Vasconcelos 2011). It is a tumor promoter and probable carcinogen that creates the cyclic pentapeptide hepatotoxin nodularin, which is like microcystins, significantly inhibits protein phosphatases

1 and 2A (Lopes and Vasconcelos 2011). Although nodularin intoxication has not resulted in any human deaths, it has been linked to a range of stock and domestic animal poisoning incidents (Lopes and Vasconcelos 2011).

5.1.3.3 Oscillatoria

Oscillatoria is a common species of bacteria found in freshwater systems across the globe that have been associated with disease and, in some cases, death in animals and humans. Ingestion of polluted water or food products exposes people to cyanobacteria. The presence of these bacteria in the gut exposes the body to their toxins, such as lipopolysaccharide (LPS), to B cells in the gut-associated lymphoid tissue. On the other hand, the effect of Oscillatoria sp. LPS on B cell activation is unclear (Swanson-Mungersona et al. 2017) (Table 5.2).

5.2 Materials

- Glass slides
- Forceps
- Dropper
- Coverslips
- Bacterial culture
- Inoculation loop
- 70% ethanol
- Immersion oil
- Microscope

5.3 Methods

5.3.1 Microscopic Analysis (Hayet et al. 2021)

The identification of cyanobacteria is usually carried out by using a compound light microscope. Collection of water sample containing suspected toxin producing cyanobacteria from the estuarian water body will be the preliminary step.

5.3.1.1 Experimental Protocol

1. Take a clean glass slide and wipe the surface with 70% ethanol
2. Mount the cyanobacterial culture(sample) on the slide using a dropper
3. Make a uniform smear with an inoculation loop or forceps
4. The microscopic slide culture is covered with a glass coverslip
5. View the slide through the microscope at 10×, 40× objective lens and using oil immersion at a magnification of 100× lens of the compound light microscope. The schematic representation of the protocol is given in Fig. 5.1.

Table 5.2 Identification of cyanobacterial toxic blooms of Estuarine habitat and their mode of action

Genus	Toxin produced	Bloom morphology	Biological toxicity	Mode of action	References
Microcystis	Microcystins	Surface scums and dispersed colonies 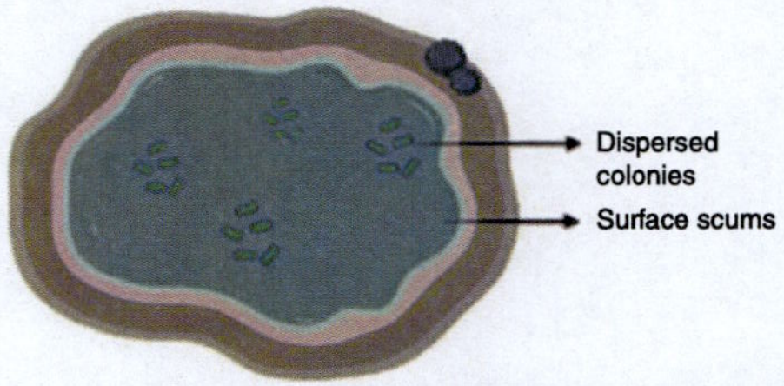	Hepatotoxic	Inhibitors of protein phosphatases 1, 2A and 3, tumor promoter, genotoxicity	Paerl (1988); Swanson-Mungersona et al. (2017)
Nodularia	Nodularin	The bloom was first noticed as a thick layer of brilliant blue-green surface scum and dispersed colonies concentrated by wind-driven advection along the lakeshore 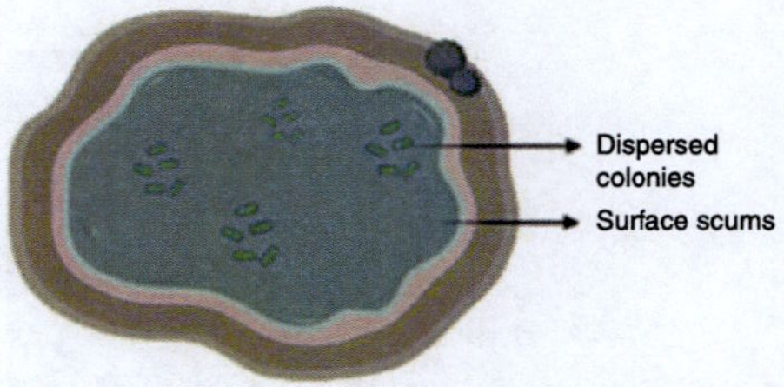	Hepatotoxic	Inhibitors of protein phosphatases 1, 2A and 3, tumor promoter	Paerl (1988); Mcgregor et al. (2012)
Oscillatoria	Lipopolysaccharides (LPS)		Cytotoxic, dermatotoxic, gastroenteritis	Dermonecrotic, protein kinase C activator, and potent tumor promoters	Paerl (1988); Swanson-Mungersona et al. (2017)

(continued)

Table 5.2 (continued)

Genus	Toxin produced	Bloom morphology	Biological toxicity	Mode of action	References
		Subsurface layers and dispersed filaments 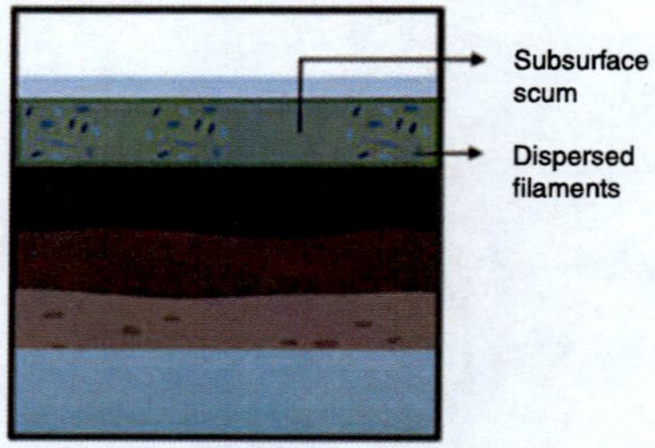			

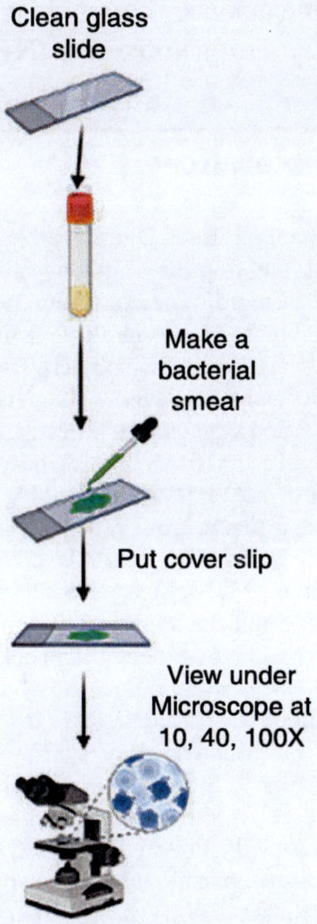

Fig. 5.1 Schematic representation of microscopic identification of cyanobacteria

5.4 Observations/Interpretations

- *Microcystis* colonies are more densely clustered. Cells are spherical shaped, enclosed in a mucilaginous sheath and non-filamentous. Each cell has gas vesicles that often appear blackish blue.
- *Nodularia* trichomes were solitary, cylindrical and straight to slightly flexuous. Akinetes were common, discoid-subspherical mostly single or in pairs. Heterocysts were also compressed discoid.
- *Oscillatoria* trichomes (threads of cells) often appear as dense, opaque ribbons, which oscillate under the microscope. The cells are short cylinders. Trichomes are straight or irregularly twisted.

5.4.1 General Interpretation of Cyanobacteria

Morphologically cyanobacteria are unicellular, Soft, green gelatinous, membranous, filamentous with a true or false type of branching, repeatedly branched, rounded cylindrical, heterocysts or non-heterocysts, trachomatous, perennation (hormogones and akinetes) and multicellular with carbon dioxide fixation (CO_2) in vegetative cells and nitrogen fixation (N_2) in heterocysts (Shiels et al. 2019).

References

Acuña S, Baxa D, Lehman P, Teh FC, Deng DF, Teh S (2020) Determining the exposure pathway and impacts of microcystis on Threadfin Shad, Dorosoma petenense, in San Francisco estuary. Environ Toxicol Chem 39:787–798

Albrecht M, Pröschold T, Schumann R (2017) Identification of cyanobacteria in a eutrophic coastal lagoon on the southern Baltic coast. Front Microbiol 8:1–16

Bormans M, Savar V, Legrand B, Mineaud E, Robert E, Lance E, Amzil Z (2020) Cyanobacteria and cyanotoxins in estuarine water and sediment. Aquat Ecol 54:625–640. https://doi.org/10.1007/s10452-020-09764-y

Brown A, Foss A, Miller MA, Gibson Q (2018) Detection of cyanotoxins (microcystins/nodularins) in livers from estuarine and coastal bottlenose dolphins (Tursiops truncatus) from Northeast Florida. Harmful Algae 76:22–34

Elliott M, McLusky DS (2002) The need for definitions in understanding estuaries. Estuar Coast Shelf Sci 55:815–827

Fristachi A, Sinclair JL, Hall S et al (2008) Occurrence of cyanobacterial harmful algal blooms workgroup report. Adv Exp Med Biol 619:45–103

Gibble CM, Kudela RM (2014) Detection of persistent microcystin toxins at the land-sea interface in Monterey Bay, California. Harmful Algae 39:146–153

Hayet S, Sujan KM, Mustari A, Miah MA (2021) Hemato-biochemical profile of Turkey birds selected from Sherpur district of Bangladesh. Int J Adv Res Biol Sci 8:1–5

Kaur HP, Prasad B, Kaur S (2015) A review on applications of biosurfactants produced from unconventional inexpensive wastes in food and agriculture industry. World J Pharm Res 4:827

Lehman PW, Teh SJ, Boyer GL, Nobriga ML, Bass E, Hogle C (2010) Initial impacts of Microcystis aeruginosa blooms on the aquatic food web in the San Francisco estuary. Hydrobiologia 637:229–248

Lehman PW, Kurobe T, Teh SJ (2022) Impact of extreme wet and dry years on the persistence of Microcystis harmful algal blooms in San Francisco estuary. Quat Int 621:16–25

Lopes VR, Vasconcelos VM (2011) Planktonic and benthic cyanobacteria of European brackish waters: a perspective on estuaries and brackish seas, pp 292–304

Mcgregor GB, Stewart I, Sendall BC, Sadler R, Reardon K, Carter S, Wruck D, Wickramasinghe W (2012) First report of a toxic Nodularia spumigena (Nostocales/Cyanobacteria) bloom in sub-tropical Australia. I. Phycological and public health investigations. Int J Environ Res Public Health 9:2396–2411

Miller MA, Kudela RM, Mekebri A et al (2010) Evidence for a novel marine harmful algal bloom: cyanotoxin (microcystin) transfer from land to sea otters. PLoS One 5:1–11

Paerl HW (1988) Nuisance phytoplankton blooms in coastal, estuarine, and inland waters. Limnol Oceanogr 33:823–843

Perillo GM (1995) Geomorphology and sedimentology of estuaries. Developments in sedimentology. Geomorphol Sedimentol Estuaries 53:17–47

Rastogi RP, Madamwar D, Incharoensakdi A (2015) Bloom dynamics of cyanobacteria and their toxins: environmental health impacts and mitigation strategies. Front Microbiol 6:1–22

Sellner KG, Doucette GJ, Kirkpatrick GJ (2003) Harmful algal blooms: causes, impacts and detection. J Ind Microbiol Biotechnol 30:383–406

Shiels K, Browne N, Donovan F, Murray P, Saha SK (2019) Molecular characterization of twenty-five marine cyanobacteria isolated from coastal regions of Ireland. Biology (Basel) 8:59. https://doi.org/10.3390/biology8030059

Swanson-Mungersona M, Incroccia R, Subramaniamb V, Williamsc P, Halld ML, Mayer AMS (2017) Effects of cyanobacteria Oscillatoria sp. lipopolysaccharide on B cell activation and Toll-like receptor 4 signaling. Physiol Behav 275:101–107

United States Environmental Protection Agency (2020) Cyanobacteria and cyanotoxins. In: Cyanobacteria and cyanotoxins, pp 1–11

Identification of Toxic Blooms of Cyanobacteria in Saltpan Habitat

6

Ramasamy Thangaraj, Mohamed Arief Mohideen Pitchai, and N. Thajuddin

Abstract

Cyanobacteria are photosynthetic prokaryotic which are found in a wide variety of habitats. Excess salt concentration in sea water is said to be hypersaline environments. Saltpans are one of the hypersaline extreme environment that exhibit a wide range of environment stress through salinity changes. Cyanobacterial toxins recreational the development of numerous method to detect the toxin and their product as well as identification and quantification of toxins. Cyanotoxins are chemical compounds produced by cyanobacterial mats grown in aquatic ecosystems. These may threaten human health and aquatic organisms. Extraction of these toxins is usually associated with many difficulties due to their concentration in aquatic ecosystems. In current study, we overview numerous methods that have been developed for the cyanotoxin analysis.

Keywords

Cyanobacteria · Saltpan · Extraction and cyanotoxin

R. Thangaraj (✉) · M. A. M. Pitchai
Department of Microbiology, Ayya Nadar Janaki Ammal College, Sivakasi, Tamil Nadu, India

N. Thajuddin
B.S. Abdur Rahman Crescent Institute of Science & Technology (Deemed to be University), Chennai, Tamil Nadu, India

Department of Microbiology, National Repository for Microalgae and Cyanobacteria - Freshwater (NRMC-F), School of Life Sciences, Bharathidasan University, Tiruchirappalli, Tamil Nadu, India

N. Thajuddin et al. (eds.), *Protocols for Cyanobacteria Sampling and Detection of Cyanotoxin*, https://doi.org/10.1007/978-981-99-4514-6_6

6.1 Introduction

Cyanobacteria found in various environment such as sea, river, lake, and ponds, but some algae like cyanobacteria and blue green algae present in the saltpan. Saltpan are constructed by human because of production of raw salt along the side of sea. In the hypersaline environment of saltpan diversity of microalgae community is very low, but among the halophilic microalgae, diatoms, cyanobacteria, and green algae occupied or found in the saltpan. Saltpan represents the majority simplified ecosystems for the simple reason that the number of species at any trophic level is low. In the hypersaline extreme environment microalgae community and diversity is very low (Nagasathya and Thajuddin 2008).

Cyanobacteria are practical grouping of photosynthetic organisms growing using with the sunlight; it includes a diverse array of colour and morphologies from single cell to multicellular (Ashok Kumar et al. 2011). Cyanobacteria have high morphological and ecological variability (Whitton and Potts 2012), and due to this, its taxonomical classification is relatively intricate and is in constant change (Komarek et al. 2014). New cyanobacteria families, genera, and species, as Oculatellaceae and Trichocoleaceae (Mai et al. 2018), Aliinostoc (Bagchi et al. 2017), Halotia (Genuario et al. 2015), and Lusitaniella, and Hyellapatelloides (Brito et al. 2017) and Compactonostoc shennongjiaensis (Cai et al. 2019) have been reported in recent years. These taxonomic redefinitions are challenging for the identification and distribution of toxins in many of the new genera (Osterholm et al. 2020), thus increasing the lack of knowledge on toxic strains. Although cyanobacteria are widely distributed through many environments, most cyanobacteria and cyanotoxins studies are focused on planktonic species from inland freshwater systems (Cires et al. 2017), with less information in extreme environments (e.g., hot springs, caves, hypersaline lakes, polar deserts), especially regarding their toxicity.

Some cyanobacteria produced many more toxins, such as cyanotoxins and microcystin. These cyanotoxins very harmful ecological, cause health social and economic effect. It reduced the water quality and aquatic environment ecosystem (Plaas and Paerl 2021). Several efficient sensitive chemical analytical methods and bioassay have been developed to determine the concentration of cyanotoxins in environmental samples. Some of the methods are currently applied for the routine monitoring; the others are mainly used as scientific tool for the harmful algae. The aim of the various assessment methods developed for cyanotoxin analysis and to discuss the advance and shortages of their application. These cyanotoxins are structurally cyclic heptapeptide amino acids (Shimizu et al. 2014).

There are few certified standards for cyanotoxins variant also. To determine cyanobacterial toxins in animal tissues, the reversed phase high performance liquid chromatography method coupled with UV, photodiode array, and mass spectrometer detectors was applied. Different microcystins have been isolated from several species of cyanobacteria, and several of them can be produced by a single bloom (Zastepa et al. 2017). Cyanobacteria produce the level of cyanotoxins in blooms rather than characterizing the water-soluble fractions of the toxins (Ai et al. 2020).

6.2 Materials Required

- Polythene bags
- Whatman filter
- Distilled water
- Plastic bottle
- Magnetic stirrer
- BG 11 medium
- Centrifuge
- Sonicator
- NaCl
- Methanol
- Ethyl acetate
- Petri plate
- Conical flask
- Spectrometer

6.3 Methods

6.3.1 Collection of Cyanobacterial Mats

- Cyanobacterial mats along with their aqueous media are collected from Sivakasi temple.
- (Enhhal et al., 2013). The samples are transfer into the laboratory and is maintained at temperature of 22 °C ± 2 °C during the experimental work in the laboratory.
- The sample is transferred to a glass aquarium 16 L capacity and left for 2 days in the laboratory condition to allow the mats to reform again.
- Then the cyanobacterial mat is collected and transferred to a glass plate and left in the laboratory for air-drying. The air-dried materials are kept in plastic bottles for further analysis (Fig. 6.1).

6.3.2 Extraction of Cyanotoxins from Media by Activated Charcoal

- The media (one litre cyanobacterial mats growth media) is filtered through cyanobacterial mats and growth media collection.
- Use Whatman filter paper to remove the suspended particulate matter.
- Series of two litre capacity, quick fit round bottom flasks are washed with distilled water and subdivided into four groups. Group 1 containing 10 g of activated charcoal, Group 2 containing 10 g of clay (bentonite), Group 3 containing 10 g of clay-HDTMA complex (Hexadecyltrimethylammonium bromide), and Group 4 contained 10 mL (hexane: ethyl acetate mixture 9:1).

Fig. 6.1 Sampling site view of saltpan and microscopic view of cyanobacteria

- One litre of filter growth media is added to each flask. Then the flask is kept under continuous magnetic stirring for 48 h. The magnetic stirrers are stopped to allow precipitation charcoal, clay, and clay-HDTMA. Group 4 contents are transferred to a glass cylinder 4 L capacity. Then the cylinder is left for 2 h to allow the organic layer to form.

- Then the organic layer is collected and the organic solvent is removed by gentle reduced pressure using rotary evaporation (Safi et al. 2019; Nahhal et al. 2020). The content of each flask is separated by filtration. Each supernatant is re-extracted again by the same materials mentioned above.
- At the end of the extraction processes, the solid materials are collected separately and allowed to air-drying in the lab for 24 h.
- Then the collected solid materials are transferred to a glass flask contained 20 mL solvent mixture (hexane: ethyl acetate 9:1 v/v) and transferred to ultrasonic device for a period of 15 min at high speed.
- The solvent mixture is separated by filtration and the extraction procedure is repeated again with another 10 mL solvent mixture as mentioned above.
- The collected solvent mixture is evaporated under gentle reduced pressure to re-concentrate the cyanotoxin to allow the formation of cyanotoxin crystals.
- The collected dried sediments (charcoal, clay, and clay-HDTMA) (about 20 g) was transferred to a 100 mL conical flask containing 50 mL absolute methanol and put under sonication at high speed for 15 min.
- Then the content is filtered and the charcoal is extracted again by the volume and procedure mentioned above. The supernatant is collected and evaporated under reduced gentle pressure using a rotary evaporator up to complete dryness. This allows the formation of crude cyanotoxins crystals.
- This extraction procedure is repeated using clays and organo-clays to extract the cyanotoxins. The extraction procedure of cyanotoxins from aqueous media (growth media) and cyanobacterial mats (sediment) was done using different adsorbing materials (solid phase extraction).

References

Ai Y, Lee S, Lee J (2020) Drinking water treatment residuals from cyanobacteria bloom-affected areas: investigation of potential impact on agricultural land application. Sci Total Environ 706: 1–11

Ashok Kumar N, Viji D, Baluswami M (2011) A short term study of algal flora of a salt pan near Chennai. Int J Recent Sci Res 2(12):292–296

Bagchi SN, Dubey N, Singh P (2017) Phylogenetically distant clade of Nostoc-like taxa with the description of Aliinostoc gen. nov. and Aliinostocmorphoplasticum sp. nov. Int J Syst Evol Microbiol 67:3329–3338

Brito A, Ramos V, Mota R, Lima S, Santos A, Vieira J, Vieira CP, Kastovsky J, Vasconcelos VM, Tamagnini P (2017) Description of new genera and species of marine cyanobacteria from the Portuguese Atlantic coast. Mol Phylogenet Evol 111:18–34

Cai F, Li X, Yang Y, Jia N, Huo D, Li R (2019) Compactonostoc shennongjiaensis gen. & sp. nov. (Nostocales, Cyanobacteria) from a wet rocky wall in China. Phycologia 58:200–210

Cires S, Casero MC, Quesada A (2017) Toxicity at the edge of life: a review on cyanobacterial toxins from extreme environments. Mar Drugs 15:233

Genuario DB, Vaz MGMV, Hentschke GS, Sant Anna CL, Fiore MF (2015) Halotia gen. nov., a phylogenetically and physiologically coherent cyanobacterial genus isolated from marine coastal environments. Int J Syst Evol Microbiol 65:663–675

Komarek J, Kastovsky J, Mares J, Johansen JR (2014) Taxonomic classification of cyanoprokaryotes (cyanobacterial genera) 2014, using a polyphasic approach. Preslia 86:295–335

Mai T, Johansen JR, Pietrasiak N, Bohunicka M, Martin MP (2018) Revision of the Synechococcales (Cyanobacteria) through recognition of four families including Oculatellaceae fam. nov. and Trichocoleaceae fam. nov. and six new genera containing 1410 species. Phytotaxa 365:1–59

Nagasathya A, Thajuddin N (2008) Diatom diversity in hypersaline environment. J Fish Aquat Sci 3:328–333

Nahhal Y, Lubbad R, Al-Agha MR (2020) Toxicity evaluation of Chlorpyrifos and Diuron below maximum residue limits in rabbits. Toxicol Environ Health Sci 12:177–190

Osterholm J, Popin RV, Fewer DP, Sivonen K (2020) Phylogenomic analysis of secondary metabolism in the toxic cyanobacterial genera Anabaena, Dolichospermum and Aphanizomenon. Toxins 12:248

Plaas HE, Paerl HW (2021) Toxic cyanobacteria: a growing threat to water and air quality. Environ Sci Technol 55(1):44–64

Safi JM, Yassin MM, El-Nahhal YZ, Abed YA, Safi MJ, Suleiman HD (2019) Childhood Lead poisoning in Gaza strip, The Palestinian Authority. J Trace Elem Med Biol 54:118–125

Shimizu K, Sano T, Kubota R, Kobayashi N, Tahara M, Obama T, Sugimoto N, Nishimura T, Ikarashi Y (2014) Effects of the amino acid constituents of microcystin variants on cytotoxicity to primary cultured rat hepatocytes. Toxins 6:168–179

Whitton B, Potts M (2012) Introduction to the cyanobacteria. In: Whitton BA (ed) The ecology of cyanobacteria II–their diversity in time and space. Springer, Dordrecht, pp 1–13. isbn:978-94-007-3854-6

Zastepa A, Watson SB, Kling H, Kotak B (2017) Spatial and temporal patterns in microcystin toxins in Lake of the Woods surface waters. Lake Reservoir Manag 33(4):433–443

Sampling and Identification of Toxic Cyanobacteria in the Landfill Leachate

7

Vishnu Chitthan, Nisha Banu, N. Thajuddin, and D. Dhanasekaran

Abstract

The knowledge on anthropogenic environments have shown increasing interest around worldwide nowadays. Dumping of solid waste creates landfill leachate when it is exposed to rain. Landfill leachate proliferates the toxic photosynthetic cyanobacteria which create harmful algal blooms (cyanoHABs). The toxic cyanobacteria produce a variety of cyanotoxins which are microcystins, saxitoxins, cylindrospermopsin, and anatoxin-a; these toxins have different mode of action such as hepatotoxins, neurotoxins, cytotoxins, and gastrointestinal toxins. Cyanotoxins affects the growth or leads to mortality to the biodiversity which dependent of that particular environment. This is important to take care because the overall food chain is affected due to cyanoHABs in the aquatic environment. The cyanobacteria genera occur in the toxic environment can be identified by through its morphological structure using light microscopy, but the confirmatory way of detection till species level is only through PCR amplification of that specific gene which codes for the production of cyanotoxin. Thus, the molecular technique is always reliable for detection of cyanotoxin synthesising gene through specific primers (e.g., mcyE-F2 and mcyE-R4 is used in detecting

V. Chitthan (✉) · N. Banu
Department of Microbiology, Bharathidasan University, Tiruchirappalli, Tamil Nadu, India

N. Thajuddin
B.S. Abdur Rahman Crescent Institute of Science & Technology (Deemed to be University), Chennai, Tamil Nadu, India

Department of Microbiology, National Repository for Microalgae and Cyanobacteria - Freshwater (NRMC-F), School of Life Sciences, Bharathidasan University, Tiruchirappalli, Tamil Nadu, India

D. Dhanasekaran
Department of Microbiology, National Repository for Microalgae and Cyanobacteria - Freshwater (NRMC-F), Bharathidasan University, Tiruchirappalli, Tamil Nadu, India

N. Thajuddin et al. (eds.), *Protocols for Cyanobacteria Sampling and Detection of Cyanotoxin*, https://doi.org/10.1007/978-981-99-4514-6_7

mycE gene). The most widespread of the cyanotoxins are the peptide toxins in the class called microcystins; there are around 80 known microcystins, including Microcystin-LR, which is generally considered one of the most toxic.

Keywords

Cyanobacteria · Microcystins · CyanoHABs and landfill leachate

7.1 Introduction

The aggregation of organic and inorganic waste from industries and agricultural sectors creates anthropogenic environment. The landfill leachate formed by percolation of rainwater and intervention of ground water through the solid waste which causes extraction of suspended and soluble solids in it. This waste combines and contaminates both the soil and aquatic environments thus leading to an interruption in the food chain of aquatic organisms. According to landfill leachate, the biggest problem is there are no standard assessment practice is followed till date due to the different types of organic and inorganic substance combines at a wide range of combinations (Montvydienė et al. 2020).

Landfill leachate contains high level of anthropogenic nutrients which lead to formation of cyanobacterial harmful blooms (cyanoHABs) (Beversdorf et al. 2015; Huisman et al. 2005). These cyanoHABs produce toxic secondary metabolites that affect the growth and lead to mortality to the zooplankton, fish, and even mammals, which depends on the aquatic ecosystem for consumption and habitat (Bartram and Chorus 1999). The cyanobacterial toxicity is expressed after the cell is collapsed and its internal bioactive compounds are released in the environment due to increased concentration in the metals, etc. (Fauziah et al. 2013; Osada et al. 2011). The cyanobacteria produce a variety of toxins such as microcystins, saxitoxins, cylindrospermopsin, and anatoxin-a; these toxins are specific and affect undesirably in the body from skin rashes to lung failure (Cheung et al. 2013). These cyanotoxins are classified based on their mode of action, such as hepatotoxins, neurotoxins, cytotoxins, and gastrointestinal toxins (Carmichael 1997). The *Microcystis aeruginosa* is a species which is mainly known to produce a ubiquitous cyanotoxins called microcystins that have hepatotoxic effect which could be considered as a cyclic hepato peptide (Oberholster et al. 2003). Still now the number of cyanobacterial taxa which is a cause for the toxins remains unclear globally, and this data will increase when the exploration of new environments and toxins progress (Shams et al. 2015). The identification of genus responsible for cyanotoxin can be quickly assessed by visualizing it under light microscopy, but a complete and reliable assessment of risk is evaluated through molecular and analytical techniques to detect under species level by knowing its genetic and chemotypic differences. Therefore, the method suitable for characterization of its genetic and chemotypic difference is amplification of DNA by PCR method and the amplicons can be further analysed for the presence of genes that codes for cyanotoxin synthesis (Meriluoto and Codd n.d.; Meriluoto et al. 2017). This chapter defines the way of sample

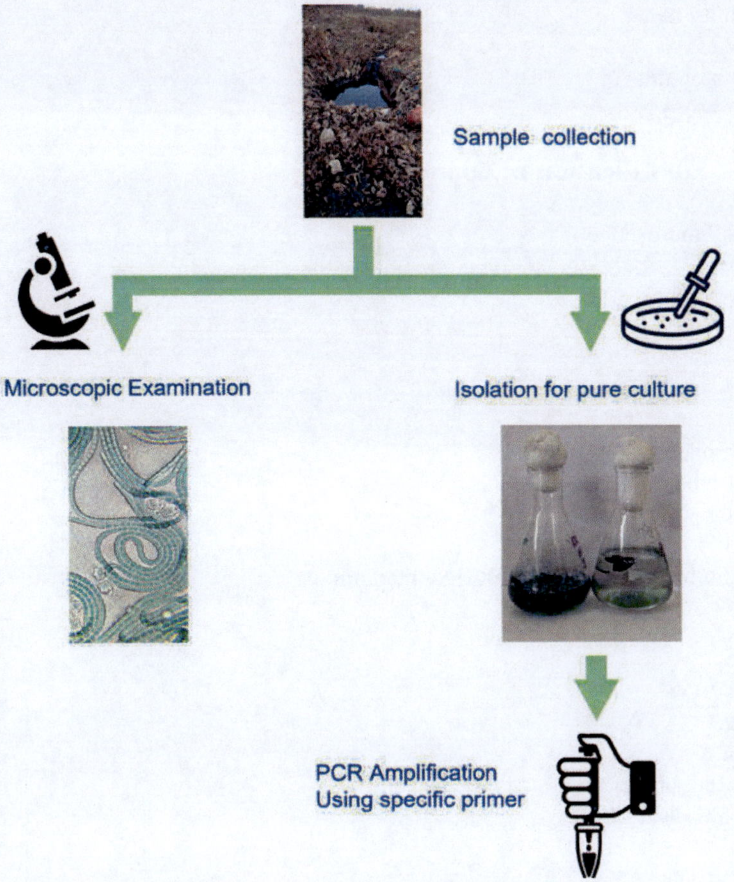

Fig. 7.1 Isolation and identification of toxic cyanobacteria from landfill leachate

collection from landfill leachate and preliminary screening to identify the genus through morphological identification under light microscope. Then it provides information on pure culturing a single cyanobacterial strain for extraction of DNA to identify its sequence to study its gene which codes for the cyanotoxin (Fig. 7.1).

7.2 Materials

7.2.1 Cyanobacteria Sample Collection

- Plastic bag
- Forceps
- pH meter

- Salinity meter
- Gloves
- Plankton net

7.2.2 Identification of Cyanobacteria

- Light microscope
- Slides
- Cover slips

7.2.3 Isolation of Culture

- Petri plates
- Test tubes
- Flasks
- BG11+ medium
  Medium composition for BG11+ medium

$NaNO_3$	1.5 g
K_2HPO_4	0.04 g
$MgSO_4 \cdot 7H_2O$	0.075 g
$CaCl_2 \cdot 2H_2O$	0.036 g
Citric acid	0.006 g
Ferric ammonium citrate	0.006 g
EDTA (disodium salt)	0.001 g
Na_2CO_3	0.02 g
Trace metal mix A5	1.0 mL
Agar (if needed)	10.0 g
Distilled water	1.0 L

The pH should be 7.1 after sterilization.
Trace metal mix A5

H_3BO_3	2.86 g
$MnCl_2 \cdot 4H_2O$	1.81 g
$ZnSO_4 \cdot 7H_2O$	0.222 g
$NaMoO_4 \cdot 2H_2O$	0.39 g
$CuSO_4 \cdot 5H_2O$	0.079 g
$Co(NO_3)_2 \cdot 6H_2O$	49.4 mg
Distilled water	1.0 L

7.2.4 DNA Isolation

- Eppendorf tube
- 50 mM Tris
- 5 mM EDTA
- 50 mM NaCl
- Lysozyme
- Proteinase K
- 10% SDS
- Phenol: Chloroform: Isoamyl alcohol (25:24:1)
- 4 M ammonium acetate
- Isopropanol
- 70% ethanol
- TE Buffer

7.2.5 PCR Amplification

- 10× Taq polymerase buffer
- dNTPs
- Thermocycler
- mcyE-F2-Forward Primer (Sequence: GAA ATT TGT GTA GAA GGT GC)
- mcyE-R4 -Reverse Primer (Sequence: AAT TCT AAA GCC CAA AGA CG)

7.3 Methods

7.3.1 Cyanobacterial Sample Collection

7.3.1.1 Cyanobacterial Mat Collection from Leachate

1. This procedure should be carried out with gloves for safety purposes.
2. Collect the cyanobacterial mat using forceps and transfer into a plastic cover or vial containing little quantity of sterilised BG11+ medium.

7.3.1.2 Cyanobacteria Sample Collection from Leachate

Use Plankton nets with mesh size 42 μm, rinse and tilt the samples to collect the cyanobacteria present freely in the leachate.

7.3.2 Morphological Identification of Cyanobacteria Using Light Microscopy

1. Cyanobacterial diversity should be identified based on its morphological structure using light microscope.

2. To identify the genus and species with the help of published reports and standard monographs.

7.3.3 Isolation of Pure Culture

1. Spread plate method-Pipette out 50µl of the leachate sample into BG11+ agar plate with agar concentration 1.2% and spread it over using an L rod.
2. Incubate the plate in 2000 lux light intensity for 12:12 h on and off cycle.

7.3.4 Culture Purification

1. After 7 days of incubation, magnify the plate under light microscope for the presence of pure filamentous cyanobacteria and pick gently using a sterile needle to inoculate it in a test tube with 1 mL of sterilised BG11+ medium.
2. Incubate the test tube in 25 °C at 2000 lux light intensity.
3. It takes around 3–4 weeks for a visible growth in the medium.

7.3.5 Molecular Identification

7.3.5.1 DNA Isolation

1. Suspend the cyanobacterial cells in TE buffer and grind it for at least 5 min.
2. Transfer it into a fresh 1.5 mL centrifuge tube.
3. Add lysozyme and RNase A to the tube for pre-treatment and incubate it for 30 min.
4. Then add 30 µL of 10% SDS and 200 µL of phenol: chloroform: isoamyl alcohol (25:24:1) into the tube and vortex it.
5. Centrifuge the tubes at 15,000 rpm for 5 min on 4 °C and collect the supernatant.
6. Perform the purification of DNA by adding 0.1 volume of 3 M ammonium acetate solution, with that add 2.5 volume of 100% ethanol.
7. Maintain the tubes in −20 for 1 h and then centrifuge it for 20 min at 15,000 rpm.
8. Wash the pellet with 70% ethanol after discarding the supernatant and dry it before being resuspended in 100 µL TE buffer.

7.3.5.2 Detection of Cyanotoxin Synthetase Gene by PCR Amplification Method

The most widespread of the cyanotoxins are the peptide toxins in the class called microcystins. The procedure to detect microcystin synthetase gene is given below:

1. For PCR detection mcyE-F2 and mcyE-R4 primers are suitable for determining mcyE gene.
2. Perform the reaction mixtures in 20-µL volumes with 1µL of DNA template.

3. PCR mixture with 1 × Mg (+) PCR buffer, 250 μM dNTPs, 0.5 μM of forward and reverse primers, 0.5 U of Ex Taq DNA polymerase and sterile water must be used (Vaitomaa et al. 2003).

PCR Cycles

- Initial denaturation −95 °C for 3 min.
- Followed by 24 cycles of 94 °C for 30 s, 60 °C for 30 s, and 72 °C for 1 min.
- Final extension −72 °C for 10 min.

Confirm the presence of mcyEgene through 1% agarose gel electrophoresis.

7.4 Observation

In this chapter, the detection of only microcystin synthetase gene is covered; there are a variety of cyanotoxins, and each has its specific primers for detection. Since microcystin is the most common cyanotoxin present in the wastewater, it is taken for primary detection.

References

Bartram J, Chorus I (eds) (1999) Toxic cyanobacteria in water. CRC Press, Boca Raton. https://doi.org/10.1201/9781482295061

Beversdorf LJ, Miller TR, McMahon KD (2015) Long-term monitoring reveals carbon-nitrogen metabolism key to microcystin production in eutrophic lakes. Front Microbiol 6:456. https://doi.org/10.3389/fmicb.2015.00456

Carmichael W (1997) The cyanotoxins. In: Incorporating in plant pathology—classic papers. Elsevier, pp 211–256. https://doi.org/10.1016/S0065-2296(08)60282-7

Cheung MY, Liang S, Lee J (2013) Toxin-producing cyanobacteria in freshwater: a review of the problems, impact on drinking water safety, and efforts for protecting public health. J Microbiol 51(1):1–10. https://doi.org/10.1007/s12275-013-2549-3

Fauziah SH, Izzati MN, Agamuthu P (2013) Toxicity on Anabas Testudineus: a case study of sanitary landfill leachate. Procedia Environ Sci 18:14–19. https://doi.org/10.1016/j.proenv.2013.04.003

Huisman J, Matthijs HCP, Visser PM, editors (2005) Harmful cyanobacteria, vol. 3. Springer. https://doi.org/10.1007/1-4020-3022-3

Meriluoto J, Codd G (n.d.) Toxic: cyanobacterial monitoring and cyanotoxin analysis

Meriluoto J, Spoof L, Codd GA (2017) Handbook of cyanobacterial monitoring and cyanotoxin analysis

Montvydienė D, Šulčius S, Jurgelėnė Ž, Makaras T, Kalcienė V, Taraškevičius R, Kazlauskas M, Kazlauskienė N (2020) Contrasting ecotoxic effects of landfill leachate and cyanobacterial biomass on aquatic organisms. Water Air Soil Pollut 231(7):323. https://doi.org/10.1007/s11270-020-04684-x

Oberholster P, Botha A-M, Grobbelaar J (2003) Microcystis aeruginosa: source of toxic microcystins in drinking water. Afr J Biotechnol 3(3):159–168. https://doi.org/10.5897/AJB2004.000-2029

Osada T, Nemoto K, Nakanishi H, Hatano A, Shoji R, Naruoka T, Yamada M (2011) Analysis of ammonia toxicity in landfill leachates. ISRN Toxicol 2011:1–6. https://doi.org/10.5402/2011/954626

Shams S, Capelli C, Cerasino L, Ballot A, Dietrich DR, Sivonen K, Salmaso N (2015) Anatoxin-a producing Tychonema (Cyanobacteria) in European waterbodies. Water Res 69:68–79. https://doi.org/10.1016/j.watres.2014.11.006

Vaitomaa J, Rantala A, Halinen K, Rouhiainen L, Tallberg P, Mokelke L, Sivonen K (2003) Quantitative real-time PCR for determination of microcystin synthetase E copy numbers for *Microcystis* and *Anabaena* in lakes. Appl Environ Microbiol 69(12):7289–7297. https://doi.org/10.1128/AEM.69.12.7289-7297.2003

Handling and Processing of Cyanobacterial Toxin Containing Pond and Lake Water Samples

8

Tamalika Chakraborty, Sutripto Ghosh, Sumana Roy, and Dipanjan Mandal

Abstract

Blue-green algae, or cyanobacteria, are a kind of prokaryotic cell that can photosynthesize because it contains the chlorophyll molecule. They were the first cells on Earth thought to be capable of photosynthetic metabolism. Their absence of membrane-bound organelles is a defining feature. The thylakoids are present, dispersed throughout the cytoplasm. Microcystins, on the other hand, are toxic chemicals made by cyanobacterial cells that have been shown to have a direct negative impact on aquatic ecosystems and the human population. The purpose of this study was to identify cyanobacterial cells and examine the presence of microcystin in lake and pond water samples. Both Gram staining and PCR based on 16S rDNA detection verified the presence of cyanobacterial cells. The purest form of the toxin is obtained using column chromatography followed by high-performance liquid chromatography. Furthermore, microcystin was found in both of the separated water samples using an ELISA designed specifically for microcystin detection.

T. Chakraborty (✉) · S. Ghosh
Division of Life Science, Guru Nanak Institute of Pharmaceutical Science and Technology, Kolkata, West Bengal, India
e-mail: tamalika.chakraborty@gnipst.ac.in

S. Roy
Division of Pharmaceutics, Guru Nanak Institute of Pharmaceutical Science and Technology, Kolkata, West Bengal, India

D. Mandal
Division of Pharmacology, Guru Nanak Institute of Pharmaceutical Science and Technology, Kolkata, West Bengal, India

N. Thajuddin et al. (eds.), *Protocols for Cyanobacteria Sampling and Detection of Cyanotoxin*, https://doi.org/10.1007/978-981-99-4514-6_8

Keywords

Cyanobacteria · Cyanotoxin · Microcystin · Blue-green algae · HPLC · ELISA

8.1 Introduction

Blue-green algae, or cyanobacteria, can be found in both fresh and salt water. In particular, cyanotoxins, which are generated by toxic species of cyanobacteria during times of eutrophication, pose a threat to human and animal health (Lawton and Codd 1991). It has been reported that both humans and animals have been poisoned by cyanotoxins. Moreover, hepatocyte apoptosis and necrosis are caused by cyanotoxins because of their role as inhibitors of protein phosphatases 1 and protein phosphatases 2A (PP1 and PP2A, respectively) (Dawson 1998). Further, they are responsible for digestive disorder and rashes (Chernoff et al. 2002; Martins et al. 2005).

Microcystins, a class of cyclic heptapeptides, are poisonous cyanotoxins that display their noxious effects because of the presence of the odd amino acid Adda (3-amino-9-methoxy-2,6,8-trimethyl-10-phenyl-4,6-dienoic acid) (Codd et al. 1999). More than 70 distinct microcystin analogues have been identified (MacElhiney and Lawton 2005).

While microcystins have been studied extensively in both fresh and brackish water, marine settings have received less attention (Lawton and Codd 1991). There have been detections of microcystin in seawater all around the world, including the Atlantic, Pacific, Caribbean, Indian, Arabian, and Mediterranean seas (Martins et al. 2005; MacElhiney and Lawton 2005). Microcystins were detected in marine waters with concentrations ranging from 0.003 to 19.8 ng/L in one Greek study, from Amvrakikos Gulf (Taş et al. 2006; Vareli et al. 2012).

The presence of cyanotoxins have been reported in various parts of the world, especially in areas utilized for human activities like fishing or tourism. However, non-adequate information is available on marine cyanobacteria that are hazardous. On the other hand, illegitimate use of fertilizers in agriculture has resulted in the release of phosphate along with nitrate salts in waterbodies (Poulos et al. 2000). Cyanobacterial blooms are the end result of a chain of events that have led to eutrophication. It has been observed that people engaging in recreational and other professional activities in waterbodies polluted with cyanotoxins from cyanobacteria suffer immediate deleterious effects. People who come into contact with cyanotoxin-contaminated water have been documented to suffer from a variety of adverse health effects, including but not limited to skin infections, eye irritations, allergies, and a number of water-borne disorders (Sangolkar et al. 2006; Stewart et al. 2009). As a result, divers, fish farmers, water sport instructors, people who clean the shore, those who harvest fish and shellfish, and fishermen all run the danger of being exposed to harmful marine cyanobacteria while doing their jobs (Stewart et al. 2009). Ingestion of toxic marine cyanobacteria is thought to be detrimental to zooplankton as well as crustaceans (Figueiredo et al. 2004; Carmichael 1992). In addition, research using mussel embryos has demonstrated that extracts of the cyanobacteria *Synechocystis* spp. and *Synechococcus* spp. totally block development of the embryos (Carmichael 1992; Martins et al. 2005).

The objective of this study was to develop a methodology for handling as well as sampling of cyanobacterial toxin.

8.2 Materials

8.2.1 Materials to Prepare Pringsheim's Broth

- Potassium nitrate
- Magnesium sulphate
- Ammonium nitrogen phosphate
- Calcium chloride
- Iron chloride
- Weight balance
- Autoclave

8.2.2 Materials Required for the Isolation of Cyanobacteria

- Pond water
- Lake water
- Distilled water
- Sodium chloride

8.2.3 Materials Required for Gram Staining

- Glass slides
- Crystal violet
- Gram's iodine
- Ethanol (70%)
- Safranin
- Light microscope

8.2.4 Materials Required for Measuring the Total Protein Concentration by Folin Lowry Method

- Sodium carbonate 0.1 N
- Copper sulphate solution
- Sodium potassium tartrate
- Lowry reagent

- Folin Ciocalteu reagent
- Spectrophotometer

8.2.5 Materials Required for Structural Analysis

8.2.5.1 Microtiter Plate

- Sodium acetate buffer (pH 4.6)
- Sodium chloride (20%)
- Vacuum grease
- Cover slip
- Forcep
- X-ray diffractometer

8.2.6 Other Materials Required

- Brij35 buffer
- Sodium dodecyl sulphate
- Agarose
- Superdex resin
- Sephadex-LH20 resin
- Acetonitrile
- Double distilled water
- Measuring cylinder
- Conical flask 1000 mL
- Glass bottle
- Micropipette
- Microtips
- Spirit lamp
- Laminar hood
- Test tubes
- Tris-HCl
- EGTA

8.3 Procedure

8.3.1 Preparation of Pringsheim's Broth for Cyanobacteria Isolation (Pringsheim 1946; Thakare et al. 2018)

1. Weigh 0.2 g of Potassium nitrate, 0.010 g of Magnesium sulphate, 0.02 g of Ammonium nitrogen phosphate, 0.005 g of Calcium chloride, and 0.005 g of Iron chloride, and add them to a conical flask.

2. Add 1000 mL of distilled water to the conical flask containing the required weighed substances.
3. Autoclave the conical flask containing the desired substances for 15 min at 121 ° C and 15 psi pressure for sterilization.

8.3.2 Isolation of Cyanobacteria

1. Collect pond water and lake water in two different clean glass bottles.
2. Serially dilute the sample from two different sources from 10^{-1} through 10^{-6} using 0.9% NaCl solution.
3. Transfer 1 mL of each dilution each into separate test tubes containing 20 mL of Pringsheim's broth.
4. Incubate at 25–28 °C for 4 weeks in sunlight, for the photosynthetic cyanobacteria to grow.
5. The samples are then stained using Gram staining.
6. Confirm the presence of cyanobacterial cells by PCR analysis using primers, based on the amplification of 16S rDNA of cyanobacterial cells.

8.3.3 Staining Using Gram Staining Method (Kalaitzidou et al. 2015; Hoiczyk and Hansel 2000)

1. Take a small amount of water sample from both the sources into two different slides
2. Stain both the slides by using crystal violet as primary stain
3. Add Gram's iodine to both the slides which acts as mordant.
4. Add ethanol to both the slides to discolorize.
5. Add two drops of Safranin to both the slides.
6. Observe and examine under the light microscope.

8.3.4 Extraction of Intracellular Microcystin from Cyanobacterial Cells (Phelan and Downing 2007; Weidner et al. 2003; Botes et al. 1982)

1. Take a small amount of cyanobacterial cell culture.
2. Centrifuge at 10,000 ×*g* for a time period of 10 min and freeze after centrifugation in liquid nitrogen overnight.
3. Add 50 μL of 70% methanol per mg of dried cellular mass.
4. Break the cell wall by sonication for eight burst of 3 min each; at 40% power on ice and 50% duty cycle.
5. Remove the cell debris by centrifuging for 10 min at 2000 ×*g* and desiccate the supernatant in a speed vacuum.

8.3.5 Purification of Microcystin

8.3.5.1 Purification and Isolation of Intracellular Microcystin Using Gel Permeation Chromatography and HPLC (Phelan and Downing 2007)

1. Fill a column with a resin (Sephadex- LH20) used for purification of natural substances.
2. Add 80% methanol to the column for elution.
3. Load 3 mL of the extract into the eluted column at a flow rate of 0.5 mL/min.
4. The eluent is 4 mL which was collected to determine the absorbance at 238 nm.
5. Confirm the presence of microcystin using ELISA specific for microcystin identification.
6. Resuspend the toxin in 30% methanol.
7. Load the resuspended toxin in C_{18} column and run at a rate of 3 mL/min.
8. Use methanol and Milli-Q water as mobile phase.
9. Measure the absorbance at 238 nm and examine the peak.

8.3.5.2 Purification and Isolation of Intracellular Microcystin Using SDS PAGE and Gel Permeation Chromatography (Hummert et al. 2000; Lowry et al. 1951)

1. Perform the extraction process to extract intracellular microcystin as in Sect. 3.4.
2. Precipitate the extracted protein using ammonium sulphate.
3. Concentrate the precipitated microcystin by removing ammonium sulphate by dialysis.
4. Estimate the total protein by Folin Lowry method.
5. Separate the protein according to size using gel filtration chromatography by loading inert beads (Superdex) into a column, load the extracted protein into the column and use a mixture of 50% acetonitrile and 50% distilled water as the mobile phase. The flow rate is adjusted 1 mL per minute.
6. Separate the different fractions obtained by SDS PAGE and then identify the band of microcystin by comparison to a known microcystin ladder.
7. Estimate the total protein (Microcystin) again by Folin Lowry method and calculate the difference between the two values.

Estimation of Total Protein (Microcystin) by Folin Lowry Method (Lowry et al. 1951)

1. Prepare Solution A consisting of sodium carbonate (2%) in 0.1 N sodium hydroxide.
2. Prepare solution B consisting of copper sulphate solution (0.5%) in 1% sodium potassium tartrate.
3. Prepare Lowry reagent (Solution C) by adding 50 mL of solution A to 1 mL of Solution B.
4. Add 0.4 mL of working solution in five test tubes and add distilled water to make up the volume to 1 mL by adding distilled water.
5. Add 4.5 mL of solution C to all the test tubes and incubate at RT for 10 min.

6. Add 0.5 mL of Folin Ciocalteu reagent to all the test tubes and incubate at RT for 30 min.
7. Prepare a blank by adding distilled water in place of working solution.
8. Note the absorbance at 660 nm and prepare a curve by plotting the amount of protein on X axis and absorbance at Y axis to determine the amount of microcystin present in the sample.

8.3.6 Storage of Cyanobacterial Microcystin (Phelan and Downing 2007)

1. Collect the purified toxin eluted from HPLC and evaporate under vacuum.
2. Resuspend in 100% methanol or ethanol, add 0.1 mM EGTA, 0.03% Brij 35 Buffer, 50 mM Tris-HCl, and double distilled water make the final concentration to 500 ng/mL.
3. Store in dark at 4 °C.

8.3.7 Other Analytical Method of Determining Cyanobacterial Toxin (Kalaitzidou et al. 2015)

1. Take 10 mL of pond water sample and lake water sample and filter it using filter of 0.5 μm pore size separately. Store the two different filter papers in freezer at − 80 °C until microcystin detection.
2. Cut the frozen filters into small pieces of 2–3 mm size carefully and extract using 10 mL 75% methanol by dipping the filters into it overnight.
3. After the overnight incubation, discard the supernatant and resuspend the pellet in 10 mL of 75% methanol, heat at 55 °C for 12–15 min.
4. Centrifuge the solution at 3000 rpm for 20 min at 4 °C. Carefully isolate the supernatant in another test tube.
5. Repeat the entire procedure once more, collect the extracts, concentrate using rotary evaporation until completely dry, and again resuspend in 1 mL of 75% methanol.
6. Filter the resuspended solution using syringe filters.
7. Screen the extracts for the presence of microcystin using ELISA kit specific for microcystin identification. Perform the ELISA test according to the protocol of the manufacturer.
8. Finally, measure the absorbance (440 nm) and read using microplate ELISA Photometer.
9. Draw a standard curve for determining microcystin concentration by plotting against concentrations of 0.15 ppb, 0.40 ppb, 1.00 ppb, 2.00 ppb, and 5.00 ppb of microcystin LR analogues. Repeat the above procedural steps for both, lake water and pond water samples separately.

8.3.8 Structural Elucidation Using X Ray Crystallography (Junius et al. 2016)

1. Take a microtiter plate.
2. Add 10 μL of distilled water, sodium acetate buffer (pH 4.6), and sodium chloride (20%) each.
3. Mix the contents in the well.
4. Apply vacuum grease along the top edge of the well in order to create a closed environment for crystallization.
5. Take a cover slip and pipette out 2 μL of the solution from the solution from the well on the cover slip.
6. Add equal amount of protein (Microcystin) solution (2 μL) on the same cover slip.
7. Carefully pick up the cover slip with a forcep and set it on the well so that the droplet is towards the well.
8. Gently press the cover slip, such that the well becomes completely airtight.
9. Once crystals start appearing gently view the crystals under a microscope.
10. Pick a crystal that is the right size, has good dimensions, and has no rough edges.
11. Gently isolate the crystal with the tip of the pin.
12. Mount the pin into the X-ray Diffractometer such that the X-ray beam is focused directly on the crystal.
13. Obtain the X-ray diffraction images using a wavelength of around 0.98 Å.
14. Note/examine the data generated using the X-ray detector software.

8.4 Observations

8.4.1 Identification of Cyanobacteria

On Gram staining, pink long grain thread-like structures were observed which confirmed the presence of Gram-negative cyanobacterial cells in both, pond as well as lake water, when observed under the light microscope. Further, PCR amplification confirmed the presence of 16S rDNA in cyanobacterial cells.

8.4.2 Isolation and Purification of Microcystin

Cells were completely lysed using sonication for eight burst of 3 min each; at 40% power on ice and 50% duty cycle. Isolation and purification of microcystin using Sephadex-LH20 yielded 74% of toxin recovery whereas HPLC yielded 33% microcystin.

8.4.3 Identification and Confirmation of Presence of Cyanobacterial Toxin

We confirmed the presence of cyanobacteria in the water samples obtained from lake as well as pond water. ELISA confirmed the presence of microcystin, a toxin produced by cyanobacterial cells in both, the lake as well as pond water.

8.4.4 Structural Elucidation Using X-Ray Crystallography

The coordinates of the diffraction patterns generated by the X-ray detector software were noted and examined further for structural model building. ARP/wARP version 8 was used further for microcystin structural model building using the X-ray diffraction data generated.

8.5 Future Aspects

The ability to recognize and manage microcystin is becoming an essential aspect of cutting-edge research. It allowed us to explore various previously inaccessible avenues. Microcystins have also shown antibacterial, antimycobacterial, as well as cytotoxic activities (Ramos et al. 2015; Benegas et al. 2021). Microcystin structure determination by NMR, MS, or X-ray crystallography can aid in the elucidation of previously unknown structural features by comparing them to known microcystin structures found in publicly accessible protein databases. As such, they may be used to find new therapeutic targets and develop novel medications for use in treating people, animals, and plants.

Acknowledgement The authors are very much thankful to Guru Nanak Institute of Pharmaceutical Science and Technology for providing the necessary resources and guidance during the course of this work.

References

Benegas GRS, Bernal SPF, de Oliveira VM, Passarini MRZ (2021) Antimicrobial activity against Microcystis aeruginosa and degradation of microcystin-LR by bacteria isolated from Antarctica. Environ Sci Pollut Res Int 28(37):52381–52391. https://doi.org/10.1007/s11356-021-14458-5

Botes DP, Kruger H, Viljoen CC (1982) Isolation and characterization offour toxins from the blue-green alga, Microcystis aeruginosa. Toxicon 20(6):945–954

Carmichael WW (1992) Cyanobacteria secondary metabolites—the cyanotoxins. J Appl Bacteriol 72:445–459

Chernoff N, Hunter ES, Hall LL, Rosen MB, Brownie CF, Malarkey D, Marr M, Herkovits J (2002) Lack of teratogenicity of microcystin-LR in the mouse and toad. J Appl Toxicol 22(1):13–17

Codd G, Bell S, Kaya K, Ward C, Beattie K, Metcalf J (1999) Cyanobacterial toxins, exposure routes and human health. Eur J Phycol 34:405–415

Dawson RM (1998) The toxicology of microcystins. Toxicon 36(7):953–962

Figueiredo DR, Azeiteiro UM, Esteves SM, Gonçalves FJM, Pereira MJ (2004) Microcystin-producing blooms-a serious global public health issue. Ecotoxicol Environ Saf 59(2):151–163

Hoiczyk E, Hansel A (2000) Cyanobacterial cell walls: news from an unusual prokaryotic envelope. J Bacteriol 182(5):1191–1199. https://doi.org/10.1128/JB.182.5.1191-1199.2000

Hummert C, Reichelt M, Luckas B (2000) New strategy for the determination of microcystins and diarrhetic shellfish poisoning (DSP) toxins, two potent phosphatases 1 and 2A inhibitors and tumor promoters. Fresenius J Anal Chem 366(5):508–513. https://doi.org/10.1007/s002160050101

Junius N, Oksanen E, Terrien M, Berzin C, Ferrer JL, Budayova-Spano M (2016) A crystallization apparatus for temperature-controlled flow-cell dialysis with real-time visualization. J Appl Crystallogr 49(3):806–813. https://doi.org/10.1107/S1600576716004635

Kalaitzidou M, Filliousis G, Petridou EJ, Economou V, Theodoridis A, Aggelidis P (2015) Isolation of toxic marine cyanobacteria and detection of microcystins in Thermaikos gulf in Central Macedonia in Greece. HAICTA

Lawton LA, Codd GA (1991) Cyanobacterial (blue-green algal) toxins and their significance in UK and European waters. Water Environ J 5:460–465

Lowry OH, Rosebrough NJ, Farr AL, Randall RJ (1951) Protein measurement with the Folin phenol reagent. J Biol Chem 193(1):265–275

MacElhiney J, Lawton LA (2005) Detection of the cyanobacterial hepatotoxins microcystins. Toxicol Appl Pharmacol 203(3):219–230

Martins R, Pereira P, Welker M, Fastner J, Vasconcelos VM (2005) Toxicity of culturable cyanobacteria strains isolated from the Portuguese coast. Toxicon 46:454–464

Phelan RR, Downing TG (2007) Optimization of laboratory scale production and purification of microcystin-LR from pure cultures of Microcystis aeruginosa. Afr J Biotechnol 6(21): 2451–2457. https://doi.org/10.5897/ajb2007.000-2388

Poulos SE, Chronis GT, Collins MB, Lykousis V (2000) Thermaikos gulf coastal system, NW Aegean Sea: an overview of water/sediment fluxes in relation to air–land–ocean interactions and human activities. J Mar Syst 25(1):47

Pringsheim E (1946) Pure cultures of algae. Cambridge University Press, London, p 119

Ramos DF, Matthiensen A, Colvara W, de Votto AP, Trindade GS, da Silva PE, Yunes JS (2015) Antimycobacterial and cytotoxicity activity of microcystins. J Venom Anim Toxins Incl Trop Dis 21:9. https://doi.org/10.1186/s40409-015-0009-8

Sangolkar LN, Maske SS, Chakrabarti T (2006) Methods for determining microcystins (peptide hepatotoxins) and microcystin producing cyanobacteria. Water Res 40(19):3485–3496

Stewart I, Carmichael WW, Sadler R, McGregor GB, Reardon K, Eaglesham GK, Wickramasinghe WA, Seawright AA, Shaw GR (2009) Occupational and environmental hazard assessments for the isolation, purification and toxicity testing of cyanobacterial toxins. Environ Health 8(52): 1–12

Taş S, Okuş E, Yılmaz AA (2006) The blooms of a cyanobacterium, Microcystis cf. aeruginosa in a severely polluted estuary, the Golden Horn, Turkey. Estuar Coast Shelf Sci 68(3–4):593–599

Thakare KT, Nadare M, Parbat JM (2018) Effect of culture medium on growth of blue green algae. Int J Curr Microbiol Appl Sci 7(10):2685–2690. https://doi.org/10.20546/ijcmas.2018.710.312

Vareli K, Zarali E, Zacharioudakis GSA, Vagenas G, Varelis V, Pilidis G, Briasoulis E, Sainis I (2012) Microcystin producing cyanobacterial communities in Amvrakikos Gulf (Mediterranean Sea, NW Greece) and toxin accumulation in mussels (Mytilus galloprovincialis). Harmful Algae 15:109–118

Weidner C, Visser P, Fastner J, Metcalf JS, Codd GA, Mur LR (2003) Effects of light on the microcystin content of Microcystis strain PCC7806. Appl Environ Microbiol 69:1475

9 Isolation of Toxin-Producing Cyanobacteria from Aquatic Samples with *Anabaena*

Ramasamy Thangaraj, SakthiThangavel Ramnath, and N. Thajuddin

Abstract

Cyanobacteria are known to produce hepatotoxic substances, the functional and ecological role of these toxins, however, remains largely unclear. In freshwater, such as lakes and ponds, harmful blooms are most commonly caused by cyanobacteria (also called blue-green algae), which are a kind of single-celled organism called phytoplankton. Some cyanobacteria produce toxins (poisons) called cyanotoxins. The freshwater cyanotoxins like anabaena may produce toxins such as anatoxin, microcystin, and saxitoxin. Saxitoxin is a potent neurotoxin produced by several species of *Anabaena*. The identification of different saxitoxin analogues is essential to later determine its toxicity. This present protocol gives a clear view about the toxin-producing cyanobacteria from freshwater pond respected with anabaena.

Keywords

Cyanobacteria · Cyanotoxins · Anabaena · Saxitoxin

R. Thangaraj (✉) · S. Ramnath
Department of Microbiology, Ayya Nadar Janaki Ammal College, Sivakasi, Tamil Nadu, India

N. Thajuddin
B.S. Abdur Rahman Crescent Institute of Science & Technology (Deemed to be University), Chennai, Tamil Nadu, India

Department of Microbiology, National Repository for Microalgae and Cyanobacteria - Freshwater (NRMC-F), School of Life Sciences, Bharathidasan University, Tiruchirappalli, Tamil Nadu, India

N. Thajuddin et al. (eds.), *Protocols for Cyanobacteria Sampling and Detection of Cyanotoxin*, https://doi.org/10.1007/978-981-99-4514-6_9

Abbreviations

FLD	Fluorescence Detection
HILIC	Hydrophilic Interaction Liquid Chromatography
LCMS	Liquid Chromatography Mass Spectrometry
MC	MicroCentrifuge
RPLC	Reverse Phase Liquid Chromatography

9.1 Introduction

Cyanobacteriaare blue-green algae, Gram-negative bacteria, found in both marine and freshwater environments. Cyanobacteria is an ancient group of photosynthetic microbes that occur in most inland waters and that can have major effects on the water quality and functioning of aquatic ecosystems. Cyanobacteria can fix atmospheric nitrogen, and they are not dependent on a fixed source of carbon and are widely distributed throughout aquatic environments. These include freshwater and marine environments and in some soils. Such organisms are known to produce a number of toxins including neurotoxins, such as the *saxitoxins* and *anatoxin-a* (ATX-a), and *hepatotoxins*, such as *cylindrospermopsin* (CYN) and *microcystins*. Five types of cyanobacteria have been identified as toxin producers, including two strains of *Anabaena flosaquae*, *Aphanizomenon flosaquae*, *Microcystisaeruginosa*, and *Nodularia* species. Five morphologically distinct sections have been recognized: unicellular colonies with binary fission, unicellular colonies with multiple fission, multicellular colonies, multicellular colonies with differentiated cells, and branched multicellular colonies with differentiated cells (Schirrmeister et al. 2011). The cyanotoxins are a diverse group of natural toxins, both from the chemical and the toxicological points of view). *Anabaena* may produce a few different toxins, including anatoxin, microcystin, and saxitoxin. However, the wide range of structures and charge states of most cyanobacterial toxins make it difficult to resolve all of the toxins in one analysis. Colonies are often large enough to be seen with the naked eye and, when numerous, may change the color of the water or form scums at the water surface. Dense proliferations of cyanobacteria are referred to as algal blooms. The environmental health of aquatic ecosystems is threatened by the global proliferation of harmful cyanobacterial blooms (Burford et al. 2020; Paerl and Barnard 2020). They are dominated by toxigenic cyanobacterial genera, for example, Cylindrospermopsis, *Dolichospermum* (formerly *Anabaena*), *Microcystis*, and *Planktothrix*, characterized by gene sequences encoding the production of toxic metabolites known as cyanotoxins. When blooms have the potential to cause harmful effects, they are called harmful algal blooms (HABs). The events negatively impact water quality, degrade ecosystem integrity, and pose a threat to human health (Carmichael and Boyer 2016; Chapra et al. 2017; Bullerjahn et al. 2016; Qin et al. 2010; Westrick and Szlag 2018). The incidence of HABs is increasing in most regions and is usually linked to excess nutrient inflow into surface water from agricultural and industrial sources (Hudnell 2010; Gkelis and Zaoutsas 2014). Severe and often prolonged HABs now plague some of the world's largest waterbodies, such as Lakes Erie, Ontario, Okeechobee, and Winnipeg (North America), Taihu (China), Kasumigaura (Japan), Kinneret (Israel), Victoria (Africa), and the Baltic and Caspian Seas (Europe and West Asia, respectively) (Paerl and Paul 2011; Kling et al. 2011; Zohary et al. 2012). Blooms are often comprised of species that are inedible or low in nutritional quality to invertebrate grazers, negatively affecting foodweb biodiversity and production (Paerl and Otten 2013). The LC–MS methods that have been reported for cyanotoxins are usually based on reversed-phase liquid chromatography (RPLC) separations. However, the wide range of structures and charge states of most cyanobacterial toxins make it difficult to resolve all of the toxins in one analysis.

9.2 Saxitoxin

Saxitoxins are produced in freshwater and marine environments. In marine environments, they are often referred to as paralytic shellfish poisons (PSPs) associated with their bioaccumulation in shellfish such as *Saxidomus giganteus*, from which the first identification was made. Saxitoxins are tricyclic purine derivatives (Wiegand and Pflugmacher 2005), with a molecular weight around 299 (Svrcek and Smith 2004). So far, 27 variants of saxitoxins have been found (Ho et al. 2012). Within the *Dolichospermum* genus, saxitoxin production has been only reported in *D. circinalis* (*Anabaena circinalis*) blooms in Australian freshwater systems, such as dams, weirs on rivers, and reservoirs (Beltran and Neilan 2000). The gene cluster responsible for saxitoxin biosynthesis (genes sxtA-E and sxtG-X spanning 28 kbp) has been described for *D. circinalis* (Mihali et al. 2009). Most human saxitoxin toxicoses have been associated with the ingestion of marine shellfish, which accumulate saxitoxins produced by marine dinoflagellates (Cusick and Sayler 2013). In freshwaters, saxitoxins are produced by cyanobacteria in the genera *Anabaena, Aphanizomenon, Planktothrix, Cylindrospermopsis, Lyngbya*, and *Scytonema* (Smith et al. 2012; Wiese et al. 2012). A specific quantitative PCR (qPCR) method based on SYBR green chemistry was developed to quantify saxitoxin-producing *D. circinalis* (Al-Tebrineh et al. 2010), and saxitoxin concentrations were positively correlated with sxtA gene copy numbers, indicating that the latter can be used as a measure of potential toxigenicity in *D. circinalis* and possibly other cyanobacterial blooms. This assay was expanded and combined into a multiplex qPCR assay to detect sxtA, cyrA, mcyE-ndaF, and 16S rRNA from most cyanobacterial species known to produce the *saxitoxin, cylindrospermopsin, microcystin*, and *nodularin toxins* (Al-Tebrineh et al. 2012). It also inhibits calcium and potassium channels in excitable cells thereby affecting the production of action potentials which can cause fatal cardiac arrhythmias (Fig. 9.1).

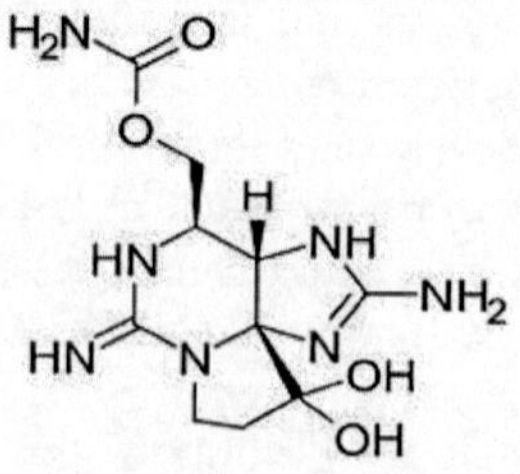

Fig. 9.1 Chemical Structure of Saxitoxins (STXs)

9.3 Materials Required

- Polythene bags
- Filtering cloth
- Distilled water
- BG 11 medium
- Centrifuge
- Sonicator
- Petriplate
- Conical flask

9.4 Sample Collection and Site Description

The algal sample is collecting from the pond Periyakulam located in Kaliappa Nagar, Sivakasi, Tamil Nadu. The latitude and longitude of Periyakulam pond is 9°27′10.3″ N 77°47′07.7″E.

9.5 Cultivation

The collected algal sample is inoculating in the plates using BG11 and Chu 10 medium. This medium is used for the cultivation of freshwater algae. After inoculation, it is used to freeze dried by lyophilization method.

9.6 Extraction of Toxin

1. Floating cyanobacterial mat is collected from the pond in the laboratory by filtration and it is freeze dried.
2. To lyophilize the cells of 20 mg, dry biomass sample is suspended in 80 μL of water at room temperature (about 15 °C) stirring at medium speed in the dark for 2 h.
3. Then add 125 μL of acetonitrile–water–formic acid (80:19.9:0.1)
4. Mix well and allow it to sit for 10 min.

5. The mixture is filtering using an Ultrafree-MC 0.45 μm membrane (Millipore) at 628.3 rad.s^{-1} for 10 min.
6. The filter is wash twice with 125 μL of the same extracting solvent.
7. The filtrates are combined, the volume is adjusted to 500 μL with extracting solvent, and is analysed directly by RPLC–fluorescence detection (FLD) and/or HILIC–MS.
8. For quantitation of toxins present in the samples of strains of *A. circinalis* test, the above extraction procedure is modified slightly; accurately weigh the samples of material (20–40 mg) after suspension in water are extracted with 4 × 800 μL aliquots of acetonitrile–water–formic acid (80:19.9:0.1) and the combined extracts made to a final volume of 4 mL (Dell'Aversano et al. 2004).

References

Al-Tebrineh J, Mihali TK, Pomati F, Neilan BA (2010) Detection of saxitoxin-producing cyanobacteria and Anabaena circinalis in environmental water blooms by quantitative PCR. Appl Environ Microbiol 76:7836–7842

Al-Tebrineh J, Merrick C, Ryan D, Humpage A, Bowling L, Neilan BA (2012) Community composition, toxigenicity, and environmental conditions during a cyanobacterial bloom occurring along 1100 kilometers of the Murray river. Appl Environ Microbiol 78:263–272

Beltran EC, Neilan BA (2000) Geographical segregation of the neurotoxin-producing cyanobacterium Anabaena circinalis. Appl Environ Microbiol 66:4468–4474

Bullerjahn GS, McKay RM, Davis TW, Baker DB, Boyer GL, D'Anglada LV, Doucette GJ, Ho JC, Irwin EG, Kling CL, Kudela RM, Kurmayer R, Michalak AM, Ortiz JD, Otten TG, Paerl HW, Qin B, Sohngen BL, Stumpf RP, Visser PM, Wilhelm SW (2016) Global solutions to regional problems: collecting global expertise to address the problem of harmful cyanobacterial blooms. A Lake Erie case study. Harmful Algae 54:223–238

Burford M, Carey C, Hamilton DP, Huisman J, Paerl H, Wood S, Wulff A (2020) Perspective: advancing the research agenda for improving understanding of cyanobacteria in a future of global change. Harmful Algae 91:101601

Carmichael WW, Boyer GL (2016) Health impacts from cyanobacteria harmful algae blooms: implications for the north American Great Lakes. Harmful Algae 54:194–212

Chapra SC, Boehlert B, Fant C, Bierman VJ, Henderson J, Mills D, Mas DML, Rennels L, Jantarasami L, Martinich J, Strzepek KM, Paerl HW (2017) Climate change impacts on harmful algal blooms in U.S. freshwaters: a screening level assessment. Environ Sci Technol 51(16): 8933–8943

Cusick KD, Sayler GS (2013) An overview on the marine neurotoxin, saxitoxin: genetics, molecular targets, methods of detection and ecological functions. Mar Drugs 11(4):991–1018

Dell'Aversano C, Eaglesham GK, Quilliam MA (2004) Analysis of cyanobacterial toxins by hydrophilic interaction liquid chromatography–mass spectrometry. J Chromatogr A 1028: 155–164

Gkelis S, Zaoutsas N (2014) Cyanotoxin occurrence and potentially toxin producing cyanobacteria in freshwaters of Greece: a multi-disciplinary approach. Toxicon 78:1–9

Ho L, Sawade E, Newcombe G (2012) Biological treatment options for cyanobacteria metabolite removal—a review. Water Res 46:1536–1548

Hudnell HK (2010) The state of US freshwater harmful algal blooms assessments, policy and legislation. Toxicon 55(5):1024–1034

Kling HJ, Watson SB, McCullough GK, Stainton MP (2011) Bloom development and phytoplankton succession in lake Winnipeg: a comparison of historical records with recent data. Aquat Ecosyst Health Manag 14:219–224

Mihali TK, Kellmann R, Neilan BA (2009) Characterisation of the paralytic shellfish toxin biosynthesis gene clusters in Anabaena circinalis AWQC131C and Aphanizomenon sp. NH-5. BMC Biochem 10:8

Paerl HW, Barnard MA (2020) Mitigating the global expansion of harmful cyanobacterial blooms: moving targets in a human- and climatically-altered world. Harmful Algae 96:101845

Paerl HW, Otten TG (2013) Harmful cyanobacterial blooms: causes, consequences, and controls. Microb Ecol 65:995–1010

Paerl HW, Paul V (2011) Climate change: links to global expansion of harmful cyanobacteria. Water Res 46:1349–1363

Qin B, Zhu G, Gao G, Zhang Y, Li W, Paerl HW, Carmichael WW (2010) A drinking water crisis in Lake Taihu, China: linkage to climatic variability and lake management. Environ Manag 45(1): 105–112

Schirrmeister BE, Anisimova M, Antonelli A et al (2011) Evolution of cyanobacterial morphotypes: taxa required for improved phylogenomic approaches. Comm Integr Biol 4(4): 424–427

Smith FMJ, Wood SA, Wilks T et al (2012) Survey of Scytonema (cyanobacteria) and associated saxitoxins in the littoral zone of recreational lakes in Canterbury, New Zealand. Phycologia 51(5):542–551

Svrcek C, Smith DW (2004) Cyanobacteria toxins and the current state of knowledge on water treatment options: a review. J Environ Eng Sci 3:155–185

Westrick JA, Szlag D (2018) A cyanotoxin primer for drinking water professionals. J Am Water Works Assoc 110:E1–E16

Wiegand C, Pflugmacher S (2005) Ecotoxicological effects of selected cyanobacterial secondary metabolites: a short review. Toxicol Appl Pharm 203:201–218

Wiese M, D'Agostino PM, Mihali TK et al (2012) Neurotoxic alkaloids: saxitoxin and its analogs. Mar Drugs 8(7):2185–2211

Zohary T, Nishri A, Sukenik A (2012) Present-absent: a chronicle of the dinoflagellate Peridinium gatunense from lake Kinneret. Hydrobiologia 698:161–174

Isolation of Toxin Producing Cyanobacteria from Aquatic Samples—*Anabaenopsis* sp. 10

M. Veerapagu, K. R. Jeya, and A. Sankara narayanan

Abstract

Climate change, eutrophication, and cyanobacteria's capacity to adapt to and flourish in shifting environmental circumstances are all contributing to an increase in their prevalence in freshwaters. The public's health is increasingly in danger due to the spread of dangerous cyanobacterial blooms (CyanoHABs) on a worldwide scale. Production of cyanotoxins, which are harmful by-products, is a defining characteristic of cyanoHABs. The greatest health risk posed by cyanoHABs is the release of cyanotoxins into recreational, fishable, and potable water sources. Humans that consume cyanotoxins over an extended period of time may have serious health problems. Across a range of habitats, several cyanobacterial taxa generate a variety of poisons, including anatoxin, cylindrospermopsin, microcystin, nodularin, and saxitoxin. Heterocytous cyanobacteria of the genus *Anabaenopsis* are mostly found in tropical, subtropical, and temperate lakes and ponds. In warm weather, nearly all of the species that contribute to algal blooms generate cyanotoxin. As a result, it is important to isolate *Anabaenopsis* sp. from various aquatic environments.

M. Veerapagu (✉)
Department of Biotechnology and Microbiology, National College (Autonomous), Tiruchirappalli, Tamilnadu, India

K. R. Jeya
Department of Biotechnology, Government Arts and Science College (Women), Veppur, Perambalur, Tamilnadu, India

A. Sankara narayanan
Department of Life Sciences, Sri Sathya Sai University for Human Excellence, Kalaburagi, Karnataka, India

N. Thajuddin et al. (eds.), *Protocols for Cyanobacteria Sampling and Detection of Cyanotoxin*, https://doi.org/10.1007/978-981-99-4514-6_10

Keywords

Cyanobacteria · Eutrophication · CyanoHABs · Cyanotoxin · Microcystin · *Anabaenopsis* sp.

10.1 Introduction

Water is the most significant natural resource that is necessary for home, agricultural, and industrial uses. Water should therefore be safe enough to be used or consumed with little chance of immediate or long-term harm, making it a crucial problem for public health. Cyanobacterial blooms, which produce cyanotoxins, are a rapidly expanding form of water contamination. Globally, reports of cyanobloom-related water contamination in seas, lakes, rivers, lagoons, streams, wells, and water reservoirs have been made (Meriluoto et al. 2017; Svircev et al. 2019; Zhang et al. 2019). Water sources may get contaminated by cyanotoxins and cyanoblooms, making them unfit for use by humans and other living things.

It is well known that water eutrophication encourages cyanobacterial growth, which in turn leads to cyanobacterial blooms (Paerl et al. 2018). Public health is affected by cyanobacterial harmful algal blooms, which can include several genera like *Microcystis*, *Anabaena*, *Anabaenopsis*, *Oscillatoria*, and *Nostoc*. These blooms are caused when cyanotoxins deplete dissolved freshwater oxygen, which makes it impossible for other aquatic and sediment species to survive (Ndlela et al. 2016). Negative impacts include genotoxicity, liver, intestinal, cardiovascular, and central nervous system damage when humans consume water or fish from these waterbodies (Cao et al. 2019; Wen et al. 2019).

Cyanobacteria are prokaryotic organisms that are very diverse and are present in almost every ecosystem on Earth. In contrast to other prokaryotes, they perform oxygenic photosynthesis and have chlorophyll-a (bacteria and archaea). Their nearest cousins are the chloroplasts of higher plants and purple bacteria (Moore et al. 2019). The size and structure of the multicellular aggregates that the majority of cyanobacterial species create can be used to distinguish cyanobacteria from other microorganisms in freshly collected field samples. Identification is greatly aided by the kind of cell division and the following cell separation, or lack thereof.

Divided cells in unicellular forms (like *Synechococcus* sp.) totally separate from one another and fail to produce (real) filaments. However, certain "unicellular" species can create microbial colonies or mats by encasing solitary cells in a mucous matrix (mucilage). In cultures, organisms that normally form colonies frequently develop as solitary cells or aggregates with a morphology that is distinct from that of naturally occurring colonies.

It is generally known that when combined with human-persuaded action, natural circumstances may quickly expand to produce a mass large enough to be seen with the unaided eye. Cyanobloom is the name of this occurrence (Massey et al. 2020). Among the species that frequently generate blooms are *Microcystis* sp., *Anabaena* sp., *Nostoc* sp., *Anabaenopsis*, *Cylindrospermopsis* sp., and *Planktothrix* sp. In both industrialized and developing nations, the occurrence of hazardous cyanoblooms is crucial for the production of drinking water. This is owing to the fact that toxic cyanoblooms in drinking water sources may cause a decrease in water quality and increase the danger of toxin exposure for water users, both of which are now major health concerns on a worldwide scale.

Algal blooms are caused when algae growth is accelerated by specific environmental factors in an aquatic environment. HABs, or hazardous algal blooms, are blooms that have the potential to endanger aquatic ecosystems or human health. Cyanobacteria, sometimes referred to as blue-green algae, are microorganisms that may generate HABs in freshwater ecosystems. Toxins can be produced by certain cyanobacterial HABs, or cyanoHABs. CyanoHABs and the toxins they produce have been shown to be harmful to humans, animals, aquatic ecosystems (Chatziefthimiou et al. 2021), the economy, drinking water sources, property values, and recreational activities like swimming and fishing for both sport and profit.

The advancement of cyanoblooms has promoted natural substances that are harmful to living things. The poisons known as cyanotoxins are hazardous to humans, animals, plants, and crops. The most often reported cyanotoxins include microcystin, nodularin, cylindrospermopsin, anatoxin-a, and saxitoxin in this area (Miglione et al. 2021). Several cyanobacterial species, including those from the genera *Aphanizomenon*, *Anabaena*, *Anabaenopsis*, *Cylindrospermopsis*, *Microcystis*, *Nostoc*, *Nodularia*, and *Oscillatoria*, among others, generate cyanotoxins (Trung et al. 2018).

Cyanobacterial cells generate and contain cyanotoxins (intracellular). Instead of being continuously excreted from the cyanobacterial cells, these toxins are mostly released upon cell death and lysis (i.e., cell break) during an algal bloom. Some cyanobacteria species can, however, release extracellular poisons into the water without rupturing or dying. A group of filamentous, heterocystous cyanobacteria known as *Anabaenopsis* is divided into akinetes and fragments. Microcystins, which are harmful to both people and animals, can be produced by *Anabaenopsis*. The majority of the species in this genus are tropical and subtropical, while a few produce blooms throughout the warmer months in temperate climates. Bloom-formers are among the kinds of planktonic organisms in *Anabaenopsis*.

There are several methods for the isolation of cyanobacteria *Anabaenopsis*. To get pure cultures of cyanobacteria from various sources, methods like the streak plate approach and microscope assisted unialgal isolation by dilution (MAUD) are crucial.

10.2 Materials

- Sample container bottle
- Ice chest box
- Test tube
- Distilled water
- Pipette
- Laminar air flow cabinet

- Vortex mixer (Eppendorf)
- Water bath
- Autoclave
- Petriplate (100 × 15 mm, Hi media)
- White fluorescent lamp (Philips)
- Inoculation loop (Hi media)
- Microscopic slide (Hi media)
- BG11 medium (Hi media)
- Agar (Hi media)
- Algal growth chamber
- Microscope (Olympus)
- 0.2 μm filter (Merck)
- Antibiotics: Trimethoprim, Cycloheximide (Hi media)

10.2.1 Preparation of Antibiotic Solutions

Dissolve 150 mg of trimethoprim and cycloheximide separately in 50 ml of dimethyl sulfoxide to get a concentration of 3 mg/mL.

Filter sterilize antibiotic solutions through 0.2 μm filter (Noroozi et al. 2017).

10.2.2 Preparation of BG11 Agar Medium Plate

1. Measure the required quantity of BG 11 broth medium according to manufacturer's recommendation.
2. Add 1.2–1.5% agar concentration to the medium (Videau and Cozy 2019).
3. Boil the medium for complete dissolution in a water bath or hot plate, then adjust the pH of the medium to 7.1 ± 0.2.
4. Sterilize the medium in an autoclave at 121 °C for 15 min.
5. Add 30 μg/mL of filter sterilized antibiotic solutions trimethoprim and cycloheximide to the BG11 medium (Bairwa et al. 2021).
6. Pour approximately 20–25 mL of sterile BG11 medium onto sterile petriplates.
7. Allow the medium to cool and solidify for 15–20 min.

10.3 Methods

10.3.1 Collection of Sample

1. Collect water samples in a screw cap bottles aseptically.
2. Keep the water samples in a container at 4 °C (Noroozi et al. 2017).
3. Transport the samples to the lab as early as possible.

10.3.2 Streak Plate Method (Lee et al. 2014)

1. Prepare and sterilize a series of test tubes containing 9 mL of distilled water in an autoclave.
2. Aseptically transfer 1 mL of the sample to the first tube with the help of a sterile pipette.
3. Mix well the sample in a vortex and make the dilution up to 10^{-8} (Parthiban and Jambulingam 2023)
4. Spread 1 mL from each dilution onto BG11 agar medium plate.
5. Incubate the plates in an algal growth chamber at 25 ± 2 °C, illumination at 150 μmol photons m^{-2} s^{-1}under a 12 h light/dark cycle for about 14 days (Wei et al. 2020).
6. After visible growth, select the colonies from mixed culture in the BG11 medium plate.
7. Streak the cyanobacterial colonies onto BG11 agar medium plate and incubate under controlled conditions as above.
8. Repeat the streak plate method procedure until isolated cyanobacterial colonies are obtained (Lee et al. 2014).
9. Identify the isolated colonies for the presence of *Anabaenopsis* sp. by using the standard key [19, 20].

10.3.3 Microscope-Assisted Uni Algal through Dilution (MAUD) Method (Verma et al. 2021)

10.3.3.1 Enrichment of Cyanobacteria

1. Collect the samples as described above.
2. Add 10 mL of water samples to a sterile 90 mL of BG11 medium aseptically.
3. Incubate it in an algal growth chamber at 25 ± 2 °C, illumination at 150 μmol photons m^{-2} s^{-1} under a 12 h light/dark cycle for about 14 days (Wei et al. 2020).

10.3.3.2 Microscope-Assisted Unialgaldilution (MAUD)

1. Wipe and clean a series of microscopic slides with 70% alcohol.
2. After incubation transfer 20 μL of mixed cyanobacterial culture to a microscopic slide.
3. Add 15 μL of sterile water to a series of microscopic slides.
4. Perform four-fold serial dilution by transferring 5 μL of cyanobacterial culture with the help of micropipette to 15 μL of sterile water droplets in a slide.
5. Mix well the suspension.
6. Observe each slide under microscope for the presence of unialgal culture.
7. Transfer the droplets containing unialgal culture to a 2 mL of sterile BG11 medium tube (Verma et al. 2021).
8. Incubate the cyanobacterial culture under controlled conditions as described above.

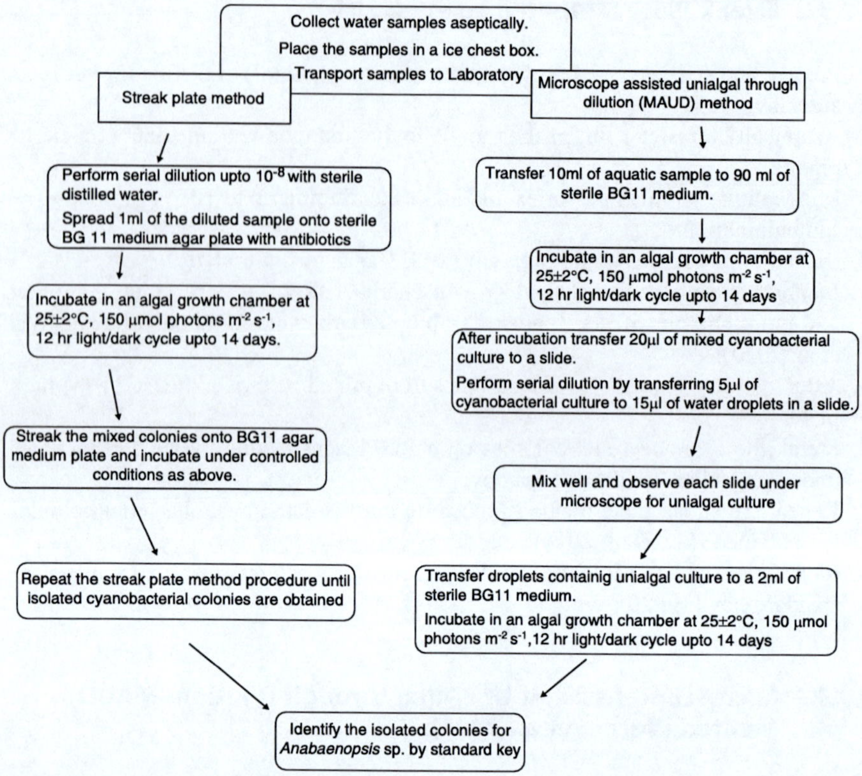

Fig. 10.1 Isolation of toxin producing cyanobacteria from aquatic samples—*Anabaenopsis* sp.

9. After incubation identify the culture for *Anabaenopsis* sp. with the help of standard key (Desikachary 1959; Komarek 2005) (Fig. 10.1).

References

Bairwa HK, Prajapat G, Jain S, Khan IA, Ledwani L, Yadav P, Agrawal A (2021) Evaluation of UV-B protection efficiency of mycosporine like amino acid extracted from the cyanobacteria *Anabaenopsis* sp. SLCyA isolated from a hypersaline lake. Bioresour Technol Rep 15:00749

Cao L, Huang F, Massey IY, Wen C, Zheng S, Xu S, Yang F (2019) Effects of microcystin-LR on the microstructure and inflammation-related factors of jejunum in mice. Toxins 11(9):482

Chatziefthimiou AD, Banack SA, Metcalf JS (2021) Harmful algal and cyanobacterial harmful algal blooms in the Arabian seas: current status, implications, and future directions. In: The Arabian seas: biodiversity, environmental challenges and conservation measures. Springer, Cham, pp 1083–1101

Desikachary TV (1959) Cyanophyta. Indian Council of Agriculture Research, New Delhi

Komarek J (2005) Phenotypic diversity of the heterocytous cyanoprokaryotic genus *Anabaenopsis*. Fottea 5(1):1–35

Lee K, Eisterhold ML, Rindi F, Palanisami S, Nam PK (2014) Isolation and screening of microalgae from natural habitats in the midwestern United States of America for biomass and biodiesel sources. J Nat Sci Biol Med 5(2):333

Massey IY, Al Osman M, Yang F (2020) An overview on cyanobacterial blooms and toxins production: their occurrence and influencing factors. Toxin Rev 41(1):326–346

Meriluoto J, Blaha L, Bojadzija G, Bormans M, Brient L, Codd GA, Visser PM (2017) Toxic cyanobacteria and cyanotoxins in European waters-recent progress achieved through the CYANOCOST action and challenges for further research. Adv Oceanogr Limnol 8(1):161–178

Miglione A, Napoletano M, Cinti S (2021) Electrochemical biosensors for tracing cyanotoxins in food and environmental matrices. Biosensors 11(9):315

Moore KR, Magnabosco C, Momper L, Gold DA, Bosak T, Fournier GP (2019) An expanded ribosomal phylogeny of cyanobacteria supports a deep placement of plastids. Front Microbiol 10:1612

Ndlela LL, Oberholster PJ, Van Wyk JH, Cheng PH (2016) An overview of cyanobacterial bloom occurrences and research in Africa over the last decade. Harmful Algae 60:11

Noroozi M, Amozegar MA, Rahimi R, Shahzadeh Fazeli SA, Bakhshi Khaniki G (2017) The isolation and preliminary characterization of native cyanobacterial and microalgal strains from lagoons contaminated with petroleum oil in Khark Island. Biol J Microorganism 5(20):33–41

Paerl HW, Otten TG, Kudela R (2018) Mitigating the expansion of harmful algal blooms across the freshwater-to-marine continuum. Environ Sci Technol 52:5519–5529

Parthiban J, Jambulingam R (2023) Enhancing the biodiesel production potential of *Synechococcus elongatus* and *Anabaena* cyanobacterial strain isolated from saline water using different media composition and organic carbon sources. Sustainability 15(1):870

Svircev Z, Lalic D, Bojadzija Savic G, Tokodi N, Drobac Backovic D, Chen L, Codd GA (2019) Global geographical and historical overview of cyanotoxin distribution and cyanobacterial poisonings. Arch Toxicol 93(9):2429–2481

Trung B, Dao TS, Faassen E, Lurling M (2018) Cyanobacterial blooms and microcystins in southern Vietnam. Toxins 10(11):471

Verma S, Bagul SY, Choudhary P, Chakdar H, Das S, Siddiqui N, Saxena AK (2021) Microscope assisted uni-algal isolation through dilution (MAUD): a simple modified technique for tapping diverse cyanobacteria. 3 Biotech 11(7):343

Videau P, Cozy LM (2019) *Anabaena* sp. strain PCC 7120: laboratory maintenance, cultivation, and heterocyst induction. Curr Protoc Microbiol 52(1):e71

Wei J, Xie X, Huang F, Xiang L, Wang Y, Han T, Yang F (2020) Simultaneous *Microcystis* algicidal and microcystin synthesis inhibition by a red pigment prodigiosin. Environ Pollut 256: 113444

Wen C, Zheng S, Yang Y, Li X, Chen J, Wang X, Yang F (2019) Effects of microcystins-LR on genotoxic responses in human intestinal epithelial cells (NCM460). J Toxicol Environ Health A 82(21):1113–1119

Zhang C, Massey IY, Liu Y, Huang F, Gao R, Ding M, Yang F (2019) Identification and characterization of a novel indigenous algicidal bacterium *Chryseobacterium* species against *Microcystis aeruginosa*. J Toxicol Environ Health A 82(15):845–853

Isolation of Toxin Producing Cyanobacteria from Aquatic Samples with *Cylindrospermopsis* sp. 11

O. Archana and Lokesh Ravi

Abstract

Cyanotoxins problems have puzzled researchers for the last century because of odor issues, decreased water purity, aquatic animal death due to toxicity of cyanotoxins, and many more things. *Cylindrospermopsis* sp. is also one such freshwater planktonic cyanobacteria responsible for producing acute and chronic toxic effects in many countries. Hence, it is essential to isolate and purify the *Cylindrospermopsis* sp. from its natural habit and to study its morphology and chemical compounds responsible for toxicity effects on the environment in a laboratory condition (*in vitro*), so to isolate this cyanobacterium, certain methods and tools are required which are described in this chapter. And also, this chapter is focused on cyanotoxins produced by this species and their effects on the environment and animals.

Keywords

Cylindrospermopsin · Paralytic shellfish poison · Microspade · *Cylindrospermopsis raciborskii* · Saxitoxin

O. Archana
Department of Botany, St Joseph's University, Bengaluru, Karnataka, India

L. Ravi (✉)
Department of Food Technology, Faculty of Life and Allied Health Sciences, MS Ramaiah University of Applied Sciences, Bengaluru, India
e-mail: lokesh.ravi@sjc.edu.in

N. Thajuddin et al. (eds.), *Protocols for Cyanobacteria Sampling and Detection of Cyanotoxin*, https://doi.org/10.1007/978-981-99-4514-6_11

11.1 Introduction

Cyanobacteria are also known as blue–green algae because of the presence of Phycocyanin (Blue) and Chlorophyll (Green) pigment. Cyanobacteria are reported to be the first photosynthetic organisms existed on Earth that originated almost 2.6 ~ 3.5 billion years ago. And also believe that photosynthetic organelles in eukaryotic organisms originated from the endosymbiosis process between Cyanobacterium and the Phagotrophic host. These prokaryotic algae exhibit diverse morphology and appear in different forms like filamentous, unicellular, planktonic, benthic, or colonial. Majority of the cyanobacteria perform photosynthesis in the presence of oxygen but very few use sulfur-dependent anoxygenic approach for their photosynthesis. Filamentous cyanobacteria have a specialized structure called heterocyst which helps in fixing atmospheric nitrogen (Lau et al. 2015).

Cyanobacteria grows very well in waterbodies that contain important nutrients such as phosphorous, nitrogen, and other life-supporting nutrients and elements (eutrophic reservoirs). Some cyanobacteria will produce cyanotoxins (biotoxins). Based on their chemical structure, which is commonly categorized into alkaloids, cyclic peptides, and lipopolysaccharides. And also, based on the toxicity mechanism of target organs, cyanotoxins are classified into neurotoxins, hepatotoxins, cytotoxins, irritant toxins, and dermatotoxins (Somdee et al. 2013).

Some of the species of the genus *Cylindrospermopsis* are *C. raciborskii*, *C. gangetica*, *C. philippinensis*, *C. africana*, *C. cuspis*, *C. sinuosa*, *C. helicoidea*, *C. acuminatocrispa*, *C. catemaco*, *C. taverae*, *C. curvispora*. *Cylindrospermopsis* sp., are cosmopolitan and occur in temperate, trophic, and sub-trophic freshwater regions (Li et al. 2017).

Some of the key characteristic features, which are required for identifying planktic *Cylindrospermopsis* species are

1. Presence of gas vacuoles
2. Pointed and attenuated trichome end
3. Presence or absence of akinetes near one or both the ends of the trichome
4. Presence of 1–3 vegetative cells between the akinetes and the terminal heterocyst
5. Heterocyst develops primarily from terminal cells

A few species of *Cylindrospermopsis* lack akinete, for example, *C. curvispora*, *C. catemaco*, *C. taverae*, *C. cuspis*, *C. helicoidea* (Li et al. 2017; Saker and Neilan 2001). Filamentous diazotrophic *Cylindrospermopsis raciborskii* shows two different morphologies in its shape such as curly and straight filament. Figure 11.1 shows the different forms of this species. These differences make variations in the size and number of cells per trichome. The straight filament tends to produce more cell toxins than curved filaments (Jones and Sauter 2005).

In a study by González-Blanco et al. (2020) researchers reported *Cylindrospermopsis raciborskii* strain 11K and its toxins. This species produces zwitterion toxins such as cylindrospermopsin, 7-deoxy-cylindrospermopsin, and 7-epi-cylindrospermopsin. These zwitterionic compounds contain positively

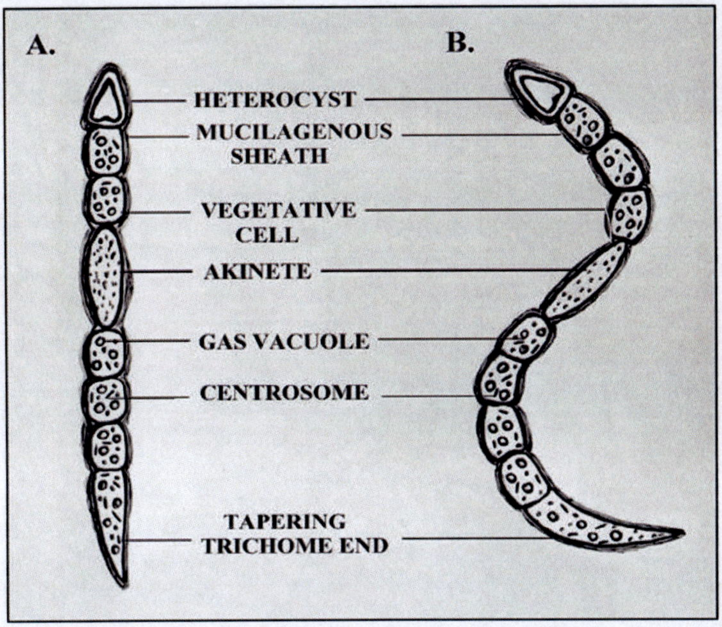

Fig. 11.1 Diagramatic representation of *Cylindrospermopsis raciborskii*. (**a**) Straight and (**b**) Curved filament

charged tricyclic guanidine moiety and a negatively charged sulfate group and also biologically active uracil moiety. In Cylindrospermopsin and 7-epi-cylindrospermopsin the uracil group is attached to the heterocyclic rings via hydroxylated asymmetric carbon. But this kind of hydroxylation is not there in 7-deoxy-cylindrospermopsin, 7-deoxy-desulfo-cylindrospermopsin, and 7-deoxy-desulfo-12-acetyl-cylindrospermopsin produced by *Cylindrospermopsis raciborskii* ITEP 18. The lack of sulfate moiety in 7-deoxy-desulfo-cylindrospermopsin and 7-deoxy-desulfo-12-acetyl-cylindrospermopsin compounds makes them more hydrophobic as contrasted with the zwitterionic compounds. 5-chloro-cylindrospermopsin and Cylindrospermic acid, two innocuous oxidation products. Desulfo cylindrospermopsin is a degradation substance that can be produced by acid hydrolysis (González-Blanco et al. 2020).

Cylindrospermopsin is a cyclic guanidine alkaloid produced by *Cylindrospermopsis raciborskii,* which is associated with the *cyr*gene cluster (Antunes et al. 2015). This species is also known to produce Carbamate alkaloid neurotoxin which is Paralytic Shellfish Poison (PSP). These PSPs can inhibit the activity of muscle cells, as a result, paralysis can occur and also causes respiratory arrest which leads to the death of the organisms (Fabre et al. 2017). The most toxic of PSP is Saxitoxins, which are neurotoxic alkaloids possessing many isoforms like Neosaxitoxin, and decarbamoylsaxitoxin (Antunes et al. 2015).

Table 11.1 Chemical structure of toxins produced by *Cylindrospermopsis raciborskii*

Toxin name	Chemical structure	Molecular weight (g/mol)	PubChem ID
Cylindrospermopsin	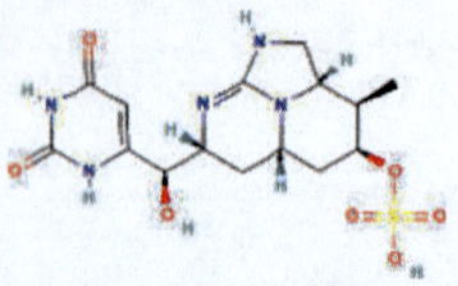	415.4	135565888
7-deoxy-cylindrospermopsin	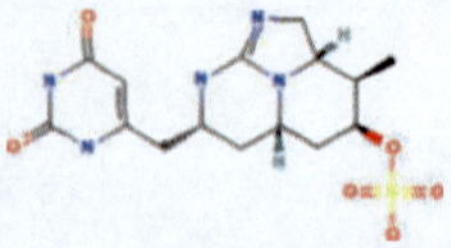	–	440700993
7-epi-cylindrospermopsin	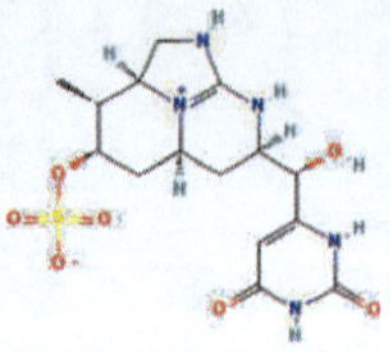	415.4	10341781
7-deoxy-desulfo-cylindrospermopsin	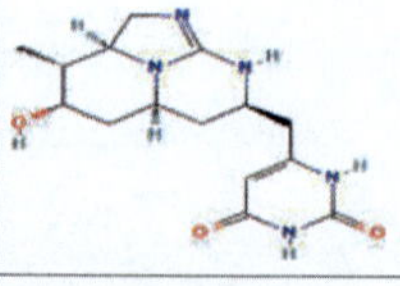	319.36	11247498
Saxitoxin	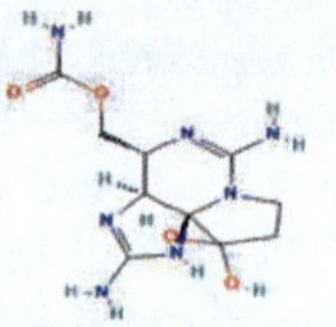	299.29	56947150
Neosaxitoxin	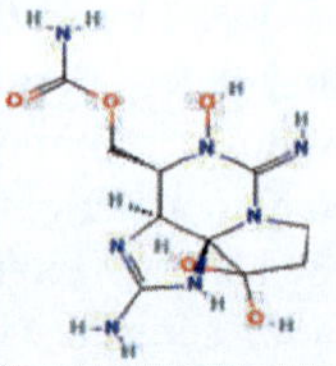	315.29	21117946

Exposure to *Cylindrospermopsis raciborskii* causes liver damage and related kidneys, adrenal gland, lungs, heart, thymus, and small intestine problems in mice. Because of the production of alkaloid cylindrospermopsin. This alkaloid can inhibit protein synthesis (Bernard et al. 2003) (Table 11.1).

11.2 Material and Methods

11.2.1 Sample Collection and Maintenance

Sample collection can be done by using a plankton net.

1. Plankton net size and mesh size are important while collecting samples with a plankton net. Usually, a 10-μm mesh phytoplankton net size is used to collect Cylindrospermopsis sp.
2. Samples can be stored in Lugol's solution for further morphological analysis
3. Samples can also be stored using freezing technique at −20 °C for cyanotoxin identification.
4. Mass culture can be maintained at 28 °C in 1 L of Jaworski's liquid medium in 3 L Erlenmeyer flasks with natural or artificial light (cool white fluorescent) at an intensity of 100–200 ktE m^2/s.
5. After 14 days, successful cultures in the liquid medium, the sample can be streaked on agar plates (0.5% agar) (Somdee et al. 2013).
6. Culture can also be maintained in the laboratory by utilizing the precipitate-free medium known as ASM culture medium. It is composed of sodium nitrate, magnesium sulfate, magnesium chloride, calcium chloride, potassium hydrogen phosphate, disodium phosphate, ferric chloride, boric acid, manganese chloride, zinc chloride, cobalt chloride, copper chloride, ethylenedinitrilotetraacetic acid disodium salt dihydrate (Gorham et al. 1964).
7. Culture should be maintained at 22 °C, in12-h light/dark intervals, and the intensity of light should be between 30 and 50 μmol (González-Blanco et al. 2020).

11.2.2 Micropipette–Isolation Method

The main aim of the micropipette isolation method is to collect cells from the sample without any damage and to transfer them into a sterile droplet. The micropipette method is useful to obtain a single algal cell. This is an effective method to pick unicellular soft microcolony or fine filamentous cyanobacteria, it can also be utilized to pick single filaments which are short and sturdy. Figure 11.2 represents the preparation of the micropipette from the Pasteur pipette from step 1 to step 5 mentioned below.

1. Hold the commercially available Pasteur pipette tip of one end to the flame and rotate the pipette gradually with the help of a hand and support the flamed tip of the pipette through forceps.
2. As soon as the pipette becomes soft, remove the pipette from the flame and gently pull the tip of the pipette to produce a thin pipette
3. Relocate the forceps on the thin pipette
4. Quickly bend the pipette so that it will break

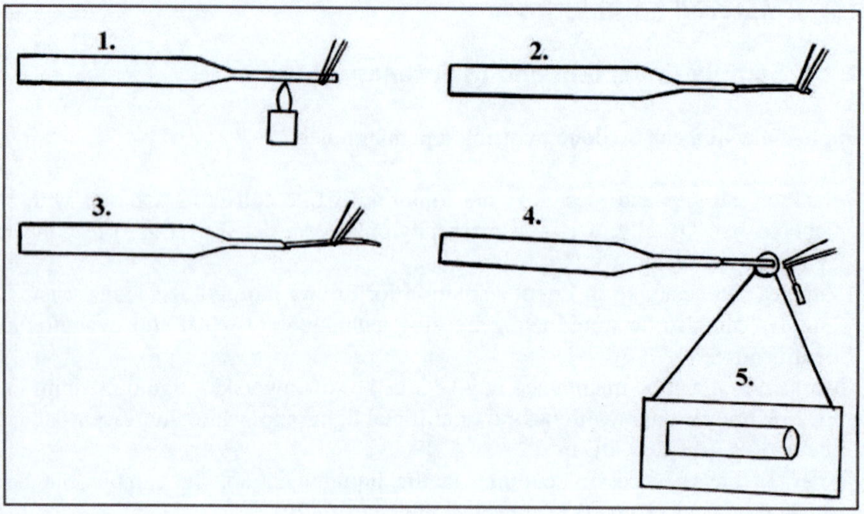

Fig. 11.2 Micropipette preparation from Pasteur pipette

5. Tip with a smooth edge is suitable for collecting samples (Hoshaw and Rosowski 1973).

   NOTE: Before the use of the pipette, it is suggested to humidify the tip of the tube by quickly touching the sterile agar plate surface so that the capillary tube gets moisture from the agar plate.
6. Hold the micropipette with one hand under the dissection scope and above the sample plate, due to the capillary action sample/filament entering into the capillary tube.
7. Once the capillary tube is filled, gently blow the Pasteur micropipette onto the new sterilized agar plate surface.
8. Prepare a fresh Pasteur micropipette for each isolation because the capillary tube would melt when it is re-sterilizing under the Bunsen burner (Rippka 1988).

11.2.3 Isolation of Algal Cells Using Streaking Method

This method is usually used to dilute the culture/inoculum on a solid agar medium to obtain a pure colony. 0.5–2.0% of Agar can be used to grow algal culture. And inoculation has to be done inside the Laminar Air Flow (LAF) to control contamination of the microorganisms.

[**Note:** Strictly follow the rules and regulations of a microbiology lab.]

Inoculation loop can be either purchased commercially or made from a platinum/nichrome wire.

1. Take 0.7 mm diameter and 10 cm long platinum/nichrome wire.
2. Bend one end of the platinum/nichrome wire to make a circle of diameter 3–5 mm.
3. And the other end is anchored in a glass rod or any other commercially available handles.
4. Heat the circle end of the loop until it turns red with the help of a Bunsen burner and leave that for cooling.
5. Once the loop is cooled down take a loopful culture from the culture plate.
6. Spread the inoculum over the first quarter of the agar plate, passing over it many times, and try to homogenize aggregates of filaments.
7. After every quarter streaking, flame the loop, leave it for cooling and use it to make several parallel streaks traversing at an angle of 90° to the previously spread filament.
8. Repeat this process another two times, and care should be taken at the fourth quadrant streaks because the loop should not touch the first streak to get isolated colonies.
9. After streaking, the agar plate is incubated until colonies of cells exist on the plate.

The incubation time varies from a few days to several months based on the source of collection. Usually, oceanic algal culture takes months to grow on culture media. Figure 11.3 shows quadrant streaking of cyanobacterial culture (Rippka 1988).

11.2.4 Isolation from Microspade Method

This method is specifically used to pick the colonies or filaments which are compact and dry.

1. Take 0.7–1 mm thickened and 8–10 cm long nichrome or platinum wire for constructing a microspade.
2. Heat the nichrome or platinum wire until it becomes red with the help of a Bunsen burner.
3. Immediately after the wire became red, make approximately 5 mm length of one end of the wire flat as much as possible with the help of a hammer.
4. After cooling, file the flattened part with two parallel cutting sides of about 2-3 mm in length to get a rhomboidal spade.
5. And the other end is anchored in a glass rod or any other commercially available handles.
6. Place the agar plate which has culture under a dissection microscope in such a way that the filaments are easily accessible to the microspade.
7. Make shallow incisions into the agar plate in a rectangle or square shape with the help of microspade cutting edges.
8. Take rectangle or square agar block which has culture and transfer the block with the help of a microspade broadside to the fresh agar plate (Rippka 1988).

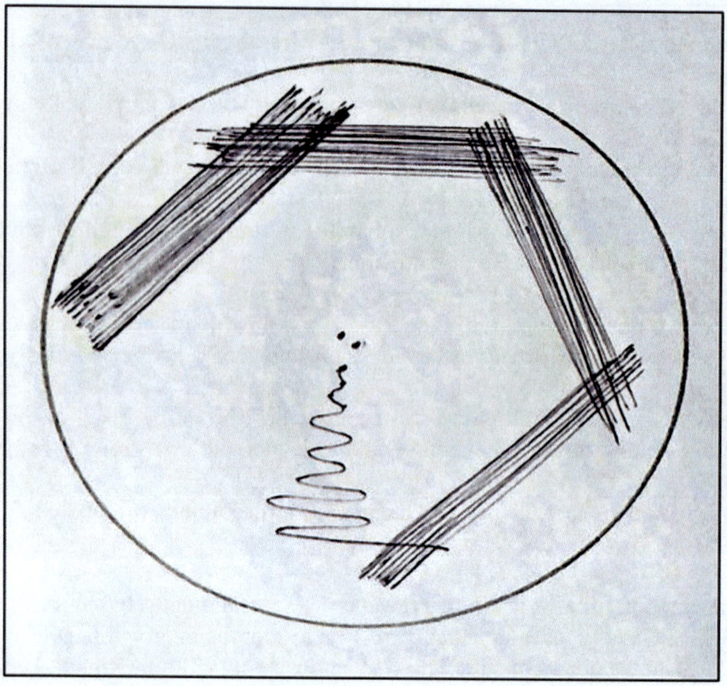

Fig. 11.3 Quadrant streaking of cyanobacterial culture

11.3 Management of Toxin Producing Cyanobacteria

1. Algaecides can be applied to lakes and reservoirs to control cyanobacterial blooms. And chemical oxidation and ultraviolet irradiation are also helpful in removing cyanobacterial blooms.
2. Clays and commercially available aluminum sulfate have been utilized for the control of cyanobacterial blooms in water. Aluminum sulfate treatment efficiency depends on the concentration and the flocculant type.
3. Destratification and aeration have also been utilized to treat cyanobacterial bloom.
4. The use of sonication or ultrasound waves to disturb the cells of cyanobacteria.
5. Avoid nutrient pollution sources in waterbodies.
6. Traditional treatments like coagulation, flocculation, rapid granular filtration, and sedimentation are effective in eliminating intact cells (United States Environmental Protection Agency 2015).

11.4 Conclusion

Cylindrospermopsis sp. is known to produce a cyanotoxin called Cylindrospermopsin. This toxin has adverse effects on humans as well as on animal health. This is generally regarded as an undesirable property of the organism; however, it is important to notice the applications of this organism and its secondary metabolites in medical fields such as antimicrobial and anticancer agents. Recent studies support the pharmaceutical applications of *Cylindrospermopsis* and future investigations are in demand to prove and successfully use these *Cylindrospermopsis* sp. to their maximum potential for human benefits. The protocols described in this chapter would help towards isolation and investigation of such microorganisms for exploiting their significance.

References

Antunes JT, Leão PN, Vasconcelos VM (2015) Cylindrospermopsis raciborskii: review of the distribution, phylogeography, and ecophysiology of a global invasive species. Front Microbiol 6:473

Bernard C, Harvey M, Briand JF, Biré R, Krys S, Fontaine JJ (2003) Toxicological comparison of diverse Cylindrospermopsis raciborskii strains: evidence of liver damage caused by a French C. raciborskii strain. Environ Toxicol 18:176–186

Fabre A, Lacerot G, de Paiva RR, Soares MCS, de Magalhães VF, Bonilla S (2017) South American PSP toxin-producing Cylindrospermopsis raciborskii (Cyanobacteria) decreases clearance rates of cladocerans more than copepods. Hydrobiologia 785:61–69

González-Blanco C, Dörr FA, Albuquerque R, Onuki J, Pinto E (2020) Alternative isolation protocol for desulfo and zwitterionic cylindrospermopsin alkaloids and comparison of their toxicity in HepG2 cells. Molecules 25:3027

Gorham PR, McLachlan J, Hammer UT, Kim WK (1964) Isolation and culture of toxic strains of Anabaena flos-aquae (Lyngb.) de Bréb. SIL Proceedings, 1922–2010. 15:796–804

Hoshaw RW, Rosowski JR (1973) Culture methods and growth measurements. In: Stein JR (ed) Algal culturing techniques. Cambridge University Press, Cambridge, pp 53–68

Jones WW, Sauter S (2005) Distribution and abundance of Cylindrospermopsis raciborskii in Indiana lakes and reservoirs

Lau NS, Matsui M, Abdullah AAA (2015) Cyanobacteria: photoautotrophic microbial factories for the sustainable synthesis of industrial products. Biomed Res Int 2015:754934

Li XC, Huo SL, Cai FF, Yang YM, Xi BD, Li RH (2017) The taxonomy and phylogeny of the genus Cylindrospermopsis (Cyanobacterium) evaluated by adding five new records from China. Phytotaxa 316:224–238

Rippka R (1988) Isolation and purification of cyanobacteria. Methods Enzymol 167:3–27

Saker ML, Neilan BA (2001) Varied diazotrophies, morphologies, and toxicities of genetically similar isolates of Cylindrospermopsis raciborskii (Nostocales, Cyanophyceae) from northern Australia. Appl Environ Microbiol 67:1839–1845

Somdee T, Kaewsan T, Somdee A (2013) Monitoring toxic cyanobacteria and cyanotoxins (microcystins and cylindrospermopsins) in four recreational reservoirs (KhonKaen, Thailand). Environ Monit Assess 185:9521–9529

United States Environmental Protection Agency (2015) Drinking water health advisory for the cyanobacterial toxin cylindrospermopsin

Isolation of Toxin Producing Cyanobacteria from Aquatic Sample with *Nodularia* sp. 12

Gomathi Selvam, Ambikapathy Varatharaju, and Panneerselvam Annamalai

Abstract

Cyanobacteria commonly known as blue-green algae are not truly eukaryotic algae. They are Gram-negative prokaryotes, perform oxygenic photosynthesis, and also fix atmospheric N_2. Nodularin toxin is commonly detected in cyanobacterial blooms containing *Nodularia* sp. Toxic blooms of *Nodularia* have been associated with poisonings of domestic animals in different parts of the world. Traditionally, *Nodularia* have been classified on the basis of the morphology of the different types of cells, on *Nodularin* production, on ultra-structural features of the cells and on ecological characteristics. In this study, we practise isolation of toxin producing cyanobacteria from aquatic sample of *Nodularia*. The results were discussed in detail.

Keywords

Cyanobacteria · Nodularia · Aquatic sample · Isolation · Identification

12.1 Introduction

The present world population of about 7.2 billion is expected to cross 9.6 billion by the end of year 2050. Nowadays, sustainable agriculture practices have envisaged an important role of these tiny microorganisms in achieving food security without creating environmental problems. The recent trends of using the bio-inoculants containing beneficial soil microbes over synthetic fertilizers, insecticides, and pesticides for enhancing crop productivity is a welcome step (Singh 2011; Singh

G. Selvam (✉) · A. Varatharaju · P. Annamalai
PG and Research Department of Botany, A.V.V.M Sri Pushpam College (Autonomous), Poondi (Affiliated to Bharathidasan University, Tiruchirappalli), Thanjavur, Tamilnadu, India

N. Thajuddin et al. (eds.), *Protocols for Cyanobacteria Sampling and Detection of Cyanotoxin*, https://doi.org/10.1007/978-981-99-4514-6_12

et al. 2011). Very recently, it has been proposed that cyanobacteria could be the vital bio-agents in ecological restoration of degraded lands (Singh 2014). Cyanobacteria are also useful for wastewater treatment and have the ability to degrade the various toxic compounds even the pesticides (Cohen 2006).

They are ubiquitous in ponds, lakes, water streams, rivers, and wetlands. They can easily survive extreme environments such as hot springs, hyper-saline waters, freezing environments, and arid deserts. Cyanobacteria produce a variety of biologically active compounds of antibacterial, antifungal, antialgal, and antiviral potential (Teuscher et al. 1992; Dahms et al. 2006).

The genus *Nodularia* (Nostocales) consists of filamentous heterocystous nitrogen-fixing cyanobacteria, which are found in brackish water and freshwater, as well as in terrestrial environments, worldwide. Cyanobacteria of the genus *Nodularia* form toxic blooms in brackish waters worldwide (Laamanen et al. 2001).

12.2 Materials

- Glassware
- Chemicals
- Stock solutions for BG-11 medium
- Stock 1:
  - Na_2Mg EDTA 0.1 g/L
  - Ferric ammonium citrate 0.6 g/L
  - Citric acid $1H_2O$ 0.6 g/L
  - $CaCl_2$. $2H_2O$ 3.6 g/L
- Stock 2:
  - $MgSO_4$ $7H_2O$ 7.5 g/L
- Stock 3:
  - K_2HPO_4 $3H_2O$ 4.0 g/L
- Stock 4 (Microelements):
  - H_3Bo_3 2.86 g/L
  - $MnCl_2$ $4H_2O$ 1.81 g/L
  - $ZnSo_4$ $7H_2O$ 0.222 g/L
  - $CuSo_4$ 5H2O 0.079 g/L
  - $CoCl_2$ $6H_2O$ 0.050 g/L
  - $NaMoO_4$ $2H_2O$ 0.391 g/L

12.3 Methods (Kızılkaya et al. 2016)

12.3.1 Study Area

1. The *Nodularia* aquatic samples are collected from cyanobacterial bloom.

12.3.2 Sampling Collection

1. Collect the sample from surface water.
2. Collect the sample in 3 L bucket and equally divide into three different 1 L bottles.
3. Add formaldehyde solution (40%) for microscopic analysis.
4. The second 1 L sample will be for molecular analysis.
5. The third 1 L sample for toxin analyses.

12.3.3 Medium Preparation

1. For basic BG11 medium, combine the following stock solutions:

Stock solution	Per liter of medium
Stock 1	10 mL
Stock 2	10 mL
Stock 3	10 mL
Na_2CO_3	0.02 g
Stock 4	1.0 mL
$NaNO_3$	1.5 g

2. Combine stocks and adjust pH to 7.5
3. After autoclaving and cooling the pH adjusted to 7.1.
4. For solid media, add 1% noble agar.

12.3.4 Isolation (Ferris and Hirsch 1991; Roger and Kulasooriya 1980)

1. Transfer water sample to sterile 250 mL conical flasks containing 90 mL BG11 medium.
2. Observe under microscopic growth appearance, during 3–4 weeks of incubation at 30 °C.
3. Streaking method BG11 agar medium to get single culture from isolates.
4. Incubate the plates at 30 °C (400–500 lux).

12.3.5 Identification (Komarek et al. 1993; Desikachary 1959)

1. Presence or absence of gas vesicles, size and shape of trichomes, dimensions and shapes of vegetative cells, akinetes and heterocyst.

2. Examine the bloom sample under light microscope three different slides for quantification.
3. Measure 20 filaments in each slide.

12.4 Observation

Observe the morphology of trichomes, heterocyst size, shape of vegetative cells and akinetes in isolated cyanobacterial species.

Acknowledgment The authors are thankful to PG and Research Department of Botany, A.V.V.M Sri Pushpam College (Autonomous), Poondi, Thanjavur, and also to director and staff members of Indian Biotrack Research Institute, Thanjavur for providing the lab facilities.

References

Cohen RRH (2006) Uses of microbes for cost reduction of metal removal from metals and mining industry waste streams. J Clean Prod 14:1146–1157. https://doi.org/10.1016/j.jclepro.2004.10.009

Dahms HU, Xu Y, Pfeiffer C (2006) Antifouling potential of cyanobacteria: a mini-review. Biofouling 22:317–327. https://doi.org/10.1080/08927010600967261

Desikachary IV (1959) Cyanophyta. ICAR, New Delhi, 685 pp

Ferris MJ, Hirsch CF (1991) Method for isolation and purification of cyanobacteria. Appl Environ Microbiol 57(5):1448–1452

Kızılkaya IT, Demirel Z, Kesici K, Kesici E, Sukatar A (2016) Morphological, molecular and toxicological characterization of *Nodularia spumigena* Mertens in Jungens (1822) from Brackishwater Lake Bafa (Turkey). Sinop Univ J Nat Sci 1(1):39–52

Komarek J, Hubel M, Hubel H, Smarda J (1993) The Nodularia studies 2. Taxonomy. Algol Stud 68:1–25

Laamanen MJ, Gugger MF, Lehtimaki JM, Haukka K, Sivonen K (2001) Diversity of toxic and nontoxic Nodularia isolates (cyanobacteria) and filaments from the Baltic Sea. Appl Environ Microbiol 67(10):4638–4647. https://doi.org/10.1128/AEM.67.10.4638-4647.2001

Roger PA, Kulasooriya SA (1980) Blue green algae and rice. International Rice Research Institute, Los Baños

Singh JS (2011) Methanotrophs: the potential biological sink to mitigate the global methane load. Curr Sci 100:29–30

Singh JS (2014) Cyanobacteria: a vital bio-agent in eco-restoration of degraded lands and sustainable agriculture. Climate Change Environ Sustain 2:133–137

Singh JS, Pandey VC, Singh DP (2011) Efficient soil microorganisms: a new dimension for sustainable agriculture and environmental development. Agric Ecosyst Environ 140:339–353. https://doi.org/10.1016/j.agee.2011.01.017

Teuscher E, Lindequist U, Mundt S (1992) Cyanobakterien, Quellen biogener Wirkstoffe. Pharm Ztg Wiss 137:57–69

Isolation of Toxin Producing Cyanobacteria from Aquatic Samples with *Nostoc* sp.

13

Rajasekar Thirunavukkarasu, P. Gayathri, and V. Jeyanthi

Abstract

Cyanotoxins are secondary metabolites produced by cyanobacteria, a group of photosynthetic prokaryotes that are mostly found in freshwater. Cyanobacteria form the algae blooms in the aquatic ecosystems and potential release of cyanotoxins. Bioaccumulate toxins and spread them up the food chain to people and other animals, toxins pose a new threat to aquatic life. Cyanotoxin poisoning is most frequently transported on by consuming contaminated water by human. Hepatotoxins known as microcystins (MCs) are widely distributed. The most dangerous toxins are found in water supplies. Several cyanobacterial genera, including Microcystis, Anabaena, Nostoc, and Planktothrix, were found to produce MCs. In the present aim to isolation of toxin producing *Nostoc sp.* from aquatic samples with using molecular studies.

Keywords

Nostoc · Cytotoxin · Microcystins

R. Thirunavukkarasu (✉)
Centre for Drug Discovery and Development, Sathyabama Institute of Science and Technology, Chennai, Tamil Nadu, India

P. Gayathri
Department of Medical Laboratory Technology (MLT), School of Allied Health Sciences (AHS), Chennai, Tamil Nadu, India

V. Jeyanthi
Department of Biotechnology, SRM Arts and Science College, Chennai, Tamil Nadu, India

N. Thajuddin et al. (eds.), *Protocols for Cyanobacteria Sampling and Detection of Cyanotoxin*, https://doi.org/10.1007/978-981-99-4514-6_13

13.1 Introduction

There are a great number of viruses in natural waters, in both marine and freshwater environments (Bergh et al. 1989), and it is suspected that a large proportion of these viruses are infectious for bacteria or cyanobacteria. The first isolation of freshwater cyanophages was reported about 40 years ago, and during the following two decades numerous cyanophage strains were isolated (Safferman et al. 1969).Cyanobacteria belong to an ancient group of photosynthetic prokaryotes presenting a broad range of cellular strategies, physiological capacities, and adaptations that support their colonization of diverse environments worldwide. Cyanobacteria can even exist in extreme habitats and are able to settle in diverse biotopes such as hot springs (Demay et al. 2019). They are also known for their production of potent toxins (microcystins, anatoxins, saxitoxins) (Te and Gin 2011). Cyanobacteria of the genus Nostoc are found on every continent on earth in a wide range of terrestrial and aquatic ecosystems. Although numerous strains of Nostoc have been identified, only a few Nostoc species have been characterized in detail. Episodes of toxic cyanobacterial blooms have recently been rather well documented in subtropical and temperate regions (Herdman et al. 2001). However, information on toxic cyanobacteria and cyanotoxins from tropical regions like Southeast Asia is limited.

Nostoc is a cosmopolitan nitrogen-fixing cyanobacterial genus with some 200 described species (Komárek and Hauer 2009). Species of this genus can occur in aquatic, terrestrial, aerial ecosystems and in symbioses with fungi, mosses, liverworts, ferns, lichens, and vascular plants. According to the current taxonomy of the genus Nostoc, several morphotypes are genetically heterogeneous and fall outside of the main Nostoc cluster in phylogenetic analysis (Lukešová et al. 2009; Papaefthimiou et al. 2008; Rajaniemi et al. 2005). Therefore, some Nostoc morphotypes may be split into new taxonomic units in the future. In Brazil, only 11 species of Nostoc were identified (Fiore et al. 2009; Sant'Anna et al. 2007), and there is no report of microcystin production by this genus.

Microcystin-LR (MC-LR) has been assigned a provisional guideline value of 1 g/L in drinking water by the World Health Organization (WHO). Even in countries that are members of the European Union, the concentration of cyanotoxins in drinking water is not yet adequately regulated due to the lack of comprehensive toxicological data for a variety of cyanotoxins (EU). In this report, attention is focused on the methodologies commonly used to detect cyanotoxins in water environments using *Nostoc sp.*

In this study, an investigation of microcystin production by the *Nostoc sp.* isolated from a eutrophic freshwater reservoir. The isolated Nostoc strain was identified by morphological and molecular analyses. The microcystin production by the cyanobacterium under culture condition was evaluated using ELISA immunoassay. Fragments of mycD, mycE, and mcyG genes involved in microcystin biosynthesis were sequenced. Sequence analysis of genes encoding the small subunit of ribosomal RNA (16S rDNA) is currently the most promising approach to the phylogenetic classification of cyanobacteria.

13.2 Material

- BG-110 media
- Conical flasks
- A bright field microscope and digital camera
- Microscope slide
- Ph-meter
- Phase contrast light microscope
- Cetyl-trimethyl-ammonium bromide (CTAB)
- PCR thermal cycler
- Oligonucleotide primer
- dNTP
- $MgCl_2$
- PCR buffer
- *Taq* DNA polymerase

13.3 Methods

13.3.1 Sample Collection

- Cyanobacteria isolated by the serial dilution method from aquatic samples (Lukešová 1993).
- From the obtained water suspensions, dilutions (10–2, 10–3) were made, and 0.1 mL aliquots were spread on solidified (1.5%) BG-110 media (Rippka et al. 1979).
- Plates were incubated at 24 °C with a 16:8 (light:dark) photoperiod and 2000–3000 lx of light intensity maintained
- The isolated, pure cyanobacterial strains were kept alive in the BG-1 medium.
- All inoculations were performed in an aseptic environment, and all cultures were shaken twice to prevent cell clumping and speed up the growth process.
- The cultures were routinely checked for biological interfering.
- Subculturing the axenic cultures into fresh medium on a regular basis while maintaining the same culture conditions, the cultures were kept in an exponential growth phase.
- 1-mL water samples were inoculated without nitrogen in 100-mL sterilized BG-11 medium (pH 7.5). By using the serial dilution and plating method, the organism was isolated and purified (Stanier et al. 1971).
- *Nostoc sp.* was isolated in its purest form, and cultures were kept alive in 250 mL flasks filled with nitrogen-free BG-11 medium (pH 7.5) at 25 °C under an illumination of 2000 lux over a 16/8-h photoperiod. To maintain homogeneity, the culture vessels were vigorously shaken four to five times per day.
- Repeated subculturing of each strain was done by transferring into fresh medium every 20 days in order to maintain the axenic culture.
- Laboratory stock cultures of the test organisms were kept.

13.3.2 Morphology Identification

13.3.2.1 Light Microscopy

- The morphology of the gathered algal members was examined using a bright field microscope with a digital camera.
- Morphological identify *Nostoc sp.* variation studies and taxonomical methods described by (Desikachary 1959) and (Anand 1989).
- To identify cyanobacterium, morphological parameters were used (Mishra et al. 2019).
- The morphological characteristics of cyanobacterium were examined under 45× and 10× Olympus microscopes.
- Magnus Image Projector and image analysis using standards were used to analyse morphological characters.
- In accordance with the taxonomic descriptions of Rippka et al. (1979), the stain was attributed to cyanobacterial species.
- The isolated cyanobacteria's fresh unialgal culture was examined using an Olympus BX 40 Phase contrast light microscope.

13.3.3 Analyses of Microcystin

13.3.3.1 Growth and Cell Collection

Nostoc sp. was grown in 2 L of BG-110 liquid medium for 30 days at 25 ± 1 °C with gentle orbital shaking (100 rpm) and constant white, fluorescent illumination (40 mol photon μm^{-2} s^{1}). The culture was centrifuged at 12,000 ×*g* for 5 min, and the cell pellet was lyophilized for further investigation.

13.3.4 Genetic Analyses

13.3.4.1 DNA Extraction, PCR Amplification, and Sequencing

- Cetyl-trimethyl-ammonium bromide (CTAB)-based extraction technique used isolated total genomic DNA from cyanobacteria.
- Invesntigate the relationship between the microcystin synthetase genes from cyanobacteria sp using mcyD, mcyE and mcyG genes from cyanobacteria sp. (Table 13.1)
- In a PCR thermal cycler, PCR amplification was carried out with 25 L reactions containing 10 ng of genomic DNA, 0.2 M of each oligonucleotide primer, 200 M dNTP, 3.0 mM $MgCl_2$, 1 PCR buffer, and 1.5 U Platinum Taq DNA polymerase.

Table 13.1 List of primer for amplification

Name of the primer	Gene	Bp
mcyD	mcyD F/mcyD R	818 bp
mcyE	mcyEF2/mcyER4	809 bp
mcyG	mcyGF/mcyGR	534 bp

- The protocol used was as follows: 95 °C for 3 min, then 30 cycles of 94 °C for 30 s of denaturation, 56 °C for 30 s of annealing, and 72 °C for 1 min of elongation. In accordance with the supplier's instructions, the PCR product produced was cloned into a pGEM®-T Easy Vector System.
- Quimiocompetent Escherichia coli DH5 cells were transformed using the heat-shock method (Sambrook et al. 1989) and incubated at 37 °C overnight on solid LB medium containing UltraPuraTM X-Gal (50 g mL^1) (Invitrogen) and ampicillin sodium salt (100 g mL^1).
- Alkaline lysis technique, recombinant plasmids were isolated from white colonies after growth (Birnboim and Doly 1979).
- DYEnamic ET Terminator Cycle Sequencing kit and the primers T7 and M13 anchored in the pGEM®-T Easy Vector, the cloned gene fragments were sequenced (GE Healthcare, Little Chalfont, Buckinghamshire, UK).
- A PCR system was used to perform the cycle sequencing reaction, and the conditions were 25 cycles of 95 °C for 20 s, 50 °C for 15 s, and 60 °C for 60 s.
- Following the sequencing reaction, the DNA was precipitated using 60 L of 100% ethanol and 2 L of sodium acetate buffer (1.5 M sodium acetate, pH 9.0, and 250 mM EDTA, pH 8.0).
- Tube was centrifuged at 12,000 ×*g* for 15 min at 4 °C, and the supernatant was thrown away. The supernatant was collected after centrifuging the DNA pellet for 5 min after washing it with 150 L of 70% ethanol.
- Purified reaction was resuspended in HiDi formamide (Applied Biosystems), and the sample was placed in an ABI PRISM 3100 Genetic Analyzer.
- The pellet was air-dried overnight at room temperature in the dark (Applied Biosystems).
- Only bases with a quality >20 were taken into consideration when assembling the sequenced fragments into a single contig using the programme Phred/Phrap/Consed (Philip Green, University of Washington, Seattle, WA, USA).
- PCR amplification using the primer set NMT-F/NMT-R (1600 bp), created to target a portion of the C-terminal adenylation domain and the full NMT domain from Anabaena and Nostoc strains, was used to confirm the presence of the N-methyltransferase (NMT) domain in the first module of McyA, responsible for the N-methylation of the microcystins (Fewer et al. 2008).
- A PCR gradient annealing temperature (48.0, 48.2, 48.7, 49.6, 50.7, 52.0, 53.4, 54.8, 56.1, 57.1, 58.0, and 58.4 °C) was used to improve the specificity of the PCR reaction.
- Anabaena circinalis CENA191, a strain that produces microcystins, was used as a positive control.
- Using the oligonucleotide primers 27F1 and 1494Rc (Neilan et al. 1997), which correspond to the *E. coli* 16S rRNA gene positions 27–1494, the almost complete 16S rRNA gene from the genomic DNA of the *Nostoc sp.* CENA88 was amplified by PCR (Fiore et al. 2005).

Phylogenetic Analyses

- A phylogenetic tree was created using the 16S rRNA gene sequence that was obtained for this study along with reference sequences retrieved from GenBank.

- Kimura 2-parameter model of sequence evolution was used to reconstruct the tree using the neighbour-joining (NJ) method, which was implemented by the MEGA version 4.0 software package.
- Using 1000 replications, bootstrap percentages were used to estimate the tree's robustness.
- The three mcy genes' DNA sequences were converted into protein sequences, and the evolutionary history was also recreated by assembling a data set of 695 amino acids made up of McyG, McyD, and McyE in accordance with the positions in which they were arranged in the mcy gene cluster.
- The NJ method, implemented in the MEGA 4.0 programme, produced the unrooted dendrogram using the P-distance model with bootstrap test (1000 replications) (Tamura et al. 2007).

13.4 Observation/Any Special Note (If Any)

Acknowledgement The authors extend their appreciation to the Deanship of Scientific Research, Vice Presidency for Graduate Studies and Scientific Research, King Faisal University, Saudi Arabia (Project# GRANT3868). The authors acknowledge the Management of Sathyabama Institute of Science and Technology, Chennai-600 119, Tamil Nadu, India for infrastructural support.

Funding: This work was supported by the Deanship of Scientific Research, Vice Presidency for Graduate Studies and Scientific Research, King Faisal University, Saudi Arabia (Project# GRANT3868).

References

Anand N (1989) Hand book of blue-green algae. Bishen Singh Mahendra Pal Singh, Dehra Dun, p 77

Bergh Ø, Børsheim KY, Bratbak G, Heldal M (1989) High abundance of viruses found in aquatic environments. Nature 340(6233):467–468

Birnboim HC, Doly J (1979) A rapid alkaline extraction procedure for screening recombinant plasmid DNA. Nucleic Acids Res 7:1513–1523

Demay J, Bernard C, Reinhardt A, Marie B (2019) Natural products from cyanobacteria: focus on beneficial activities. Mar Drugs 17(6):320

Desikachary TV (1959) Cyanophyta. Indian Council of Agricultural Research, New Delhi, 669 pp

Fewer DP, Tooming-Klunderud A, Jokela J, Wahlsten M, Rouhiainen L, Kirstensen T, Rohrlack T, Jakobsen KS, Sivonen K (2008) Natural occurrence of microcystin synthetase deletion mutants capable of producing microcystins in strains of the genus Anabaena (Cyanobacteria). Microbiology 154:1007–1014

Fiore MF, Neilan BA, Copp JN, Rodrigues JLM, Tsai SM, Lee H, Trevors J (2005) Characterization of nitrogen-fixing cyanobacteria in the Brazilian Amazon floodplain. Water Res 39:5017–5026

Fiore MF, Genuário DB, da Silva CSP, Shishido TK, Moraes LAB, Neto RC, Silva-Stenico ME (2009) Microcystin production by a freshwater spring cyanobacterium of the genus Fischerella. Toxicon 53(7–8):754–761

Herdman M, Castenholz RW, Rippka R (2001) Form-genus VIII. Nostoc Vaucher 1803. In: Boone DR, Castenholz RW (eds) Bergey's manual of systematic bacteriology, vol 1, 2nd edn. Springer, New York, pp 575–580

Komárek J, Hauer T (2009) Worldwide electronic publication. University of South Bohemia and Institute of Botany AS CR. CyanoDB. cz-On-line database of cyanobacterial genera

Lukešová A (1993) Soil algae in four secondary successional stages on abandoned fields. Algol Stud 81–102

Lukešová A, Johansen JR, Martin MP, Casamatta DA (2009) Aulosira bohemensis sp. nov.: further phylogenetic uncertainty at the base of the Nostocales (cyanobacteria). Phycologia 48(2): 118–129

Mishra SK, Singh J, Pandey AR, Dwivedi N (2019) Indole-3-acetic acid production by the cyanobacterium Fisherella muscicola NDUPC001. Curr Sci 116(7):1233–1237

Neilan BA, Jacobs D, Dot TD, Blackall LL, Hawkins PR, Cox PT, Goodman AE (1997) RNA sequences and evolutionary relationships among toxic and nontoxic cyanobacteria of the genus Microcystis. Int J Syst Bacteriol 47:693–697

Papaefthimiou D, Hrouzek P, Mugnai MA, Lukesova A, Turicchia S, Rasmussen U, Ventura S (2008) Differential patterns of evolution and distribution of the symbiotic behaviour in nostocacean cyanobacteria. Int J Syst Evol Microbiol 58(3):553–564

Rajaniemi P, Hrouzek P, Kaštovska K, Willame R, Rantala A, Hoffmann L et al (2005) Phylogenetic and morphological evaluation of the genera Anabaena, Aphanizomenon, Trichormus and Nostoc (Nostocales, Cyanobacteria). Int J Syst Evol Microbiol 55(1):11–26

Rippka R, Deruelles J, Waterbury JB, Herdman M, Stanier RY (1979) Generic assignments, strain histories and properties of pure cultures of cyanobacteria. Microbiology 111(1):1–61

Safferman RS, Schneider IR, Steere RL, Morris ME, Diener TO (1969) Phycovirus SM-1: a virus infecting unicellular blue-green algae. Virology 37(3):386–395

Sambrook J, Fritsch EF, Maniatis T (1989) Molecular cloning. A laboratory manual, 2nd edn. New York, Cold Spring Harbor Laboratory Press

Sant'Anna CL, Azevedo MTDP, Branco LHZ, Komárek J (2007) New aerophytic morphospecies of Nostoc (cyanobacteria) from São Paulo state, Brazil. Hoehnea 34:95–101

Stanier RY, Kunisawa R, Mandel MCBG, Cohen-Bazire G (1971) Purification and properties of unicellular blue-green algae (order Chroococcales). Bacteriol Rev 35(2):171–205

Tamura K, Dudley J, Nei M, Kumar S (2007) MEGA4: molecular evolutionary genetics analysis (MEGA) software version 4.0. Mol Biol Evol 24:1596–1599

Te SH, Gin KYH (2011) The dynamics of cyanobacteria and microcystin production in a tropical reservoir of Singapore. Harmful Algae 10(3):319–329

14 Isolation of Toxins Producing Cyanobacteria from Aquatic Sample with Schizothrix

Vajiha Banu Habeeb Mohamed, Sumithra Pasumalaiarasu, and Kavitha Parangusadoss

Abstract

Cyanotoxins is an intracellular toxic substance that can be produced by a wide variety of planktonic cyanobacteria. Most commonly, it occurs in a genera of Microcystis, Dolichospermum (previously Anabaena), and Planktothrix. Cyanotoxins are released from the intracellular components of the cyanobacterial cells (algal bloom) into the surrounding water mostly during cell lysis (i.e., cell rupture). However, some cyanobacterial species are capable of releasing toxins (extracellular) into the water without cell rupture or death. Cyanobacteria were present on the earth 3.5 billion years ago since it has been colonized almost all terrestrial and aquatic ecosystems. They produce a high number of bioactive molecules, among which some of them are cyanotoxins. Excessive growth of algae in on freshwater ecosystem to form algal blooms in an anthropogenic activities and climatic changes might be arising a major problem of human and animal health. These cyanotoxins include cyclic peptides and alkaloids. Among the cyclic peptides the microcystins and the nodularins and alkaloids contain anatoxin-a, anatoxin-a(S), cylindrospermopsin, saxitoxins (STXs), aplysiatoxins, and lyngbyatoxin. Cyanotoxins may cause adverse health effects from mild skin rash to serious illness (death). When people are exposed to the drinking water containing microcystin and cylindrospermopsin, they could cause serious illness of liver and kidney damage. Acute illnesses like hay fever, skin rashes, respiratory and gastrointestinal distress were also raised by short-term exposure of

V. B. H. Mohamed (✉)
PG and Research Department of Microbiology, Jamal Mohamed College (Autonomous), Tiruchirappalli, Tamilnadu, India

S. Pasumalaiarasu · K. Parangusadoss
PG and Research Department of Microbiology, Srimad Andavan Arts and Science College (Autonomous), Tiruchirappalli, Tamilnadu, India

N. Thajuddin et al. (eds.), *Protocols for Cyanobacteria Sampling and Detection of Cyanotoxin*, https://doi.org/10.1007/978-981-99-4514-6_14

cyanotoxins in recreational activities. Cyanotoxins can be grouped according to their biological effects as hepatotoxins, neurotoxins, cytotoxins, and dermatotoxin. During water treatment cyanobacteria produce the algal toxins. Aplysiatoxin produced by *Schizothrix*, which comes under dermatotoxins. Lipopolysaccharide endotoxin is also produced by certain Cyanobacteria including *Schizothrix calcicola*. Climate change, as well as progressive eutrophication, is responsible for the global expansion and increased frequency of cyanobacterial blooms in aquatic environments. This chapter presents the knowledge that has evolved on the topic of toxins produced by *Schizothrix*.

Keywords

Schizothrix · Aplysiatoxin · Cyanotoxins · Lipopolysaccharide · Eutrophication

14.1 Introduction

The Schizothrix is a genus of cyanobacteria in the family Schizotrichaceae. Based on the structure of filaments and thallus, more ensheathed and trichomes are enveloped by common sheaths. The toxin-producing genera include Anabaena, *Aphanizomenon, Cylindrospermopsis, Gloeotrichia, Hapalosiphon, Lyngbya, Microcystis, Nodularia, Nostoc, Oscillatoria, Schizothrix, Spirulina,* and *Synechocystis (*Komarek et al. 2006*)*.

Aplysiatoxins can cause severe skin dermatitis and are potent tumour promoters and *protein kinase C activators*. Cyanobacterial blooms have a significant impact on water quality.

Lipopolysaccharides (LPS) were isolated from *Schizothrix* species by hot phenol-water extraction. The polysaccharide moiety was composed of glucosamine, galactose, glucose, mannose, xylose, and rhamnose. The lipid A part contained beta-hydroxylauric, myristic, pentadecanoic, palmitic, beta-hydroxypalmitic, stearic, oleic, and linoleic acids (Keleti et al. 1979).

Cyanobacteria can lead to harmful effect for humans when water is used for drinking consumption; in fact, they can create many problems in drinking water treatment plants (increase of solids load, bacterial growth in sand and GAC filters, low efficiency of disinfection) and in the distribution system (growth in reservoir tanks and pipes).

The cyanobacteria occur in a variety of environments and have been found globally. Toxin-producing cyanobacterial harmful algal blooms (CHABs) have been increasing worldwide and cause a threat to drinking and recreational water (Cheung et al. 2013).

The progressive eutrophication and climate changes is responsible for the global expansion and increased cyanobacterial blooms in aquatic environments. This phenomenon has received a lot of attention worldwide, as it is a serious problem for drinking water management. The massive occurrence of cyanobacteria has a significant impact on the physical and chemical properties of water such as reduced transparency of water column, oxygen deficit, and they are to produce varieties of bioactive compounds, that compounds affect the taste and odour of water, toxins like hepatotoxic microcystins, neurotoxic anatoxin-a, cytotoxic cylindrospermopsin and oligopeptides of unclear biological function (Bober et al. 2022).

New methods are being developed for the purification of the toxins and also for their recovery and quantification of waters. These include the use of chemical, cytotoxicity, and immunological methods to complement the mouse bioassay which has hitherto been used in cyanobacterial toxin studies with laboratory cultures and water samples. Information on the regulation of cyanobacterial toxin production and on the possible biological significance of the toxins in aquatic environments is also presented. A greater awareness of cyanobacterial toxins in water are destined for human use (Codd et al. 1989).

The endotoxin producing few cyanobacterial strains were characterized and belonging to the following genera include *Synechococcus*, *Synechocystis* and *Microcystis*, filamentous heterocystous Anabaena, and filamentous non-nitrogen fixing *Phormidium*, *Oscillatoria*, and *Schizothrix* (Rapala et al. 2002).

Lyngbya, *Schizothrix*, and *Oscillatoria* are marine toxic forms of filamentous cyanobacteria. Toxins producing cyanobacteria are responsible for the contact dermatitis called "swimmers itch" (Carmichael 1986).

14.2 Materials

- Water sample
- GF/C glass fibre papers
- Medium—(BG-11 Medium and Chu's-Medium)
- Lugol's iodine solution
- Distilled water
- Conical flask
- Slides
- Light microscope

14.3 Methods

14.3.1 Sample Collection (Mohamed 2008)

The aquatic water sample is collected from cyanobacterial water blooms. Collect samples for the bacteriological analyses in sterilized glass bottles and transport, refrigerate to the laboratory, and process within 24 h after sampling.

14.3.2 Isolation and Identification of *Schizothrix*

1. Collect the water samples and cyanobacterial mats and transport them to the laboratory.
2. Prepare any one of the cyanobacterial isolation medium in conical flask and sterilize at 121 °C for 30 min.
3. Inoculate water sample into the cyanobacterial isolation medium.
4. Incubate the conical flask under light at 28–30 °C for 2–3 weeks.
5. Algal growth appears in broth after 2–3 weeks.
6. Perform the wet mount and observe the *Schizothrix*.

14.3.3 Toxin Analysis (Mohamed 2008)

- Filter the sample through GF/C glass fibre paper.
- An aliquot of the filtrates is used for chemical analysis, while the remaining part is kept at 20 °C for toxin analysis from *Schizothrix*.
- Mix the mats and homogenize it with 1 L of sterile distilled water.
- Preserve the aliquot (500 mL) of the homogenized pool sample in Lugol's iodine solution and store it in the dark condition for 24 h.
- Remove the supernatant, after sedimentation process and mix the remaining solution (100 mL) and use it for the identification and enumeration of cyanobacteria.
- Use another aliquot (50 mL) of the homogenized pool sample for isolation and purification of the species constituting the cyanobacterial mats.
- Freeze and dry the remaining volume of homogenized mats and keep it at −20 °C for toxin analysis.

Acknowledgement The authors acknowledge the PG and Research Department of Microbiology, Jamal Mohamed College (*Autonomous*), Tiruchirappalli.

References

Bober B, Bialczyk J, Chrapusta-Srebrny E (2022) Effect of abiotic factors on the stability of chosen oligopeptides isolated from the freshwater cyanobacterium *Woronichinia naegeliana*. Int J Environ Sci Technol 20:7057. https://doi.org/10.1007/s13762-022-04474-4

Carmichael WW (1986) Algal toxins. In: Advances in botanical research, vol. 12. Academic, pp 47–101

Cheung MY, Liang S, Lee J (2013) Toxin-producing cyanobacteria in freshwater: a review of the problems, impact on drinking water safety, and efforts for protecting public health. J Microbiol 51:1–10. https://doi.org/10.1007/s12275-013-2549-3

Codd GA, Bell SC, Brooks WP (1989) Cyanobacterial toxins in water. Water Sci Technol 21(3): 1–13

Keleti G, Sykora JL, Lippy EC, Shapiro M (1979) Composition and biological properties of lipopolysaccharides isolated from Schizothrix calcicola (Ag.) Gomont (Cyanobacteria). Appl Environ Microbiol 38(3):471–477

Komarek J, Taton A, Sulek J, Wilmotte A, Kastovska K, Elster J (2006) Ultrastructure and taxonomic position of two species of the cyanobacterial genus Schizothrix. Cryptogam Algol 27(1):53–62

Mohamed ZA (2008) Toxic cyanobacteria and cyanotoxins in public hot springs in Saudi Arabia. Toxicon 51(1):17–27

Rapala J, Lahti K, Räsänen LA, Esala AL, Niemelä SI, Sivonen K (2002) Endotoxins associated with cyanobacteria and their removal during drinking water treatment. Water Res 36(10): 2627–2635

15 Isolation of Toxin Producing Cyanobacteria from Aquatic Samples with *Lyngbya*

S. Nisha Nandhini, Kallingil Gopi Divya, and Nithaniyal Stalin

Abstract

Cyanobacteria are represented by morphologically diverse, ecologically important, phylogenetic Gram-negative prokaryotes. They constitute a phylogenetically coherent group of evolutionarily ancient phototrophic bacteria. The toxins produced by cyanobacteria are classified into endotoxins, hepatotoxins, lipopolysaccharides, and neurotoxins. Some species of *Lyngbya* can have Lyngbyatoxin-a and Aplysiatoxins, causing dermatitis with itching, burning, pain, rash, and blisters when in contact with human skin and could lead to loss of superficial layers of the skin by cell death. The effect of cyanobacterial intoxication includes epithelial cell damage in the digestive and respiratory tracts promoting tumour development. This chapter aims to elucidate the methodology for isolating the cyanotoxins from *Lyngbya*.

Keywords

Lyngbya · Lyngbyatoxin-a · Aplysiatoxins · Saxitoxins · Z8 medium · ASN-III medium

S. Nisha Nandhini
Department of Botany, St. Joseph's College (Autonomous), affiliated to Bharathidasan University, Tiruchirappalli, Tamil Nadu, India

K. G. Divya
Department of Pharmacognosy, Siddha Central Research Institute, Chennai, Tamil Nadu, India

N. Stalin (✉)
Botanical Survey of India, Central Regional Centre, Prayagraj, Uttar Pradesh, India

N. Thajuddin et al. (eds.), *Protocols for Cyanobacteria Sampling and Detection of Cyanotoxin*, https://doi.org/10.1007/978-981-99-4514-6_15

15.1 Introduction

Lyngyba is filamentous blue-green algae that flourish in freshwater. The unbranched filaments are covered by firm polysaccharide sheaths and are seen covering both water surfaces and benthic sediments (Speziale et al. 1988). The consumption of water contaminated with cyanobacteria will call gastroenteritis. Cyanobacteria can produce diverse toxins that may affect the central nervous system like anatoxin-a, anatoxin-a(s), and saxitoxins. How cyanotoxins have their toxic effects can vary among different groups. The principal toxins primarily affect liver, such microcystins, nodularin, and cylindrospermopsin (Otero and Silva 2022). The marine *Lyngbya majuscula* produces a several bioactive compounds including Dermatotoxins. Lyngbyatoxin-a a potent irritant and vesicant produce rashes and other skin reactions. Aplysiatoxins considered promoting tumour development. Similarly, *Lyngbya wollei* can produce a potent neurotoxin like saxitoxins. The toxic effects include damage to the digestive and respiratory tract's epithelial cells. The toxins produced by *Lyngbya* promote the development of a tumour and the development of fibropapillomatosis (Beasley 2020). Cylindrospermopsin is another important alkaloid cyanotoxin produced by *Lyngbya* consisting of tricyclic guanidine (Ohtani et al. 1992; Guzmán-Guillén et al. 2013). The structural variant 7-epicylindrospermopsin is highly toxic (Banker et al. 2001). Toxins released from the algal cells to the surroundings are accumulated in the organisms in the lower level of the food chain (Rücker et al. 2007; Kinnear 2010). Cyanobacterial intoxication requires accurately identifying the algal species in the contaminated water (Van der Merwe 2015). Toxicological analyses of biological specimens contracted with algal toxins are recommended. Since the toxicity of these cyanobacteria is strain specific, more than mere identification is needed to predict their hazardous level. Liquid chromatography and tandem mass spectrometry are the best methodologies adapted for the analysis (Puschner and Moore 2013).

15.2 Materials

- Z-8 medium
- Modified seawater BG-11 medium
- ASN-III medium
- Sterile agar
- Cycloheximide
- Germanium dioxide
- Cycloserine
- N-formimidoyl thienamycin monohydrate
- Distilled water
- Sterile containers
- Petri dishes
- Erlenmeyer flasks (125 mL, 500 mL, and 1 L)
- Culture vessels

- Dive knife
- Autoclave
- Incubator
- Cool white fluorescence tube

15.3 Methods

15.3.1 Sample Collection (Burja et al. 2002)

1. Collect the samples with a clean dive knife.
2. Place them in sterile sample containers and seal them underwater.
3. Ship to the laboratory within a day and process them within 1 week of collection.

15.3.2 Z-8 Medium

1. Wash the filaments with sterile Z-8 medium (Kotai 1972; Carmichael 1986).
2. Autoclave the sterile agar and wash it thrice over a period of 72 h with distilled water. Then, mix with the filter-sterilized nutrient solution.
3. Add 50 mg L^{-1} and 100 μg mL^{-1} of cycloheximide and germanium dioxide, respectively, to eliminate the growth of eukaryotes. Add cycloserine and N-formimidoyl thienamycin monohydrate (Imipenem) to prevent bacterial growth (Vaara et al. 1979).
4. Store the agar plates at 4 °C before use.
5. Cut 0.5 cm long segments and plate at the centre of Petri dishes containing 1% w/v agar and Z-8 medium. Prepare in triplicate.
6. Maintain the cultures in an incubator under 26 °C under 16/8 light and dark photoperiod.
7. Use cool white fluorescence tube with an irradiance of 22 μmol m^{-2} s^{-1} photon flux as a light source (Speziale and Dyck 1992).
8. Cut new trichomes that grow out of old ones away from the region of contamination. Transfer the trichome to another Petri dish containing the same medium. Repeat until unialgal culture is obtained.
9. Then, transfer the trichomes to 125 mL Erlenmeyer flasks containing 100 mL of Z-8 medium. Allow the filaments to adapt to the Z-8 medium for 2 weeks.
10. After 2 weeks, transfer to the 1-L Erlenmeyer flasks containing 800 mL of Z-8 medium.
11. Aerate the cultures at the rate of 43–71 Lh^{-1}.

15.3.3 Modified Seawater BG-11 Medium (Burja et al. 2002)

1. Subculture the samples in 500 mL Erlenmeyer flasks containing 200 mL of modified seawater BG-11 medium (Rossi et al. 1997).

2. Let the cultures to achieve equilibrium with their new environment at 20 °C and then inoculate in large-scale culture vessels.
3. Maintain the cultures under a 12-h light/dark regime (Lux 250) and at a temperature of 20 °C.
4. Aerate with 0.2 mm filtered air and subculture every 2 months.

15.3.4 ASN-III Medium (Rippka et al. 1979)

1. Culture the filaments in ASN-III medium and subculture repeatedly to obtain axenic culture.
2. Maintain the cultures under 25 ± 1 °C and 7.5 W/m^2 of temperature and light intensity, respectively.
3. Use cool-white, fluorescent tubes as the light source with a photoperiod of 14: 10 h.

15.4 Observation (Burja et al. 2002)

1. Use light microscopy to study the external morphology of *Lyngbya*. Autofluorescence images of filamentous *Lyngbya* visualized at the emission range of 400–500 nm under a fluorescence microscopy.

Acknowledgement S.N. acknowledge the Director, Botanical Survey of India, Kolkata; N.N. acknowledges the Department of Botany, St. Joseph's College (Autonomous), Tiruchirappalli, for infrastructural support. G.K. acknowledges the Department of Pharmacognosy, Siddha Central Research Institute, Chennai.

References

Banker R, Carmeli S, Werman M, Teltsch B, Porat R, Sukenik A (2001) Uracil moiety is required for toxicity of the cyanobacterial hepatotoxin cylindrospermopsin. J Toxicol Environ Health A 62(4):281–288

Beasley VR (2020) Harmful algal blooms (Phycotoxins)

Burja AM, Abou-Mansour E, Banaigs B, Payri C, Burgess JG, Wright PC (2002) Culture of the marine cyanobacterium, Lyngbya majuscula (Oscillatoriaceae), for bioprocess intensified production of cyclic and linear lipopeptides. J Microbiol Methods 48(2–3):207–219

Carmichael WW (1986) Isolation, culture and toxicity testing of toxic freshwater cyanobacteria (blue-green algae). In: Shilov V (ed) Fundamental research in homogenous catalysis, vol 3, pp 1249–1262

Guzmán-Guillén R, Prieto AI, Vasconcelos VM, Cameán AM (2013) Cyanobacterium producing cylindrospermopsin cause oxidative stress at environmentally relevant concentrations in sub-chronically exposed tilapia (Oreochromis niloticus). Chemosphere 90(3):1184–1194

Kinnear S (2010) Cylindrospermopsin: a decade of progress on bioaccumulation research. Mar Drugs 8(3):542–564

Kotai J (1972) Instructions for preparation of modified nutrient solution Z8 for algae. Norwegian Institute for Water Research Oslo 11(69):5

Ohtani H, Callahan RJ, Khaw BA, Fishman AJ, Wilkinson RA, Strauss HW (1992) Comparison of technetium-99m-glucarate and thallium-201 for the identification of acute myocardial infarction in rats. J Nucl Med 33(11):1988–1993

Otero P, Silva M (2022) The role of toxins: impact on human health and aquatic environments. In: The pharmacological potential of cyanobacteria, pp 173–199

Puschner B, Moore C (2013) Chapter 43—cyanobacteria. In: Small animal toxicology, pp 533–540

Rippka R, Deruelles J, Waterbury JB, Herdman M, Stanier RY (1979) Generic assignments, strain histories and properties of pure cultures of cyanobacteria. Microbiology 111(1):1–61

Rossi JV, Roberts MA, Yoo HD, Gerwick WH (1997) Pilot scale culture of the marine cyanobacterium Lyngbya majuscula for its pharmaceutically-useful natural metabolite curacin A. J Appl Phycol 9(3):195–204

Rücker J, Stüken A, Nixdorf B, Fastner J, Chorus I, Wiedner C (2007) Concentrations of particulate and dissolved cylindrospermopsin in 21 Aphanizomenon-dominated temperate lakes. Toxicon 50(6):800–809

Speziale BJ, Dyck LA (1992) Lyngbya infestations: comparative taxonomy of Lyngbya wollei comb. nov. (cyanobacteria). J Phycol 28(5):693–706

Speziale BJ, Turner EG, Dyck L (1988) Giant Lyngbya. Aquatics 10(2):4–11

Vaara T, Vaara M, Niemelä S (1979) Two improved methods for obtaining axenic cultures of cyanobacteria. Appl Environ Microbiol 38(5):1011–1014

Van der Merwe D (2015) Chapter 31—cyanobacterial (blue-green algae) toxins. In: Handbook of toxicology of chemical warfare agents, pp 421–429

16 Isolation of Toxin Producing Cyanobacteria from Aquatic Samples with *Raphidiopsis*

Nithaniyal Stalin and S. Nisha Nandhini

Abstract

Contamination of waterbodies with toxins produced by cyanobacteria affects aquatic life significantly. *Raphidiopsis* is found in fresh and brackish waters of tropical and temperate regions. It is associated with the production of deoxy-CYN, CYN, and homoanatoxin-a. Besides, they are also associated with PSP toxin production. Toxins produced by *Raphidiopsis* have yet to be studied as much as the toxins other cyanobacteria produce due to their relatively low distribution. This chapter emphasizes the culture of *Raphidiopsis* in CT and MLA media for the isolation of cyanotoxins produced by it.

Keywords

Raphidiopsis · Cyanobacteria · Cyanotoxin · CT medium · MLA medium

16.1 Introduction

Raphidiopsis is a solitary, free-floating, filamentous freshwater algae without a mucilaginous sheath. One or both the ends of the trichome are attenuated and pointed. They are characterized by the presence of akinetes and the absence of heterocysts (Fritsch et al. 1929). The lack of a heterocyst differentiates it from *Cylindrospermopsis*. *Raphidiopsis* species are less in distribution when compared with other planktonic cyanobacteria like *Anabaena* and *Oscillatoria*. Cyanobacteria

N. Stalin
Botanical Survey of India, Central Regional Centre, Prayagraj, Uttar Pradesh, India

S. Nisha Nandhini (✉)
Department of Botany, St. Joseph's College (Autonomous), affiliated to Bharathidasan University, Tiruchirappalli, Tamil Nadu, India

N. Thajuddin et al. (eds.), *Protocols for Cyanobacteria Sampling and Detection of Cyanotoxin*, https://doi.org/10.1007/978-981-99-4514-6_16

produce a diverse range of toxins. These toxins produced by freshwater cyanobacteria are classified into cytotoxins, hepatotoxins, and neurotoxins. *Raphidiopsis* was found to have cylindrospermopsin (CYN), deoxy-cylindrospermopsin (deoxy-CYN) and homoanatoxin-a. CYN and deoxy-CYN are hepatotoxins. Whereas homoanatoxin-a is a type of neurotoxin that causes intense body paralysis, dyspnea, and death within 5–10 min in mice (Li et al. 2001; Watanabe et al. 2003). *Raphidiopsis* is also a paralytic shellfish poisoning (PSP) toxin producer. These toxins are associated with harmful algal blooms in freshwater and brackish waters (Soto-Liebe et al. 2012). In species of alga that form bloom, the overall bloom toxin yield and associated risk level of the bloom vary widely in their response to abiotic drivers and toxin cell quota. Studying the cyanotoxins and predicting the bloom formation and toxin yields necessitates the isolation and culture of the naturally dominant species (Burford et al. 2014; Baxter et al. 2020). This chapter presents the protocol for the culture of *Raphidiopsis* with CT medium and MLA medium.

16.2 Materials

- CT medium
- MLA medium
- Pasteur pipette
- Screw cap test tube (18 cm × 150 mm)
- 2-L flask
- 24-well plates
- Plastic tissue culture vessels (25 cm^2 and 75 cm^2 vented)
- Plankton net (30-μm)
- Eyelash brush
- Centrifuge
- Lyophilizer
- Fluorescent lamps
- Binder growth chamber

16.3 Methods

16.3.1 Sample Collection

1. Collect water samples with plankton nets (30-μm mesh) (Bicudo and Menezes 2006).

16.3.2 Algal Culture

16.3.2.1 CT Medium (Li et al. 2001)

1. Isolate a single filament of *Raphidiopsis* by repeatedly washing it with sterile media with a Pasteur pipette.
2. Maintain the isolated strain in an 18 cm × 150 mm screw cap test tube with 10 mL of CT medium (Watanabe and Hiroki 1997).
3. Grow the cultures under a 12:12 light: dark cycle (photon flux density of 40 μmol photons m^{-2} s^{-1}) produced by fluorescent lamps at a temperature of 20 °C.
4. Culture the cells in 2-L flasks containing 1 L of CT medium.
5. Aerate the cultures with 0.22 μm filtered air at a rate of 1 L min^{-1}.
6. After 4 weeks, harvested the cultures by centrifugation at 10,000 ×*g* for 10 min.
7. Wash twice with distilled water and lyophilize.

16.3.2.2 MLA Medium (McGregor et al. 2011)

1. Streak a sub-sample onto an agar plate containing MLA medium (Bolch and Blackburn 1996; Bakr et al. 2022).
2. Transfer single trichomes from the plate to 24-well plates filled with 1.5 mL of MLA medium with an eyelash brush.
3. Sub-sample aliquots of 0.25 mL from wells and observe for growth.
4. After 6 weeks, transfer to plastic 25 cm^2 vented tissue culture vessels with 15 mL MLA medium and incubate for 18 days.
5. Maintain the cultures in plastic 75 cm^2 vented tissue culture vessels in a Binder growth chamber with 12:12 of light: dark cycle between 10 and 30 μmol m^{-2} s^{-1}.
6. Synchronize the temperature with the light cycle; 24 °C light, 18 °C dark.

16.4 Observation (Watanabe et al. 2003)

1. Observe the morphological and cytological characteristics of the cultured cells with optical and electron microscopes.

Acknowledgment S.N. acknowledges the Director, Botanical Survey of India, Kolkata. N.N. acknowledges the Department of Botany, St. Joseph's College (Autonomous), Tiruchirappalli, for infrastructural support.

References

Bakr A, Alzain MN, Alzamel NM, Loutfy N (2022) Accumulation of microcystin from *Oscillatoria limnetica* Lemmermann and *Microcystis aeruginosa* (Kützing) in two leafy green vegetable crop plants *Lactuca sativa* L. and *Eruca sativa*. Plants 11(13):1733

Baxter K, Jameson ID, Willis A (2020) Towards defining global ecotypes of the toxic cyanobacterium *Raphidiopsis raciborskii*. Appl Phycol. 1–10

Bicudo CEM, Menezes M (2006) Gêneros de algas de águas continentais do Brasil: chave para identificação e descrições. RIMA, São Carlos, p 502

Bolch CJS, Blackburn SI (1996) Isolation and purification of Australian isolates of the toxic cyanobacterium Microcystis aeruginosa Kütz. J Appl Phycol 8:5–13

Burford MA, Davis TW, Orr PT, Sinha R, Willis A, Neilan BA (2014) Nutrient-related changes in the toxicity of field blooms of the cyanobacterium *Cylindrospermopsis raciborskii*. FEMS Microbiol Ecol 89:135–148

Fritsch FE, Rich F, Stephens EL (1929) Contributions to our knowledge of the freshwater algae of Africa: 7. Freshwater algae (exclusive of diatoms) from Griqualand West. Trans R Soc South Africa 18(1):1–92

Li R, Carmichael WW, Brittain S, Eaglesham GK, Shaw GR, Liu Y, Watanabe MM (2001) First report of the cyanotoxins cylindrospermopsin and deoxycylindrospermopsin from *Raphidiopsis curvata* (cyanobacteria). J Phycol 37(6):1121–1126

McGregor GB, Sendall BC, Hunt LT, Eaglesham GK (2011) Report of the cyanotoxins cylindrospermopsin and deoxy-cylindrospermopsin from *Raphidiopsis* mediterranea Skuja (Cyanobacteria/Nostocales). Harmful Algae 10(4):402–410

Soto-Liebe K, Méndez MA, Fuenzalida L, Krock B, Cembella A, Vásquez M (2012) PSP toxin release from the cyanobacterium *Raphidiopsis brookii* D9 (Nostocales) can be induced by sodium and potassium ions. Toxicon 60(7):1324–1334

Watanabe MM, Hiroki M (1997) NIES-collection list of strains, algae and protozoa, 5th edn. National Institute for Environmental Studies. Environment Agency Japan, pp 140

Watanabe MF, Tsujimura S, Oishi S, Niki T, Namikoshi M (2003) Isolation and identification of homoanatoxin-a from a toxic strain of the cyanobacterium *Raphidiopsis* mediterranea Skuja isolated from Lake Biwa, Japan. Phycologia 42(4):364–369

17 Isolation of Toxin Producing Cyanobacteria from Aquatic Samples: *Oscillatoria*

S. Nisha Nandhini and Nithaniyal Stalin

Abstract

Freshwater resources are dwindling, creating a need to conserve the existing natural water sources. Algal blooms deteriorate the water quality of freshwater sources like lakes and ponds. This creates a need to isolate and study the cyanotoxins and the associated ill effects on the lake. *Oscillatoria* is one such genus of bloom-producing cyanobacteria which produce cyanotoxins (microcystins, anatoxins) and taste and odor metabolites (2-methylisoborneol (2-MIB), Geosmin (GSM)). This chapter elucidates the methodology of isolation and culture of *Oscillatoria*. in Z8 and BG-13 media.

Keywords

Oscillatoria · Cyanobacteria · Cyanotoxin · Z8 medium · BG-13 medium

17.1 Introduction

Oscillatoria is filamentous green algae, and the algal body is referred to as a trichome. Their cell wall comprises several biologically active bio-polymers, including cellulose and polysaccharides-bound proteins. They are found in abundance in natural ecosystems like fresh and sea waters. They can be involved in the remediation of contaminated effluents due to their uptake of nutrients and pollutants from wastewater (Hussain et al. 2017). Using water contaminated with microcystins has

S. Nisha Nandhini (✉)
Department of Botany, St. Joseph's College (Autonomous), affiliated to Bharathidasan University, Tiruchirappalli, Tamil Nadu, India

N. Stalin
Botanical Survey of India, Central Regional Centre, Prayagraj, Uttar Pradesh, India

N. Thajuddin et al. (eds.), *Protocols for Cyanobacteria Sampling and Detection of Cyanotoxin*, https://doi.org/10.1007/978-981-99-4514-6_17

led to their bioaccumulation in plant tissues which has been associated with human health risks. The estimated daily intake (EDI) of microcystins by a person of weight 60 kg who consumes 300 g of fresh plants was greater than the total daily intake of 0.04 μg kg^{-1} body weight. This raises the need to test the water before using it for both drinking and irrigation (Bakr et al. 2022). Microcystin is a hepatotoxin as it inhibits protein phosphatase leading to successive accumulation of phosphorylated proteins in the liver. It causes cell necrosis, massive hemorrhage, and death (Runnegar et al. 1995). Therefore, qualitative and quantitative monitoring of contaminated aquatic water is important and isolation of algal species remains a prerequisite for identification of toxins producing cyanobacteria. Here we provided a protocol for culture of *Oscillatoria* using Z8 and B13 media.

17.2 Materials

- Z8 media
- BG13 media
- NaCl
- Imipenem
- Cycloheximide
- Sterile syringe (0.2 μm porosity acrodisc)
- Stopper (0.2 μm porosity filter)
- GF/C glass filter paper
- Millipore membrane
- Petri dishes
- Sterilized flask
- 1-L Erlenmeyer flasks
- Distilled water
- Centrifuge
- Pressurized vacuum pump
- Lyophilizer

17.3 Methods

17.3.1 Sample Collection (Mohamed and Al Shehri 2009)

1. Selection of the study site and season of sample collection.
2. Collect the water sample in sterile bottles and filter through GF/C glass filter papers.

17.3.2 Algal Culture

17.3.2.1 Z8 Medium (Hong et al. 2010)

1. Supplement solid and liquid Z8 medium with 20 gL-L of NaCl.
2. Perform strain isolation and algal culture in these media (Kotai 1972)
3. Culture the strains for 3 weeks at 25 °C, with 28 μmol photons m^{-2} s^{-1} irradiance and 14 h/10 h of light/dark cycle.
4. After culturing, isolate different species and transfer them into a new agar medium.
5. Supplement the media with 100 μg mL^{-1} imipenem and 20 μg mL^{-1} of cycloheximide antibiotic to prevent bacterial and eukaryotic growth, respectively. This makes them axenic.
6. Take 1% of inoculum and 500 mL of liquid medium and mass culture in 1-L Erlenmeyer flasks.
7. Collect the cells after 3–4 weeks of culturing by centrifugation or filtration and then freeze-dry them.

17.3.2.2 BG-13 Medium (Mustapha et al. 2021)

1. Prepare solid and liquid BG13 media (Ferris and Hirsch 1991).
2. Autoclave for 20 min at 120 °C and 11 kg/cm^2 of temperature and pressure, respectively.
3. Add 100 mg of cycloheximide per liter to liquid and solid culture media with a sterile syringe (0.2 μm porosity acrodisc).
4. Filtered water samples through Millipore membrane (45 μm porosity) using a pressurized vacuum pump.
5. Transfer the membrane to liquid media (pH 8) and incubate for 3 weeks under the ambient temperature of 20–25 °C, an intensity of 2000 lumens, and a 12 h photoperiod.
6. Transfer the membranes into Petri dishes containing solid media for the purification phase.
7. After 3 weeks, transfer a tiny fragment of the strain to the new solid media.
8. Repeat this until you get a pure strain. Then, transfer the pure strain to 100 mL of liquid media in a sterile bottle.
9. Seal the bottle with a stopper of 0.2 μm porosity filter.
10. Repeat the steps 2–3 times to obtain axenic cultures within 12 weeks.
11. Lyophilize the cells.

17.4 Observation (Mustapha et al. 2021)

1. Subject the cells to preliminary morphological observations.
2. Observe and measure the species under a light microscope.

Acknowledgment N.N. acknowledges the Department of Botany St. Joseph's college (Autonomous), Tiruchirappalli for infrastructural support. S.N. acknowledges the Director, Botanical Survey of India, Central Regional Centre, Kolkata.

References

Bakr A, Alzain MN, Alzamel NM, Loutfy N (2022) Accumulation of microcystin from *Oscillatoria limnetica* Lemmermann and *Microcystis aeruginosa* (Kützing) in two leafy green vegetable crop plants *Lactuca sativa* L. and *Eruca sativa*. Plants 11(13):1733

Ferris JM, Hirsch CF (1991) Method for isolation and purification of cyanobacteria. Appl Environ Microbiol 57(5):1448–1452

Hong JW, Choi HG, Kang SH, Yoon HS (2010) Axenic purification and cultivation of an arctic cyanobacterium, *Nodularia spumigena* KNUA005, with cold tolerance potential for sustainable production of algae-based biofuel. Algae 25:99–104

Hussain F, Shah SZ, Zhou W, Iqbal M (2017) Microalgae screening under CO2 stress: growth and micro-nutrients removal efficiency. J Photochem Photobiol B Biol 170:91–98

Kotai J (1972) Instructions for preparation of modified nutrient solution Z8 for algae. Norwegian Institute for Water Research 11(69):5

Mohamed ZA, Al Shehri AM (2009) Microcystins in groundwater wells and their accumulation in vegetable plants irrigated with contaminated waters in Saudi Arabia. J Hazard Mater 172(1): 310–315

Mustapha O, El OK, Anass EL, Farid BA, Abdeltif EO, Mohamed I, Jamal A (2021) Characterization of cyanobacteria microcystins (cyanotoxins) blooming in the dams of Northern Morocco. Afr J Environ Sci Technol 15(3):124–141

Runnegar M, Berndt N, Kaplowitz N (1995) Microcystin uptake and inhibition of protein phosphatases: effects of chemoprotectants and self-inhibition in relation to known hepatic transporters. Toxicol Appl Pharmacol 134(2):264–272

18 Isolation of Toxin Producing Cyanobacteria from Aquatic Samples: *Planktothrix*

S. Nisha Nandhini and Nithaniyal Stalin

Abstract

Anthropogenic activities and associated climate change have led to algal blooms in several lakes around the globe. Algal blooms not only destroy the water quality but also affect the aquatic flora and fauna in the lake. The secondary metabolites produced by cyanobacteria are the chief problem creators. Toxins like microcystins (MCs) and anatoxin are produced by *Planktothrix.* MCs cause liver damage to the fishes, interrupt the embryonic development, and also inhibit eukaryotic protein phosphatase 1 and 2A. It is mandatory to quantify these toxins to ensure the drinkability and quality of the water from natural sources like ponds and lakes. This chapter focuses on the protocol for the isolation and culture of cyanotoxin producing *Planktothrix* in WC and Mineral media.

Keywords

Planktothrix · Cyanobacteria · Microcystin · WC medium · Mineral medium · ELISA

18.1 Introduction

Planktothrix is a filamentous cyanobacteria. The cells are cylindrical and a little shorter than their width. Their trichomes may be solitary, straight, or slightly curved. They are formed by the binary division of cells at right angles to the long axis in one

S. Nisha Nandhini (✉)
Department of Botany, St. Joseph's College (Autonomous), affiliated to Bharathidasan University, Tiruchirappalli, Tamil Nadu, India

N. Stalin
Botanical Survey of India, Central Regional Centre, Prayagraj, Uttar Pradesh, India

N. Thajuddin et al. (eds.), *Protocols for Cyanobacteria Sampling and Detection of Cyanotoxin*, https://doi.org/10.1007/978-981-99-4514-6_18

plane and the cells remain attached to each other with little or no constriction at the cross-walls. Trichomes are a few micrometres in diameter and contain hundreds to thousands of cells. The length of trichome varies among different species of *Planktothrix*. These filaments may be attenuated toward the ends or terminal cells may be tapered (with or without calyptras) (Komárek and Anagnostidis 2007; Kurmayer et al. 2016).

Planktothrix is a common bloom-forming cyanobacteria. It is successful in forming algal blooms in several lakes due to its ability to use phycobilins, in addition to chlorophyll-a as chromophores in addition to its adaptability and the ability to regulate buoyancy. Phycobilins, phycocyanin and phycoerythrin, enable *Planktothrix* spp. to absorb light from larger parts of the electromagnetic spectrum; chiefly, blue and green light, thus leading its red appearance. In the water column, *P. agardhii* and *P. rubescens* frequently dominate the phytoplankton (Kurmayer et al. 2016; Fournier et al. 2021). Occasionally, filaments of *P. rubescens* occasionally rise to the surface and form reddish blooms on the surface, and this is known as the "burgundy blood phenomenon" (Walsby 2005).

The metabolites produced by *Planktothrix* were found to be toxic to aquatic invertebrates and fishes. Transparency and light availability to phytoplankton and aquatic macrophytes is significantly reduced by the dense cyanobacterial blooms. Additionally, these cyanobacteria can produce a wide range of metabolites like microcystins, anatoxin which proved toxic to aquatic biota and humans (Poniedziałek et al. 2012; Meriluoto et al. 2017). Toxicity of *Planktothrix* is not only linked to microcystins, but to several other unknown toxins. This creates a need to purify and identify various cyanotoxins produced by them to assess the toxicity of *Planktothrix* and ecological assessment with respect to the lake ecosystem (Keil et al. 2002). This chapter highlights methods to collect and culture *Planktothrix* in WC and Mineral medium.

18.2 Materials

- WC medium
- Mineral medium
- Cellulose nitrate membrane filter
- 20 mm net
- Erlenmeyer flasks
- Lyophilizer
- Phytotron chamber

18.3 Methods

18.3.1 Sample Collection (Keil et al. 2002)

1. Site and season of sample collection is determined preferably during monospecific bloom.
2. Concentrate the samples with a 20 mm net.
3. Lyophilize the cyanobacteria.
4. Store at −20 °C.

18.3.2 WC Medium (Wejnerowski et al. 2020)

1. Culture the algal cells in WC medium for 70 days since reinoculation in Erlenmeyer flasks.
2. Maintain 300 mL of cultures in a walk-in phytotron chamber under 40–50 μmol photons m^{-2} s^{-1} of light intensity, photoperiod of 16:8 h light:dark cycle and temperature of 20 ± 0.5 °C.
3. The algal cultures can be used after the development of colour change in culture from greenish to yellow colour.

18.3.3 Mineral Medium (Walsby and Jüttner 2006)

1. Grow the algal cultures in the mineral medium in 300 mL Erlenmeyer flasks (Blom et al. 2001).
2. Maintain the cultures at 20 °C with continuous light of low irradiance (7 μmol m^{-2} s^{-1}).
3. Dilute the evenly suspended culture with fresh medium to obtain a low filament volume concentration (between 7 and 60 cm^3 m^{-3})
4. Keep the culture at the ambient temperature, about 20 °C and <1 μmol m^{-2} s^{-1} irradiance.
5. Mix (2 mL or 8 mL) of sample with 100 mL of medium and filter with a cellulose nitrate membrane filter (50 mm diameter and pore size 8 μm) which will retain the filaments.

18.3.4 ELISA (Hautala et al. 2013)

1. Dilute the extracts according to HPLC results in order to adjust to the microcystin concentrations within the working range of the assay.
2. Prepare two different dilutions of the same sample.
3. Analyse the samples with the QuantiPlate Microcystin Kit using the protocol provided by the manufacturer.

18.4 Observation (Walsby and Avery 1996)

1. Determine the total filament length with epifluorescence microscopy and image analysis.

Acknowledgement N.N. acknowledges the Department of Botany St. Joseph's college (Autonomous), Tiruchirappalli for infrastructural support. S.N. acknowledges the Director, Botanical Survey of India, Central Regional Centre, Kolkata.

References

Blom JF, Robinson JA, Juttner F (2001) High grazer toxicity of [D-Asp3, (E)-Dhb7] microcystin-RR of Planktothrix rubescens as compared to different microcystins. Toxicon 39:1923–1932

Fournier C, Riehle E, Dietrich DR, Schleheck D (2021) Is toxin-producing Planktothrix sp. an emerging species in Lake Constance? Toxins 13(9):666

Hautala H, Lamminmäki U, Spoof L, Nybom S, Meriluoto J, Vehniäinen M (2013) Quantitative PCR detection and improved sample preparation of microcystin-producing Anabaena, Microcystis and Planktothrix. Ecotoxicol Environ Saf 87:49–56

Keil C, Forchert A, Fastner J, Szewzyk U, Rotard W, Chorus I, Krätke R (2002) Toxicity and microcystin content of extracts from a Planktothrix bloom and two laboratory strains. Water Res 36(8):2133–2139

Komárek J, Anagnostidis K (2007) Cyanoprokaryota, 2. Teil: Oscillatoriales. Elsevier GmbH, Spektrum Akademischer Verlag, Heidelberg

Kurmayer R, Deng L, Entfellner E (2016) Role of toxic and bioactive secondary metabolites in colonization and bloom formation by filamentous cyanobacteria Planktothrix. Harmful Algae 54:69–86

Meriluoto J, Spoof L, Codd GA (eds) (2017) Introduction. Handbook of cyanobacterial monitoring and cyanotoxin analysis. Wiley, Hoboken

Poniedziałek B, Rzymski P, Kokociński M (2012) Cylindrospermopsin: water-linked potential threat to human health in Europe. Environ Toxicol Pharmacol 34(3):651–660

Walsby AE (2005) Stratification by cyanobacteria in lakes: a dynamic buoyancy model indicates size limitations met by Planktothrix rubescens filaments. New Phytol 168(2):365–376

Walsby AE, Avery A (1996) Measurement of filamentous cyanobacteria by image analysis. J Microbiol Methods 26:11–20

Walsby AE, Jüttner F (2006) The uptake of amino acids by the cyanobacterium Planktothrix rubescens is stimulated by light at low irradiances. FEMS Microbiol Ecol 58(1):14–22

Wejnerowski Ł, Falfushynska H, Horyn O, Osypenko I, Kokociński M, Meriluoto J, Jurczak T, Poniedziałek B, Pniewski F, Rzymski P (2020) In vitro toxicological screening of stable and senescing cultures of Aphanizomenon, Planktothrix, and Raphidiopsis. Toxins 12(6):400

Analysis of Cyanobacterial Species Composition and Variation Analysis Among Toxic Blooms

19

Anu Sharon and Lokesh Ravi

Abstract

Several types of cyanobacteria (blue-green algae) have the potential to cause major environmental problems due to the toxins they release. The demise of cattle and other organisms treated to these toxins provides proof that they can be acutely damaging. In this chapter, we will be looking certain protocols which are used to study different cyanobacterial composition in algal blooms and species variation among. This study will be useful in analysing the morphological variation of cyanobacteria within the toxic bloom. A broad range of social, environmental, and economic consequences are triggered by cyanobacterial blooms. A limited number of organisms that are regularly found in freshwater streams and lakes can create dangerous chemicals that can have an adverse impact on an individual's health in this book there are few methods mentioned like cell counting, molecular methods, brief note on sampling of toxic bloom, NMR methods to identify the cyanobacteria species and their morphological differences are also mentioned.

Keywords

Toxic blooms · Molecular techniques cell counting · Biochemical · Microcystis · Sampling

A. Sharon
Department of Botany, School of Life Sciences, St Joseph's University, Bengaluru, Karnataka, India

L. Ravi (✉)
Department of Food Technology, Faculty of Life and Allied Health Sciences, MS Ramaiah University of Applied Sciences, Bengaluru, India

N. Thajuddin et al. (eds.), *Protocols for Cyanobacteria Sampling and Detection of Cyanotoxin*, https://doi.org/10.1007/978-981-99-4514-6_19

19.1 Introduction

A broad range of social, environmental, and economic consequences are triggered by cyanobacterial blooms. A limited number of organisms that are regularly found in freshwater streams and lakes can create dangerous chemicals that can have an adverse impact on an individual's health. Lake and reservoir phytoplankton groups naturally contain cyanobacteria that are frequently present at a minimum rate. Eutrophication is the prime most cause that support the growth of cyanobacteria in lakes streams another factor which is also responsible to enhance the growth of cyanobacteria is less nutrient content or low production of nutrition in water; the biodiversity richness is also reduced by the growth of cyanobacteria bloom since all these cyanobacteria produces one or the other toxins that will in turn effect the marine population by the killing the fishes and other aquatic vegetation and reducing the oxygen in water (Soares et al. 2009).

Cyanobacteria (blue-green algae) are ancient microorganisms with a global distribution, and they produce oxygen by the process of photosynthesis. They were responsible for the creation of the earth's aerobic atmosphere over two thousand million years ago, and today they remain major agents in biological cycling of carbon, nitrogen, and other minerals. They are native occupants of a wide range of habitats such as littoral rocks and soils, freshwater, saline water, and marine habitats. Cyanobacteria are the leading colonizers of the ocean because of their photosynthetic development, the capability of several species to transform gaseous nitrogen to ammonia and amino acids and their tolerance to harsh and extreme climatic conditions (Chorus and Welker n.d.). Cyanobacteria may be found in a broad range of habitats in aquatic ecosystems like clumps or fine particles in the water, the water's interface, debris at the bottom, and vegetation and rocks adhered to the coastline. These regions could have high populations of cyanobacteria which appear as blooms, scums, pathogenic organisms, or as sticky mats on the water surface (Sant'Anna et al. 2008). The monophyletic yet diverse group of oxygen producing photosynthetic organisms known as cyanobacteria. The structures of cyanobacteria are extremely different, comprising single celled, colonies, filaments, and branching filamentous varieties. Moreover, the variation of cell sizes in cyanobacteria covers two orders of magnitude. It is conceivable for trichome length and aggregate size to be macroscopic and observable to the bare eyes. Cyanobacteria have endured a number of modifications in structure and function over the course of their prolonged course of evolution, and these modifications are what give them their adaptive metabolism and broad ecological compatibility. The ability of cyanobacteria to survive extreme conditions such intense heat, ultraviolet irradiation, dehydration, and water and salt pressures is a factor in their ability to survive in a diverse range of environments. Cyanobacteria may photosynthesize in low-light conditions and therefore can fix carbon dioxide by using carbonic acid at higher pH levels. Several organisms absorb nitrogen from the atmosphere as a source of nitrogen by fixing it. Additionally, cyanobacteria can survive in low-phosphate conditions because of their capacity to utilize a variety of distinct sulphur compounds and effective phosphate uptake methods. Cyanobacteria are major primary sources on a worldwide scale and are crucial to the nitrogen, carbon, and oxygen biogeochemical cycles. Cyanobacteria are estimated to provide up to 50% of total of the world's foremost supply of these types of gases.

19.2 Methods

Requirements: Plankton net, Kemmerer water bottle sampler, Van Dorn water bottle sampler, 100 mL Lugol solution, inverted microscope, or light microscope (Graham 2008).

(a) To analyse the cyanobacteria from the toxic bloom, one has to be expertise in sampling techniques now this sampling technique requires sample handling, preparation and processing of the samples that will in turn help us in identifying in various species. Sampler that are commonly used are surface samplers—handheld open mouth bottle sampler and weighted bottle sampler; horizontal bottle sampler is more specifically used for open water or surface water; Kemmerer water sampler or vertical water sampler is used in discrete sampling site (Soares et al. 2009; Hodoki et al. 2011).
(b) Once the samples are taken out from the site the subsamples are preserved in 100 mL of Lugol solution and are observed in an inverted microscope or light microscope identification and evaluation of these species are morphologically examined for further studies.
(c) For different levels of water, different samplers are used since in toxic bloom all the cyanobacteria (blue-green algae) will definitely carry toxins in them; therefore, the samplers are said to be carefully used and churns are commonly used to collect these algae from the toxic bloom while churning some of the algae that are morphologically filamentous or colonial in form might break hence individual cells have to be maintained and not to be affected.
(d) In case of collecting samples only for a selected site, one can use composite samplers. Now these collected samples are at the same depth and they are collected until they achieve a required volume (Graham 2008).
(e) Above methods are most commonly used to identify the cyanobacteria species, but there are several other methods that are rapid, quick solution methods of which some of them will be listed below.

Cyanobacterial species that are most commonly found in toxic bloom by the techniques mentioned below includes the following:

1. *Microcystis aeruginosa*
2. *Microcystis protocystis*
3. Anabeana
4. *Anabeana crassa*
5. *Lyngbya robusta*
6. Oscillatoria

7. *Radiocystis fernandoi*
8. *Cylindrospermopsis raciborskii*

19.2.1 Molecular Techniques

In modern technology, there are several methods which can assist in identifying and examining different species also molecular techniques provides a lot of advantages and instant results can be obtained. Denaturing gradient gel electrophoresis testing may assess changes in cyanobacterial community composition, and real-time PCR is helpful for examine of cyanobacterial strains (DGGE). Combining gene-specific oligonucleotide probes and a DNA chip/microarray, an elevated analytic approach could be employed to determine the structure of the cyanobacterial population (Saker et al. 2007).

19.2.2 Cell Counting

A cyanobacterial bloom is often monitored using microscopic identification and cell counts as the primary methods there are number of scientific techniques for counting cyanobacteria cells. To make it simpler to estimate the number of cells in a suspension sample, i.e., for example-Microcystis colonies alkaline hydrolyses and sonication were used although colonies were not obliterated, considerable cell dispersion was achieved in the suspension. Cyanobacteria culture samples were warmed to 80 °C that enabled the cells detached by centrifuging these techniques were matched with sonication and titanium dioxide techniques for isolating particular Microcystis colonies into suspensions of cells in both the trials it was shown on heating and centrifuging were the best techniques for liberating free cells from populations. Sonication is a simple and quick procedure, but it could cause cell damage which would cause the total number of cells to be underestimated (Pick 2016).

Requirements: Microcystis colonies, centrifuge, titanium dioxide

In contrast to microscopic identification, flow cytometry could also be used to track blooms, but this method does not offer complete information on species diversity in general cell counting assists in the quickest evaluation of cyanobacterial species that are present in toxic blooms, but this type of analysis requires more time and its purely based good laboratory skills.

19.2.3 Biochemical Methods

High performance liquid chromatography (HPLC) along with photodiode array pointer for UV absorbance is another common approach for figuring out cyanobacterial species; the wavelength that is suitable is always ranged between

200 and 300 nm; this is another method to easily identify species and also helpful in evaluating the morphological structures (Moreira et al. 2014).

19.2.4 Maldi-Assisted Laser Ionization

Another quickest method to detect the presence of cyanobacteria species, and this method is based on molecular formula in recent times; this technique has been majorly employed for the identification of many species; here, the sample of cyanobacterial species are detained on a hydrophobic chip and later they are ionized by (surface-enhanced laser desorption ionization time of flight) SELDI –TOF (Baker et al. 2001).

19.2.5 NMR Method

NMR technique has proved to be helpful in figuring out the structure of various cyanobacterial species; it is helpful in determination of morphological structure by utilizing a mixture of two-dimensional NMR, but comparatively it is very expensive than other analytical tools, and it needs a good quantity of real samples, and it mainly requires specialist assessment (Baker et al. 2001).

19.3 Morphological Studies

The morphology of the five mentioned species and the common morphological features between them are as follows:

19.3.1 Microcystis

During the sampling or by the outcome of those above-mentioned techniques, the most commonly found species are *Microcystis aeruginosa*; it is unicellular colonial form of cyanobacteria that is commonly found in stagnant water and toxic blooms. They do not have definite shape; they are considered to be free floating and covered by mucilaginous sheath; they have gas vesicles that help them for buoyancy; they secrete poisonous toxin (Meriluoto et al. 2017) (as shown in Fig. 19.1).

19.3.2 Anabaena

They are filamentous form, which commonly appear as green thick mats on the water surface; the cells are barrel shaped; presence of heterocyst in them is a characteristic feature that helps in fixation of nitrogen and also maintains symbiotic association with a pteridophyte known as azolla compared to microcystis; anabaena is

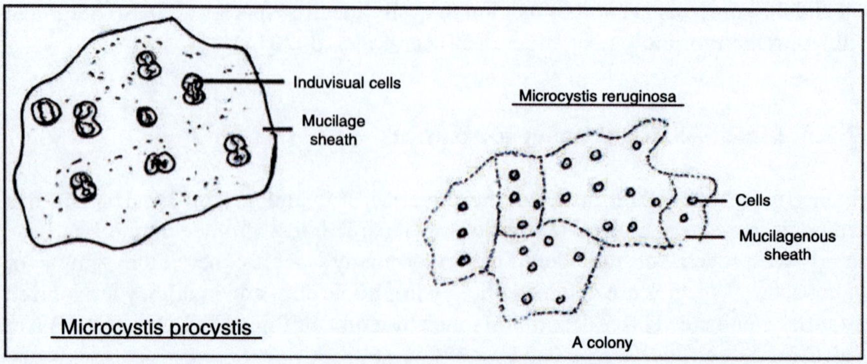

Fig. 19.1 A representation of Microcystis protocystis and Microcystis aeruginosa

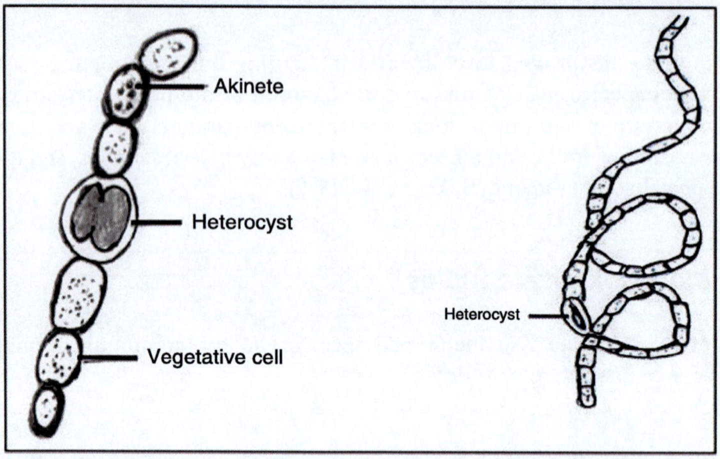

Fig. 19.2 A representation of anabaena filaments (Anabaena and Anabaena crassa)

morphologically different since it has a definite shape and also the presence of akinetes which is not seen in case of Microcystis; akinetes are thick resting spore cells. There are two species that are found, *Anabeana crassa* and *Anabeana spiroids*, morphologically; they are much similar to the genus anabaena (Lv et al. 2022) (as shown in Fig. 19.2).

19.3.3 Radiocystis fernandoi

This particular type of species is morphologically similar to Microcystis; the colonies are larger in size; the former colonies are unicellular; they have a thick mucilage surrounding them which is arranged in radial pattern; cells are all spherical in shape (Sant'Anna et al. 2008) (as shown in Fig. 19.3).

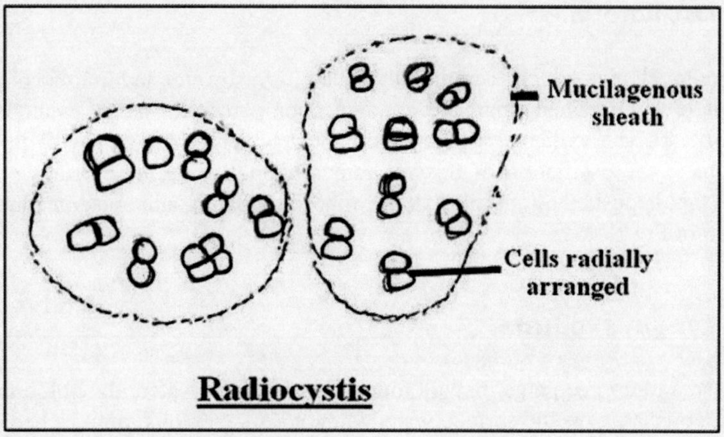

Fig. 19.3 Radiocystis cells radially arranged in a cover of mucilaginous sheath

Fig. 19.4 An illustration of Cylindrospermopsis

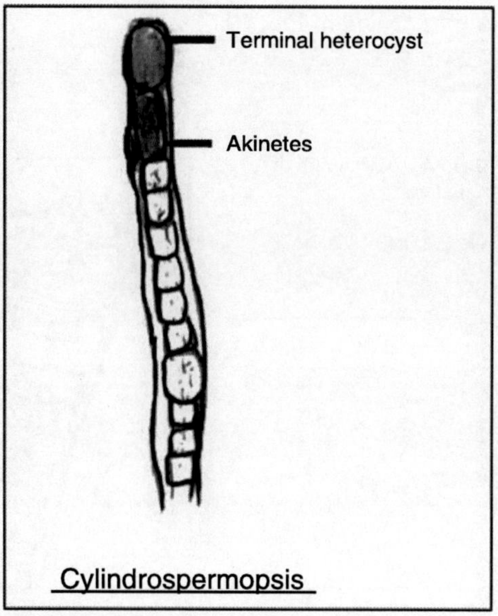

19.3.4 Cylindrospermopsis raciborskii

This species is also commonly found in toxic bloom; presence of trichomes can be observed and distinct cross walls; cells shape vary; presence of heterocyst that helps in fixing nitrogen; presence of thick resting cell akinetes are also observed (Li et al. 2017) (as shown in the below image Fig. 19.4).

19.3.5 Oscillatoria

It is a blue-green alga which is commonly found in freshwater and in toxic bloom; it is a simple unbranched filamentous alga, compared to other cyanobacteria *Oscillatoria* has a peculiar oscillatory movement which is shown by the presence of trichomes; the oscillation of this trichomes happens due to secretion of thick mucilage that is pushed out of the cell; they are free floating and uniseriate in nature (as shown in Fig. 19.5).

19.3.6 Lyngbya robusta

This is also a blue-green alga that are found growing freshwater, algal bloom and it covers a larger area on the surface water; they are filamentous, unbranched, when compared to anabaena lyngbya do not have heterocyst; therefore, they are considered as non-heterocyst form; presence of trichomes can be seen, but it is non-mobile in nature; this is a morphological difference when compared to *Oscillatoria*. (as represented in Fig. 19.6).

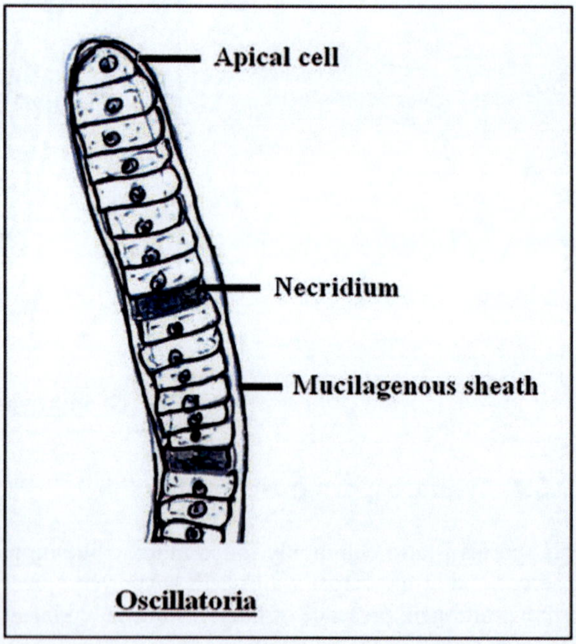

Fig. 19.5 A Diagram of Oscillatoria

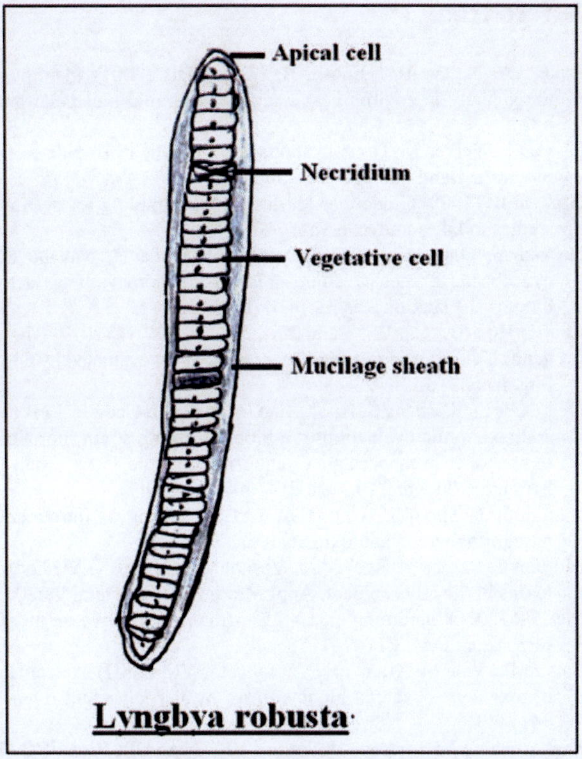

Fig. 19.6 A representation of Lyngbya robusta

19.4 Conclusion

Cyanobacteria which are considered to be the most ancient microorganism with a global distribution; they were actually responsible for the earth's aerobic atmosphere; the above-mentioned techniques are the most common ones that assist in identifying various species in the toxic bloom; this particular analysis gives us a brief outcome as on to figure out different species of cyanobacteria in toxic blooms; while analysing them it was found to be that there are few species that are always present in vast number; and morphologically they are different from the other cyanobacteria that are present in the same toxic bloom, for example, the most common blue-green algae that are found in toxic blooms are *Microcystis which* does not have any specific morphology or shape and always appear in the amorphous form and are enclosed in a mucilaginous sheath while on the other hand *Oscillatoria* which is morphologically different from that of the *Microcystis* have trichomes and shows a peculiar oscillatory movement by secretion of mucilage. *Anabaena* has heterocyst that helps in nitrogen fixation, and there are few forms of cyanobacteria that are non-heterocyst in nature, for example, *Microcystis*, *Lyngbya*. These unique features can help identify the different distribution of cyanobacteria in a toxic bloom.

References

Baker JA, Neilan BA, Entsch B, Mckay DB (2001) Identification of cyanobacteria and their toxigenicity in environmental samples by rapid molecular analysis. Environ Toxicol 16:472–482

Chorus I, Welker M Toxic cyanobacteria in water; a guide to their public health consequences, monitoring and management, 2nd edn

Graham JL (2008) Guidelines for design and sampling for cyanobacterial toxin and taste-and-odor studies in lakes and reservoirs

Hodoki Y, Ohbayashi K, Kobayashi Y, Okuda N, Nakano SI (2011) Temporal variation in cyanobacteria species composition and photosynthetic activity in experimentally induced blooms. J Plankton Res 33:1410–1416

Li XC, Huo SL, Cai FF, Yang YM, Xi BD, Li RH (2017) The taxonomy and phylogeny of the genus Cylindrospermopsis (Cyanobacterium) evaluated by adding five new records from China. Phytotaxa 316:224–238

Lv X, Cheng Y, Zhang S, Hu Z, Xiao P, Zhang H, Geng R, Li R (2022) Polyphasic characterization and taxonomic evaluation of a bloom-forming strain morphologically resembling Radiocystis fernandoi (Chroococcales, Cyanobacteria) from Lake Erhai, China. Diversity (Basel) 14(10): 816. https://doi.org/10.3390/d14100816

Meriluoto J, Spoof L, Codd GA (2017) Section I. Introduction handbook of cyanobacterial monitoring and cyanotoxin analysis, 1st edn

Moreira C, Ramos V, Azevedo J, Vasconcelos V (2014) Methods to detect cyanobacteria and their toxins in the environment. Appl Microbiol Biotechnol 98:8073–8082

Pick FR (2016) Blooming algae: a Canadian perspective on the rise of toxic cyanobacteria. Can J Fish Aquat Sci 73:1149–1158

Saker ML, Vale M, Kramer D, Vasconcelos VM (2007) Molecular techniques for the early warning of toxic cyanobacteria blooms in freshwater lakes and rivers. Appl Microbiol Biotechnol 75: 441–449

Sant'Anna CLA, Teresade P. Werner VR, Dogo CR, Rios FRC (2008) Review of toxic species of cyanobacteria in Brazil. Arch Hydrobiol Suppl Algol Stud 126:251–265

Soares MCS, Maria MI, Marinho MM, Azevedo SMFO, Branco CWC, Huszar VLM (2009) Changes in species composition during annual cyanobacterial dominance in a tropical reservoir: physical factors, nutrients and grazing effects. Aquat Microb Ecol 57:137–149

20 Confocal Laser Scanning Microscope (CLSM) Identification of Toxic Cyanobacteria from Algal Bloom

Shakena Fathima Thajuddin, Fayaazuddin Thajuddin, N. Thajuddin, and D. Dhanasekaran

Abstract

Cyanobacteria are known for their hazardous toxicity and are capable of producing heavy algal blooms in freshwater ecosystems. They are produced in great amounts by the cyanoHABs and pose a serious threat to humans, livestock as well as aquatic life. Cyanobacteria emit natural fluorescence due to the presence of two pigments namely chlorophyll a and phycobilin. Physiological processes occurring in photosynthetic cyanobacteria can be studied elaborately in vivo using the Confocal Laser Scanning Microscope (CLSM), which is unquestionably the most effective method as it offers live-cell imaging and is constructed with super resolution technology. They are reliable, time efficient and plays a vital role in identification of toxic cyanobacteria.

Keywords

Confocal microscope · Laser illumination · Fluorophores

S. F. Thajuddin (✉)
Department of Microbiology, School of Life Sciences, Bharathidasan University, Tiruchirappalli, Tamil Nadu, India

F. Thajuddin
School of Life Sciences, Bharathidasan University, Tiruchirappalli, Tamil Nadu, India

N. Thajuddin
B.S. Abdur Rahman Crescent Institute of Science & Technology (Deemed to be University), Chennai, Tamil Nadu, India

Department of Microbiology, National Repository for Microalgae and Cyanobacteria - Freshwater (NRMC-F), School of Life Sciences, Bharathidasan University, Tiruchirappalli, Tamil Nadu, India

D. Dhanasekaran
Department of Microbiology, National Repository for Microalgae and Cyanobacteria - Freshwater (NRMC-F), Bharathidasan University, Tiruchirappalli, Tamil Nadu, India

N. Thajuddin et al. (eds.), *Protocols for Cyanobacteria Sampling and Detection of Cyanotoxin*, https://doi.org/10.1007/978-981-99-4514-6_20

20.1 Introduction

Cyanobacteria are prokaryotic photoautotrophic picoplankton that can thrive in a variety of environments, and it is one of the planet's oldest living things, according to fossil records (Jakubowska and Szeląg-Wasielewska 2015; Leao et al. 2009; Sánchez-Baracaldo et al. 2022). They possess an ability to develop symbiotic connections with a variety of diverse hosts. They also had a significant effect on biological diversity (Sánchez-Baracaldo et al. 2022). Cyanobacteria are known for their hazardous toxicity and are capable of producing heavy algal blooms in freshwater ecosystems (Fig. 20.1). One of the most significantly affecting cyanotoxins includes hepatotoxins, neurotoxins, genotoxins, inflammatory, and cytotoxic agents. They are produced in great amounts by the cyanoHABs and pose a serious threat to humans, livestock as well as aquatic life (Głowacka et al. 2011). It has recently come to light that the frequency of blooms may rise dramatically under future climate change (Kalsoom et al. 2018; Paerl and Huisman 2008).

Cyanobacteria emit natural fluorescence due to the presence of two pigments namely chlorophyll a and phycobilin (Campbell et al. 1998). Due to the occurrence of these pigments, cyanobacteria impact fluorescence at the wavelength of 543–633 nm (Garcia-Pichel et al. 1995). Physiological processes occurring in photosynthetic cyanobacteria can be studied elaborately in vivo using the Confocal Laser Scanning Microscope (CLSM), which is unquestionably the most effective method as it offers live-cell imaging and is constructed with super resolution technology (Fig. 20.2) (Elliott 2020). The high spatial and temporal resolution sets

Fig. 20.1 Microalgal blooms in freshwater ecosystems

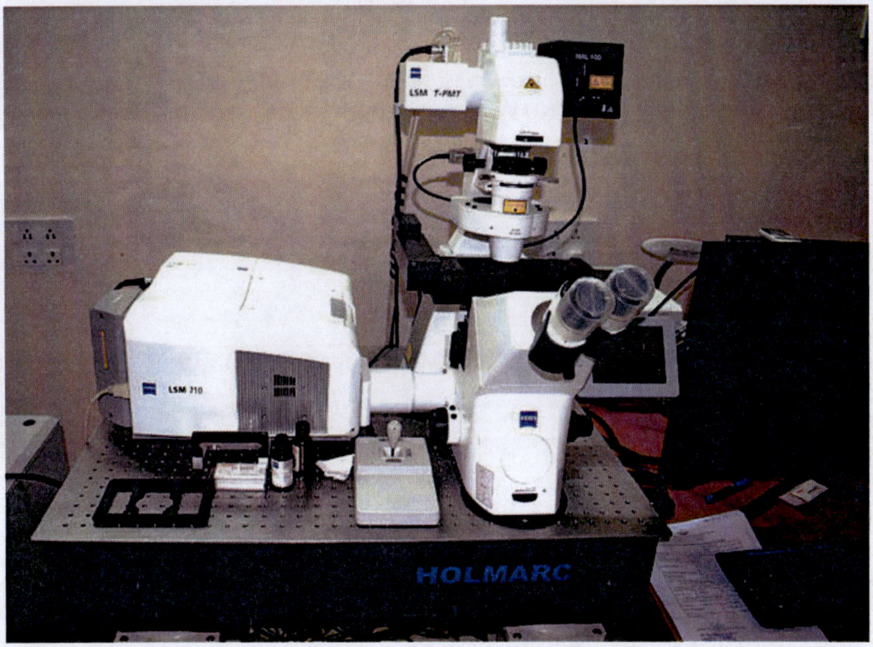

Fig. 20.2 Carl Zeiss—Confocal Laser Scanning Microscope 710 (Bharathidasan University, Tiruchirappalli)

confocal laser scanning microscopes (CLSMs) unique from other imaging technologies (Draaijer et al. 1995).The fundamental components of a typical confocal microscope involve an objective lens, out-of-focus plane, in-focus plane, beam splitters, detector, confocal pinhole (aperture), laser and oscillator mirrors. Modern confocal microscopes still share the same fundamental elements as the original design such as pinholes, objective lens, and low-noise detectors, but they also frequently come with quick scanning mirrors, filter for wavelength selection, and laser illumination (Campbell et al. 1998).

Confocal microscopy works on the fundamental premise that the illumination and detection optics are focused on the same diffraction-limited spot, which is moved over the sample to create the whole image on the detector. Confocal imaging illuminates the whole field of view, but objects outside the focal plane add nothing to the image, reducing the haze that can be seen in normal light microscopy on thick, heavily scattering substances and enabling optical sectioning (Elliott 2020; Draaijer et al. 1995). There are several different light sources that can be used for confocal microscopy. Depending on the needed power and wavelength, mercury lamps or metal halide can be utilized (Castellano-Muñoz et al. 2012). To create an image of the optical section, a laser is focused onto a pair of scanning mirrors, sweeping the beam in a single field of view in both the x and y directions. The sample is then moved progressively across the entire object (Garcia-Pichel et al. 1995). The capacity to create high-resolution, three-dimensional images of live samples of cyanobacteria using confocal microscopy is one of its key benefits (Hameed et al. 2016; LewisOscar et al. 2015, 2021). Confocal microscopy can also be used to view various fluorophores in a single sample, enabling the simultaneous viewing of various processes or pathways (Rocco et al. 2021).

20.2 Material Required

1. Cyanobacteria sample
2. Carl Zeiss—Confocal Laser Scanning Microscope (CLSM) 710
3. Glass slide
4. Micropipette
5. Mortar and pestle
6. Coverslip

20.3 Procedure

20.3.1 Startup of the System

1. In the power remote, turn on the following three switches: MAIN, SYSTEM/PC, and COMPONENT Switch (Fig. 20.3).

20.4 Handling of the Argon Laser

1. First, switch the argon laser's power supply's key switch from standby to ON. (Fig. 20.4).
2. The laser is subsequently kept in standby mode, independent of the run-idle-switch, for around 5 min.
3. Change the run-idle-switch from RUN to IDLE.
4. Within 50 s of entering RUN mode, the laser's power grows to its maximum level.
5. If the set power is stable and the laser is ready to use, a green LED light will be on.

20.5 Sample Preparation

1. Cyanobacterial sample is grinded with the help of mortar and pestle.
2. 10 μL of sample is placed in a glass slide and coverslip is fixed.
3. Prepared slide is mounted on the stage of CLSM.
4. Cyanobacteria does not require staining technique, due to their unique autofluorescence property.

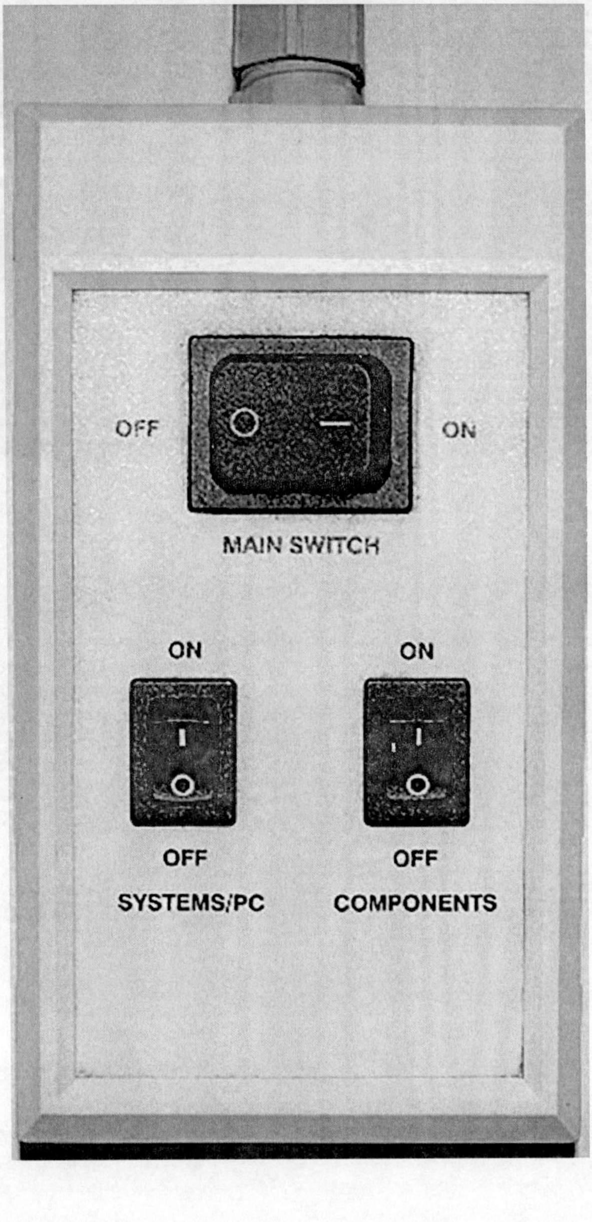

Fig. 20.3 Power remote switch of CLSM 710

Fig. 20.4 Laser control switch of CLSM 710

20.6 Starting Zen Software

1. Double click the ZEN icon on desktop to start the Carl Zeiss LSM software.
2. In a small startup window, choose START THE SYSTEM to observe freshly prepared sample.
3. To alter already-existing photos, choose IMAGE PROCESSING MODE.
4. Click on CAPTURE option to snap the cyanobacteria of best resolution.
5. Microscopic images are stored in specific folders and copied in CD or DVDs.
6. Avoid the use of pen drives and hard disks to prevent virus infestation in the CLSM system.

20.7 Observations

Freshwater samples were collected in the pond located near Bharathidasan University, Tiruchirappalli. Samples visualized under Confocal Laser Scanning Microscope (CLSM) revealed the presence of toxic cyanobacteria. Identification of cyanobacteria under CLSM is easier, due to their unique autofluorescence property. The following cyanobacteria species have been identified in the freshwater algal bloom samples: *Phormidium* sp., *Anabaena* sp., *Oscillatoria* sp., *Oscillatoria princeps, Oscillatoria* sp., *Oscillatoria amphigranulata, Oscillatoria chalybea, Chlorella* sp. (Fig. 20.5). Thus, Confocal Laser Scanning Microscope (CLSM) are reliable, time efficient, and plays a vital role in identification of toxic cyanobacteria from the collected water samples.

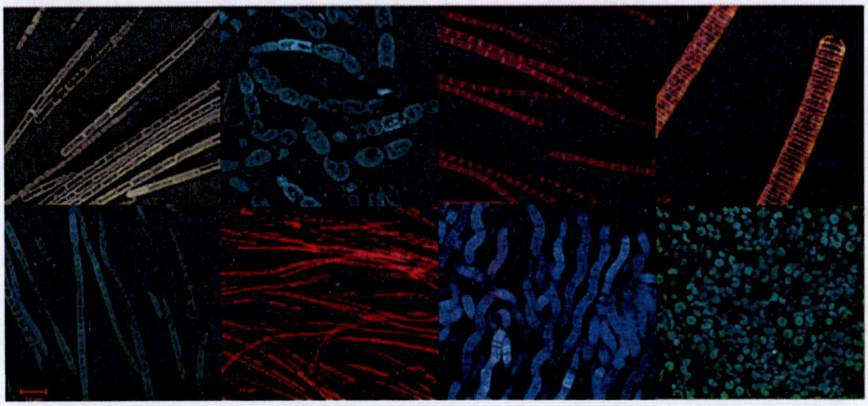

Fig. 20.5 Confocal microscopic images of microalgae and cyanobacteria: (1) *Phormidium* sp.; (2) *Anabaena* sp.; (3) *Oscillatoria* sp.; (4) *Oscillatoria princeps*; (5) *Oscillatoria* sp.; (6) *Oscillatoria amphigranulata*; (7) *Oscillatoria chalybea*; (8) *Chlorella* sp.

Acknowledgement The authors would like to thank Mr. C. Arutselvan, Research Scholar, Department of Microbiology, Bharathidasan University, Tiruchirappalli, for providing the confocal microscopic images of microalgae and cyanobacteria. We sincerely acknowledge the financial support from the DBT (Govt. of India), for the establishment of National Repository for Microalgae and Cyanobacteria—Freshwater NRMC-F (Phase II) (BT/PR29901/PBD26/694/2018), DST PURSE-Scheme DST/SR/PURSE Phase 2/16(C) and School of Life Sciences, Bharathidasan University, Tiruchirappalli, Tamil Nadu, India.

References

Campbell D, Hurry V, Clarke AK, Gustafsson P, Öquist G (1998) Chlorophyll fluorescence analysis of cyanobacterial photosynthesis and acclimation. Microbiol Mol Biol Rev 62(3): 667–683

Castellano-Muñoz M, Peng AW, Salles FT, Ricci AJ (2012) Swept field laser confocal microscopy for enhanced spatial and temporal resolution in live-cell imaging. Microsc Microanal 18(4): 753–760

Draaijer A, Sanders R, Gerritsen HC (1995) Fluorescence lifetime imaging, a new tool in confocal microscopy. In: Handbook of biological confocal microscopy, pp 491–505

Elliott AD (2020) Confocal microscopy: principles and modern practices. Curr Protoc Cytom 92(1): e68

Garcia-Pichel F, Belna J, Sus ANNE (1995) Estimates of global cyanobacterial biomass. Archiv Für Hydrobiologie: MonographischeBeiträge 148:213

Głowacka J, Szefel-Markowska M, Waleron M, Łojkowska E, Waleron K (2011) Detection and identification of potentially toxic cyanobacteria in Polish water bodies. Acta Biochim Pol 58(3): 321–333

Hameed ASH, Karthikeyan C, Ahamed AP, Thajuddin N, Alharbi NS, Alharbi SA, Ravi G (2016) In vitro antibacterial activity of ZnO and Nd doped ZnO nanoparticles against ESBL producing Escherichia coli and Klebsiella pneumoniae. Sci Rep 6(1):1–11

Jakubowska N, Szeląg-Wasielewska E (2015) Toxic picoplanktonic cyanobacteria. Mar Drugs 13(3):1497–1518

Kalsoom U, Hasan CK, Tedone L, Desire C, Li F, Breadmore MC, Nesterenko PN, Paull B (2018) Low-cost passive sampling device with integrated porous membrane produced using multimaterial 3D printing. Anal Chem 90(20):12081–12089

Leao PN, Vasconcelos MTS, Vasconcelos VM (2009) Allelopathy in freshwater cyanobacteria. Crit Rev Microbiol 35(4):271–282

LewisOscar F, MubarakAli D, Nithya C, Priyanka R, Gopinath V, Alharbi NS, Thajuddin N (2015) One pot synthesis and anti-biofilm potential of copper nanoparticles (CuNPs) against clinical strains of Pseudomonas aeruginosa. Biofouling 31(4):379–391

LewisOscar F, Nithya C, Vismaya S, Arunkumar M, Pugazhendhi A, Nguyen-Tri P, Alharbi SA, Alharbi NS, Thajuddin N (2021) In vitro analysis of green fabricated silver nanoparticles (AgNPs) against Pseudomonas aeruginosa PA14 biofilm formation, their application on urinary catheter. Prog Org Coat 151:106058

Paerl HW, Huisman J (2008) Blooms like it hot. Science 320(5872):57–58

Rocco B, Cimadamore A, Sarchi L, Bonetti LR, Bertoni L, Azzoni P, Assumma S, Turri F, Bozzini G, Eissa A, Micali S (2021) Current and future perspectives of digital microscopy with fluorescence confocal microscope for prostate tissue interpretation: a narrative review. Transl Androl Urol 10(3):1569

Sánchez-Baracaldo P, Bianchini G, Wilson JD, Knoll AH (2022) Cyanobacteria and biogeochemical cycles through earth history. Trends Microbiol 30(2):143–157

21 Determination of Cyanotoxin Patterns in Strains and Species of Cyanobacteria

Thangaraj Ramasamy, Nizhanthini Chellapandian, Ashok Muthumari, Ponmanickam Ponnirul, and N. Thajuddin

Abstract

Cyanobacteria are a group of photosynthetic prokaryotes, especially found in freshwater. They produce cyanotoxin as their secondary metabolite. In favorable conditions (i.e., high nutrient levels, light intensity, and water temperature), cyanobacteria can form blooms, a natural phenomenon characterized by algal biomass accumulation and the possible release of cyanotoxins in water ecosystems. Toxins represent an emerging threat for the aquatic organisms which can bioaccumulate in the food chain and affect wildlife and humans. The consumption of contaminated water and skin contact during recreational activities are among the most possible reasons for human poisonings caused by cyanotoxins. In this protocol, attention is focused on the methodologies commonly used to detect cyanotoxins in environments mainly in water sources.

Keywords

Cyanotoxins · Bioassays · LC-MS · ELISA and 16S rRNA

T. Ramasamy (✉) · N. Chellapandian · A. Muthumari
Department of Microbiology, Ayya Nadar Janaki Ammal College, Sivakasi, Tamil Nadu, India

P. Ponnirul
Department of Zoology, Ayya Nadar Janaki Ammal College, Sivakasi, Tamil Nadu, India

N. Thajuddin
B.S. Abdur Rahman Crescent Institute of Science & Technology (Deemed to be University), Chennai, Tamil Nadu, India

Department of Microbiology, National Repository for Microalgae and Cyanobacteria - Freshwater (NRMC-F), School of Life Sciences, Bharathidasan University, Tiruchirappalli, Tamil Nadu, India

N. Thajuddin et al. (eds.), *Protocols for Cyanobacteria Sampling and Detection of Cyanotoxin*, https://doi.org/10.1007/978-981-99-4514-6_21

21.1 Introduction

Cyanobacteria are eubacteria which have a cell structure similar to prokaryotic. Cyanobacteria are structurally bacteria like but functionally they are algal like. Cyanobacteria possess a photosynthetic pigment chlorophyll a, characteristics feature of eukaryotic algae and higher plants. Cyanobacteria form a natural part of phytoplankton assemblages in lakes and reservoirs. As a result of changes in temperature of water, solar irradiance, metrological conditions, hydrology, nutrient supply, cyanobacterial abundance, and community composition varies seasonally. While referring to cyanobacteria, the term bloom commonly is associated with accumulation of cyanobacteria at water surface; however, cyanobacteria blooms are more typically mixed throughout the photic zone, water column.

Cyanobacteria are the only member of freshwater phytoplankton communities to produce toxin. Cyanobacterial toxins are naturally produced by-products. Their functions include primary role in cellular processes, secondary metabolites, allelopathy, or defense mechanism. Cyanobacterial toxins are chemically and bioactively diverse. The three main classes of cyanobacterial toxins, defined by their effects on human health, are dermatoxins, neurotoxins, and hepatotoxins (Chorus 1999; Chorus et al. 2000). Dermatotoxins cause acute, often severe, dermatitis after contact with cyanobacteria and also may cause inflammation of the gastro-intestinal tract. Neurotoxins, which affect the central nervous system and hepatoxins, are extremely toxic and affect the liver.

Cyanobacterial toxins can be determined by a variety of analytical techniques. Techniques include bioassays such as enzyme-linked immunosorbent assays (ELISA), protein phosphate inhibition assays (PPIA), neurotransmitter inhibition assays, radioassays, gas chromatography flame ionization detection (GC-FID), and liquid chromatography (LC) coupled with UV visible spectrometry and mass spectrometry (LC/uv-vis) (Fig. 21.1).

21.2 Materials Required

- Glass bottle
- BG 11 medium
- Methanol
- Chloroform
- Sea water
- Agarose gel
- Distilled water

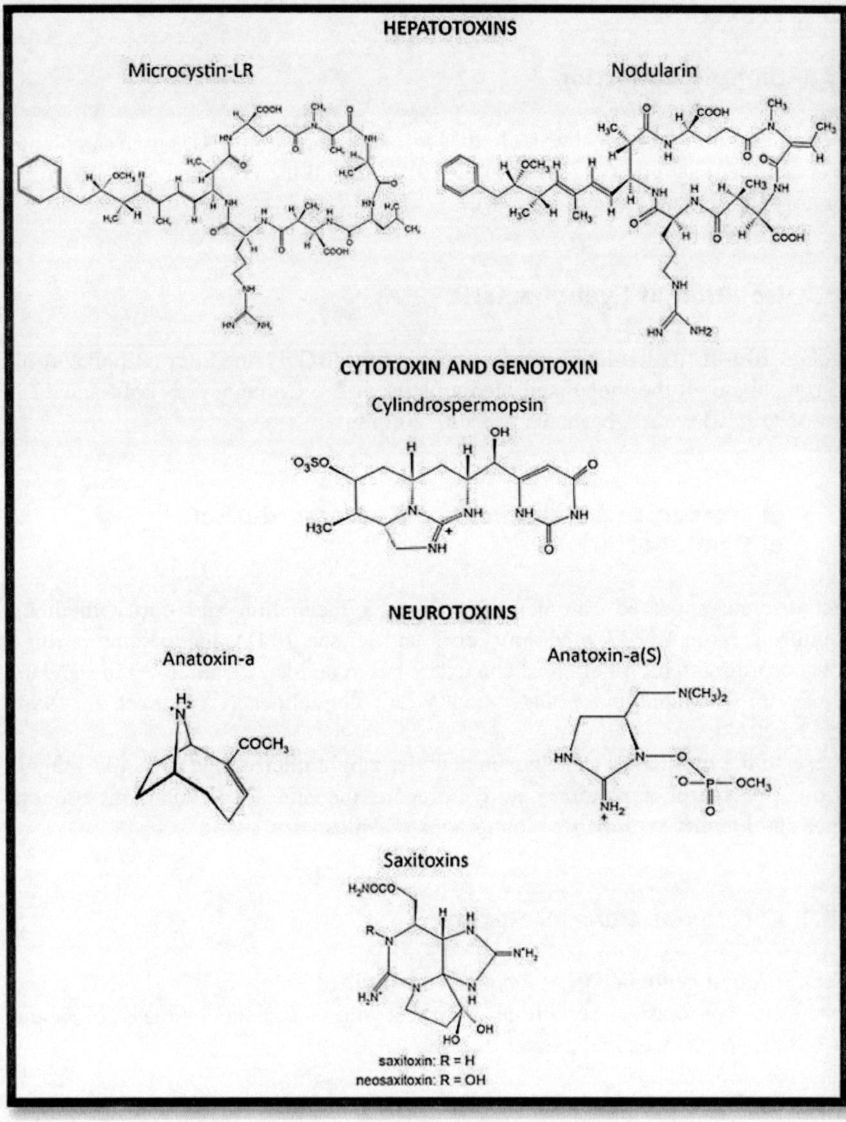

Fig. 21.1 Categories of cyanotoxin

21.3 Procedure

21.3.1 Sample Collection

Collect the sample from local water bodies in a dark glass bottle. Use grab sampler to collect sediment sample (Gozari et al. 2019). Take all the collected samples to the laboratory immediately (within 2–3 h).

21.3.2 Isolation of Cyanobacteria

Inoculate 100 μL (0.1 mL) of each sample on the BG 11 medium (Rippika et al. 1979). Incubate all the inoculated media plates at 28 °C under 12 h light and 12 h darkness (12:12) with light intensity 1500–2000 lux.

21.3.3 Microscopic Observation and Characterization of Cyanobacteria

Select discrete appeared colonies after 2 week incubation and purify them by subculturing on to BG 11 medium (Ferris and Hirsch 1991). Identify the purified isolate according to morphological characters based on identification key in standard methods for examination of water and WHO Guidelines (Lawton et al. 1999; Carranzo 2012).

Observe the morphological characters under a light microscope with 40X magnification. For morphotaxonomic study, prepare the slides and take microscopic images and identify with paper monographs (Desikachary 1959).

21.3.4 Cytotoxins Genes Detection

For extraction of genomic DNA, inoculate the purified isolate in BG 11 medium and incubate in appropriate conditions. Harvest the sufficient biomass of about (50–100 mg) of isolates and extract the DNA.

21.3.5 PCR (Polymerase Chain Reaction)

After DNA extraction, the cyanotoxin biomarkers, including microcystin, nodularine, and cylindrospermopsin gene detection is to be done by PCR.

Perform agarose gel electrophoresis, for the detection of results.

21.3.6 Extraction of Cyanotoxin

After cultivation period, harvest the cells by centrifugation. Wash the harvested cells with distilled water to remove salts.

Then obtain cyanobacterial extracts by homogenization using methanol: chloroform (1:1) in the ratio of 1:10 (biomass: solvent) and keep at 25 °C for 24 h in a photo incubator.

Centrifuge the solutions at 10,000 rpm for 10 min. Recover the supernatant, dry, and store at −20 °C.

21.3.7 Bioassays

Dissolve 50 mg of the obtained cyanobacterial crude extracts in a 1 mL of dimethyl sulfoxide for bioassays (Maruthanayagam et al. 2013).

For rapid monitoring of cyanotoxins concentration, perform the most commonly used assays such as ELISA and PPIA.

21.3.8 Confirmation of Cyanotoxin Patterns

Perform Liquid Chromatography-Mass Spectrometry (LC-MS) for identification, detection, and confirmation of cyanotoxin in the samples. Liquid chromatography-mass spectrometry with electrospray ionization is a standard method for detection, identification, and confirmation of cyanotoxins in environmental samples. Liquid chromatography-mass spectrometry enables the simultaneous separation and identification of cyanotoxins in mixture. (Banack et al. 2010).

21.3.9 Molecular Identification of Cyanobacteria

For phylogenetic analysis, amplify 16S rRNA gene using PCR reaction with the help of universal primers including 9F and 1541R described by Heuer et al. (1997).

Then analyze the 16S rRNA gene sequences by BLAST program at NCBI. Then construct the phylogenetic tree based on neighbor joining algorithm using MEGA X program (Kumar et al. 2018).

21.4 Conclusion

Cyanotoxins are considered an emerging hazard to natural reservoirs and drinking water supply. This protocol provides the methodologies commonly used to detect cyanotoxins in environment mainly in water sources. However, many efforts have still to be made in order to improve the existing available techniques and to develop new tools for detecting toxic cyanobacteria species. The lack of comprehensive toxicity data referred to all cyanotoxins is an important gap to overcome so as to improve cyanobacterial regulations and to make a decisive step towards recognizing cyanotoxins as a risk to human health and environment.

References

Banack SA, Caller TA, Stommel EW (2010) The cyanobacteria derived toxin beta n –methylamino-L-alanine and amyotrophic lateral sclerosis. Toxins (Basel) 2:2837–2850. https://doi.org/10.3390/toxins2122837

Carranzo IV (2012) Standard methods for examination of water and wastewater. In: Anales De Hidrología Médica, vol 5. Universidad Complutense de Madrid, pp 100–185. 978-087553-013-0

Chorus I (1999) Toxic cyanobacteria in water. A guide to their public health consequences, monitoring and management. E&FN Spon, London

Chorus I, Falconer IR, Salas HJ, Bartram J (2000) Health risks caused by freshwater cyanobacteria in recreational water. J Toxicol Environ Health 3:323–347

Desikachary TV (1959) Indian council of agricultural research. ICAR, New Delhi, pp 71–619

Ferris MJ, Hirsch C (1991) Method for isolation and purification of cyanobacteria. Appl Environ Microbiol 57:1448–1452. https://doi.org/10.1128/aem.57.5.1448-1452.1991

Gozari M, Zaheri A, Jahromi ST, Gozari M, Karimzadeh R (2019) Screening and characterization of marine actinomycetes from the northern Oman Sea sediments for cytotoxic and antimicrobial activity. Int Microbiol 22:521–530. https://doi.org/10.1007/s10123-019-00083-3

Heuer H, Krsek M, Baker P, Smalla K, Wellington E (1997) Analysis of actinomycete communities by specific amplification of genes encoding 16S rRNA and gel -electrophoretic separation in denaturing gradients. Appl Environ Microbiol 63:3233–3241

Kumar S, Stecher G, Knyaz C, Tamura K (2018) MEGA X: molecular evolutionary genetics analysis across computing platforms. Mol Biol Evol 35:1547–1549. https://doi.org/10.1093/molbev/msy096

Lawton L, Marsalek B, Padisák J, Chorus I (1999) Determination of cyanobacteria in the laboratory. Toxic cyanobacteria in water: A guide to their public health consequences, monitoring and management. World Health Organization, pp 1–28

Maruthanayagam V, Nagarajan M, Sundararaman M (2013) Cytotoxicity assessment of cultivable marine cyanobacterial extracts in Artemia salina (brine shrimp) larvae and cancer cell lines. Toxin Rev 32:1–9. https://doi.org/10.3109/15569543.2012.754772

Rippika R, Deruelles J, Waterbury JB, Stainer R, Y (1979) Genetic assignment, strain histories and properties of pure culture of cyanobacteria. J Gen Microbiol 111:1–61

22 Chemical and Toxicological Studies of the Toxic Freshwater Cyanobacterium: *Microcystis aeruginosa*

K. R. V. Bandara, P. M. Samudrani, H. A. S. N. Abeysiri, G. Y. Liyanage, P. M. Sandupama, D. M. P. I. Dissanayake, M. M. Pathmalal, and S. A. Kulasooriya

Abstract

Globally, cyanobacterial hazardous algal blooms are becoming more frequent and intense due to eutrophication and climate change. *Microcystis aeruginosa*, that cause hazardous algal blooms, produce Microcystins (MCs), which are linked to hepatotoxicity in humans and animals. Sensitive, quick, and reliable technologies that detect MCs are needed to effectively manage and regulate these toxins and minimize health concerns due to them. The chemical and toxicological analytical techniques, including biological (mouse bioassay) and chemical (High-Performance Liquid Chromatography) research, used to identify MCs are summarized in this chapter.

K. R. V. Bandara · H. A. S. N. Abeysiri · M. M. Pathmalal
Centre for Water Quality and Algae Research, Department of Zoology, University of Sri Jayewardenepura, Nugegoda, Sri Lanka

Present Address: Faculty of Graduate Studies, University of Sri Jayewardenepura, Nugegoda, Sri Lanka

P. M. Samudrani · P. M. Sandupama · D. M. P. I. Dissanayake
Centre for Water Quality and Algae Research, Department of Zoology, University of Sri Jayewardenepura, Nugegoda, Sri Lanka

G. Y. Liyanage
Centre for Water Quality and Algae Research, Department of Zoology, University of Sri Jayewardenepura, Nugegoda, Sri Lanka

Department of Aquatic Bioresources, Faculty of Urban and Aquatic Bioresources, University of Sri Jayewardenepura, Nugegoda, Sri Lanka

S. A. Kulasooriya (✉)
National Institute of Fundamental Studies, Kandy, Sri Lanka
e-mail: ananda@kulasooriya.com

N. Thajuddin et al. (eds.), *Protocols for Cyanobacteria Sampling and Detection of Cyanotoxin*, https://doi.org/10.1007/978-981-99-4514-6_22

Keywords

Microcystis aeruginosa · Cyanobacteria · Microcystins (MCs) · Mouse bioassay · High-Performance Liquid Chromatography

22.1 Introduction

As the population increases, agricultural and urban development are placed under enormous pressure (Hassan and Nazem 2016). The excessive input of exogenous nutrients causes eutrophication (Hassan and Nazem 2016). With the increasing impacts of climate change and eutrophication, occurrence of cyanobacterial blooms is one of the global environment problems, causing pollution and health problems (Nazari-Sharabian et al. 2018). Numerous scientists have been concerned about their causes. Microcystins (MCs), most often found in water and to a lesser extent in desert environments, are primarily produced by species of cyanobacteria belonging to the genera *Microcystis, Anabaena, Aphanizomenon, Nostoc, Cylindrospermopsis, and Planktothrix* (Nowruzi and Porzani 2021*;* Svirčev et al. 2019*). M. aerugino* sa is a photosynthetic cyanobacterium that plays an important role in global oxygenation (Qu et al. 2018) and cyanobacterial toxin production (Mazard et al. 2016).

Microcystin hepatotoxin can accumulate in the liver and kidney of terrestrial mammals mainly through the bile acid transport system and exert its toxic activity by inhibiting protein synthesis (Zurawell et al. 2005; Zanchett and Oliveira-Filho 2013). In 1996, the most severe case of human poisoning by microcystin in hospital hemodialysis water resulted in the death of 76 patients, representing the first report of human death caused by microcystin contamination (Massey et al. 2018). Based on toxicity, Microcystin-LR (MC-LR) is by far the most potent hepatotoxin among the different variants of MCs and has become a global focus (Li et al. 2021; Zhou et al. 2015). The International Agency for Research on Cancer classified this toxin as a group 2B carcinogen (Ostry et al. 2017), and the World Health Organization (WHO) recommended provisional 1 μg/L MC-LR guidelines for drinking water quality (World Health Organization 1998). For plants, MCs adversely affect physiological processes, including tissue development, enzyme activation, gene expression, photosynthesis, and other functions necessary for a plant to grow (Campos et al. 2021).

To effectively manage and control MCs and prevent or minimize their health risks, sensitive, fast, and reliable screening methods capable of detecting these toxins are urgently required. Early detection of MCs can help to counteract these deadly toxins and avoid further poisoning of ecosystems and human health threats. An important consideration in analyzing water samples for MCs is to determine the differences between intracellular and extracellular toxins produced by *M. aeruginosa* (Sun et al. 2013). To successfully determine the toxin level, there should be cell lysis to release intracellular toxins, primarily by freeze-thawing and ultrasonication bath (Massey et al. 2020). Therefore, the first step towards MCs hazard detection must contain developing sensitive, fast, and reliable screening methods to identify these toxins. This chapter aims to describe the chemical and toxicological studies used for MC detection in terms of their novel development and usage.

Various MCs detection studies have been developed, such as Protein Phosphatase Inhibition Assays (PPIA), Enzyme Linked Immunosorbent Assays (ELISAs), High-Performance Liquid Chromatography (HPLC), Liquid Chromatography Mass Spectrometry (LC-MS), and biosensors (Kumar et al. 2020). Chemical methods, such as HPLC and LC-MS, can identify various variants more sensitively. The most commonly and widely used laboratory technique to analyze MCs using different stationery and aqueous mobile phases containing methanol or acetonitrile is HPLC (Li et al. 2014). It is worth noting that the confined range of MCs detection is associated with concentration factors attained and sample volume. Parameters such as mobile phase composition and HPLC conditions, including flow rate, temperature, and column features (including stationary phase, silanol activity, and length) may account for an excellent separation and sensitivity of HPLC (Lawton et al. 1994). A globally certified reference material to purify and quantify MCs has been accepted for the efficient implementation of this technique. This will help to ensure standardization of routine laboratory analysis of these toxins. Mouse Bioassay (MBA) was performed as a toxicological study to detect the toxicity of MCs in animals. Generally, toxins extracts are administered via intraperitoneal injection into mice (Massey et al. 2020). MBA may also apply microbes, invertebrate and vertebrate animals, cell cultures or plants and plant extracts to detect the toxicity of MCs (Bláha et al. 2016).

22.2 Materials

- 50 μm mesh size plankton net
- 100 mL plankton collecting bottles
- Autoclaved amber color glass bottles
- 50 mL beakers
- Whatman 110 mm GF/C filters (diameter of 0.22 μm)
- HPLC vials
- C18 column
- Male mice with animal house facilities for toxicological study
- Milli-Q water
- 0.9% NaCI
- Methanol
- Trifluoroacetic Acid (TFA)

22.3 Methodology

22.3.1 Mouse Bioassays and LD_{50}

22.3.1.1 Collection of Cyanobacteria

1. Collect cyanobacteria by concentrating with a 50 μm mesh size plankton net.
2. Lyophilize the concentrated samples and keep at −20 °C until extraction.

22.3.1.2 Bioassay

1. Toxicity of MCs is measured by intraperitoneal MBA using 20–30 g male mice.
2. Suspend freeze-dried cyanobacteria in 0.9% NaCl solution and prepare different concentrations following injecting two mice per dose level for each of four to eight doses.
3. Continuously observe the injected animals for 2 h and then each 30 min during the following 6 h.
4. Observe symptoms and necropsies performed to detect signs of hepatotoxicity.
5. Cyanobacteria is considered toxic if death occurs at doses below 1500 mg/kg (dried cells/animal weights).
6. Determine LD_{50} as the dose between those that produced 0% and 100% mortality (Bláha et al. 2016).

22.3.2 Extraction of MCs from Water

22.3.2.1 Intracellular MCs

1. Pass a known volume of raw water sample (500 mL) through Whatman 110 mm GF/C filters (diameter of 0.22 μm) to retain cyanobacteria cells.
2. Keep the filter discs in the freezer (at −20 °C) for 24 h, thaw and cut them into small pieces and keep them in 50 mL beakers containing 10 mL of 80% aqueous methanol.
3. Allow approximately 1 h for the extraction.
4. Repeat the extraction twice to ensure maximum recovery of cyanobacteria cells, each time with 10 mL of 80% aqueous methanol in 5 mL aliquots.
5. Collect the total extracts, sonicate for 5 min and centrifuge at 12,000 rpm for 10 min.
6. Carefully decant the supernatant into a 50 mL rotary evaporation flask and dry in vacuum at 40 °C.
7. Suspend the residue in 0.5–1 mL of 80% HPLC methanol, sonicate for 5 min, and centrifuge at 10,000 rpm for 10 min.
8. Decant the supernatant, vortex and transfer into clean HPLC vials.
9. Reverse-phase HPLC should be carried out to determine intracellular MCs in water samples (Lawton et al. 1994).

22.3.2.2 Extracellular MCs

1. Filter raw water samples (500 mL) through 110 mm GF/C filter discs to get the filtrate in a 500 mL clean glass bottle, add 5 mL methanol and mix properly.
2. Attach Solid Phase Extraction (SPE) cartridges to the vacuum manifold system and condition them with methanol and water (10 mL each).
3. Inject the water sample into the vacuum manifold.
4. Wash the cartridges with 10–30% methanol in series, air-dry and elute with 3 mL of 80% methanol mixed with TFA (Lawton et al. 1994).

22.3.3 Quantification of MCs

22.3.3.1 HPLC-Photodiode Array Method (PDA)

1. Before the analysis, filter all solvents, including Milli-Q water, through Whatman GF/C filters (diameter of 0.22 μm) and sonicate for 15 min.
2. Use the C18 column (ZORBAX ODS, 4.6 × 250 mm, 5 μm, serial number: USF0063623) and photodiode array detector at 200–300 nm with 3 nm resolution in HPLC for the quantification and identification of MCs.
3. Before HPLC of analyses, wash the C18 column with 100% methanol for 30 min. Use a mobile phase, Milli-Q water-0.05% TFA, and acetonitrile-0.05% TFA to get a gradient flow to separate MCs.
4. Maintain the temperature of the C18 column at 40 °C with a flow rate of 1 mL/min, and the system pressure at 1.5 atm.
5. Run the toxin samples against MC standards (Lawton et al. 1994).
6. Prepare a calibration curve for the different concentrations of MCs (half dilution method) using standard samples.
7. Run a standard sample at the beginning and end of a set of analyses to confirm the system's correct operation and check retention time drift.
8. Run each toxin sample in triplicate to confirm peaks in the HPLC chromatogram.
9. Determine the toxin concentrations of samples by relative peak areas and compare them with calibration plots.
10. Identify the characteristic absorption spectrum, the presence of MCs at 238 nm wavelength.
11. At the end of the analysis, wash the column with 100% methanol and Milli-Q water (Lawton et al. 1994).

Acknowledgment The authors acknowledge the Centre for Water Quality and Algae Research, Department of Zoology and the University of Sri Jayewardenepura for laboratory and infrastructural facilities.

References

Hassan MM, Nazem MNI (2016) Examination of land use/land cover changes, urban growth dynamics, and environmental sustainability in Chittagong city, Bangladesh. Environ Dev Sustain 18(3):697–716

Nazari-Sharabian M, Ahmad S, Karakouzian M (2018) Climate change and eutrophication: a short review. Eng Technol Appl Sci Res 8(6):3668

Nowruzi B, Porzani SJ (2021) Toxic compounds produced by cyanobacteria belonging to several species of the order Nostocales: a review. J Appl Toxicol 41(4):510–548

Svirčev Z, Lalić D, Bojadžija Savić G, Tokodi N, Drobac BD, Chen L, Meriluoto J, Codd GA (2019) Global geographical and historical overview of cyanotoxin distribution and cyanobacterial poisonings. Arch Toxicol 93(9):2429–2481

Qu J, Shen L, Zhao M, Li W, Jia C, Zhu H,Zhang,Q(2018) Determination of the role of *Microcystis aeruginosa* in toxin generation based on phosphoproteomic profiles. Toxins 10(7):304

Mazard S, Penesyan A, Ostrowski M, Paulsen IT, Egan S (2016) Tiny microbes with a big impact: the role of cyanobacteria and their metabolites in shaping our future. Marine Drugs 14(5):97

Zurawell RW, Chen H, Burke JM, Prepas EE (2005) Hepatotoxic cyanobacteria: a review of the biological importance of microcystins in freshwater environments. J Toxicol Environ Health B Crit Rev 8(1):1–37

Zanchett G, Oliveira-Filho EC (2013) Cyanobacteria and cyanotoxins: from impacts on aquatic ecosystems and human health to anticarcinogenic effects. Toxins 5(10):1896–1917

Massey IY, Yang F, Ding Z, Yang S, Guo J, Al-Osman M, Kamegni RB, Zeng W (2018) Exposure routes and health effects of microcystins on animals and humans: a mini-review. Toxicon 151: 156–162

Li B, Liu Y, Zhang H, Liu Y, Liu Y, Xie P (2021) Research progress in the functionalization of microcystin-LR based on interdisciplinary technologies. Coord Chem Rev 443:214041

Zhou M, Tu WW, Xu J (2015) Mechanisms of microcystin-LR-induced cytoskeletal disruption in animal cells. Toxicon 101:92–100

Ostry V, Malir F, Toman J, Grosse Y (2017) Mycotoxins as human carcinogens—the IARC monographs classification. Mycotoxin Res 33(1):65–73

World Health Organization (1998) Guidelines for Drinking-Water Quality, vol 1, 2nd edn. World Health Organization, Geneva, p 13

Campos A, Redouane EM, Freitas M, Amaral S, Azevedo T, Loss L, Máthé C, Mohamed ZA, Oudra B, Vasconcelos V (2021) Impacts of microcystins on morphological and physiological parameters of agricultural plants: a review. Plan Theory 10(4):639

Sun F, Pei HY, Hu WR, Li XQ, Ma CX, Pei RT (2013) The cell damage of Microcystis aeruginosa in PACl coagulation and floc storage processes. Sep Purif Technol 115:123–128

Massey IY, Wu P, Wei J, Luo J, Ding P, Wei H, Yang F (2020) A mini-review on detection methods of microcystins. Toxins 12(10):641

Kumar P, Rautela A, Kesari V, Szlag D, Westrick J, Kumar S (2020) Recent developments in the methods of quantitative analysis of microcystins. J Biochem Mol Toxicol 34(12):22582

Li YW, Zhan XJ, Xiang L, Deng ZS, Huang BH, Wen HF, Sun TF, Cai QY, Li H, Mo CH (2014) Analysis of trace microcystins in vegetables using solid-phase extraction followed by high performance liquid chromatography triple-quadrupole mass spectrometry. J Agric Food Chem 62(49):11831–11839

Lawton LA, Edwards C, Codd GA (1994) Extraction and high-performance liquid chromatographic method for the determination of microcystins in raw and treated waters. *Analyst 119*(7): 1525–1530

Bláha L, Cameán AM, Fessard V, Gutiérrez-Praena D, Jos Á, Marie B, Metcalf JS, Pichardo S, Puerto M, Törökné A, Vasas G (2016) Bioassay use in the field of toxic cyanobacteria. In: Handbook of Cyanobacterial Monitoring and Cyanotoxin Analysis, 4:272–279

23 Chemical and Toxicological Studies of the Toxic Freshwater Cyanobacterium: *Anabaena Flos-aquae*

Thangaraj Ramasamy, Nivetha Marimuthu, Naveena Kamalahasan, and N. Thajuddin

Abstract

Blue-green algae, more commonly known as cyanobacteria, are frequently found in freshwater systems. They can also be found in estuarine and marine waters. Some freshwater cyanobacterial blooms are able to produce highly potent toxins, known as cyanotoxins. Cyanobacterial toxins are considered as an emerging threat to natural water reservoirs and drinking water supplies. When they are ingested, these toxins are responsible for human and animal poisonings. *Anabaena flos-aquae* produces neurotoxins termed anatoxin is a potent postsynaptic depolarizing neuromuscular blocking agent. In this protocol, attention is focused on the methodologies commonly used to study and analyze the cyanotoxins in environments mainly in freshwater sources.

Keywords

Anabaena flos-aquae · Anatoxin · BG 11 medium · Microcystin · Nodularin and cylindrospermopsin gene · Polymerase chain reaction · Photo incubator · High pressure liquid chromatography

T. Ramasamy (✉) · N. Marimuthu · N. Kamalahasan
Department of Microbiology, Ayya Nadar Janaki Ammal College, Sivakasi, Tamil Nadu, India

N. Thajuddin
B.S. Abdur Rahman Crescent Institute of Science & Technology (Deemed to be University), Chennai, Tamil Nadu, India

Department of Microbiology, National Repository for Microalgae and Cyanobacteria - Freshwater (NRMC-F), School of Life Sciences, Bharathidasan University, Tiruchirappalli, Tamil Nadu, India

N. Thajuddin et al. (eds.), *Protocols for Cyanobacteria Sampling and Detection of Cyanotoxin*, https://doi.org/10.1007/978-981-99-4514-6_23

23.1 Introduction

Cyanobacteria, a large, heterogeneous group of prokaryotic, principally photosynthetic organisms. Cyanobacteria resemble the eukaryotic algae in many ways, including morphological characteristics and ecological niches and were at one time treated as algae, hence the common name of blue-green algae. Cyanobacteria range in size from 0.5 to 60 mm, which represents the largest prokaryotic organism. They are widely distributed and are extremely common in freshwater, where they occur as members of both the plankton and the benthos. They are also abundantly represented in such habitats as tide pools, coral reefs, and tidal spray zones; a few species also occur in the ocean plankton. On land, cyanobacteria are common in soil down to a depth of 1 m (39 inches) or more; they also grow on moist surfaces of rocks and trees, where they appear in the form of cushions or layers.

Cyanobacteria are Gram-negative bacteria about five genera of freshwater cyanobacteria are known to produce exotoxins. They include species of Microcystis, Anabaena, Aphanizomenon, Oscillatoria, and Synechocystis (Steven and Williams 2014). Toxins produced include: anatoxins (alkaloids and peptides of Anabaena); microcystin and microcystis-type c (peptide of Microcystis); and aphantoxin (neurotoxic compounds of Aphanizomenon similar to certain PSPs). Toxic water blooms of these species cause death of livestock and wildlife in several countries. The toxins may also be responsible for some human gastroenteritis outbreaks from municipal reservoirs (Sivonen 2009). In recreational waters, the toxins can cause contact irritation, perhaps due to a delayed hypersensitivity reaction. These phycotoxins also threaten the successful use of cyanobacteria as microalgal food since present methods of culturing do not have adequate quality control to prevent contamination by toxic strains. Methods of detection used are high pressure liquid chromatography, fluorometry, and UV absorbance.

23.2 Materials Required

- BG 11 medium
- Methanol
- Chloroform
- Agarose gel
- Distilled water
- Glass-amber material

23.3 Procedure

23.3.1 Sample Collection

For sample collection, glass-amber material should be used, do not use clear glass (anatoxin-a is light sensitive–avoid exposure to light, as this will degrade the toxin and produce inaccurate results). Use grab sampler to collect sediment sample (Gozari et al. 2019). Take all the collected samples to the laboratory immediately (within 2–3 h).

23.3.2 Isolation of Cyanobacteria

Inoculate 100 μL of each sample on the BG 11 medium (Rippika et al. 1979). Incubate all the inoculated plates at 28 °C under 12 h light and 12 h darkness (12:12) with light intensity 1500–2000 lux.

23.3.3 Characterization of Cyanobacteria

Observe the micromorphological characters under a light microscope with 40X magnification. For morphotaxonomic study, prepare the slides and take microscopic images and identify with paper monographs (Desikachary 1959).

23.3.4 Detection of Cytotoxins Gene

For extraction of genomic DNA, inoculate the purified isolate in BG 11 medium and incubate in appropriate conditions. Harvest the sufficient biomass of about (50–100 mg) of isolates and extract the DNA.

23.3.5 PCR (Polymerase Chain Reaction)

After DNA extraction, the cyanotoxin biomarkers, including microcystin, nodularin, and cylindrospermopsin gene detection is to be done by PCR.

Perform agarose gel electrophoresis, for the detection of results.

23.3.6 Extraction of Cyanotoxin

After cultivation period, harvest the cells by centrifugation. Wash the harvested cells with distilled water to remove salts.

Then obtain cyanobacterial extracts by homogenization using methanol: chloroform (1:1) in the ratio of 1:10 (biomass: solvent) and keep at 25 °C for 24 h in a photo incubator.

Centrifuge the solutions at 10,000 rpm for 10 min. Recover the supernatant, dry, and store at −20 °C.

23.3.7 Bioassays

Dissolve 50 mg of the obtained cyanobacterial crude extracts in a 1 mL of dimethyl sulfoxide for bioassays (Maruthanayagam et al. 2013).

For rapid monitoring of cyanotoxins concentration, perform the most commonly used assays such as ELISA and PPIA. Liquid chromatography-mass spectrometry enables the simultaneous separation and identification of cyanotoxins in mixture (Banack et al. 2010).

23.4 Conclusion

Cyanotoxins are regarded as the potential hazardous substances to natural reservoirs and drinking water supply. This protocol provides the methodologies commonly used to detect and study the cyanotoxins in environments mainly in freshwater sources. However, many efforts have still to be made in order to improve the existing available techniques and to develop new tools for detecting toxic cyanobacteria species. The lack of comprehensive toxicity data referred to all cyanotoxins is an important gap to overcome so as to improve cyanobacterial regulations towards recognizing cyanotoxins as a risk to human health and environment.

References

Banack SA, Caller TA, Stommel EW (2010) The cyanobacteria derived toxin beta n − methylamino-L-alanine and amyotrophic lateral sclerosis. Toxins(Basel) 2:2837–2850. https://doi.org/10.3390/toxins2122837

Desikachary, TV., 1959, Indian Council of Agricultural Research, New Delhi,71–619pp

Gozari M, Zaheri A, Jahromi ST, Gozari M, Karimzadeh R (2019) Screening and characterization of marine actinomycetes from the northern Oman Sea sediments for cytotoxic and antimicrobial activity. Int Microbiol 22:521–530. https://doi.org/10.1007/s10123-019-00083-3

Maruthanayagam V, Nagarajan M, Sundararaman M (2013) Cytotoxicity assessment of cultivable marine cyanobacterial extracts in Artemia salina (brine shrimp) larvae and cancer cell lines. Toxin Rev 32:1–9. https://doi.org/10.3109/15569543.2012.754772

Rippika R, Deruelles J, Waterbury JB, Stainer RY (1979) Genetic assignment, strain histories and properties of pure culture of cyanobacteria. J Gen Microbiol 111:1–61

Sivonen K (2009) Cyanobacterial toxins in Encyclopedia of Microbiology, 3rd edn

Steven LP, Williams DW (2014) Microbiology of waterborne diseases, 2nd edn

24 Risk Assessment of Microcystin in Water Resources

P. A. B. P. Abeyrathne, K. P. A. Imanthi, H. A. S. N. Abeysiri, G. Y. Liyanage, M. M. Pathmalal, and S. A. Kulasooriya

Abstract

Cyanobacteria naturally produce hepatotoxic substances known as microcystins in aquatic environments. When present in large quantities, they can affect the aesthetics of water quality, contaminate drinking water reservoirs and recreational rivers, obstruct normal ecosystem function, and threaten the health of animals, plants, and people. Animals and humans are often exposed to microcystin when they drink contaminated water or come into contact with it directly. According to extensive studies, oral exposure to microcystins has been linked to a wide spectrum of liver failure, from hepatocellular damage to primary liver cancer. Preliminary guidelines for microcystins in drinking and recreational water have

P. A. B. P. Abeyrathne · K. P. A. ImanthiH. A. S. N. Abeysiri
Centre for Water Quality and Algae Research, Department of Zoology, University of Sri Jayewardenepura, Nugegoda, Sri Lanka

Present Address: Faculty of Graduate Studies, University of Sri Jayewardenepura, Nugegoda, Sri Lanka

G. Y. Liyanage
Centre for Water Quality and Algae Research, Department of Zoology, University of Sri Jayewardenepura, Nugegoda, Sri Lanka

Department of Aquatic Bioresources, Faculty of Urban and Aquatic Bioresources, University of Sri Jayewardenepura, Nugegoda, Sri Lanka

M. M. Pathmalal
Centre for Water Quality and Algae Research, Department of Zoology, University of Sri Jayewardenepura, Nugegoda, Sri Lanka

Faculty of Graduate Studies, University of Sri Jayewardenepura, Nugegoda, Sri Lanka

S. A. Kulasooriya (✉)
National Institute of Fundamental Studies, Kandy, Sri Lanka
e-mail: ananda@kulasooriya.com

N. Thajuddin et al. (eds.), *Protocols for Cyanobacteria Sampling and Detection of Cyanotoxin*, https://doi.org/10.1007/978-981-99-4514-6_24

been established to prevent hazardous exposures and safeguard the public's health. The United States Environmental Protection Agency (USEPA) developed risk assessment computing models for carcinogenic and non-carcinogenic compounds using mathematical equations. This chapter addresses the potential for human exposure to microcystins found in nearby water supplies used for daily use.

Keywords

Human exposure · Mathematical model · Microcystins · Risk assessment · Water resources

24.1 Introduction

The eutrophication of water bodies and the occurrence of algal blooms result from a significant amount of anthropogenically produced nitrogen and phosphorus that has recently entered waterbodies by surface runoff, sewage systems, and atmospheric dry and wet deposition processes (Shan et al. 2019). In addition to degrading the quality of the water, cyanobacterial blooms also produce some highly toxic secondary metabolites that represent a significant risk to human health and the security of the water supply (Ge et al. 2021). Microcystins (MCs), which have a cyclic heptapeptide structure, are the most common algal toxins. An uncommon 3-amino-9-methoxy-2,6,8-trimethyl-10-phenyl-4,6-decadienoic acid (ADDA) with two variable L amino acid residues is present in these biotoxins. At least 246 variants have been discovered since the 1980s when the structure of MCs was initially identified. Globally, Microcystine-LR (MC-LR), Microcystine-RR (MC-RR), and Microcystine-YR (MC-YR) are the variants most frequently found in water samples. As a result of its high acute toxicity and widespread occurrence, MC-LR is currently regarded as the reference compound in Asia, where numerous studies have found MC-LR, MC-RR, and MC-YR (Kholafazad Kordasht et al. 2020). Among them, MC-LR is the most frequently reported and thoroughly investigated compound. A toxico-kinetic investigation found that MCs cause serine and threonine protein phosphatases to be inhibited after being taken up by hepatocytes through a carrier-mediated transport pathway. In addition, recent research examined the impact of 21 MCs analogues on the activity of protein phosphatase 2A and discovered that MC-LR was the most potent inhibitor. Inhibiting protein phosphatases causes a rise in protein phosphorylation, which impacts several procedures and causes various cellular reactions, including apoptosis, diminished DNA repair, and tumour promotion (Zong et al. 2022).

Environmental health issues are assessed to identify toxic cyanobacteria as a threat to human and animal health. However, a careful examination of human illnesses and fatalities brought on by MCs exposure is necessary (Shan et al. 2019; Mokoena et al. 2016). The health hazards presented by exposure to MCs are difficult to measure because the actual exposure and the effects, particularly concerning humans, have not been established with certainty. Algal scums can also be consumed or touched, exposing one to MCs. It is possible to distinguish between three different sources of MC exposure, direct contact between exposed body parts, such as the mouth, throat, ears, and areas covered by swimming suits, as well as unintentional ingesting and water spray inhalation among them (Mokoena et al. 2016). Additional exposure paths, though less frequent, include consuming contaminated fruits, vegetables, aquatic organisms, dietary supplements, and algae (Melaram et al. 2022). The disease that broke out at a haemodialysis centre in Caruaru, Pernambuco State, Brazil, was the most widely documented human MCs intoxication. Some patients also experienced abrupt liver failure, visual abnormalities, nausea, vomiting, and muscle weakness (Mokoena et al. 2016; Lins et al. 2016). Researchers concluded that humans were especially exposed to MC-LR, YR, and AR after comparing the symptoms of humans with those of the studied animals. According to a study, drinking tainted water exposes humans to MCs, which causes stomach and intestinal irritation, liver cancer, and spleen sickness. MCs have been related to encouraging the growth of tumours and has been shown to impact the liver and the colon. Long-term consumption of MCs contaminated water can result in liver damage, liver cancer, intestine cancer, and genotoxicity. According to epidemiological research, the high prevalence of primary liver cancer in various southern Chinese cities (such as Jiangsu Taixing, and Haimen) is caused by MCs contaminating the water supply (Kholafazad Kordasht et al. 2020). Over the past 80 years, more than 80% of Serbia's water supply reservoirs have witnessed algae blooms, with MC-LR levels in some of these reservoirs up to 650 mg L^{-1}. Studies had shown that in the 10 years before their examination, the prevalence of primary liver cancer significantly increased in areas where these waters were used for human consumption. It has been demonstrated that aquatic species exposed to MCs exhibit altered behaviour, bioaccumulation in tissues, growth inhibition, and decreased fertility and survivability. Worldwide, toxic MC occurrences have led to wild animals, cattle, and aquatic biota deaths. The ecosystem's significant factors of MCs toxicity include resident organisms' biological detoxifying systems. Numerous aquatic organisms retain the biotransformation of MCs into conjugate intermediates via the glutathione route, which may protect them from the toxicologic effects of MCs. MCs are removed from the body through the metabolic process and released into the environment. As MCs go up the food chain to higher trophic levels, they can cause toxin release from exposed creatures that can be harmful to both humans and other aquatic organisms. Additionally, MCs can migrate into soils used for agriculture and leach into groundwater (Lins et al. 2016).

Carcinogenic and non-carcinogenic toxins are categorized and said to have either acute or chronic health consequences when consumed. Since there has not yet been research to establish whether MCs causes cancer, only theories suggesting it might hasten cancer progression, it is still classified as a non-carcinogen. Therefore, the hazard quotient model should be used to establish whether MCs can harm human health if present in consumable water. Direct and indirect exposure of humans to

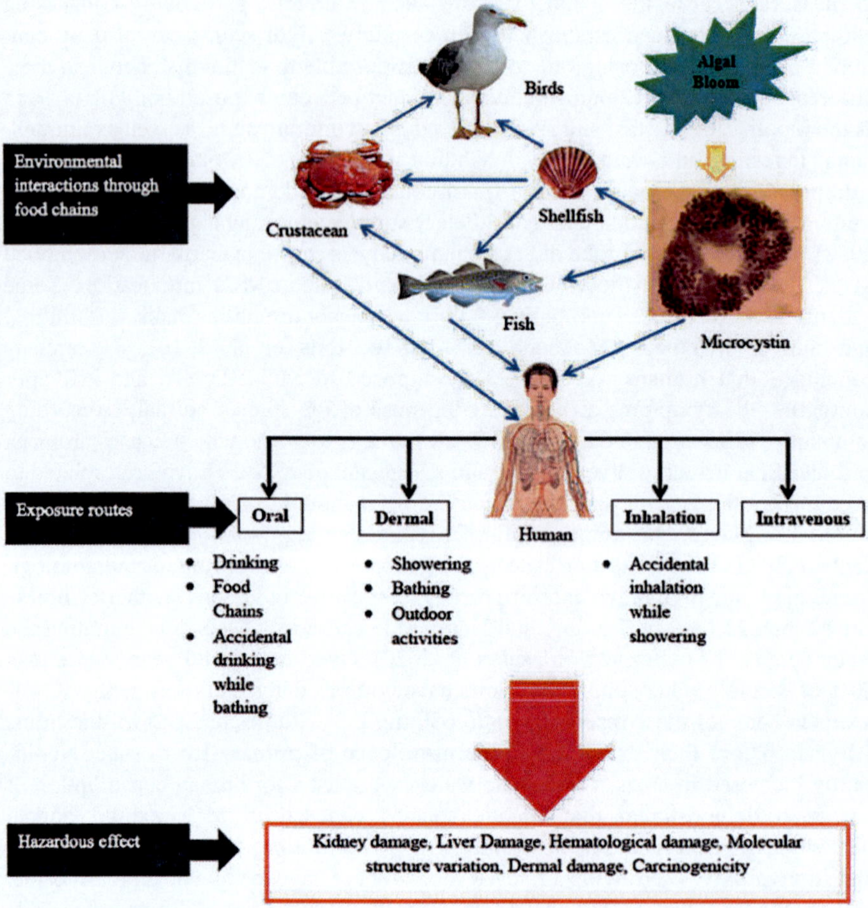

Fig. 24.1 Diagram for microcystins exposure and their toxic effects in humans, and interactions with the food chain

MCs is possible in oral, dermal, inhalation, and intravenous ways (Jaiswal et al. 2016). Drinking water intake is the primary direct exposure pathway. In contrast, other indirect exposure pathways include eating other foods of animal origin, taking dietary supplements, consuming infected crops, and eating plants that have been irrigated with MCs contaminated water (Fig. 24.1). This study assessed the calculation of the risk of facing MCs associated with the water.

24.2 Methods

24.2.1 Calculating the Hazard Quotient (HQ)

24.2.1.1 Calculating the Tolerable Daily Intake (TDI)

It was predicted that family members would be exposed to MCs orally by drinking contaminated water during their lives. Safe MCs doses in drinking water have to be established to estimate the daily dose intake. TDI was determined using toxicological data.

The dosage for non-carcinogenic chemicals can be determined using the intake equation shown below:

$$\mathrm{TDI} = \frac{\mathrm{MC} \times \mathrm{IR} \times \mathrm{EF} \times \mathrm{ED}}{\mathrm{BW} \times \mathrm{AT}}$$

TDI, MCs concentration, Ingestion Rate (IR) contact rate (L/day), Exposure Frequency (EF) (in days), Exposure Duration (ED) (in years), Body Weight (BW) (in kg), and Average Time (AT) (in days).

The HQ is based on the oral exposure to MCs, taking into account the body weight of 70 kg for adults and 15 kg for children, at daily water ingestion volumes of roughly 2 L day^{-1}, as shown in Table 24.1, for the exposure to the MCs contamination in drinking water. The oral LD_{50} or NOEAL of the pig or mice were considered for the MCs concentration.

24.2.1.2 Toxicity Assessment

Consider the possibility of creating a Reference Dose (RfD). The RfD calculates the quantity of a substance consumed in drinking water over the course of 24 h, typically given as body weight. MCs concentrations in drinking water samples were measured during the flowering and decaying seasons. The MCs concentrations were proportional to the lifetime exposure of individuals to water.

Table 24.1 Definitions of the symbols used in Eq. 1 and the assigned concentrations for the TDI calculation

Parameter	Definition	Assigned concentrations
MC	Exposure point concentration	The average concentration of contaminant (microcystins) on exposure (in mg L^{-1} if in water) $\mu\mathrm{g\ L}^{-1}$.
IR	Contact rate (in L day^{-1})	2 L/day drinking water
EF	Exposure frequency (in days per year)	350 days year^{-1}
ED	Exposure duration (in years)	Actual event duration or 30 years if chronic
BW	Body weight (in kg)	70 kg (adult), 15 kg (child)
AT	The period over which exposure is averaged (in days)	Actual event duration if not carcinogenic, or 365 days/years × 70 years if carcinogenic

$$\mathrm{RfD} = \frac{\mathrm{NOEAL\ or\ LOEAL}}{\mathrm{UF}}$$

RfD reference dose, NOEAL no-observed-adverse-effect level, LOEAL lowest-observed-adverse-effect level, and UF uncertain factor

24.2.1.3 Quantifying the Risk and Hazard

The United States Environmental Protection Agency's (USEPA) recommended exposure pathways for pollutants were followed when determining the MCs non-carcinogenic risk assessment in drinking water.

$$\mathrm{HQ} = \frac{\mathrm{TDI}}{\mathrm{RfD}}$$

HQ hazard quotient, TDI, and RfD reference dose

Additional measures must be taken for an adequate risk assessment when HQ is more than 1. These actions (options) include remediation, institutional controls, deed limitations, and source control.

24.2.2 Health Risk Assessment for Carcinogenic Substances

24.2.2.1 Calculating the Average Personal Carcinogenic Annual Risk

$$R_i^C = \frac{1 - \exp(-D_i q_i)}{70}$$

In the formula, R_i^c is the average personal carcinogenic annual risk of chemical carcinogen i through drinking water, (years); 70 indicates the average life expectancy of the population, years; D_i is the daily average exposure to chemical carcinogen i through drinking water, that is, CDI, μg/kg/day; q_i is the carcinogenic strength coefficient of the chemical carcinogen i through the drinking water, and currently, there is no recognized carcinogenic intensity coefficient of MCs. Based on the formula of carcinogenic strength coefficient of carcinogens (Jaiswal et al. 2016), this study deduced the formula as follows:

$$\mathrm{CPI} = \frac{(\mathrm{OR} - 1) \times \mathrm{LR}}{D}$$

where q_i, kg/day/g/L, is the coefficient of carcinogenic strength calculated from population data, and CPI is the Carcinogenic Potency Index. The Odds Ratio (OR) divides the exposed and non-exposed individuals in the case group into the control group. Lifetime Risk (LR) represents the lifetime risk of developing cancer in the local population. D indicated the average daily life exposure dose calculation, μg/kg/day (Wang et al. 2017).

24.2.3 Case Studies on MCs Risk Assessment

Xiao et al. *2018* studied the random sample collection in the Yongjiang river, Guangxi, China, to investigate the presence of MC-RR, MC-LR, and MC-YR. The findings showed that the Total concentrations of MC-RR (TMC-RR), MC-YR (TMC-YR), and MC-LR (TMC-LR) during the observation period varied from 0.0224 to 0.3783 μg/L, 0.0329 to 0.1433 μg/L, and 0.0341 to 0.2663 μg/L, respectively. The study investigated the carcinogenic and non-carcinogenic risk assessment using the software @Risk7.5, the risks of dietary intake of MCs. It was revealed that MC-RR had a larger carcinogenic risk in drinking water than MC-LR and MC-YR and that the presence of MCs carried significant potential health hazards, particularly for children. The International Commission on Radiological Protection (ICRP) advised a maximum allowance level of $>5 \times 10^{-5}$ for adults, whereas the USEPA recommended a maximum allowance level of $>1 \times 10^{-4}$ for children. The non-carcinogenic Hazard Index (HI) of MC-RR, MC-YR, and MC-LR all rose with time, showing that MC-LR was more harmful to human health than MC-YR and MC-RR even if its HI was <1.

Abeysiri and Manage 2022 conducted the Nile tilapia (*Oreochromis niloticus*) intake as the food was gathered from 13 freshwater reservoirs in four distinct Sri Lankan districts (Anuradhapura, Polonnaruwa, Ampara, and Badulla) to establish the concentration of MC-LR and analyse the risk assessment (Abeysiri and Manage 2022). The fish skin had the highest mean concentrations of MC-LR ((3004.25 ± 30 μg kg^{-1}) from the Padaviya, followed by the head (836.25 ± 18 μg kg^{-1}) and flesh (41.67 ± 8 μg kg^{-1}). In the skin and head of all samples, the average daily intake of MC-LR was higher than the WHO's TDI (0.04 μg kg^{-1} day^{-1}). The TDI of MC-LR in fish skin and the head in Parakrama Samudraya were 0.38 ± 0.01 μg kg^{-1} day^{-1} and 0.32 ± 0.01 μg kg^{-1} day^{-1}, respectively. The TDI of MC-LR in fish skin in Kaudulla was 0.73 ± 0.01 μg kg^{-1} day^{-1}. These numbers are above the TDI advised by the WHO. Contrarily, Parakrama Samudraya's TDI value for fish meat was 0.02 μg kg^{-1} day^{-1}, which was below the WHO standard. Fish skin and head TDI values at Halmilla, Nachchaduwa, Kalawewa, Ulhitiya, and Rathkinda exceeded WHO TDI values (Abeysiri and Manage 2022). Therefore, it is advised that people refrain from consuming *O. niloticus* head and skin parts that were obtained from reservoirs with high MC-LR levels.

Acknowledgement The authors acknowledge the Centre for Water Quality and Algae Research, Department of Zoology, University of Sri Jayewardenepura, Sri Lanka, for the immense support and guidance for the leading research to establish the water quality, and well-being of the human and environmental protection in Sri Lanka.

References

Abeysiri HASN, Manage PM (2022) Microcystin-LR contamination in Nile tilapia (Oreochromis niloticus) in some water bodies in Sri Lanka. Sri Lanka J Aquat Sci 27(2):91–105

Ge S, Qiao X, Zhao X, Li X, Liu Y (2021) Microcystin in source water: pollution characteristics and human health risk assessment. RSC Adv 11(11):6415–6422

Jaiswal KP, Yadav AS, Gothalwal R (2016) Microcystins: an emerging biomarker and toxicity prediction in human and animal. Int J Biotech Trends Technol 17(1):1–8

Kholafazad Kordasht H, Hassanpour S, Baradaran B, Nosrati R, Hashemzaei M, Mokhtarzadeh A, de la Guardia M (2020) Biosensing of microcystins in water samples; recent advances. Biosens Bioelectron 165:112403

Lins RP, Barbosa LG, Minillo A, De Ceballos BS (2016) Cyanobacteria in a eutrophicated reservoir in a semi-arid region in Brazil: dominance and microcystin events of blooms. Rev Bras Bot 39(2):583–591

Melaram R, Newton AR, Chafin J (2022) Microcystin contamination and toxicity: implications for agriculture and public health. Toxins 14(5):350

Mokoena MM, Mukhola MS, Okonkwo OJ (2016) Hazard assessment of microcystins from the household's drinking water. Appl Ecol Environ Res 14:695–710

Shan K, Shang M, Zhou B, Li L, Wang X, Yang H, Song L (2019) Application of Bayesian network including Microcystis morphospecies for microcystin risk assessment in three cyanobacterial bloom-plagued lakes, China. Harmful Algae 83:14–24

Wang Y, Xu MF, Geng MM, Li M, Chen GN (2017) Health risk assessment of microcystins from drinking water source by Monte Carlo simulation method. Huan jing ke xue= Huanjing kexue 38(5):1842–1851

Zong Z, Dang Y, Zhang Y, Yu L, Liu C, Wang J (2022) Promotion effect on liver tumour progression of microcystin-LR at environmentally relevant levels in female krasV12 transgenic zebrafish. Aquat Toxicol 252:106313

Extraction and Quantification of BMAA from Water Samples

25

Milena Bruno and Valentina Messineo

Abstract

β-N-methylamino-L-alanine (BMAA) is a non-protein amino acid produced by several types of cyanobacteria found in freshwaters, marine waters, and soil. When ingested, BMAA damages and ultimately destroys motor neurons in the spinal cord, causing the same type of damage seen in patients with ALS, and causes neurofibrillary tangles in the spinal cord and brain, similar to those seen in Alzheimer's disease. Those suffering from this BMAA-induced damage are classified as suffering from amyotrophic lateral sclerosis/parkinsonism dementia complex (ALS/PDC). The symptoms of ALS/PDC include the varying degrees of the muscular paralysis of ALS, the muscular rigidity of Parkinson's, and/or the dementia of Alzheimer's and ultimately result in death. Humans may be exposed to BMAA through the ingestion of contaminated drinking water or foods. Drinking water may become contaminated with BMAA through the proliferation of BMAA-producing cyanobacteria in drinking water sources such as lakes and reservoirs. The majority of methods used for BMAA analysis employed chromatography (either gas or liquid] while a small percentage used other methods including capillary electrophoresis (CE) and ELISA kits among others. Though the earlier studies used fluorescence detection, most of the studies since 2008 used mass spectrometry (MS) or tandem mass spectrometry (MS/MS) as the detection method. This chapter explains the detection of BMAA through a particular sample extraction and ELISA Assay. In literature, BMAA was extracted from cells using solvent which in this method were avoided because ELISA, despite of mass spectrometry, is able to see the molecule in any state (free and conjugated).

M. Bruno (✉) · V. Messineo
FAST, Core Facilities, National Institute of Health, Rome, Italy
e-mail: milena.bruno@iss.it

N. Thajuddin et al. (eds.), *Protocols for Cyanobacteria Sampling and Detection of Cyanotoxin*, https://doi.org/10.1007/978-981-99-4514-6_25

Keywords

Biotoxins · BMAA · Water · Amyotrophic lateral sclerosis, Alzheimer · Parkinson · Elisa analysis

25.1 Introduction

β-N-methylamino-L-alanine (BMAA) is a non-protein amino acid (Fig.25.1) produced by several types of cyanobacteria found in freshwaters, marine waters, and soil, discovered in 1967 from micronesian Cycas seeds on the island of Guam (Vega and Bell 1967). This neurotoxic compound (Chiu et al. 2011) was related to the high incidence of amyotrophic lateral sclerosis/Parkinsonism-dementia complex (ALS/PDC) observed among the indigenous population Chamorro.

This hypothesis was then criticized and rejected and subsequently strengthened in the early 2000s (Karamyan and Speth 2008). Since the discovery of BMAA in 1967, the link between BMAA and neurodegenerative diseases such as amyotrophic lateral sclerosis (ALS) is still under discussion (Karamyan and Speth 2008; Banack et al. 2010; Lee and Mc Geer 2012).

A research team later demonstrated that BMAA could be produced by a symbiotic cyanobacterium (*Nostoc* species) (Cox et al. 2003) and could be biomagnified within the Guam ecosystem, from cyanobacteria to the brains of people who later died due to ALS / PDC, and that large amounts of BMAA can be released from proteins after total acid hydrolysis of the samples (Cox et al. 2003; Murch et al. 2004a). After Cox et al. (Cox et al. 2005) reported that nearly all cyanobacteria can produce BMAA, other groups found BMAA in different ecosystems around the world (Metcalf et al. 2008; Esterhuizen and Downing 2008; Johnson et al. 2008; Li et al. 2010; Spáčil et al. 2010), and three biological magnification models have been suggested, respectively, in the Baltic Sea (Jonasson et al. 2010), in the Florida bay (Brand et al. 2010) and in Taihu Lake, China (Jiao et al. 2014).

There are several potential mechanisms by which BMAA can cause neurological injury (Rao et al. 2006; Lobner et al. 2007; Liu et al. 2010) and environmental

Fig. 25.1 BMAA structure

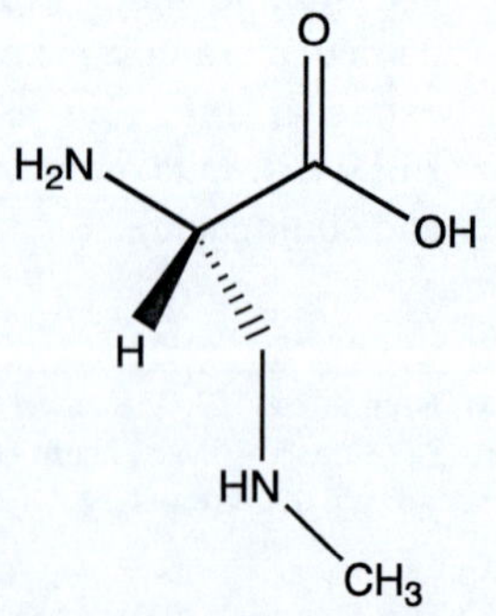

exposures can contribute to the development of neurodegenerative disorders (Murch et al. 2004b; Pablo et al. 2009).

BMAA is produced by representative species of cyanobacteria, diatoms, and dinoflagellates (Cox et al. 2005; Jiang and Ilag 2014; Lage et al. 2014). These species are distributed globally and present in terrestrial, brackish, freshwater, and marine habitats. BMAA has been detected in desert dust in the Persian Gulf (Cox et al. 2009) and in several water bodies, including urban waters in the Netherlands (Faassen et al. 2009), 11 freshwater lakes and one brackish waterbody in Great Britain (Metcalf et al. 2008); in cyanobacteria isolated from South African freshwater reservoirs (Esterhuizen and Downing 2008) in those present in an oasis of the Gobi desert (Craighead et al. 2009), and in marine and freshwater waters of Chinese ecosystems (Li et al. 2010).

Blooms of toxin-producing phytoplankton are also a recurring phenomenon in the Baltic Sea (Boesch et al. 2006; Gallon et al. 2002). Together with the revelation of BMAA in the brains of Canadian patients suffering from various neurodegenerative diseases (Pablo et al. 2009), these results suggested a possible widespread human exposure to the neurotoxin BMAA and its global implication in ALS and also in Alzheimer's disease and Parkinson. Dunlop et al. (Dunlop et al. 2013) reported that free BMAA can be mistakenly incorporated into human neuroproteins in place of serine in a cell line in vitro, and this incorporation into proteins was initially postulated as a mechanism for the bioaccumulation of BMAA in the Guam ecosystem.

Since the discovery of BMAA in 1967, the link between BMAA and neurodegenerative diseases such as amyotrophic lateral sclerosis (ALS) is still under discussion (Karamyan and Speth 2008; Lee and Mc Geer 2012).

25.2 Materials

- Ultracentrifuge 11,000 rpm, Beckman LT-55
- Eppendorf vials
- Gilson micropipette
- Bidistilled H_2O
- Vibra-Cell, Sonics & Materials Inc.
- Ultrasonic bath (Elgasonic Swiss made)
- 0.45 µm cellulose nitrate filters (Millipore, Inc.)
- ABRAXIS® BMAA, ELISA Assay, 96-test
- EZ Read 400 ELISA Reader

25.3 Methods

25.3.1 Toxin Extraction

Extract water samples as follows:

1. Obtain fresh phytoplankton aliquots (10–130 mg) by centrifugation of water samples.
2. Suspend the aliquots in 2 mL of sterile bidistilled water.
3. Stir the solution in Vortex.
4. Sonic it for 5 min. at 30–40 °C in an ultrasonic bath (Elgasonic Swiss made).
5. Centrifuge for 10 min. at 11,000 r.p.m. (Beckman L5–55 Ultracentrifuge) to eliminate debris.
6. Collect the supernatant.
7. Resuspend the pellet.
8. Repeat the whole process twice.
9. Pool the two supernatants.
10. Analyze for intracellular toxin.
11. Determine extracellular BMAA concentrations from the centrifugated water.

25.3.2 Toxin Determination with ELISA Abraxis Assay

The β-N-methylamino-L-alanine (BMAA) ELISA is an immunoassay for the quantitative and/or qualitative screening of BMAA in water samples. The BMAA ELISA allows for the analysis of 42 samples in duplicate determination. Less than 1 mL of sample is required.

The test can be performed in approximately 2 h. The test is a direct competitive ELISA based on the recognition of BMAA by specific antibodies. BMAA, when present in a sample, and a BMAA-HRP analog compete for the binding sites of the rabbit anti-BMAA antibodies in solution. The BMAA antibodies are then bound by a second antibody (goat anti-rabbit) immobilized on the wells of the microtiter plate. After a washing step and addition of the substrate solution, a color signal is generated. The intensity of the blue color is inversely proportional to the concentration of BMAA present in the sample. The color reaction is stopped after a specified time and the color is evaluated using an ELISA reader. The concentrations of the samples are determined by interpolation using the standard curve constructed with each run.

The limit of quantitation for BMAA (95% B/B_0) is approximately 4 ng/mL. The middle of the test (50% B/B_0) is approximately 100 ng/mL. Determinations closer to the middle of the calibration curve give the most accurate results. Sample concentration may be performed for samples requiring a lower limit of detection (technical bulletin available from Eurofins Abraxis by request). Sample extraction and clean-up are necessary for biological samples.

Test reproducibility: Coefficients of variation (CVs) for standards: <10%; CVs for samples: <15%.

Several literature report attempts to adapt extraction methods for instrumental analysis, in order to allow Elisa detection too (Clausi et al. 2016), but an absolute condition for ELISA analyses is the pH control of sample and extracted sample environment. Direct and indirect Elisa methods, in fact, require a pH environment usually between pH 6.0 and 9.0, or even more neutral (around 7.3), to allow antibody binding (Researchgate.net n.d.) and the ending of the reaction in algal toxins Elisa is generally performed just by adding a stop solution made by diluted sulfuric acid.

25.3.2.1 Reagents and Materials Provided

1. Microtiter plate (12 × 8 strips) coated with a secondary antibody, in a resealable aluminum pouch.
2. Lyophilized BMAA Standards: 0, 5, 25, 100, 250, 500 ng/mL (ppb), must be reconstituted before thoroughly. Reconstituted standards may be used for up to 1 month (store frozen).
3. Dilute the Wash Buffer (5×) Concentrate at a ratio of 1:5. If using the entire bottle (100 mL), add to 400 mL of deionized or distilled water and mix thoroughly.
4. The stop solution must be handled with care as it contains diluted H_2SO_4.

25.3.2.2 Working Scheme

The microtiter plate consists of 12 strips of 8 wells, which can be used individually for the test. The standards must be run with each test. Values of standards used in a previous test must never be reused.

1. Add 100 μL of the reconstituted standards or samples into the wells of the test strips according to the working scheme given. Analysis in duplicate or triplicate is recommended.
2. Add 50 μL of enzyme conjugate solution to the individual wells successively using a multi-channel pipette or a stepping pipette.
3. Add 50 μL of antibody solution to the individual wells successively using a multi-channel pipette or a stepping pipette. Cover the wells with parafilm or tape and mix the contents by moving the strip holder in a circular motion on the benchtop for 30 s. Be careful not to spill the contents.
4. Incubate the strips for 90 min at room temperature.
5. After incubation, remove the covering, decant the contents of the wells into a sink, and blot the inverted plate on a stack of paper towels.
   Wash the strips four times using the diluted wash buffer. Use at least a volume of 250 μL of 1X wash buffer for each well and each washing step. Blot the inverted plate after each wash step on a stack of paper towels. After the last wash/blot, check the wells for any remaining buffer in the wells, and if necessary, remove by additional blotting.
5. Add 150 μL of substrate (color) solution to the wells. Cover the wells with parafilm or tape and mix the contents by moving the strip holder in a circular motion on the benchtop for 30 s. Be careful not to spill the contents. Incubate the strips for 30 min at room temperature.
   Protect the strips from direct sunlight.
6. Add 100 μL of stop solution to the wells in the same sequence as for the substrate solution.
7. Read the absorbance at 450 nm using a microplate ELISA photometer within 15 min after the addition of stopping solution.

25.3.2.3 Evaluation

The evaluation of the ELISA can be performed using commercial ELISA evaluation programs such as 4-Parameter (preferred) or Logit/Log. For a manual evaluation, calculate the mean absorbance value for each of the standards.

1. Calculate the %B/B_0 for each standard by dividing the mean absorbance value for each standard by the Zero Standard (Standard 0) mean absorbance.
2. Construct a standard curve by plotting the %B/B_0 for each standard on the vertical linear (y) axis versus the corresponding BMAA concentration on the horizontal logarithmic (x) axis on graph paper. %B/B_0 for samples will then yield levels in ppb (or ng/mL) of BMAA by interpolation using the standard curve.

Results can also be obtained by using a spreadsheet macro available from Eurofins Abraxis upon request. The concentrations of the samples are determined using the standard curve run with each test. Sample extracts showing a lower concentration of BMAA than standard 1 (5 ppb) should be reported as containing <5 ppb of BMAA. Samples showing a higher concentration than standard 5 (500 ppb) must be diluted further with the provided sample diluent and re-analyzed. Semi-quantitative results can be derived by simple comparison of the sample absorbances to the absorbances of the standards. Samples with lower absorbances than a standard will have concentrations of BMAA greater than the concentration of that standard. Samples which have higher absorbances than a standard will have concentrations of BMAA less than that standard.

25.4 Conclusions

BMAA effects in human cell lines and in animals are consistent with the hypothesis that chronic exposure to this neurotoxin can trigger features of sporadic neurodegenerative diseases such as Parkinson's disease. In this regard, it is reasonable to hypothesize that BMAA may promote protein misfolding, mitochondrial dysfunction, and chronic innate immune activation in genetically susceptible individuals, leading to brain neurodegeneration (Nunes-Costa et al. 2020).

Bioaccumulation and biomagnification of BMAA has been demonstrated in various food webs. It is alarming as this intensification of BMAA will result in human exposure to higher concentrations from a direct cyanobacterial source. Many food items, fish and shellfish for example, have been identified as a source of BMAA and with the large variations in BMAA content. In a recent study (Esterhuizen-Londt and Pflugmacher 2019), some vegetables naturally irrigated with water containing a BMAA producing cyanobacterial bloom have been studied to evaluate BMAA uptake and accumulation of the toxin in edible plant tissue, emphasizing the need to further screen items for BMAA to understand the human exposure risk and to assess food safety evaluations.

References

Banack SA, Downing TG, Spácil Z, Purdie EL, Metcalf JS, Downing S, Esterhuizen M, Codd GA, Cox PA (2010) Distinguishing the cyanobacterial neurotoxin β-N-methylamino-L-alanine (BMAA) from its structural isomer 2,4-diaminobutyric acid (2,4-DAB). Toxicon 56:868–879

Boesch D, Hecky R, Chair CO, Schindler D, Seitzinger S (2006) Eutrophication of Swedish seas. Swedish Environmental Protection Agency, Valhallavägen

Brand LE, Pablo J, Compton A, Hammerschlag N, Mash DC (2010) Cyanobacterial blooms and the occurrence of the neurotoxin, beta-N-methylamino-L-alanine (BMAA), in South Florida aquatic food webs. Harmful Algae 9:620–635

Chiu AS, Gehringer MM, Welch JH, Neilan BA (2011) Does α-amino-β-methylaminopropionic acid (BMAA) play a role in neurodegeneration? Int J Environ Res Public Health 8:3728–3746

Clausi MT, Vita V, Bruno M, Franchino C, Trifirò G, Palumbo MP, Floridi F, De Pace R (2016) Validation of ELISA methods for search and quantification of β-N-methylamino-l-alanine in water and fish tissue. Int. J. Environ Anal Chem 96:1290

Cox PA, Banack SA, Murch SJ (2003) Biomagnification of cyanobacterial neurotoxins and neurodegenerative disease among the Chamorro people of Guam. Proc Natl Acad Sci U S A 100:13380–13383

Cox PA, Banack SA, Murch SJ, Rasmussen U, Tien G, Bidigare RR, Metcalf JS, Morrison LF, Codd GA, Bergman B (2005) Diverse taxa of cyanobacteria produce beta-N methylamino-L-alanine, a neurotoxic amino acid. Proc Natl Acad Sci U S A 102:5074–5078

Cox PA, Richer R, Metcalf JS, Banack SA, Codd GA, Bradley WG (2009) Cyanobacteria and BMAA exposure from desert dust: a possible link to sporadic ALS among Gulf War veterans. Amyotroph Lateral Scler 10(Suppl. 2):109–117

Craighead D, Metcalf JS, Banack SA, Amgalan L, Reynolds HV, Batmunkh M (2009) Presence of the neurotoxic amino acids beta-N-methylamino-L-alanine (BMAA) and 2,4-diamino-butyric acid (DAB) in shallow springs from the Gobi Desert. Amyotroph Lateral Scler 10(Suppl. S2): 96–100

Dunlop RA, Cox PA, Banack SA, Rodgers KJ (2013) The non-protein amino acid BMAA is misincorporated into human proteins in place of l-serine causing protein misfolding and aggregation. PLoS One 8:e75376

Esterhuizen M, Downing TG (2008) Beta-N-methylamino-L-alanine (BMAA) in novel south African cyanobacterial isolates. Ecotoxicol Environ Saf 71:309–313

Esterhuizen-Londt M, Pflugmacher S (2019) Vegetables cultivated with exposure to pure and naturally occurring β-N-methylamino-L-alanine (BMAA) via irrigation. Environ Res 169: 357–361

Faassen EJ, Gillissen F, Zweers HAJ, Lürling M (2009) Determination of the neurotoxins BMAA (β-N-methylamino-L-alanine) and DAB (α-, γ-diaminobutyric acid) by LC-MSMS in Dutch urban waters with cyanobacterial blooms. Amyotroph Lateral Scler 10:79–84

Gallon JR, Evans AM, Jones DA, Albertano P, Congestri R, Bergman B, Gundersen K, Orcutt KM (2002) Maximum rates of N2 fixation and primary production are out of phase in a developing cyanobacterial bloom in the Baltic Sea. Limnol Oceanogr 47:1514–1521

Jiang L, Ilag LL (2014) Detection of endogenous BMAA in dinoflagellate (*Heterocapsa triquetra*) hints at evolutionary conservation and environmental concern. PubRaw Sci 1:1–8

Jiao Y, Chen Q, Chen X, Wang X, Liao X, Jiang L, Wu J, Yang L (2014) Occurrence and transfer of a cyanobacterial neurotoxin β-methylamino-L-alanine within the aquatic food webs of Gonghu Bay (Lake Taihu, China) to evaluate the potential human health risk. Sci Total Environ 468-69: 457–463

Johnson HE, King SR, Banack SA, Webster C, Callanaupa WJ, Cox PA (2008) Cyanobacteria (*Nostoc commune*) used as a dietary item in the Peruvian highlands produce the neurotoxic amino acid BMAA. J Ethnopharmacol 118:159–165

Jonasson S, Eriksson J, Berntzon L, Spáčil Z, Ilag LL, Ronnevi LO, Rasmussen U, Bergman B (2010) Transfer of a cyanobacterial neurotoxin within a temperate aquatic ecosystem suggests pathways for human exposure. Proc Natl Acad Sci U S A 107:9252–9257

Karamyan VT, Speth RC (2008) Animal models of BMAA neurotoxicity: a critical review. Life Sci 82:233–246

Lage S, Costa PR, Moita T, Eriksson J, Rasmussen U, Rydberg SJ (2014) BMAA in shellfish from two Portuguese transitional water bodies suggests the marine dinoflagellate *Gymnodinium catenatum* as a potential BMAA source. Aquat Toxicol 152:131–138

Lee M, Mc Geer PL (2012) Weak BMAA toxicity compares with that of the dietary supplement beta-alanine. Neurobiol Aging 33:1440–1447

Li AF, Tian ZJ, Li J, Yu RC, Banack SA, Wang ZY (2010) Detection of the neurotoxin BMAA within cyanobacteria isolated from freshwater in China. Toxicon 55:947–953

Liu XQ, Rush T, Ciske J, Lobner D (2010) Selective death of cholinergic neurons induced by beta-methylamino-L-alanine. Neuroreport 21:55–58

Lobner D, Piana PMT, Salous AK, Peoples RW (2007) Beta-N-methylamino-L-alanine enhances neurotoxicity through multiple mechanisms. Neurobiol Dis 25:360–366

Metcalf JS, Banack SA, Lindsay J, Morrison LF, Cox PA, Codd GA (2008) Co-occurrence of beta-N-methylamino-L-alanine, a neurotoxic amino acid with other cyanobacterial toxins in British waterbodies, 1990-2004. Environ Microbiol 10:702–708

Murch SJ, Cox PA, Banack SA (2004a) A mechanism for slow release of biomagnified cyanobacterial neurotoxins and neurodegenerative disease in Guam. Proc Natl Acad Sci U S A 101:12228–12231

Murch SJ, Cox PA, Banack SA, Steele JC, Sacks OW (2004b) Occurrence of β-methylamin o-L-alanine (BMAA) in ALS/PDC patients from Guam. Acta Neurol Scand 110:267–269

Nunes-Costa D, Magalhães JD, G-Fernandes M, Cardoso SM, Empadinhas N (2020) Microbial BMAA and the pathway for Parkinson's disease neurodegeneration. Front Aging Neurosci 12: 26. https://doi.org/10.3389/fnagi.2020.00026

Pablo J, Banack SA, Cox PA, Johnson TE, Papapetropoulos S, Bradley WG, Buck A, Mash DC (2009) Cyanobacterial neurotoxin BMAA in ALS and Alzheimer's disease. Acta Neurol Scand 120:216–225

Rao SD, Banack SA, Cox PA, Weiss JH (2006) BMAA selectively injures motor neurons via AMPA/kainate receptor activation. Exp Neurol 201:244–252

Researchgate.net (n.d.). https://www.researchgate.net/post/Any_tips_for_ELISA_at_multiple_pHs#:~:text=pH%20affects%20the%20reactions%20in,before%20carrying%20out%20the%20ELISA.&text=The%20reaction%20between%20antibody%20and,fragment%20will%20bind%20your%20protein

Spáčil Z, Erkisson J, Jonasson S, Rasmussen U, Ilag LL, Bergman B (2010) Analytical protocol for identification of BMAA and DAB in biological samples. Analyst 135:127–132

Vega A, Bell EA (1967) α-amino-β-methylaminopropionic acid, a new amino acid from seeds of Cycas circinalis. Phytochemistry 6:759–762

26 Extraction and Quantification of BMAA from Fish Tissue

Milena Bruno and Valentina Messineo

Abstract

β-N-methylamino-L-alanine (BMAA) is a non-protein amino acid produced by several types of cyanobacteria found in freshwaters, marine waters, and soil. When ingested, BMAA damages and ultimately destroys motor neurons in the spinal cord, causing the same type of damage seen in patients with ALS, and causes neurofibrillary tangles in the spinal cord and brain, similar to those seen in Alzheimer's disease. Those suffering from this BMAA-induced damage are classified as suffering from amyotrophic lateral sclerosis/parkinsonism dementia complex (ALS/PDC). The symptoms of ALS/PDC include the varying degrees of the muscular paralysis of ALS, the muscular rigidity of Parkinson's, and/or the dementia of Alzheimer's and ultimately result in death. Humans may be exposed to BMAA through the ingestion of contaminated drinking water or foods. Drinking water may become contaminated with BMAA through the proliferation of BMAA-producing cyanobacteria in drinking water sources such as lakes and reservoirs. The majority of methods used for BMAA analysis employed chromatography (either gas or liquid) while a small percentage used other methods including capillary electrophoresis (CE) and ELISA kits among others. Though the earlier studies used fluorescence detection, most of the studies since 2008 used mass spectrometry (MS) or tandem mass spectrometry (MS/MS) as the detection method. This chapter explains the detection of BMAA through a particular sample extraction and ELISA Assay. In literature, BMAA was extracted from cells using solvent which in this method were avoided because ELISA, despite of mass spectrometry, is able to see the molecule in any state (free and conjugated).

M. Bruno (✉) · V. Messineo
FAST, Core Facilities, National Institute of Health, Rome, Italy
e-mail: milena.bruno@iss.it; valentina.messineo@iss.it

N. Thajuddin et al. (eds.), *Protocols for Cyanobacteria Sampling and Detection of Cyanotoxin*, https://doi.org/10.1007/978-981-99-4514-6_26

Keywords

Toxin · Neurotoxic compound · Amyotrophic lateral sclerosis · Alzheimer · Parkinson · Contaminated fish and seafood

26.1 Introduction

The presence of BMAA in cyanobacteria and other matrices has been controversial (Kruger et al. 2012; Faassen 2014). The existence of at least three natural isomers (Fig. 26.1), 2,4 diaminobutyric acid (DAB), N-2-amino-ethyl-glycine (AEG), and amino-N-methyl-alanine (BAMA), can be partially involved in the BMAA controversy and highlighted the need for highly selective methods to unambiguously qualify and quantify BMAA from its isomers. DAB is a neurotoxic isomer of BMAA (Kruger et al. 2012) that was first found in cyanobacteria in 2008 (Rosen and Hellenas 2008), but has also been widely reported in prokaryotes and eukaryotes (Banack et al. 2010a).

Among all known isomers of BMAA, AEG, and BAMA were selected from two groups for the development of detection methods because they could potentially interfere with the analysis of BMAA (Banack et al. 2011; Jiang et al. 2012). In fact, AEG has been found in cyanobacteria, and its production has been suggested to be highly conserved (Banack et al. 2010b) while BAMA has been observed in Baltic Sea molluscs (Jiang et al. 2014).

The toxicity of these two isomers has not been studied so far, especially for AEG. To date, two papers have reported BMAA in cyanobacteria by selective MS/MS methods using a derivatization of samples (Jiang et al. 2012; Berntzon et al. 2013).

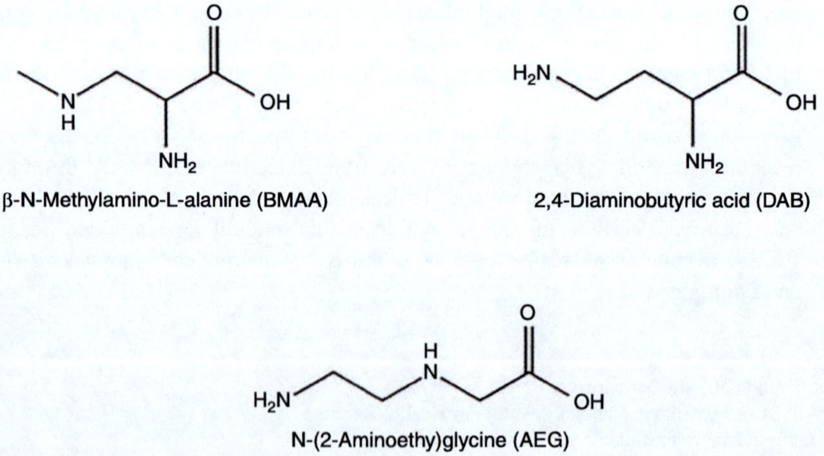

Fig. 26.1 BMAA and his isomers structure

Consumption of contaminated aquatic organisms and seafood is a possible route of human exposure to BMAA. Potential associations of sporadic ALS with BMAA present in aquatic organisms have been hypothesized in Chesapeake Bay, Maryland, USA (Field et al. 2013) and in the Thau lagoon, France (Masseret et al. 2013) with methods that use the derivatization of analytes.

Jonasson et al. (Jonasson et al. 2010) found that zooplankton, which naturally feeds on cyanobacteria, contains clearly higher levels of BMAA than BMAA producers.

In addition, several fish tissues were analyzed and contained concentrations up to 200 times higher of BMAA than cyanobacteria. The results obtained in this Baltic Sea study revealed, for the first time, that BMAA has been biotransferred to an aquatic ecosystem outside Guam.

The Baltic Sea is not the only source of commercial fish in Sweden, a country that contains many lakes, including some of the largest in Europe.

In the Baltic Sea, BMAA has been revealed in several commercial fish species, raising the question of BMAA bioaccumulation in Swedish lake systems. Here the presence of BMAA in water samples from Lake Finjasjon was revealed and bioaccumulation patterns were identified in both planktivorous/benthivorous and piscivorous fish, according to fish species, total weight, sex, and harvest season. For the first time, large numbers of fish individuals were used to draw conclusions on the bioaccumulation of BMAA in a closed ecological community based on a statistical approach.

It can be concluded that the feeding patterns (planktivorous/benthivorous) and the increase in the age of the fish can lead to a higher concentration of BMAA tissue. The suggested link between BMAA and the onset of neurodegenerative diseases and the identification of this cyanotoxin in aquatic organisms has stimulated research on the potential risk of human exposure associated with the supply of food products from eutrophic waterbodies around the world.

In the study of Clausi et al. (Clausi et al. 2016), the ELISA method for BMAA detection both in water and in fish tissue was validated and indicated as an easy, fast, and cheaper screening way to monitor this dangerous toxin. In this work, the fish tissue extraction was performed using TCA (Trichloroacetic acid) 0.1 N, then cleaned up (Water-Oasis 6 cc cartridge) to eliminate the acid solution and purified to avoid the interference of a low pH in the Elisa test.

In the following method, no acid solution is used since ELISA test is able to detect both free and bound BMAA.

26.2 Materials

- Volumetric Rotavapor flask (100 mL)
- Volumetric flask (5 mL)
- Paper filters
- Eppendorf vials
- Gilson micropipette

- Bidistilled H_2O
- 100% MeOH (HPLC grade)
- Homogenizer Ultra-Turrax T8, IKA Werke, Staufen, Germany
- Rotavapor Buchi
- Ultracentrifuge 11,000 rpm, Beckman LT-55
- Oasis HLB 3 cc Cartridge (Waters Corporation)
- ABRAXIS® BMAA, ELISA Assay, 96-test
- EZ Read 400 ELISA Reader

26.3 Methods

26.3.1 Tissue Extraction

1. Immediately homogenize tissues for 15'with 10 mL of 100% MeOH (HPLC grade).
2. Centrifuge it at 5000 g for 5′.
3. Filter the supernatant on a paper filter and collect it in a Rotavapor flask.
4. Repeat extraction on the solid sample (point 1).
5. Filter the sample directly on a paper filter and collect it in the same Rotavapor flask of the first extraction.
6. Wash the paper filter with 5 mL of 100% MeOH for three times (tot. 15 mL).
7. Bring the methanolic extract to a little volume of 1–2 mL by drying it in Rotavapor.
8. Put the 1–2 mL in a 5 mL volumetric flask by washing the Rotavapor flask with MeOH to reach the total amount (sample + MeOH) of 5 mL.
9. Pick up 1 mL of the extract.
10. Dilute it in 1 mL of bidistilled water.
11. Condition the cartridge as follows:
12. Apply 1 mL of 100% MeOH.
13. Apply 1 mL of bidistilled H_2O.
14. Then transfer the sample on the head of the cartridge.
15. Elute the sample with 1 mL of a 5% MeOH/H_2O solution.
16. Collect the sample with 1 mL of 100% MeOH.
17. Dry this fraction in Rotavapor.
18. Reconstitute the sample washing the Rotavapor flask with 2 mL of bidistilled H_2O.
19. The sample is ready to be tested with ELISA Assay.

26.3.2 Toxin Determination with ELISA Abraxis Assay

The β-N-methylamino-L-alanine (BMAA) ELISA is an immunoassay for the quantitative and/or qualitative screening of BMAA in water samples. The BMAA ELISA allows for the analysis of 42 samples in duplicate determination. Less than 1 mL of sample is required.

The test can be performed in approximately 2 h. The test is a direct competitive ELISA based on the recognition of BMAA by specific antibodies. BMAA, when present in a sample, and a BMAA-HRP analogue compete for the binding sites of the rabbit anti-BMAA antibodies in solution. The BMAA antibodies are then bound by a second antibody (goat anti-rabbit) immobilized on the wells of the microtiter plate. After a washing step and addition of the substrate solution, a color signal is generated. The intensity of the blue color is inversely proportional to the concentration of BMAA present in the sample. The color reaction is stopped after a specified time and the color is evaluated using an ELISA reader. The concentrations of the samples are determined by interpolation using the standard curve constructed with each run.

The limit of quantitation for BMAA (95% B/B_0) is approximately 4 ng/mL. The middle of the test (50% B/B_0) is approximately 100 ng/mL. Determinations closer to the middle of the calibration curve give the most accurate results. Sample concentration may be performed for samples requiring a lower limit of detection (technical bulletin available from Eurofins Abraxis by request). Sample extraction and clean-up are necessary for biological samples.

Test reproducibility: Coefficients of variation (CVs) for standards: <10%; CVs for samples: <15%.

Several literature report attempts to adapt extraction methods for instrumental analysis, in order to allow Elisa detection too (Clausi et al. 2016), but an absolute condition for ELISA analyses is the pH control of sample and extracted sample environment. Direct and indirect Elisa methods, in fact, require a pH environment usually, between pH 6.0 and 9.0, or even more neutral (around 7.3), to allow antibody binding (Researchgate.net n.d.), and the ending of the reaction in algal toxins Elisa is generally performed just by adding a stop solution made by diluted sulfuric acid.

26.3.2.1 Reagents and Materials Provided

1. Microtiter plate (12 × 8 strips) coated with a secondary antibody, in a resealable aluminum pouch.
2. Lyophilized BMAA Standards (6): 0, 5, 25, 100, 250, 500 ng/mL (ppb), must be reconstituted before thoroughly. Reconstituted standards may be used for up to 1 month (store frozen).
3. Dilute the Wash Buffer (5×) Concentrate at a ratio of 1:5. If using the entire bottle (100 mL), add to 400 mL of deionized or distilled water and mix thoroughly.
4. The stop solution must be handled with care as it contains diluted H_2SO_4

Working Scheme

The microtiter plate consists of 12 strips of 8 wells, which can be used individually for the test. The standards must be run with each test.

1. Add 100 μL of the reconstituted standards or samples into the wells of the test strips according to the working scheme given. Analysis in duplicate or triplicate is recommended.
2. Add 50 μL of enzyme conjugate solution to the individual wells successively using a multi-channel pipette or a stepping pipette.
3. Add 50 μL of antibody solution to the individual wells successively using a multi-channel pipette or a stepping pipette. Cover the wells with parafilm or tape and mix the contents by moving the strip holder in a circular motion on the benchtop for 30 s. Be careful not to spill the contents.
4. Incubate the strips for 90 min at room temperature.
5. After incubation, remove the covering, decant the contents of the wells into a sink, and blot the inverted plate on a stack of paper towels. Wash the strips four times using the diluted wash buffer. Use at least a volume of 250 μL of 1 × wash buffer for each well and each washing step. Blot the inverted plate after each wash step on a stack of paper towels. After the last wash/blot, check the wells for any remaining buffer in the wells, and if necessary, remove by additional blotting.
6. Add 150 μL of substrate (color) solution to the wells. Cover the wells with parafilm or tape and mix the contents by moving the strip holder in a circular motion on the benchtop for 30 s. Be careful not to spill the contents. Incubate the strips for 30 min at room temperature. Protect the strips from direct sunlight.
7. Add 100 μL of stop solution to the wells in the same sequence as for the substrate solution.
8. Read the absorbance at 450 nm using a microplate ELISA photometer within 15 min after the addition of stopping solution.

26.3.2.2 Evaluation

The evaluation of the ELISA can be performed using commercial ELISA evaluation programs such as 4-Parameter (preferred) or Logit/Log. For a manual evaluation, calculate the mean absorbance value for each of the standards.

1. Calculate the $\%B/B_0$ for each standard by dividing the mean absorbance value for each standard by the Zero Standard (Standard 0) mean absorbance.
2. Construct a standard curve by plotting the $\%B/B_0$ for each standard on the vertical linear (y) axis versus the corresponding BMAA concentration on the horizontal logarithmic (x) axis on graph paper. $\%B/B_0$ for samples will then yield levels in ppb (or ng/mL) of BMAA by interpolation using the standard curve.

Results can also be obtained by using a spreadsheet macro available from Eurofins Abraxis upon request. The concentrations of the samples are determined using the standard curve run with each test. Sample extracts showing a lower concentration of BMAA than standard 1 (5 ppb) should be reported as containing <5 ppb of BMAA. Samples showing a higher concentration than standard 5 (500 ppb) must be diluted further with the provided sample diluent and re-analyzed. Semi-quantitative results can be derived by simple comparison of the sample absorbances to the absorbances of the standards. Samples with lower absorbances than a standard will have concentrations of BMAA greater than the concentration of that standard. Samples which have higher absorbances than a standard will have concentrations of BMAA less than that standard.

26.4 Conclusions

The need to establish fast and simple methods (like this precise Elisa) for an ordinary monitoring of BMAA presence in the food web, is gradually reaching an increasing importance due to the more and more frequent detections of commercial fish and seafood contamination, with concurrent epidemiological evidences of Parkinson's, ALS, and Alzheimer disease diffusion in areas close to contaminated waterbodies (Fiore et al. 2020).

References

Banack SA, Caller TA, Stommel EW (2010a) The cyanobacteria derived toxin beta-N-methylamino-L-alanine and amyotrophic lateral sclerosis. Toxins 2:2837–2850

Banack SA, Downing TG, Spácil Z, Purdie EL, Metcalf JS, Downing S, Esterhuizen M, Codd GA, Cox PA (2010b) Distinguishing the cyanobacterial neurotoxin β-N-methylamino-L-alanine (BMAA) from its structural isomer 2,4-diaminobutyric acid (2,4-DAB). Toxicon 56:868–879

Banack SA, Metcalf JS, Spacil Z, Downing TG, Downing S, Long A, Nunn PB, Cox PA (2011) Distinguishing the cyanobacterial neurotoxin beta-N-methylamino-L-alanine [BMAA] from other diamino acids. Toxicon 57:730

Berntzon L, Erasmie S, Celepli N, Eriksson J, Rasmussen U, Bergman B (2013) BMAA inhibits nitrogen fixation in the cyanobacterium *Nostoc* sp. PCC 7120. Mar Drugs 11:3091–3108

Clausi MT, Vita V, Bruno M, Franchino C, Trifirò G, Palumbo MP, Floridi F, De Pace R (2016) Validation of ELISA methods for search and quantification of β-N-methylamino-l-alanine in water and fish tissue. Int J Environ Anal Chem 96(13):1290–1299. https://doi.org/10.1080/03067319.2016.1249480

Faassen EJ (2014) Presence of the neurotoxin BMAA in aquatic ecosystems: what do we really know? Toxins 6:1109–1138

Field NC, Metcalf JS, Caller TA, Banack SA, Cox PA, Stommel EW (2013) Linking beta-methylamino-L-alanine exposure to sporadic amyotrophic lateral sclerosis in Annapolis, MD. Toxicon 70:179–183

Fiore M, Parisi R, Filippini T, Mantione V, Platania A, Odone A, Signorelli C, Pietrini V, Mandrioli J, Teggi S (2020) Living near waterbodies as a proxy of cyanobacteria exposure and risk of amyotrophic lateral sclerosis: a population based case-control study. Environ ResEnviron Res 186:109530–109530

Jiang L, Aigret B, Borggraeve W, Spacil Z, Ilag L (2012) Selective LC-MS/MS method for the identification of BMAA from its isomers in biological samples. Anal Bioanal Chem 403:1719–1730

Jiang LY, Eriksson J, Lage S, Jonasson S, Shams S, Mehine M, Ilag LL, Rasmussen U (2014) Diatoms: a novel source for the neurotoxin BMAA in aquatic environments. PLoS One 9: e84578

Jonasson S, Eriksson J, Berntzon L, Spáčil Z, Ilag LL, Ronnevi LO, Rasmussen U, Bergman B (2010) Transfer of a cyanobacterial neurotoxin within a temperate aquatic ecosystem suggests pathways for human exposure. Proc Natl Acad Sci U S A 107:9252–9257

Kruger T, Oelmuller R, Luckas B (2012) The origin of β-N-methylamino-L-alanine (BMAA): cycads and/−or cyanobacteria? J Endocytobiosis Cell Res 22:29–36
Masseret E, Banack S, Boumédiène F, Abadie E, Brient L, Pernet F, Juntas-Morales R, Pageot N, Metcalf J, Cox P (2013) The French network on A.L.S.: investigation, dietary BMAA exposure in an amyotrophic lateral sclerosis cluster from Southern France. PLoS One 8:e83406
Researchgate.net (n.d.). https://www.researchgate.net/post/Any_tips_for_ELISA_at_multiple_pHs#:~:text=pH%20affects%20the%20reactions%20in,before%20carrying%20out%20the%20ELISA.&text=The%20reaction%20between%20antibody%20and,fragment%20will%20bind%20your%20protein
Rosen J, Hellenas KE (2008) Determination of the neurotoxin BMAA (beta-N-methylamino-L-alanine) in cycad seed and cyanobacteria by LC-MS/MS (liquid chromatography tandem mass spectrometry). Analyst 133:1785–1789

Postmortem Sampling of *Cyanobacteria* in the Case of Suspected Drowning

27

Viktor Soma Poór, Vivien Fejes, Dominika Szűcs, and Gábor Simon

Abstract

Drowning is one of the leading causes of accidental deaths worldwide. Diagnosis of drowning is often not possible based solely on the autopsy findings. Witness reports, camera records, or ancillary methods can help to evaluate the cause of death in these cases. The presence of *Cyanobacteria* or diatoms in the samples acquired from the victim during autopsy can provide useful information for the forensic pathologist. The amplification of Cyanobacteria-specific DNA offers superior sensitivity of the traditional diatom test.

In this protocol, we describe the necessary steps of sampling soft tissues (spleen, kidney, lung) and femoral bone marrow. Sternal bone marrow aspiration is also described. Compared to a routine forensic autopsy, some additional precautions should be taken to avoid contamination, which is also discussed. The collected tissue samples can be lysed using proteinase K digestion.

Keywords

Forensic · Drowning · Proteinase K · Autopsy · Cyanobacteria · Diatom · Bacillariophyceae

27.1 Introduction

Drowning is the third leading cause of unintentional injury death worldwide, accounting for 236.000 annual death according to WHO (WHO n.d.). Forensic pathological investigations of drowning-related fatalities are challenging, especially

V. S. Poór (✉) · V. Fejes · D. Szűcs · G. Simon
Department of Forensic Medicine, Medical School, University of Pécs, Pécs, Hungary
e-mail: viktor.s.poor@aok.pte.hu

N. Thajuddin et al. (eds.), *Protocols for Cyanobacteria Sampling and Detection of Cyanotoxin*, https://doi.org/10.1007/978-981-99-4514-6_27

regarding differentiating antemortem and postmortem submersion; hence, the former usually indicates an accidental death (drowning), but the latter suggests homicide. There are a lot of well-known and quite reliable macroscopic autopsy signs of drowning, such as foam in the airways, inflated lungs (emphysema aquosum), free liquid in the sphenoid sinuses (Svechnikov's sign), dilution of intestinal content (Schneppe et al. 2021). However, putrefaction and resuscitation attempts can make macroscopic autopsy findings uninterpretable. Therefore, establishing the diagnosis of drowning is often not possible based solely on the autopsy findings. Witness reports, camera records, or ancillary methods can help to evaluate the cause of death in these cases. One of the most used ancillary methods is the examination of the presence of microorganisms unique to certain drowning environments in the samples acquired from the victim during autopsy. Comparing the microorganisms detected in postmortem samples with those collected from the water at the possible death sites can also help to determine where the victim possibly died (which can be different in an arranged crime scene).Originally, these techniques were based on microscopic detection of diatom shells after nitric acidic digestion of the human tissues (Revenstorf 1904). Classic diatom test is known for a high level of false-positive results. Detection of *Cyanobacteria* offers several advantages: they are more abundant; their smaller size facilitates their entry through the alveolar membrane. Sensitivity is further increased by PCR amplification of specific DNA fragments (Nübel et al. 1997; Rácz et al. 2016).

Contamination leading to false positives, and technical issues leading to false negative results frequently raise concerns of misdiagnosis (Armstrong and Erskine 2018). Proper postmortem sampling methods must therefore apply during these investigations.

Organs mostly used for the forensic evaluation of the presence of waterborne organisms are shown in Fig. 27.1. Bone marrow is the most conventional tissue for testing, as it prevents contamination even in the case of advanced decay. However, the number of algae cells recovered from femoral bone marrow is generally lower than that of other organs. Recently, sternal bone marrow aspiration was suggested instead of femoral sampling because it yields mostly red bone marrow, with higher cell numbers (Szűcs et al. 2022). Aspirate sampling can be performed at the beginning of the autopsy procedure, decreasing the chance of accidental contamination. While the algae collected from the lungs will not prove drowning, their composition might help to connect the cadaver to the scene of drowning (Zhao et al. 2016; Coelho et al. 2016).

Several methods were developed for tissue lysis, mainly based on the classic diatom test, including nitric acid plus hydrogen peroxide or Soluene-350. These methods do not damage the silica shell of diatoms, but destroy *Cyanobacteria* cells, so we adopted a more gentle digestion protocol based on proteinase K (Ming et al. 2007).

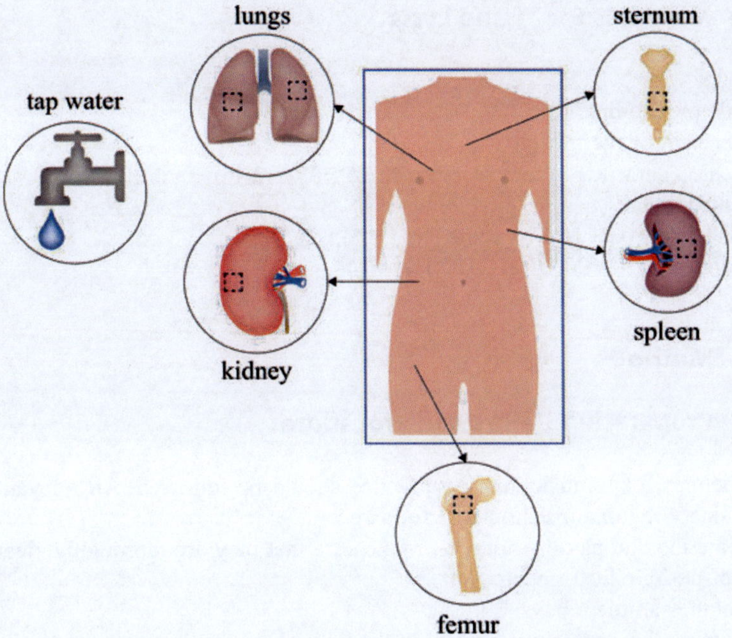

Fig. 27.1 Recommended samples for postmortem investigation of *Cyanobacteria*

27.2 Materials

27.2.1 Materials for Traditional Autopsy Sampling

- Sterilized or single-use autopsy scalpels
- Sterilized tweezers
- Clean oscillating bone saw
- Sterilized spatula
- 15 mL centrifuge tubes

27.2.2 Materials for Sternal Aspirate Sampling

- Sternal aspiration needle
- Syringe (at least 20 mL)
- 15 mL centrifuge tube
- Lysis buffer described in 27.2.3

27.2.3 Materials for Tissue Lysis

- Pipettes
- Sterile pipette tips
- Water bath or thermostat
- Benchtop centrifuge (compatible with 15 mL centrifuge tubes)
- Distilled water
- Lysis buffer: 0.01 M Tris buffer (pH 7.5), 2% SDS
- Proteinase K (~ 600 U/mL, ~20 mg/mL)

27.3 Methods

27.3.1 Preparation Before the Procedures

1. Guidelines for forensic autopsy practice should be followed. All relevant legal and safety regulations should be followed.
2. Prepare the autopsy instruments. Make sure that they are thoroughly cleaned or use disposable instruments.
3. Label the sample tubes.
4. Following the documentation, wash the body thoroughly, removing most of the potential contaminants.
5. Sample the water used in the autopsy room, in certain regions, tap water can contain *Cyanobacteria* or *Bacillariophyceae*, which can contaminate the samples.

27.3.2 Sample Collection During Autopsy

27.3.2.1 Sternal Aspirate Sampling

1. Sternal aspirate sampling should be performed at the beginning of the autopsy before the skin is cut above the sternum.
2. Clean the skin area over the sternum thoroughly with a clean sponge or tissue.
3. Palpitate the body of the sternum.
4. Set the depth limit of the sternal aspiration needle according to the thickness of the soft tissue over the sternum. It is important to ensure that the needle will reach the cavity of the sternum but should not go through it.
5. With a firm movement push it into the bone.
6. Remove the central trocar from the biopsy needle.
7. Attach the syringe to the end of the needle.
8. Generate a strong suction force by pulling the plunger of the syringe.
9. Wait until the red bone marrow enter the syringe.
10. Remove the syringe.
11. Eject the red bone marrow into the 15 mL centrifuge tube.

12. Wash the syringe with lysis buffer and eject it into the centrifuge tube.
13. Until further analysis, store the samples at 4 °C or – 20 °C.

27.3.2.2 Soft Tissue (Lung, Spleen, Kidney (Cortical Region))

1. Remove and separate the organs with conventional autopsy techniques (Pomara et al. 2010).
2. Excise the tissue sample from the lung (all lobes), spleen (convexity), and kidney (cortex) with a sterile scalpel.
3. The sample size for each organ should be at least 2 g.
4. To avoid contamination of algae from the skin, a new instrument or new blade should be used when cutting the internal organs.
5. Until further analysis, store the samples at 4 °C or – 20 °C.

27.3.2.3 Bone Marrow (Femur)

1. Make a longitudinal cut on the skin in the anterior femoral region.
2. Remove the soft tissues from the diaphysis of the femur without applying water for cleaning.
3. Make two transverse cuts on the diaphysis of the femur 5 cm apart with an oscillating autopsy saw. The cuts should reach the point of the largest diameter but should not exceed it significantly.
4. Make two longitudinal cuts connecting the transverse ones.
5. To avoid contamination, a new saw blade should be used.
6. Remove the cut-out bone, and scrape bone marrow with a spatula into a 15 mL tube.
7. Until further analysis, store the samples at 4 °C or – 20 °C.

27.3.3 Tissue Lysis

1. Prepare the lysis buffer.
2. Add ~2 g of tissue sample to 6 mL digestion buffer. (Fig. 27.2a)
3. Add 20 μL Proteinase K.
4. Incubate at 56 °C for 2 days or until it is completely digested (Fig. 27.2b). Shake it regularly.
5. Save the lysate for downstream analysis.
6. Certain *Cyanobacteria* and *Bacillariophyceae* taxa can withstand the proteinase K lysis (Gaget et al. 2017). In this case, they can be concentrated by centrifugation.
7. Centrifuge at ~1100 g for 15 min.
8. The pellet contains the intact cells (Fig. 27.2c).

Acknowledgment We would like to thank Józsefné Bedőcs for her excellent technical assistance.

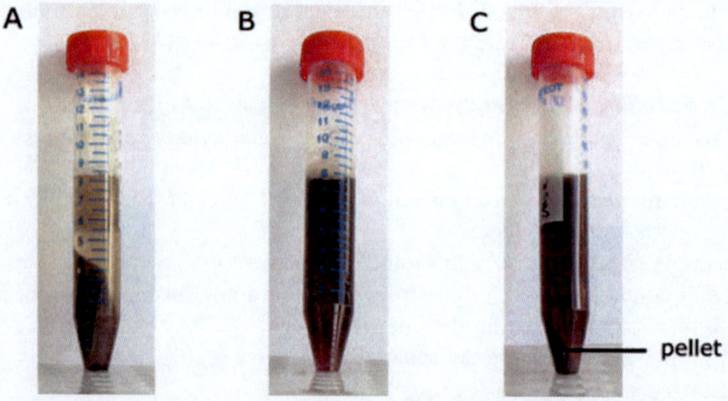

Fig. 27.2 Tissue lysis with proteinase K. (**a**) Tissue sample is added to 6 mL lysis buffer. (**b**) Completely digested tissue. (**c**) The digested tissue after centrifugation

References

Armstrong EJ, Erskine KL (2018) Investigation of drowning deaths: a practical review. Acad Forensic Pathol 1:8–43

Coelho S, Ramos P, Ribeiro C, Marques J, Santos A (2016) Contribution to the determination of the place of death by drowning - a study of diatoms' biodiversity in Douro river estuary. J Forensic Legal Med 41:58–64

Gaget V, Keulen A, Lau M, Monis P, Brookes JD (2017) DNA extraction from benthic cyanobacteria: comparative assessment and optimization. J Appl Microbiol 1:294–304

Ming M, Meng X, Wang E (2007) Evaluation of four digestive methods for extracting diatoms. Forensic Sci Int 1:29–34

Nübel U, Garcia-Pichel F, Muyzer G (1997) PCR primers to amplify 16S rRNA genes from cyanobacteria. Appl Environ Microbiol 8:3327–3332

Pomara C, Karch SB, Fineschi V (eds) (2010) Forensic autopsy. CRC Press

Rácz E, Könczöl F, Tóth D, Patonai Z, Porpáczy Z et al (2016) PCR-based identification of drowning: four case reports. Int J Legal Med 5:1303–1307

Revenstorf V (1904) Der Nachweis der aspirierten Ertränkungsflüssigkeit als Kriterium des Todes durch Ertrinken. Vierteljahrsschrift für Gerichtl Medizin und öffentliches Sanitätswes 27:274–289

Schneppe S, Dokter M, Bockholdt B (2021) Macromorphological findings in cases of death in water: a critical view on "drowning signs". Int J Legal Med 1:281–291

Szűcs D, Fejes V, Kozma Z, Poór VS, Sipos K (2022) Sternal aspirate sampling of Bacillariophyceae (diatoms) and cyanobacteria in suspected drowning cases. J Forensic Legal Med 85:102298

WHO n.d. WHO fact sheets - Drowning

Zhao J, Ma Y, Liu C, Wen J, Hu S et al (2016) A quantitative comparison analysis of diatoms in the lung tissues and the drowning medium as an indicator of drowning. J Forensic Legal Med 42: 75–78

Part II

Toxicity Analysis of Cyanotoxin

Detection of Cyanotoxins Using Vertebrate Bioassays in Mice

28

K. T. Dilrukshi, H. A. S. N. Abeysiri, K. P. A. Imanthi, P. A. K. C. Wijerathna, P. M. Manage, and S. A. Kulasooriya

Abstract

Cyanotoxins are diverse groups of secondary metabolites such as peptides, alkaloids, and lipopolysaccharides formed by numerous genera of cyanobacteria which are highly toxic to many organisms, including animals, plants, algae, and humans. Cyanotoxins, bioactive compounds, can be accumulated in marine and freshwater ecosystems and are responsible for potential environmental health threats. The detection of cyanotoxins in the aquatic environment is substantial to maintain a healthy ecosystem, and the biological assay method can be used as a cyanotoxin detection method. The bioassay is responsible for measuring the effect of cyanotoxins qualitatively and quantitatively on biological tissues, including vertebrate animals. Moreover, the most preferred bioassay for detecting cyanotoxins is mice bioassay due to total toxicity in drinking water can be estimated using a mice model as an in vivo study. Hence, this chapter summarizes the methods used to detect cyanotoxins using mice as a vertebrate animal model, and the toxicity measures based on clinical, biochemical, behavioural, and histopathological changes in mice tissues after ingestion of cyanotoxins.

Keywords

Cyanotoxins · Bioassay · Mice · Vertebrate · Toxicity

K. T. Dilrukshi · H. A. S. N. Abeysiri · K. P. A. Imanthi · P. A. K. C. Wijerathna · P. M. Manage
Centre for Water Quality and Algae Research, Department of Zoology, University of Sri Jayewardenepura, Nugegoda, Sri Lanka

Present Address: Faculty of Graduate Studies, University of Sri Jayewardenepura, Nugegoda, Sri Lanka

S. A. Kulasooriya (✉)
National Institute of Fundamental Studies, Kandy, Sri Lanka
e-mail: ananda@kulasooriya.com

N. Thajuddin et al. (eds.), *Protocols for Cyanobacteria Sampling and Detection of Cyanotoxin*, https://doi.org/10.1007/978-981-99-4514-6_28

28.1 Introduction

Cyanobacteria are photosynthetic bacteria frequently found in freshwater ecosystems (Saad and Atia 2014; Amezaga et al. 2014). Cyanobacteria play a key role in preserving the ecological balance and the biodiversity of microorganisms and higher organism communities (Sukenik and Kaplan 2021). The intense proliferation of cyanobacteria in waterbodies occurs due to anthropogenic eutrophication and climate changes, consequently triggering a phenomenon identified as bloom, which may compromise the quality of drinking and recreational water (Hallegraeff 2010; Sultana et al. 2022). Cyanobacteria produce cyanotoxins, secondary metabolites with significant toxicity levels during blooms (Pathmalal 2009). Cyanotoxins can impact humans, animals, and plants and are classified according to their toxicological target in hepatotoxins, cytotoxins, neurotoxins, and dermatoxins (Rastogi and Sinha 2009; Wijewickrama and Manage 2019). During a bloom, some cyanobacteria known to be toxic and others that do not produce toxins may be present together (Pathmalal 2019).

Cyanotoxins have been associated with adverse health effects in humans, which may be exposed to these compounds through oral, inhalation or dermal (Pathmalal 2019). Cyanotoxins bioaccumulate in the aquatic food chain and seriously threaten aquatic organisms across the trophic level (Ugya et al. 2019). The exposure of man and other terrestrial organisms to cyanotoxins results from the ingestion of drinking water contaminated with these toxins. However, man can also be exposed to cyanotoxins after feeding on seafood that contains an elevated amount of cyanotoxin accumulated in their tissues (Imam et al. 2020).

Furthermore, molecular, biochemical, and chemical techniques are the most frequent methodological approaches to detecting toxic cyanobacteria and their cyanotoxins (Bouteiller et al. 2022). Although numerous methods are available for uncovering cyanotoxins in aquatic environments, most methods are not easily accessible and require sophisticated laboratory expertise for usage (Imam et al. 2020; Zanchett and Oliveira-Filho 2013; Haque et al. 2017). The biological assay method is a type of cyanotoxin detection method that involves using living cells or tissue to determine cyanotoxin concentration in an aquatic environment (Imam et al. 2020). The assay is based on the qualitative and quantitative determination of the effect of cyanotoxin on biological tissue, including invertebrate animals, vertebrate animals, plants, and microorganisms (Imam et al. 2020).

Mouse bioassay has been intensively used during the last two decades and still is the most preferred bioassay due to total toxicity in drinking water can be estimated using a mice model as an in vivo study (Lawton and Edwards 2008; Lawton and Codd 1991). The first major drawback of using mouse assay is the need for an animal house facility for rearing the animals for routine experiments. Secondly, the use of animals in toxicity studies is against scientific ethics and is banned in most countries. Moreover, where more than one kind of cyanotoxin is present in the environment, the more rapid-acting toxin, such as microcystin-LR, may mask other symptoms (Imam et al. 2020). But the overall toxicity due to cyanobacteria can be estimated in drinking water supplies using mouse bioassays. Hence, this chapter summarizes the methods which can be used to detect cyanotoxins using mice as a vertebrate animal model.

28.2 Materials

- Distilled water
- Standards of cyanotoxins
- Sterilized centrifuge tubes
- Sterilized microcentrifuge tubes
- Metabolic cages
- Standard mouse cages
- Sterilized Sondhi needles
- 27 gauge needles with sterilized syringes
- Cotton wool
- Formalin
- Isopropyl alcohol
- CO_2
- Tissue papers
- Paraffin vax
- Haematoxylin, eosin stains
- Petri dishes

28.3 Methodology

28.3.1 Exposure to the Cyanotoxins in Mice Utilized in Vertebrate Bioassay

28.3.1.1 Ethics Approval

1. Obtain ethical approval for the experimental procedure and follow the experimental animal procedures following their guidelines.

28.3.1.2 Mice Model and Standard Conditions

1. Employ Swiss albino mouse grown under microbiologically controlled conditions.
2. Allow to acclimatize the mice to the animal house settings for 8–10 days.
3. Maintain the animal house conditions during the study period as: a 12:12 h light-dark cycle, a temperature of 20–26 °C, and a humidity of 40–70%.
4. Use standard mouse cages of 1500 $cm^2 \times 24$ cm, with autoclaved wooden shavings as bedding material.
5. Prepare and feed with a standard pellet diet for the Swiss albino mice with locally available food ingredients.

28.3.1.3 Design of the Experiment

1. Utilize distilled water and laboratory-made water samples to evaluate the effects of cyanotoxins on the organs of mice.
2. Prepare concentration series (at least three concentrations) from different cyanotoxins. For example, Microcystin-LR (MC-LR), Microcystin-RR (MC-RR), Nodularin (NOD), Cylindrospermopsin (CYN). Three concentrations as follows:
   - Concentration similar to WHO recommended level.
   - Concentration higher than the WHO recommended level.
   - Concentration lower than the WHO recommended level.

28.3.1.4 Sample Size Calculation

1. Sample size per group = 2 $SD^2 (Z_{\alpha/2} + Z_{\beta})^2/d^2$ (Charan and Kantharia 2013).
   SD - Standard deviation from the previous study by Thammitiyagodage et al. (2017).

   $Z_{\alpha/2} = Z_{0.05/2} = Z_{0.025} = 1.96$ (From Z table) at type 1 error of 5%.
   $Z_{\beta} = Z_{0.20} = 0.842$ (From Z table) at 80% power.
   d = effect size = difference between mean value.
   Hence,
   Sample size = $2\ SD^2\ (1.96 + 0.842)^2/d^2$.
   $= 2\ (10.37)^2\ (1.96 + 0.842)^2/17^2$.
   $= 5.84 \sim 6$.
   Expected attrition or death of animals: 10%.
   For 10% attrition ($6/0.9 = 6.7$) ~ 7
2. Recruit 7 mice per each group mentioned above.
3. Feed the mice from different groups with water samples prepared by diluting the toxin standards with distilled water for 90 days.
4. Feed the mice belonging to the control group with distilled water.

28.3.1.5 Sample Collection

Blood

1. Obtain the final blood sample after performing cardiac punctures on the animals at the day of sacrifice.
2. After centrifugation at 5000 rpm for 5 min, separate the serum from the blood and keep it at −20 °C until used for serum biochemistry analysis.

Urine

1. Place each mouse in a metabolic cage for 24 h to collect urine at days 0, 7, 14, 28, 42, 60, and 90 and store at −80 °C until the analysis is completed.

28.3.2 Determination of Cytotoxicity in Mice

28.3.2.1 Bodyweight Gain, Absolute Weight, and Relative Organ Weights (Clinical Assays)

1. Record the initial body weight of all individual mice before treatment with water samples.
2. Measure the body weights of each group of mice twice a week during the experiment.
3. Measure the final body weight of each mouse on the day of sacrifice and calculate the bodyweight gain by subtracting the final body weight from the initial body weight.
4. Determine the absolute and relative organ weights (liver and kidney).

$$\text{Relative organ weight} = \frac{\text{Fresh organ weight (g)}}{\text{Body weight (g)}} \times 100$$

28.3.2.2 Observation of Behavioural Changes

1. During the study period, record the daily water consumption of individual mice.
2. Examine each mouse routinely for indications such as decreased appetite, emaciation, hair loss, and behavioural abnormalities such as salivation, cannibalism, and lethargy.

28.3.2.3 Urine Analysis and Serum Biochemistry (Biochemical Assays)

Detection of Nephrotoxicity of Cyanotoxins

1. Measure serum creatinine levels in the blood samples using commercially available diagnostic kits.
2. Measure the creatinine levels in each mouse's urine using commercially available reagents.
3. Test the Kidney Injury Molecule-1 (KIM-1) level as a urinary biomarker to diagnose acute kidney damage using the collected urine samples.
4. Evaluate the expression of KIM-1 using an enzyme-linked immunosorbent assay (ELISA) kit.

Detection of Hepatotoxicity of Cyanotoxins

1. Investigate the levels of Glutamate Pyruvate Transaminase (GPT), Glutamate Oxaloacetate Transaminase (GOT), Alkaline Phosphatase (ALP), and Lactate Dehydrogenase (LDH) in serum using commercially available diagnostic kits

28.3.2.4 Histopathology

1. After 90 days, euthanize the mice and collect the final blood samples after sacrificing the animals by performing cardiac punctures.
2. To collect the organs (kidney and liver), dissect each mouse and then weigh the organs on a digital scale.

3. Bi-valve each kidney and examine both cut surfaces for colour changes, necrosis, or fibrosis.
4. Slice the liver and examine both cut surfaces for colour changes, necrosis, or fibrosis.
5. Fix the kidney and liver tissues in 10% formalin for 24–48 h, dehydrate in a series of ethanol concentrations, and clear by using xylene.
6. Then embed the tissues in paraffin and use the Haematoxylin and eosin stains to stain kidney and liver tissues.
7. Examine the slides under a light microscope.
8. Grade the significant levels of renal and liver histological lesions using scoring classification defined by (Toblli et al. 1999) and classify the lesions as mild, moderate, and severe.
9. Dispose of the animals according to the guidelines.

28.3.3 Statistical Analysis

1. Present the data as a mean and standard deviation.
2. Use one-way analysis of variance (ANOVA) to detect significant differences between groups with a p 0.05 limit.

References

Amezaga JM, Amtmann A, Biggs CA, Bond T, Gandy CJ, Honsbein A, Karunakaran E, Lawton L, Madsen MA, Minas K, Templeton MR (2014) Biodesalination: a case study for applications of photosynthetic bacteria in water treatment. Plant Physiol 164(4):1661–1676

Bouteiller P, Lance E, Guérin T, Biré R (2022) Analysis of Total-forms of Cyanotoxins Microcystins in biological matrices: a methodological review. Toxins 14(8):550

Charan J, Kantharia N (2013) How to calculate sample size in animal studies? J Pharmacol Pharmacother 4(4):303–306

Hallegraeff GM (2010) Ocean climate change, phytoplankton community responses, and harmful algal blooms: a formidable predictive challenge 1. J Phycol 46(2):220–235

Haque F, Banayan S, Yee J, Chiang YW (2017) Extraction and applications of cyanotoxins and other cyanobacterial secondary metabolites. Chemosphere 183:164–175

Imam TS, Tsagero MH, Ari HA, Aminu SB, Ugya AY (2020) Emerging approach to uncover cyanotoxins in aquatic environment: a concise review. Current World Environment 15(1):100

Lawton LA, Codd GA (1991) Cyanobacterial (blue-green algal) toxins and their significance in UK and European waters. Water and Environment Journal 5(4):460–465

Lawton LA, Edwards C (2008) Conventional laboratory methods for cyanotoxins. In: Cyanobacterial harmful algal blooms: state of the science and research needs. Springer, New York, NY, pp 513–537

Pathmalal, M., 2009. Hepatotoxic effects of Microcystis aeruginosa (PCC 7820) on Wister rats

Pathmalal, M., 2019. Cyanotoxins: a hidden cause of chronic kidney disease of unknown etiology (CKDu) in Sri Lanka-a review

Rastogi RP, Sinha RP (2009) Biotechnological and industrial significance of cyanobacterial secondary metabolites. Biotechnol Adv 27(4):521–539

Saad A, Atia A (2014) Review on freshwater blue-green algae (cyanobacteria): occurrence, classification and toxicology. Biosci Biotechnol Res Asia 11(3):1319–1325

Sukenik A, Kaplan A (2021) Cyanobacterial harmful algal blooms in aquatic ecosystems: a comprehensive outlook on current and emerging mitigation and control approaches. Microorganisms 9(7):1472

Sultana S, Awal S, Shaika NA, Khan S (2022) Cyanobacterial blooms in earthen aquaculture ponds and their impact on fisheries and human health in Bangladesh. Aquac Res 53(15):5129–5141

Thammitiyagodage MG, Gunatillaka MM, Ekanayaka N, Rathnayake C, Horadagoda NU, Jayathissa R, Gunaratne UK, Kumara WG, Abeynayake P (2017) Ingestion of dug well water from an area with high prevalence of chronic kidney disease of unknown etiology (CKDu) and development of kidney and liver lesions in rats. Ceylon Med J 62(1):20

Toblli JE, Ferder L, Angerosa M, Inserra F (1999) Effects of amlodipine on tubulointerstitial lesions in normotensive hyperoxaluric rats. Hypertension 34(4):854–858

Ugya AY, Imam TS, Ma J (2019) Mini-review on the efficacy of aquatic macrophytes as mosquito larvicide. J Appl Bot Food Qual 92:320–326

Wijewickrama MM, Manage PM (2019) Accumulation of microcystin-LR in grains of two rice varieties (Oryza sativa L.) and a leafy vegetable, Ipomoea aquatica. Toxins 11(8):432

Zanchett G, Oliveira-Filho EC (2013) Cyanobacteria and cyanotoxins: from impacts on aquatic ecosystems and human health to anticarcinogenic effects. Toxins 5(10):1896–1917

Detection of Cyanotoxin Toxicity in Fish

29

Ramakrishnan Kumar and D. Dhanasekaran

Abstract

Cyanobacteria can bloom under the right circumstances, a process marked by the increase of algal biomass and the potential production of cyanotoxins in aquatic habitats. Many cyanobacterial toxins can have a serious negative impact on health or even result in death in both humans and animals. One of the most prevalent causes of cyanotoxin poisoning in humans is drinking contaminated water. Other common causes include skin contact and ingesting water while participating in recreational activities. This protocol demonstrates how to determine the lethal concentration (LC_{50}) of acute toxicity, as well as the experimental design, growth parameters, histopathology, hematology, and serum biochemistry.

Keywords

Cyanobacteria · Cyanotoxin · Lethal concentration (LC_{50}) · Acute toxicity · Histopathology · Hematology

29.1 Introduction

Freshwater ecosystems are home to the majority of the photosynthetic microorganisms known as cyanobacteria. Cyanobacteria are generally known as blue-green bacteria. These are prokaryotic microorganisms that resemble bacteria,

R. Kumar (✉)
Department of Microbiology, School of Life Sciences, Bharathidasan University, Tiruchirappalli, Tamil Nadu, India

D. Dhanasekaran
Department of Microbiology, National Repository for Microalgae and Cyanobacteria - Freshwater (NRMC-F), Bharathidasan University, Tiruchirappalli, Tamil Nadu, India

N. Thajuddin et al. (eds.), *Protocols for Cyanobacteria Sampling and Detection of Cyanotoxin*, https://doi.org/10.1007/978-981-99-4514-6_29

but unlike bacteria, they are oxygen-evolving photosynthetic organisms rather than chemosynthetic organisms. Because of their lengthy evolutionary history, they have evolved to cope with climatic, geochemical, and human-caused changes (Duy et al. 2000). Cyanobacteria play a critical role in preserving the ecological balance as well as the biodiversity of lower and higher order microbial communities. Moreover, they have the ability to colonize salty or brackish water and can withstand temperatures that are both very low and very high. The high growth of cyanobacteria in waterbodies is being intensified by human eutrophication and climate change, leading to a condition known as bloom that may impair the quality of drinking and recreational water. It has been noted that cyanobacteria create poisonous secondary metabolites known as cyanotoxins during blooms (Sanseverino et al. 2017). Due to the nutrient enrichment of the water throughout the summer, cyanobacterial blooms have a tendency to develop, and a thick oily coating is deposited on the water's surface. These algal blooms give the water an unpleasant flavor and smell. The poisons are discharged into the water as this bloom degenerates, and the quantity of toxins may be too high for human consumption. The cyanobacterial species that produce toxins and are usually found in freshwaters are Lyngbya, Microcystis, Anabaena, Nodularia, Planktothrix, and Aphanizomenon. Cyanotoxins are classified based on their toxicological target and can affect humans, animals, and plants. These colonial or filamentous cyanobacteria produce a wide range of toxic substances that are diverse in their chemical and biological composition, including hepatotoxins (microcystin and nodularin), neurotoxins (anatoxins and saxitoxins), paralytic shellfish toxins (produced by Aphanizomenon), cytotoxins (cylindrospermopsin), and dermatotoxins (lipopolysaccharides) (Agrawal et al. 2012). The cytoplasm of cells, where the bulk of cyanotoxins are present, is intracellular. When an algal bloom decays, cyanobacteria often release the toxins contained inside their cells into the water, but in some species, living cells can also produce poisons (extracellular toxins). All of the poisons that are discharged into the water can bioaccumulate in the ecosystem and in creatures that are waterborne, where they can then be transferred to aquatic life and people. As poisons build up in fish, those who consume them may have symptoms that range from serious illnesses to death (Falfushynska et al. 2023). The invention of efficient extraction and analytical techniques capable of examining edible fish, shellfish, plants, and animals is required. To minimize health concerns and the frequency of blooms, national laws and guidelines that attempt to restrict cyanobacterial biomass in our rivers are required. In order to support regulatory policies and stop new risks to human and environmental health, it is also necessary to enhance early warning systems for cyanobacteria.

29.2 Material Required

29.2.1 Designing of the Experiment

1. Fish
2. Fiber, glass, or plastic tank

3. Ph meter
4. Purified water

29.2.2 Histopathology

1. Formalin
2. PBS
3. Automatic Embedding Machine
4. Microtome
5. Paraffin wax
6. Slide
7. Microscope

29.2.3 Hematology and Serum Biochemistry

1. Chloroform
2. Centrifuge
3. Syringe
4. EDTA tube
5. Freeze drier

29.3 Methods

29.3.1 Design of the Experiment

1. Freshwater fish are collected from the fishpond.
2. Transfer the fish into 1000-L fiber, glass, or plastic tanks where they are acclimatized and controlled diet is fed for 2 weeks.
3. Maintain the photoperiod in this experiment 16 h light to 8 h dark.
4. Check the water quality every 7 days during this experimental period.
5. Change the water in the experimental tank every morning 20–30%

29.3.2 Acute and Chronic Toxicity

1. Determine the lethal concentration (LC_{50}) value of acute toxicity in 96 h period.
2. A stratified random approach is used to different concentration cyanotoxin to assign fish to tanks at the beginning of each test among the five treatments and one control, each with two repetitions.
3. Observe the fish tank for the occurrence of dead fish and removed from tanks at 12-h intervals.

4. Find out the LC50 value to fix the higher concentrations and lower concentrations for the chronic toxicity experiment (Adelman et al. 2009).
5. The chronic toxicity assay is performed for the period of 30 days.
6. To examine the toxin accumulation in the fish, every 10 days, dissect the fish and collected the gills, intestine, liver, kidney, and muscle tissue and store in 10× PBS for histology, hematology examination.

29.3.3 Growth Rate

1. Growth evaluations is calculated the test chemical concentration that impacted fish development during a 30-day period.
2. During measurement, food is withheld for 24 h prior to the fish.
3. Fish are given 100 mg/L of chloroform to make them unconscious for the measurement.
4. Using the scale to measure the length and weight balance to measure the weight of the fish.

29.3.4 Histopathology

1. Collect the organs and store the tissue using 10% formalin.
2. The organ is fixed in paraffin wax using automatic embedding equipment.
3. Keep the blocks in the freezer for at least 24 h before cutting; use a microtome to cut the block in the size (5 μm).
4. Use hematoxylin-eosin stain in the organs and fix the slide.
5. After preparation of slide, use microscope to examine the organ for histology (Martins et al. 2018).

29.3.5 Hematology

1. Anesthetize fish using chloroform at a concentration of 100 mL/L.
2. Use 1 mL syringe to collect the blood from the caudal vein of the fish
3. Anticoagulant 10% ethylene diamine tetra-acetate (EDTA) tube is used to collect blood.
4. Analyze the number of white blood cells (WBC), red blood cells (RBC), hematocrit (Hct), hemoglobin (HB), mean corpuscular hemoglobin (MCH), mean corpuscular volume (MCV), and mean corpuscular hemoglobin concentration (MCHC) (Abdel-Aziz et al. 2020).

29.3.6 Serum Biochemistry

1. Anesthetize fish using chloroform at a concentration of 100 mL/L.
2. Use 1 mL syringe to collect the blood from the caudal vein of the fish.
3. After collection wait for 10 min for clotting of blood, then centrifuge at 9000 rpm for 10 min.
4. Collect the serum and freeze dry in −20°C for analyzing the biochemical test.
5. Test the serum to identify the total protein (g/dL), albumin (g/dL), serum globulin, plasma glucose, and creatinine using colorimetric method (Abdel-Aziz et al. 2020).

29.4 Observation

The Histopathology study shows normal in the control and experimental due to toxicity, the liver's architecture change. Hepatocytes frequently detached after 20–30 days of treatment. Parenchyma cells are damaged after the long exposure of toxicity.

Evaluation of growth rate is measured using the following equation:

$$\text{Survival of Percentage} = \frac{\text{Final no.of fish}}{\text{Initial no.of fish}} \times 100$$

Specific Growth Rate

$$= \frac{\text{Final average weight of the fish} - \text{Initial average weight of the fish}}{\text{Experimental trials duration}} \times 100$$

$$\text{Weight Gain Percentage} = \frac{\text{Final weight} - \text{Initial weight}}{\text{Initial weight}} \times 100$$

$$\text{Length Gain Percentage} = \frac{\text{Final length} - \text{Inital length}}{\text{Initial length}} \times 100$$

References

Abdel-Aziz M, Bessat M, Fadel A, Elblehi S (2020) Responses of dietary supplementation of probiotic effective microorganisms (EMs) in Oreochromis niloticus on growth, hematological, intestinal histopathological, and antiparasitic activities. Aquac Int 28:947–963

Adelman IR, Kusilek LI, Koehle J, Hess J (2009) Acute and chronic toxicity of ammonia, nitrite, and nitrate to the endangered Topeka shiner (Notropis topeka) and fathead minnows (Pimephales promelas). Environ Toxicol Chem 28(10):2216–2223

Agrawal M, Yadav S, Patel C, Raipuria N, Agrawal MK (2012) Bioassay methods to identify the presence of cyanotoxins in drinking water supplies and their removal strategies. Eur J Exp Biol 2(2):321–336

Duy TN, Lam PK, Shaw GR, Connell DW (2000) Toxicology and risk assessment of freshwater cyanobacterial (blue-green algal) toxins in water. In: Reviews of environmental contamination and toxicology: continuation of residue reviews, pp 113–185
Falfushynska H, Kasianchuk N, Siemens E, Henao E, Rzymski P (2023) A review of common cyanotoxins and their effects on fish. Toxics 11(2):118
Martins ML, Cardoso L, Furtado W, Tancredo KR, Lehmann NB, Figueiredo AB, Steckert LD, Silva KAG, Padua SB, Ferreira TH (2018) Histopathology guide for freshwater fish. Editora UFSC, Florianopolis
Sanseverino I, António DC, Loos R, Lettieri T (2017) Cyanotoxins: methods and approaches for their analysis and detection. Joint Research Centre

Detection of Cyanotoxins Using Invertebrate Bioassays in Brine Shrimp

30

V. Jeyanthi, Rajasekar Thirunavukkarasu, and Kumaran Subramanian

Abstract

Cyanobacteria are a group of Gram-negative bacteria form blooms in polluted water bodies. The presence of cyanobacterial blooms release toxins against vertebrate or invertebrate aquatic animals. It is essential to monitor microcystins (MCs) in aquatic species in order to assess the risk associated with human ingestion. The aim of this chapter is to assess the toxic effects of cyanotoxins including microcystins in brine shrimp *Artemia salina*. The brine shrimp lethality assay is performed for preliminary assessment of toxicity. Further, the sensitive and fast cyanotoxin detection methods such as enzyme-linked immunosorbent assay (ELISA) and protein phosphatase inhibition assay (PPIA) were also performed to determine the concentration of cyanotoxins.

Keywords

Cyanotoxins · Brine shrimp · Lethality · ELISA · PPIA

V. Jeyanthi
Department of Biotechnology, SRM Arts and Science College, Chennai, Tamil Nadu, India

R. Thirunavukkarasu (✉)
Centre for Drug Discovery and Development, Sathyabama Institute of Science and Technology, Chennai, Tamil Nadu, India
e-mail: rajasekar.cddd@sathyabama.ac.in

K. Subramanian
PG and Research Department of Microbiology, Sri Sankara Arts and Science College (Autonomous) Enathur, Kanchipuram, India

N. Thajuddin et al. (eds.), *Protocols for Cyanobacteria Sampling and Detection of Cyanotoxin*, https://doi.org/10.1007/978-981-99-4514-6_30

30.1 Introduction

Cyanobacteria (blue-green algae) are photosynthetic bacteria extensively found in both aquatic and terrestrial ecosystems. They frequently occur in brackish waters, rivers, lakes, and ponds. They are Gram-negative, free-living group of prokaryotes that belong to the family Cyanophyceae. They develop symbiotic relationships with lichens and plants. When conditions are favorable, cyanobacteria exist in vegetative forms; however, when conditions are unfavorable and anoxic, they produce heterocysts. Globally, aquatic habitats have seen an increase in cyanobacterial blooms due to eutrophication and continuous climate change (Lopes et al. 2010). This growth is hazardous because excessive cyanobacteria can affect the function of ecosystem. Numerous studies have shown that global warming and increased CO_2 levels due to the prevalence of cyanobacterial blooms around the world. Cyanobacterial species such as Microcystis, *Nodularia*, *Anabaena*, *Planktothrix*, *Aphanizomenon*, *Cylindrospermopsin*, and *Lyngbya* were commonly involved in the production of wide range of secondary metabolites known as toxins (Xie and Park 2007; Manubolu et al. 2018). Cyanobacteria are able to adapt to the environmental changes such as high temperature, pH, dissolved phosphorous, and nitrogen by the release of secondary metabolites. These metabolites may be toxic or non-toxic. Non-toxic secondary metabolites include phytohormones, siderophores, and mycosporine-like amino acids (MAAs) and scytonemin (Roy-Lachapelle et al. 2021). Toxic secondary metabolites are usually classified depending on the organs they affected. It includes microcystin, nodularin, neurotoxins, anatoxins, saxitoxins, paralytic shellfish toxins, cytotoxins, and dermatotoxins. Cyanotoxins are being responsible for both acute and chronic poisonings of animal and human. These toxins polluted the clean water and affected the ecosystem of aquatic animals, such as shrimps, fish, mussels, and snails. During cyanobacterial blooms, microcystins are typically the most common cyanotoxins found in drinking, fishing, and recreational waters (Schmidt et al. 2013). Different kinds of aquatic animals contaminated with microcystins have been reported by many investigators. Microcystin-RR and Microcystin-LR have received the most attention by researchers. Approximately 80 different microcystin variants have been discovered so far. Many bioassays have been employed for cyanotoxin detection and biochemical screening in cyanobacteria (Metcalf et al. 2001). There are several analytical techniques are available to evaluate microcystins. Most methods measure toxin concentrations, rather than cytotoxicity. Sensitive techniques include HPLC, ELISA, and protein phosphatase inhibition assays (Metcalf et al. 2001). They differ from each other in their sensitivity, specificity, usability, and their capacity to provide both qualitative and quantitative information of in vitro toxicity of cyanotoxins. ELISA provides a quick and accurate response for microcystins. Protein phosphatase enzymes control cell cycle events and dephosphorylate proteins. The activity of these enzymes is inhibited by microcystins. The protein phosphatase inhibition assays used for the detection of cyanotoxins. The brine shrimp *Artemia salina* is a representative species of toxicity studies. Nowadays, the cytotoxic effect of toxins is commonly assessed using the lethality assay to analyze the impact of cyanobacteria in aquatic habitats (Oh et al. 2022; Preece et al. 2015). This bioassay provides a valuable tool and serves as an inexpensive test. But, a precise and accurate method for detecting microcystins from zooplankton samples is still required. This chapter explains the in vitro cytotoxicity assays such as lethality, ELISA, and protein phosphatase inhibition assay for the microcystin detection in brine shrimp.

30.2 Methods

30.2.1 Preparation of Standard and Test Samples

Prepare the standard Microcystin-LR and cyanobacterial extract samples individually in the concentration of 1 mg/mL in 80% methanol.

Perform the serial dilution of standard stock and test samples in artificial sea water to give final concentrations of 40, 20, 10, 5, 2.5, 1.25 μg/mL Microcystin-LR and test samples separately.

30.2.1.1 *A. Salina* Lethality Assay (Sarah et al. 2017)

Materials Required

- Brine shrimp eggs
- Artificial sea water
- Pasteur pipette
- Microtiter plate
- Microscope

Procedure

1. Collect the *A. salina* brine shrimp eggs.
2. Allow to hatch in artificial sea water at 25 °C under continuous illumination and aeration.
3. After 24–26 h of incubation, the nauplii (first larval stage) will hatch.
4. Add 100 μL of fresh artificial sea water to the 96-well microtiter plate.
5. Transfer the hatched larvae using a Pasteur pipette (25 individuals/well).
6. Add 100 μL of the serially diluted standard Microcystin-LR and test cyanobacterial extracts to the microtiter plate.
7. Perform the experiment in triplicates. Use artificial sea water alone as the negative control.
8. Cover the microplates and incubate at 25 °C for 24 h.
9. Count the number of live larvae under the microscope and calculate the percentage of death using the mean of three replicates.

30.2.1.2 ELISA Assay (Carmichael and An 1999)

Materials Required

- Standard Microcystin-LR
- Brine shrimp tissue test samples
- Microcystin-LR Antibody Solution
- Microcystin-LR secondary antibody HRP Conjugate Solution
- Wash buffer (5×) Concentrate
- Substrate (Color) Solution (TMB)
- Stop Solution

Procedure

1. Homogenization protocol for the extraction of microcystins from brine shrimp as followed by Ohio (2007).
2. Add 50 μL of the standard Microcystin-LR and 50 μL of Brine shrimp test samples to the 96-well microtiter plate separately. Perform the experiment in triplicates.
3. Add 50 μL of the Microcystin-LR antibody solution to the wells individually using a multi-channel pipette.
4. Mix the contents of the microtiter plate in a circular motion for 30 s.
5. Incubate the plate for 90 min at room temperature.
6. Decant the contents of the wells and wash the well three times with 250 μL of 1 × wash buffer and blot it dry.
7. Add 100 μL of the Microcystin-LR secondary antibody-HRP enzyme conjugate solution. Again mix the contents in a circular motion for 30 s.
8. Incubate the plate for 30 min at room temperature.
9. Decant the contents of the wells and wash the well three times with 250 μL of 1 × wash buffer and blot it dry.
10. Add 100 μL of substrate (color) solution to the individual wells.
11. Mix the contents of the microtiter plate in a circular motion for 30 s.
12. Incubate the plate for 20–30 min at room temperature.
13. Add 50 μL of stop solution to the wells using a multi-channel pipette.
14. Measure the absorbance at 450 nm in an ELISA Plate reader within 15 min after the addition of the stopping solution.

30.2.1.3 Protein Phosphatase Inhibition Assay (Heresztyn and Nicholson 2001)

Materials Required

- Component A: pNPP (1 vial)
- Component B: Assay buffer
- Component C: 10× Lysis buffer
- Component D: Triton X-100
- Component E: Stop solution
- Component F: 1 M DTT

Procedure

Prepare Tissue Extract for Protein Phosphatase

1. Collect the brine shrimp having exposure to different concentrations of microcystins.
2. Homogenize the brine shrimp tissues in 1 × lysis buffer and 20 μL of Triton X-100.
3. Centrifuge the homogenized sample at 10,000 × g for 10 min at 4 °C.
4. Collect the supernatant to perform protein phosphatase assay.
5. For positive control, add unique protease inhibitors in different concentration before homogenization.

Prepare pNPP Reaction Mixture

6. Prepare the stock solution by dissolving pNPP with 250 μL of deionized water (component A). Store at −20 °C up to 3–4 weeks. Add 3 μL of 1 M DTT (component F) per mL of assay buffer (component B). Dilute pNPP stock solution (component A) 1:100 in DTT-containing assay buffer.

Protein Phosphatase Inhibition Assay

7. Add 50 μL of protein phosphatase containing brine shrimp sample per well in 96-well microtiter plate. Use buffers as a negative control.
8. Add 50 μL of pNPP reaction mixture per well in 96-well microtiter plate.
9. Then, mix the reagents for 30 s by gently shaking the microplate.
10. Measure the absorbance at 405 nm continuously and record data every 5 min for 30 min.

References

Carmichael WW, An J (1999) Using an enzyme linked immunosorbent assay (ELISA) and a protein phosphatase inhibition assay (PPIA) for the detection of microcystins and nodularins. Nat Toxins 7:377–385

Heresztyn T, Nicholson BC (2001) A colorimetric protein phosphatase inhibition assay for the determination of cyanobacterial peptide hepatotoxins based on the dephosphorylation of phosvitin by recombinant protein phosphatase. Environ Toxicol 16(3):242–252

Lopes VR, Fernandez N, Martins RF, Vasconcelos V (2010) Primary screening of the bioactivity of Brackish water cyanobacteria: toxicity of crude extracts to *Artemia salina* larvae and Paracentrotus lividus embryos. Mar Drugs 8:471–482

Manubolu M, Lee J, Riedl KM, Kua ZX, Collart LP, Ludsin SA (2018) Optimization of extraction methods for quantification of microcystin-LR and microcystin-RR in fish, vegetable, and soil matrices using UPLC–MS/MS. Harmful Algae 76:47–57

Metcalf JS, Bell SG, Codd GA (2001) Colorimetric immuno-protein phosphatase inhibition assay for specific detection of microcystins and nodularins of cyanobacteria. Appl Environ Microbiol 67(2):904–909

Oh KH, Beak KM, Shin Y, Cho YC (2022) Assessing the microcystins concentration through optimized protein phosphatase inhibition assay in environmental samples. J Microbiol 60(6): 602–609

Ohio EPA (2007) Preparation of tissue samples for extraction of organic and metals determination. Ohio EPA method 581.10, Cincinnati, OH, USA

Preece EP, Moore BC, Swanson ME, Hardy FJ (2015) Identifying best methods for routine ELISA detection of microcystin in seafood. Environ Monit Assess 187:12

Roy-Lachapelle A, Solliec M, Sauve S, Gagnon C (2021) Evaluation of ELISA-based method for total anabaenopeptins determination and comparative analysis with on-line SPE-UHPLC-HRMS in freshwater cyanobacterial blooms. Talanta 223(Pt 2):121802

Sarah QS, Anny FC, Misbahuddin M (2017) Brine shrimp lethality assay. Bangladesh J Pharmacol 12:186–189

Schmidt JR, Shaskus M, Estenik JF, Oesch C, Khidekel R, Boyer GL (2013) Variations in the microcystin content of different fish species collected from a eutrophic Lake. Toxins 5:992–1009

Xie L, Park H (2007) Determination of microcystins in fish tissues using HPLC with a rapid and efficient solid phase extraction. Aquaculture 271:530–536

31 Identification of Cyanobacterial Toxins from Lichen Thalli

Ulla Kaasalainen

Abstract

Lichens are symbioses with a lichenized fungus and photosynthetic green algae and/or cyanobacteria as main partners. Over 10% of lichen species are estimated to associate with cyanobacterial symbionts, most commonly of the genus *Nostoc*. Lichen-symbiotic *Nostoc* are known to produce an abundance of bioactive compounds, including hepatotoxic microcystin and nodularin. This chapter describes how to screen lichen specimens for the bioactive microcystin and nodularin producing gene cluster using PCR and how to identify microcystin and nodularin variants from lichen thalli using LC-MS/MS.

Keywords

Lichen · Symbiosis · *Nostoc* · Microcystin · Nodularin · *McyE* · Liquid chromatography · Mass spectrometry · LC-MS

31.1 Introduction

Lichensare symbiotic associations with a lichen-forming fungus and photosynthetic green algae and/or cyanobacteria (cyanobionts) as the most dominant organisms. In addition to being significant primary producers in many lichen-rich environments, lichens contribute to various ecosystems by, for example, fixing nitrogen, inhibiting erosion, and providing food and shelter for insects and other animals (Gotsch et al. 2016; Seaward 2001). It is estimated that over 10% of all lichen species have cyanobacterial symbionts, most commonly of the genus *Nostoc* (Rikkinen 2002).

U. Kaasalainen (✉)
Finnish Museum of Natural History, University of Helsinki, Helsinki, Finland
e-mail: ulla.kaasalainen@helsinki.fi

N. Thajuddin et al. (eds.), *Protocols for Cyanobacteria Sampling and Detection of Cyanotoxin*, https://doi.org/10.1007/978-981-99-4514-6_31

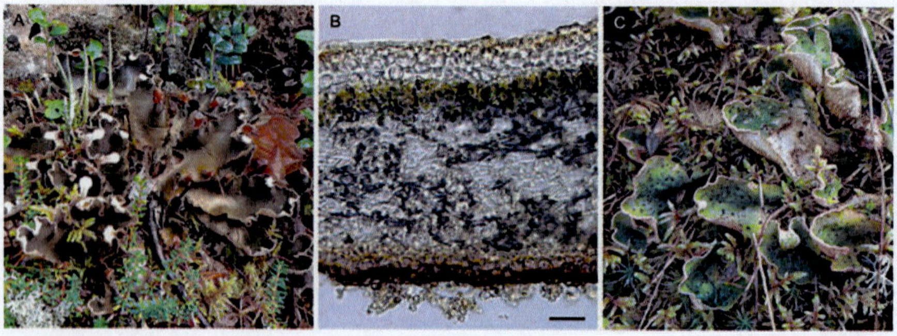

Fig. 31.1 Lichens with cyanobacterial symbionts. (**a**) *Peltigera malacea* has a cyanobacterial (*Nostoc*) photobiont. (**b**) Cross-section of a layered cyanolichen thallus with *Nostoc* cyanobacteria in a dark green photobiont layer between the dense cortex above and the looser medullar hyphae below, formed mainly by the lichenized fungus (scale 30 μm). (**c**) *Peltigera aphthosa* has a green algal primary photobiont and additionally cyanobacteria in separate structures (cephalodia), showing as dark spots on the light green upper surface

Most macro lichens have a layered (heteromerous) thallus structure where the photobionts, like cyanobacteria, are situated in a separate photobiont layer below the upper cortex (Fig. 31.1b). Some lichens with a green algal primary photobiont additionally have cyanobacteria in separate structures called cephalodia (Fig. 31.1c).

Cyanobacteria originating from various terrestrial symbioses, including several *Nostoc* strains isolated from lichens, produce an abundance of bioactive compounds (Hrouzek et al. 2011; Magarvey et al. 2006; Kaasalainen et al. 2009; Kaasalainen et al. 2012; Oksanen et al. 2004). Especially microcystins are relatively frequently present in cyanolichens, and concentrations over 0.2 mg/g and more than 50 structural variants of microcystins and three variants of nodularin have been detected from lichen thalli (Kaasalainen et al. 2012). Microcystins and nodularins are cyclic peptides produced by a number of cyanobacterial genera, especially in aquatic environments (Sivonen 2009). They are potent inhibitors of eukaryotic protein phosphatases and highly toxic (MacKintosh et al. 1990; Honkanen et al. 1991). Microcystins are heptapeptides with a typical structure of D-Ala^1-X^2-D-MeAsp^3-Z^4-Adda^5-D-Glu^6-Mdha^7, denoted as MC-XZ, where X and Z are varying L-amino acids, most commonly leucine^2 and arginine^4 (MC-LR) or double arginine (MC-RR), and Adda the characteristic 3*S*-amino-9*S*-methoxy-2*S*,6,8*S*-trimethyl-10-phenyldeca-4*E*,6*E*-dienoic acid moiety. Nodularins, on the other hand, comprise of five amino acids and lack the residues in microcystin positions 1 and 2.

The synthesis of microcystins and nodularins is directed by a hybrid non ribosomal peptide synthetase–polyketide synthase (NRPS-PKS) enzyme complex, encoded by the *mcy* gene complex (Rouhiainen et al. 2004; Tillett et al. 2000; Fewer et al. 2013; Christiansen et al. 2003). The sporadic distribution of microcystin and nodularin production among cyanobacteria is thought to be explained by multiple losses of the gene cluster but also horizontal gene transfer has been discussed as a possible mechanism (Rantala et al. 2004; Mikalsen et al. 2003; Christiansen et al. 2008; Jungblut and Neilan 2006). Additionally, similar microcystin variants have evolved independently in different lineages of cyanobacteria (Shishido et al. 2013). The genes encoding nodularin synthetase seem to be derived from those encoding microcystin synthetase (Rantala et al. 2004; Moffitt and Neilan 2004). The characteristic Adda is essential for the toxicity and show only little variation between microcystin and nodularin variants (Kaasalainen et al. 2012; Sivonen and Jones 1999; Neffling 2010). Hence, the microcystin synthetase gene E (*mcyE*) and the corresponding nodularin synthetase gene F (*ndaF*), involved in the synthesis of Adda and the formation of the bond between Adda and *D*-glutamate (Rouhiainen et al. 2004; Tillett et al. 2000), have been used to screen for potentially hepatotoxin-producing cyanobacteria (Rantala et al. 2008). However, the frequency of inactive *mcy* genotypes, i.e., strains that do not detectably produce the toxins even though they possess the biosynthetic genes, seems to vary between cyanobacterial genera (Mikalsen et al. 2003; Christiansen et al. 2008; Christiansen et al. 2006).

For the qualitative and quantitative analysis of microcystin and nodularin variants from different matrices, liquid chromatography-mass spectrometry (LC-MS) in an established and effective method (Neffling 2010; Schreidah et al. 2020; Massey et al. 2020). However, LC-MS analysis of many samples may become laborious and hence a preliminary screening for the presence of the biosynthetic genes is recommendable. When over 800 cyanolichen specimens from different parts of the world were screened for the *mcyE /ndaF* gene and analysed for microcystins and nodularins using LC-MS/MS, the gene was obtained from 12% of the specimens of which approximately half also contained detectable amounts of the toxins (Kaasalainen et al. 2012).

31.2 Materials

- Drying oven
- Freezer (−18 °C)
- Paper bags
- Sealable plastic bags
- Preparation microscope
- Basic DNA laboratory facilities for PCR (incl. micropipettes, centrifuge, thermocycler, gel electrophoresis, and imaging equipment and materials)
- Sterile needles or a scalpel
- 1.5 mL micro centrifuge tubes
- DNA extraction kit (e.g., GeneJet Genomic DNA Purification Kit, Thermo Fisher Scientific)
- Small pestles (fitted for 1.5 mL micro centrifuge tubes)
- dNTPs
- DNA polymerase (e.g., Phusion High-Fidelity DNA Polymerase from Thermo Fisher Scientific)
- DMSO
- Primers (for amplification): mcyEdgnF (5′-tcaacaggaaayccyaaaggag-3′) mcyEdgnR (5′-gaccaaccatcdraaatgatatggtgcat-3′) (Fewer et al. 2007)

- Primers (for sequencing): mcyE-F2 (5′-gaaatttgtgtagaaggtgc-3′), mcyE-R4 (aattctaaagcccaaagacg) (Rantala et al. 2004); pEintF (5′-aatataactatcaagaacc-3′), pEintR (5′-cctaaatctccggttctaaa-3′) (Kaasalainen et al. 2012)
- Positive control (e.g., HAMBI/UHCC 0152: *Nostoc* sp. 152 (Sivonen et al. 1990))
- 1 kb DNA ladder
- Freeze drier
- Cell disruptor
- HPLC-MS/MS facilities (e.g., ion trap mass detector with electrospray ionization)
- C18 HPLC column (e.g., Phenomenex Luna C18, 150 by 2.0 mm, 5 μm)
- Small mortar
- Methanol
- Glass beads (~0.5 mm in diam.)
- Syringes
- 0.2-μm-pore-size syringe filters
- Formic acid
- 2-propanol

31.3 Methods

31.3.1 Specimen collection and storage

1. Collect specimens in paper bags. To provide sufficient material for the analyses, at least several square centimetres of thallus should be collected.
2. Especially in humid conditions, dry the specimens, for example, in a drying oven, in 30 °C with good ventilation.
3. Store the dry specimens in a freezer (−20 °C) in sealed plastic bags.

(Note: With dry specimens and in relatively dry conditions, specimens can also be stored in paper bags in room conditions. With moist specimens and/or humid conditions, efficient drying and storage in a freezer is highly recommendable.)

31.3.2 Screening for the microcystin synthetase gene E (*mcyE*) and the nodularin synthetase gene F (*ndaF*)

31.3.2.1 DNA extraction

1. Of a lichen with a cyanobacterial main symbiont (Fig. 31.1a, b), select a small fragment (~2 × 2 mm) of lichen thallus for DNA extraction under a preparation microscope using a needle or a scalpel. Remove any additional structures (rhizines, tomentum) from the lower side. With a species with cephalodia (Fig. 31.1c), select 1–5 cephalodia. Place the material in a 1.5 mL micro centrifuge tube.
2. Extract DNA using the GeneJet Genomic DNA Purification Kit following the manufacturer's protocol for Gram-negative bacteria (protocol D). Start from step 2 and incorporate the following modifications:

3. After adding the adding of Digestion Solution (step 2), use a pestle to mash the sample and to produce the needed suspension. (Note: In case large intact pieces remain, these should be removed as they may cause issues with the filters in later steps.) Proceed with adding the Proteinase K.
4. Elute (step 10) with 100 μL of Elution Buffer or, in case of small cephalodia, with 50 μL.

31.3.2.2 Screening of the *mcyE/ndaF* (Kaasalainen et al. 2012)

1. Prepare the screening PCR for the *mcyE/ndaF* gene: With a total volume of 30 μL, include 1 μL of genomic DNA, 200 μM dNTPs, 0.5 μM of each of the primers mcyEdgnF and mcyEdgnR, 0.5 U of Phusion high-fidelity DNA polymerase, and 3% (vol/vol) of DMSO. Use, for example, a *Nostoc* strain known to produce microcystins as a positive control.
2. Thermocycle: Initial denaturation of 1 min at 98 °C follower by 35 cycles of 10 s at 98 °C, 30 s at 59 °C (annealing), and 1 min 45 s at 72 °C, with a final extension of 10 min at 72 °C.
3. Check success with gel electrophoresis.
4. For the sequencing of the samples interpreted as microcystin/nodularin positive based on the screening, repeat the PCR in 60 μL total volume and with 2 μL of genomic DNA and 1 U of enzyme. Sequencing can be performed with the primers mcyE-F2, mcyE-R4, pEintF, and pEintR.

(Note: (1) Optimize the PCR using the positive control if necessary. (2) For the molecular identification of the lichen species, the same DNA extraction can be used. Nuclear internal transcribed spacer (ITS) is an efficient identifying genetic marker for most large cyanolichen species. ITS can most often be amplified and sequenced with primers ITS1 (5′-tccgtaggtgaacctgcgg-3′) or ITS5F (5′--ggaagtaaaagtcgtaacaagg-3′) paired with ITS4 (5′-tcctccgcttattgatatgc-3′) (White et al. 1990).)

31.3.3 HPLC-MS/MS analysis of microcystins and nodularins (Kaasalainen et al. 2009)

1. Select a good-sized piece of clean lichen thallus (or cephalodia) for extraction, removing any unnecessary parts (e.g., rhizines, tomentum).
2. Freeze dry the sample, homogenize with a mortar, and store in −18 °C.
3. Use 10–150 mg of the freeze-dried sample with 1.0 mL of methanol and shake with glass beads in a cell disruptor three times for 20 s.
4. Centrifuge at 10600 × g for 5 min and filter the supernatant with a 0.2-μm-pore-size syringe filter.
5. Analyse the extracts using LC-MS/MS with a C18 column:
    - Mobile phase of 0.1% formic acid in water (A) and 0.1% formic acid in 2-propanol (B) (gradient from 5% B to 100% B over 50 min, flow rate of 0.15 mL/min at 40 °C).
    - Ionization in positive mode.
    - Spectra should be recorded from 50 to 1200 m/z.

Table 31.1 The main microcystin and nodularin variants detected from lichens. Variants often detected from the same lichen thalli are listed together in the order of relative intensity; only variants with a relative intensity commonly >10% are included. *N* = number of specimens containing at least the main variant (Kaasalainen et al. 2009; Kaasalainen et al. 2012)

	m/z [M + H]$^+$	N	Lichen genera	Geography
[Leu^1]MC-LR	1037	19	*Peltigera*	Finland, USA, Norway, Sweden, Japan
[Leu^1]MC-HtyrR	1101			
[ADMAdda^5]MC-RR	1066	10	*Nephroma, Peltigera*	Scotland, Norway
[Asp^3, ADMAdda^5] MC-LR	1009			
[Asp^3, ADMAdda^5] MC-RR	1052			
[Asp^3, DMAdda^5] MC-LR	967	3	*Nephroma, Lobaria, Peltigera*	Scotland, Svalbard
Nod-R	825	3	*Sticta, Peltigera*	Kenya, Argentina
[Leu^1, ADMAdda^5] MC-LR	1065	3	*Nephroma, Peltigera*	Argentina, China
MC-RR	1038	2	*Sticta*	USA
[Asp^3]MC-RR	1024			
[ADMAdda^5]MC-LR	1023	2	*Peltigera*	Finland, Kenya
[DMAdda^5]mc-LR	981			
[Asp^3, ADMAdda^5] mc-LR	1009			
[Leu^1, Asp^3]MC-LR	1023	1	*Peltigera*	China

(Note: Optimize the analysis using the positive control. Further analysis details available in (Kaasalainen et al. 2009).)

31.4 Observation

Especially in humid tropical conditions, active drying and freezing significantly enhances the preservation of specimens in comparison to specimens stored in normal indoor conditions. For example, the percentage of successfully sequenced fungal ITS regions rose from as low as <20% to up to over 95% with the suggested preservation and storage among several hundred lichen specimens collected from East Africa. However, the photobiont (cyanobacteria, green algae) genetic regions seem to be more persistent and the sequencing success less affected by storage conditions than the genetic regions of the lichenized fungi.

Microcystin variants present in lichens often differ from the basic MC-LR microcystin structure. Typical modifications include ADMAdda (9-*O*-acetylDMAdda) or DMAdda (9-*O*-desmethylAdda) in the fifth position, leucine in the first position, aspartic acid in the third position, and arginine in the variable second position (Kaasalainen et al. 2009; Kaasalainen et al. 2012; Kaasalainen 2012; Kaasalainen et al. 2013) (Table 31.1; Fig. 31.2).

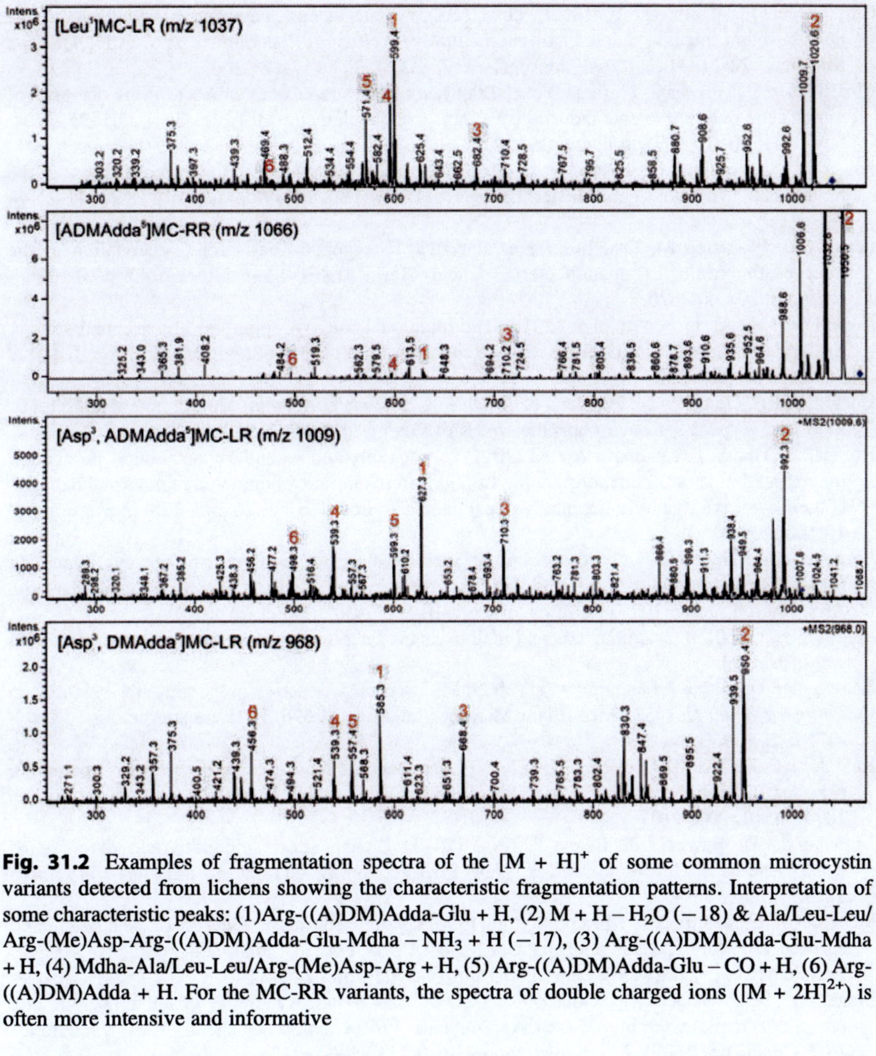

Fig. 31.2 Examples of fragmentation spectra of the $[M + H]^+$ of some common microcystin variants detected from lichens showing the characteristic fragmentation patterns. Interpretation of some characteristic peaks: (1)Arg-((A)DM)Adda-Glu + H, (2) M + H − H_2O (−18) & Ala/Leu-Leu/Arg-(Me)Asp-Arg-((A)DM)Adda-Glu-Mdha − NH_3 + H (−17), (3) Arg-((A)DM)Adda-Glu-Mdha + H, (4) Mdha-Ala/Leu-Leu/Arg-(Me)Asp-Arg + H, (5) Arg-((A)DM)Adda-Glu − CO + H, (6) Arg-((A)DM)Adda + H. For the MC-RR variants, the spectra of double charged ions ($[M + 2H]^{2+}$) is often more intensive and informative

Acknowledgement The described protocols are largely based on work done for my doctoral thesis, and I would like to acknowledge the support received especially from Jouni Jokela, David Fewer, Matti Wahlsten, Kaarina Sivonen, and Jouko Rikkinen (University of Helsinki).

References

Christiansen G, Fastner J, Erhard M et al (2003) Microcystin biosynthesis in planktothrix: genes, evolution, and manipulation. J Bacteriol 185:564–572. https://doi.org/10.1128/JB.185.2.564-572.2003

Christiansen G, Kurmayer R, Liu Q et al (2006) Transposons inactivate biosynthesis of the nonribosomal peptide microcystin in naturally occurring *Planktothrix* spp. Appl Environ Microbiol 72:117–123. https://doi.org/10.1128/AEM.72.1.117-123.2006

Christiansen G, Molitor C, Philmus B et al (2008) Nontoxic strains of cyanobacteria are the result of major gene deletion events induced by a transposable element. Mol Biol Evol 25:1695–1704. https://doi.org/10.1093/molbev/msn120

Fewer DP, Rouhiainen L, Jokela J et al (2007) Recurrent adenylation domain replacement in the microcystin synthetase gene cluster. BMC Evol Biol 7:183. https://doi.org/10.1186/1471-2148-7-183

Fewer DP, Wahlsten M, Österholm J et al (2013) The genetic basis for O-acetylation of the microcystin toxin in cyanobacteria. Chem Biol 20:861–869. https://doi.org/10.1016/j.chembiol.2013.04.020

Gotsch SG, Nadkarni N, Amici A (2016) The functional roles of epiphytes and arboreal soils in tropical montane cloud forests. J Trop Ecol 32:455–468. https://doi.org/10.1017/S026646741600033X

Honkanen RE, Dukelow M, Zwiller J et al (1991) Cyanobacterial nodularin is a potent inhibitor of type 1 and type 2A protein phosphatases. Mol Pharmacol 40:577–583

Hrouzek P, Tomek P, Lukešová A et al (2011) Cytotoxicity and secondary metabolites production in terrestrial Nostoc strains, originating from different climatic/geographic regions and habitats: is their cytotoxicity environmentally dependent? Environ Toxicol 26:345–358. https://doi.org/10.1002/tox.20561

Jungblut A-D, Neilan BA (2006) Molecular identification and evolution of the cyclic peptide hepatotoxins, microcystin and nodularin, synthetase genes in three orders of cyanobacteria. Arch Microbiol 185:107–114. https://doi.org/10.1007/s00203-005-0073-5

Kaasalainen U (2012) Cyanobacteria and their toxins in lichen symbiosis. Doctoral thesis,. University of Helsinki

Kaasalainen U, Jokela J, Fewer DP et al (2009) Microcystin production in the tripartite cyanolichen *Peltigera leucophlebia*. Mol Plant-Microbe Interact 22:695–702. https://doi.org/10.1094/MPMI-22-6-0695

Kaasalainen U, Fewer DP, Jokela J et al (2012) Cyanobacteria produce a high variety of hepatotoxic peptides in lichen symbiosis. Proc Natl Acad Sci U S A 109:5886–5891. https://doi.org/10.1073/pnas.1200279109

Kaasalainen U, Fewer DP, Jokela J et al (2013) Lichen species identity and diversity of cyanobacterial toxins in symbiosis. New Phytol 198:647–651. https://doi.org/10.1111/nph.12215

MacKintosh C, Beattie KA, Klumpp S et al (1990) Cyanobacterial microcystin-LR is a potent and specific inhibitor of protein phosphatases 1 and 2A from both mammals and higher plants. FEBS Lett 264:187–192. https://doi.org/10.1016/0014-5793(90)80245-E

Magarvey NA, Beck ZQ, Golakoti T et al (2006) Biosynthetic characterization and chemoenzymatic assembly of the cryptophycins. Potent anticancer agents from cyanobionts. ACS Chem Biol 1:766–779. https://doi.org/10.1021/cb6004307

Massey IY, Wu P, Wei J et al (2020) A mini-review on detection methods of microcystins. Toxins (Basel) 12(10):641. https://doi.org/10.3390/toxins12100641

Mikalsen B, Boison G, Skulberg OM et al (2003) Natural variation in the microcystin synthetase operon mcyABC and impact on microcystin production in Microcystis strains. J Bacteriol 185:2774–2785. https://doi.org/10.1128/JB.185.9.2774-2785.2003

Moffitt MC, Neilan BA (2004) Characterization of the nodularin synthetase gene cluster and proposed theory of the evolution of cyanobacterial hepatotoxins. Appl Environ Microbiol 70:6353–6362. https://doi.org/10.1128/AEM.70.11.6353-6362.2004

Neffling MR (2010) Fast LC-MS detection of cyanobacteria peptide hepatotoxins–method development for determination of total contamination levels in biological materials. Doctoral thesis,. Åbo Akademi University

Oksanen I, Jokela J, Fewer DP et al (2004) Discovery of rare and highly toxic microcystins from lichen-associated cyanobacterium *Nostoc* sp. strain IO-102-I. Appl Environ Microbiol 70:5756–5763. https://doi.org/10.1128/AEM.70.10.5756-5763.2004

Rantala A, Fewer DP, Hisbergues M et al (2004) Phylogenetic evidence for the early evolution of microcystin synthesis. Proc Natl Acad Sci U S A 101:568–573. https://doi.org/10.1073/pnas.0304489101

Rantala A, Rizzi E, Castiglioni B et al (2008) Identification of hepatotoxin-producing cyanobacteria by DNA-chip. Environ Microbiol 10:653–664. https://doi.org/10.1111/j.1462-2920.2007.01488.x

Rikkinen J (2002) Cyanolichens: An Evolutionary Overview. In: Rai AN, Bergman B, Rasmussen U (eds) Cyanobacteria in Symbiosis. Springer, Dordrecht, pp 31–72

Rouhiainen L, Vakkilainen T, Siemer BL et al (2004) Genes coding for hepatotoxic heptapeptides (microcystins) in the cyanobacterium *Anabaena* strain 90. Appl Environ Microbiol 70:686–692. https://doi.org/10.1128/AEM.70.2.686-692.2004

Schreidah CM, Ratnayake K, Senarath K et al (2020) Microcystins: Biogenesis, Toxicity, Analysis, and Control. Chem Res Toxicol 33:2225–2246. https://doi.org/10.1021/acs.chemrestox.0c00164

Seaward M (2001) Environmental role of lichens. In: Nash TH (ed) Lichen biology, Repr. Cambridge University Press, Cambridge, pp 274–298

Shishido TK, Kaasalainen U, Fewer DP et al (2013) Convergent evolution of D-Leucine(1) microcystin-LR in taxonomically disparate cyanobacteria. BMC Evol Biol 13:86. https://doi.org/10.1186/1471-2148-13-86

Sivonen K (2009) Cyanobacterial Toxins. In: Schaechter M (ed) Encyclopedia of Microbiology. Elsevier, Oxford, pp 290–307

Sivonen K, Jones G (1999) Cyanobacterial toxins. In: Bartram J, Chorus I (eds) Toxic Cyanobacteria in Water: A Guide to their Public Health Consequences, Monitoring and Management, 1st edn. CRC Press, London, pp 41–111

Sivonen K, Carmichael WW, Namikoshi M et al (1990) Isolation and characterization of hepatotoxic microcystin homologs from the filamentous freshwater cyanobacterium *Nostoc* sp. strain 152. Appl Environ Microbiol 56:2650–2657. https://doi.org/10.1128/aem.56.9.2650-2657.1990

Tillett D, Dittmann E, Erhard M et al (2000) Structural organization of microcystin biosynthesis in *Microcystis aeruginosa* PCC7806: an integrated peptide-polyketide synthetase system. Chem Biol 7:753–764. https://doi.org/10.1016/S1074-5521(00)00021-1

White TJ, Bruns T, Lee S et al (1990) Amplification and direct sequencing of fungal ribosomal RNA genes for phylogenetics. In: Innis MA, Gelfand DH, Sninsky JJ et al (eds) PCR protocols: a guide to methods and applications. Academic Press, pp 315–322

Detection of Cyanotoxins Using Bacterial Bioassays

32

Thenmozhli Geetha Saravanan, Kishore Kumar Annamalai, Abirami Baskaran, Ranjani Singaraj, Radhakrishnan Manikkam, and Manigundan Kaari

Abstract

Cyanobacteria is a group of photosynthetic prokaryotes primarily found in freshwater and marine environment. A natural phenomenon known as cyanobacterial blooms, which are characterized by a build-up of algal biomass and the potential production of cyanotoxins in aquatic environments. Cyanotoxins pose a threat to aquatic life because they can bio-accumulate on them and spread them up the food chain to humans and other animals. The oral, dermal, and inhalation routes are additional human exposure routes. Although many different analytical techniques have been used to identify and measure the presence of cyanotoxins in aquatic environments, the majority of these techniques are not readily available. Instead, agencies that monitor drinking water supplies are better served by relatively simple, low-cost techniques that can assess the potential health hazard and enable management decisions. Because cyanotoxins harm aquatic biota, it is crucial to find them in aquatic environments. Due to a number of benefits, including a shorter test period, sensitivity, cost-effectiveness, and convenience of use, the bioluminescence inhibition bioassay has been used extensively to evaluate the toxicity. Additionally, this bioassay was found to be equally applicable to all matrices like river water, sewage, sewage sludge, landfill leachate, herbicides, treated wastewater, organic and inorganic chemicals, metals, etc. for toxicity monitoring. The Microtox Acute Toxicity Test, sometimes known as Microtox, has been significant in the advancement of straightforward microscale toxicity testing. A specific clonal strain of bioluminescent bacteria is cultured in a distinctive lyophilized vial format is used in this test. This method can process a lot of samples quickly, easily, cheaply, and sensitively. Microtox is

T. Geetha Saravanan · K. K. Annamalai · A. Baskaran · R. Singaraj · R. Manikkam (✉) · M. Kaari
Centre for Drug Discovery and Development, Sathyabama Institute of Science and Technology, Chennai, Tamil Nadu, India

N. Thajuddin et al. (eds.), *Protocols for Cyanobacteria Sampling and Detection of Cyanotoxin*, https://doi.org/10.1007/978-981-99-4514-6_32

a screening technology that offers an alternative to conventional whole animal testing on invertebrates and fish, which is difficult, expensive, and time-consuming.

Keywords

Cyanotoxin · Bioluminescence · Microtox · *Vibrio fischeri*

32.1 Introduction

Cyanobacteria are found ubiquitous in nature, and they proliferate rapidly to produce blooms, in the presence of phosphorus concentration. Toxins produced by cyanobacteria are known as cyanotoxins. Cyanotoxins produced by blooming cyanobacteria have the potential to harm and even kill humans and animals (Ramya et al. 2020; Van Apeldoorn et al. 2007). Additionally, cyanotoxins can build up in fish and shellfish, which can lead to poisonings like shellfish poisoning. They are very toxic in terms of endotoxins, cytotoxins, neurotoxins, and hepatotoxins. These toxic compounds have a negative impact on the survival of aquatic organisms, wild and/or farm animals, and humans. Living in toxic bloom-rich environment, aquatic organisms such as plants and animals, phytoplankton, and zooplankton are all directly exposed to the negative effects of different cyanotoxins (Imam et al. 2020). The use of drinking water polluted with cyanotoxins or the direct eating of cells of cyanobacteria that produce toxins causes intoxication in animals and human beings. It has been demonstrated that several environmental stresses like ambient light, humidity, pH, and nutrition have a significant impact on the proliferation of various cyanobacteria and the manufacture of their toxic substances (Imam et al. 2020).

There are a number of techniques to determine the toxicity of cyanobacteria, including cell lines-based assays, computational methods, serological methods, and bioassays utilizing vertebrates, invertebrates, microbes, and plants (Agrawa et al. 2012; Lawton et al. 2021). By using bacterial bioassays, such as the bioluminescence inhibition method and the microbial assay for risk assessment (MARA), complex cyanotoxin substances can be evaluated. Bioluminescence inhibition assay is widely used to detect the toxicity of cyanobacteria. The underlying idea behind this assay is the reduction in bioluminescence produced by bacteria when exposed to a cyanotoxin as a result of detrimental biological activity. The naturally occurring luciferase enzyme, reduced flavin mononucleotide (FMN), long chain fatty aldehyde, and oxygen in this bacterium will undergo a chemical reaction to produce luminescence. Since the flow of electrons in the transport chain is directly related to bioluminescence, it can also be considered as a direct consequence of cellular metabolism (Abbas et al. 2018). Microtox test uses a cloned strain of the marine bacterium *Vibrio fischeri* NRRLB-11177 to detect acute toxicity (Mazzeo et al. 2018; Johnson 2005). The bacteria naturally emit light as a by-product of their metabolism. When exposed to a harmful chemical, bacteria's respiratory system is disrupted which results in reduction of light emission. This reduction is a measure of the toxicity of the sample. The Microtox photometer monitors the luminescence levels that are emitted before and after addition of the sample. Results correlate well with those from other toxicity bioassays using fish, daphnia, and shrimp. The results from Microtox are reported as EC50, i.e., the concentration at which a 50% reduction of emitted bioluminescence is observed. Such a concentration is obtained through a dose-response curve constructed by measuring the light intensity at various concentrations of the same sample. Thus when comparing the results of various tested samples, those with the lower EC50 values are considered potentially toxic (Johnson 2005; Johnson 2018). This assay requires only a few minutes (5–30 min), easy performing, requires less quantity of sample volumes and is very cost-effective.

32.2 Materials

- Microtox reagent—clonal *Vibrio fischeri* isolate
- Reconstitution solution
- Diluent—2% NaCl solution
- Osmotic adjustment solution
- Cyanotoxin extract
- DMSO
- SDI model 500 analyser (serve as both incubator and luminometer)
- MicrotoxOmni software

32.3 Methods

1. Dissolve the cyanotoxin extract in 20% DMSO to attain the required concentration.
2. Turn on the SDI model for 5 min and allow incubator to achieve optimum temperature. Maintain at 15°C for incubator well and maintain at 5 °C for reagent well.
3. Add 1 mL of reconstituent solution to reagent bacteria vials.
4. Mix well and transfer immediately to 12 × 75 mm cuvette.
5. Stir the content using vortex mixer vigorously.
6. Now place the cuvette in reagent well of the reader and hold for 5 min to stabilize the cultures emission of light prior to testing.
7. In row A, place one control and four concentration cuvette and in row B, place 5 cuvettes.
8. Add 2 mL of the diluent to cuvette A5 and 1 mL to each of the other four cuvettes in the row A.
9. In row B, add 0.5 mL of diluent to each of the five wells.
10. Pipette test cyanotoxin sample into A5 and vortex well.
11. Add 1 mL from well A5 to well A4 and thoroughly mix.
12. In the same manner, transfer 1 mL from A4 to A3, A3 to A2.

13. A1 remains as control.
14. Once the reconstituent bacterial vial has been removed from the reagent well, 10 uL of bacterial reagent should be injected into each cuvette in Row B.
15. Start the MicrotoxOmni application, enter the necessary information, and set the PC to the appropriate exposure time (5–15 min).
16. Now, create a baseline reading for emission of light using the analyser.
17. Then remove the B5 cuvette and place in reading well, thus luminometer will record light emission.
18. To record light emission, all the cuvettes from row B will be placed in reading wells and ranges from 90% to 100%.
19. Activate the incubation timer and transfer 0.5 mL of test sample to reagent culture from cuvette A5 to B5.
20. Transfer 0.5 mL in the same manner from A4 to B4, A3 to B3, and A2 to B2.
21. Place the cuvettes from row B in the reading well when the incubation period is complete, and then measure each cuvette's maximum light emission.
22. Record the data using Microtox Omni software.
23. Use the control well to correct the samples for the time-dependent drift in light output.
24. Plot the dose (concentration of the test sample) against the response of the bacteria (represented as gamma) to obtain the EC50 values from the graph.

32.4 Data Analysis

Determine the percentage of inhibition and calculate the EC50 value either by graphically using a double logarithmic co-ordinate system or by using linear regression analysis.

Acknowledgement Authors thank the management of Sathyabama Institute of Science and Technology, Tamil Nadu for their research support and encouragement. Authors Thenmozhli GS, Ranjani S and Manigundan K thank the Department of Biotechnology, New Delhi, for their support in the form of research grant (BT/PR41474/NDB/39/760/2020 dated 28.09.2021).

References

Abbas M, Adil M, Ehtisham-ul-Haque S, Munir B, Yameen M, Ghaffar A et al (2018) Vibrio fischeri bioluminescence inhibition assay for ecotoxicity assessment: a review. Sci Total Environ 626:1295–1309

Agrawa M, Yadav S, Patel C, Raipuria N, Agrawal MK (2012) Bioassay methods to identify the presence of cyanotoxins in drinking water supplies and their removal strategies. Eur J Exp Biol 2(2):321–336

Imam TS, Tsagero MH, Ari HA, Aminu SB, Ugya AY (2020) Emerging approach to uncover cyanotoxins in aquatic environment: a concise review. Curr World Environ 15(1):100

Johnson BT (2005) Microtox® acute toxicity test. In: small-scale freshwater toxicity investigations. Springer, Dordrecht, pp 69–105

Johnson BT (2018) Microtox® toxicity test system—new developments and applications. In: microscale testing in aquatic toxicology. CRC Press, pp 201–218

Lawton LA, Metcalf JS, Žegura B, Junek R, Welker M, Törökné A, Bláha L (2021) Laboratory analysis of cyanobacterial toxins and bioassays. In: Toxic Cyanobacteria in Water. CRC Press, pp 745–800

Mazzeo DE, Roberto MM, Sommaggio LR, Marin-Morales MA (2018) Bioassays used to assess the efficacy of biodegradation. In: Toxicity and biodegradation testing. Humana Press, New York, NY, pp 215–239

Ramya M, Umamaheswari A, Elumalai S (2020) Global health concern of cyanotoxins in surface water and its various detection methods. Curr Bot 11:65–74

Van Apeldoorn ME, Van Egmond HP, Speijers GJ, Bakker GJ (2007) Toxins of cyanobacteria. Mol Nutr Food Res 51(1):7–60

33 Detection of Cyanotoxins: Bioassay Using Plants

Neelam Mishra and Sheetal Gupta

Abstract

Cyanotoxins are a group of secondary metabolites which are produced by cyanobacteria like *Oscillatoria, Anabaena, Microcystis, Lyngbya, Nostoc, Nodularia, Aphanizomenon, Planktothrix*, etc. These algal species proliferate in polluted waterbodies and produce toxic secondary metabolites which further pollute the waterbodies and negatively affect the aquatic life and fish-eating birds. They often contaminate the drinking water causing health problems like gastroenteritis, hepatic hemorrhage, some allergic reactions, neurotoxic effects, effects on different organs, skin problems, carcinogenic, etc. There are various toxin detection methods like physiochemical methods, molecular methods, biochemical methods, chemical methods, and bioassays. Here in this chapter, we have discussed the bioassay methods involving plants like *Sinapis alba*, *Lepidium sativum*, *Nicotiana tabacum*, and *Solanum tuberosum* and their protocol.

Keywords

Cyanotoxins · Plant bioassay · Sinapis test · *Lepidium sativum* · *Nicotiana tabacum* · *Solanum tuberosum*

33.1 Introduction

Cyanobacteria are a group of small, microscopic photosynthetic bacteria commonly found in water ecosystems. They are also referred to as blue-green algae and phytoplankton. These microorganisms produce a wide range of secondary

N. Mishra (✉) · S. Gupta
Department of Botany, St Joseph's University, Bengaluru, Karnataka, India
e-mail: neelammishra@sju.edu.in

N. Thajuddin et al. (eds.), *Protocols for Cyanobacteria Sampling and Detection of Cyanotoxin*, https://doi.org/10.1007/978-981-99-4514-6_33

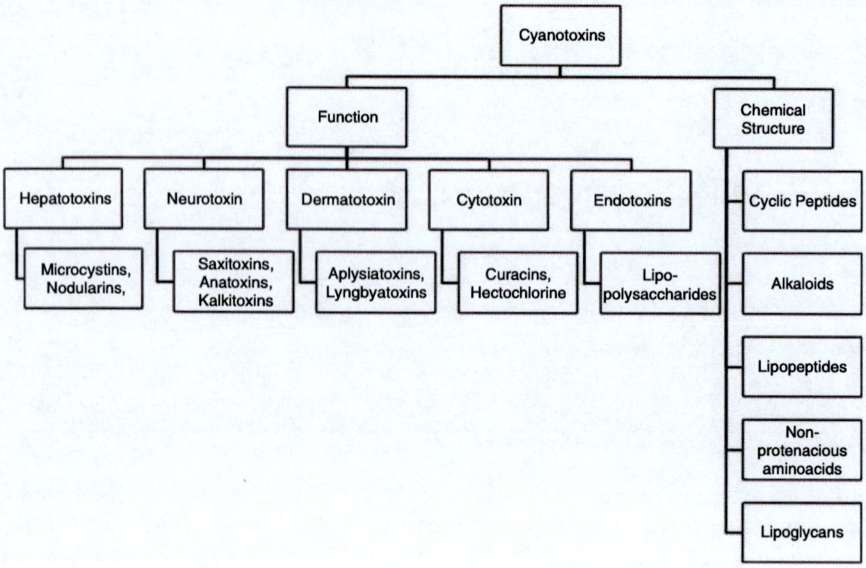

Fig. 33.1 Brief classification of cyanotoxins based on their function and chemical structure

metabolites which could either have toxic effects or non-toxic effects (Haque et al. 2017). In appropriate environmental conditions, one or few species dominate and their increased growth results in cyanobacterial bloom (Chorus and Bartram 1999). Cyanobacterial blooms are categorized as "harmful algal blooms" if they are responsible for causing adverse environmental impacts such as reduction in ecosystem stability, production of highly toxic compounds, such as cyanotoxins (Chorus 2001).

Cyanotoxins are a range of chemical compounds produced by these organisms which are known to have toxic effects on other living beings (Chorus 2012). Toxins by definition are chemical compounds that are known to negatively affect the metabolic pathways of living organisms. There are various classes of cyanotoxins produced by cyanobacteria, these being hepatotoxins, neurotoxins, dermatotoxins, and cyanotoxins based on their action (Fig. 33.1) (Du et al. 2019). Among these hepatotoxins are the most commonly occurring cyanotoxins which include microcystins, nodularins, etc. (Sivonen and Jones 2009; Carmichael 1997). Chemically cyanotoxins can be cyclic peptides (microcystins, nodularin), alkaloids (cylindrospermopsins, saxitoxins, anatoxins), lipopeptides (curacins, kalkitoxins, jamaicamides), lipoglycans (lipopolysaccharide), or non-proteinaceous like β-N-methylamino-L-alanine (Du et al. 2019; Carmichael 1997). Chemically, cyanotoxins can be cyclic peptides (microcystins, nodularins), alkaloids (cylindrospermopsins, saxitoxins, anatoxins), lipopeptides (curacins, kalkitoxins, jamaicamides), lipoglycans (lipopolysaccharide) or non-proteinaceous like β-N-methylamino-L-alanine (Du et al. 2019).

The secondary metabolites produced by the cyanobacteria are known to have both positive and negative effects on living beings, some of these chemical compounds can be of great pharmaceutical importance whereas some can show harmful effects on life, leading to bioaccumulation of toxins in higher trophic levels, cause water pollution (Shi et al. 2021; Ferrão-Filho and Kozlowsky-Suzuki 2011; Armah et al. 2013). There have been various reports of drinking water contamination due to cyanobacterial blooms producing cyanotoxins which led to human intoxication and has affected many lives (Armah et al. 2013; Kokociński et al. 2017). Microcystins are known to cause conditions like gastroenteritis, hepatic hemorrhage, primary liver cancer and some allergic reactions (Díez-Quijada et al. 2019; Ramya et al. 2020), and affect different organs (Shi et al. 2021; Dawson 1998). Cyanotoxins like nodularin and cylindrospermopsin when exposed to aquatic animals also show detrimental effects on the physiological processes of various fishes causing an overall decrease in diversity in the water bodies (Sotton et al. 2015).

Based on the bioassay methods employed cyanotoxins can be cytotoxins and biotoxins. Cytotoxins include toxins that show cytotoxic effects and bioassay methods involve culturing of cell lines whereas biotoxins include toxins that show detrimental effects on living organisms and small animals, and plants are used for their assays (Carmichael 1997).

33.2 Methods

33.2.1 Bioassay of Cyanotoxins Using Plants

Bioassay or biological assay is a method for determining and quantifying toxins using biological organisms like plants, microorganisms, invertebrates, etc. Cyanotoxins are the secondary metabolites produced by the cyanobacteria which can cause distress to several living organism because of their properties like algicidal, herbicidal, growth inhibition, changes in chlorophyll contents, abnormal development in plants, feeding inhibition, hepatotoxicity, organ failure, genotoxic effects, etc. (Imam et al. 2020; Metcalf and Codd 2012; Saqrane and Oudra 2009; Ujvárosi et al. 2019; Chen et al. 2021). All these physiological changes and the extent of these effects can be useful in determining and quantifying these cyanotoxins. Bioassay methods involving plant tissue require some more research work as not much information and effectiveness of these methods are available. Although several methods for bioassay of cyanotoxins using plant and plant extracts have been worked upon by using plants like *Sinapis alba (*Bláha et al. 2016*)*, *Lepidium sativum*, *Nicotiana tabaccum*, *Solanum tuberosum (*Sutradhar 2022*)*, *Lemna (*Bláha et al. 2016*)*, etc. Some algal members like Anacystis, Chlorella, Phormidium, and Plectonema are also used for the detection of cyanotoxins produced by *Oscillatoria* because of their algicidal effects (Ramya et al. 2020; Imam et al. 2020).

33.2.1.1 Sinapis Test

Sinapis test often referred to as the Mustard test or Blue-Green Sinapis Test (BGST) is a common plant bioassay method used for the determination and quantification of cyanotoxins, mainly microcystins and cylindrospermopsin (Vasas et al. 2002).

Procedure

1. Surface sterilize the *Sinapis alba* (white mustard) seeds using 5% hydrogen peroxide three times for 30 min (McElhiney et al. 2001) and wash it with distilled water three to four times under sterile (laminar air flow) conditions.
2. Allow the sterilized seeds to imbibe under dark conditions for 16 h.
3. The extracted and purified toxin sample mixed with ethanol is added to microtiter plates (Vasas et al. 2002). Allow the sample to evaporate under sterile conditions, leaving behind the toxin residues on the plates.
4. Prepare 1% Hoagland plant nutrient media supplemented with 0.6% bacto agar and pour it into the microtiter plates containing sample toxin residues and allow the media to solidify (Kos et al. 1995).
5. Inoculate the imbibed seeds on the microtiter plates containing the media under sterile conditions.
6. Allow the inoculated seeds to grow under dark conditions at 28 ° C.
7. Measure the length of hypocotyl after every 3–5 days and calculate the mean length and standard deviation for the same.

The presence or absence of cyanotoxins is determined by the inhibition of the growth of the hypocotyl when compared to the normal growth rate and the extent of inhibition caused by the cyanotoxins concerning their concentration (Kos et al. 1995). 50% of growth inhibition in seedlings was observed at the concentration of 3 μg/mL (Kos et al. 1995). Microcystins at a concentration ranging from 0.1 to 2.5 μg/mL show significant growth inhibition in *Sinapis alba* seedlings (Vasas et al. 2002). Cylindrospermopsin at a lower concentration of 0.01 μg/mL shows higher lateral root formation (Máthé et al. 2013).

33.2.1.2 Bioassay Using *Lepidium Sativum*

This test is also referred to as the *Lepidium sativum* seed germination test. The test involves the use of *Lepidium sativum* seed to study growth inhibition due to the toxicity effects of cyanotoxins like microcystin on the germination of seedlings. Three methods have been discussed below by which cyanotoxin toxicity detection can be performed using *L. sativum.*

Study of Growth Inhibition after Germination (Gehringer et al. 2003)

1. Surface sterilize the *Lepidium sativum* seeds using 5% hydrogen peroxide (H_2O_2) and wash the seeds with distilled water two to three times to make sure that all the H_2O_2 is washed off.
2. Allow the seeds to imbibe overnight and germinate in water.
3. Prepare a nutrient media containing the following microelements (Table 33.1) and 0.8% agar as a solidifying agent.

Table 33.1 Nutrient media composition

Elements	Percentage
Nitrogen	6.5
Phosphorous	2.7
Potassium	13.0
Calcium	7.0
Magnesium	2.2
Sulfur	7.5
Iron	0.15
Manganese	0.0024
Boron	0.0024
Zinc	0.005
Copper	0.002
Molybdenum	0.001

4. Add the extracted cyanotoxin sample into the media before pouring it into petri plates or test tubes or microtiter plates except in control.
5. Allow the media to solidify and inoculate media with germinated seeds.
6. Maintain the growth conditions as 27 °C and continuous illumination of 20 photons/m^2/s.
7. Observe the rate of growth of roots and leaves after every 2–3 days.

Study of Inhibition on Seed Germination (Šrédlová et al. 2021)

1. Take a petri dish and cut out filter paper of appropriate dimensions as the area of the petri dish and place the filter paper inside the petri dish.
2. Add 1.5 mL of distilled water in control and 1.5 mL of extracted cyanotoxin sample to the petri dish on the filter paper.
3. Place *Lepidium sativum* seeds onto the petri dish and allow them to germinate at 28 °C under dark conditions for 72 h.
4. Using the following formula, calculate the percentage of germination inhibition (IG).

$$\mathrm{IG} = 100 - \left\{ \frac{\mathrm{gs} \cdot \ \mathrm{ls}}{\mathrm{gc} \cdot \ \mathrm{lc}} \cdot 100 \right\}$$

where

gs—the percentage of seed germinated successfully in a petri dish containing sample cyanotoxin

Ls—the average length of germinated seed in mm in a petri dish containing sample cyanotoxin

gc—the percentage of seeds germinated successfully in the control petri dish

lc—the average length of germinated seed in mm in the control petri dish

Fifty percent growth inhibition has been observed at concentrations of 241 ± 13 mg/L (Šrédlová et al. 2021).

Glutathione S-Transferase and Glutathione Peroxidase Assay

1. Seeds of *L. sativum* are surface sterilized with 5% H_2O_2 for 5 min followed by three washes in sterile water for 10 min.
2. Seeds are allowed to germinate in freshwater overnight and then transfer to a nutrient medium as mentioned above.
3. Seeds are grown at 27 °C under continuous light (20 photons/m^2/s).
4. Control seeds are not treated with toxin; however, experimental seeds were treated with MCLR toxin with concentrations of 1 μg/mL and 10 μg/mL.
5. For each group, 3–5 seedlings are used to prepare homogenate using buffer (0.01 M Tris-HCl pH 7.8 and protease inhibitors: 7.5 mM PMSF, 2.5 mM EDTA, 325 mM bestatin, 3.5 mM E-64, 2.5 mM leupeptin, and 0.75 mM aprotinin).
6. The homogenate was centrifuged at 12,000 g for 2 h at 4 °C.
7. Supernatant was used to measure GST activity (Habig et al. 1974) and GPX activity (Wendel 1981).

33.2.1.3 Bioassay Using *Nicotiana Tabacum*

It is a pollen germination assay using *Nicotiana tabacum* pollen grains (Metcalf et al. 2004). Cyanotoxins like cylindrospermopsin affect the pollen tube germination in tobacco pollen grains by inhibiting the synthesis of proteins required for pollen tube formation (Metcalf et al. 2004). This effect on the pollen tube can be used as a bioassay method for the detection of cyanotoxins.

Procedure

1. Collect the pollen grains from the opened flowers of *Nicotiana tabacum* in the morning and store it.
2. Prepare a nutrient buffer containing the following components.

Components	Concentration
Sucrose	10%
Casein acid hydrolysate	0.03%
Polyethylene glycol	15.0 mM (pH 5.9)
Calcium chloride dehydrate	1.0 mM
Potassium chloride	1.0 mM
Magnesium sulfate	0.8 mM
Boric acid	1.6 mM
Copper sulfate	30 μM

3. To the nutrient buffer add 100 μL of ethanol in water for control and 100 μL of the toxin sample solution in micro-centrifuge tubes.
4. Allow the pollen grains to germinate and pollen tubes to grow for 18 h, under dark conditions at 25 °C.
5. For measuring the pollen tube growth, centrifuge the germinated pollen at 1000 g for 5 min, and remove the supernatant.

6. Add diluted Alcian blue dye to the pellet and incubate it for 30 min at room temperature.
7. Wash the pellet two to three times by adding 700 μL MilliQ water to the re-suspended pellet and centrifuging it at 1000 g for 5 min.
8. After washing three times with MilliQ water, add 500 μL citric acid 40% (w/v) solution and incubate it for 5 min then centrifuge it and remove the supernatant.
9. Calculate the absorbance of the pellet using a Versamax plate reader.

Inhibition of pollen germination was observed at concentrations of 5 μg/mL of cylindrospermopsin and 50% growth inhibition was observed at concentrations of 300 μg/mL (Metcalf et al. 2004).

33.2.1.4 Bioassay Using *Solanum Tuberosum* (McElhiney et al. 2001)

Cyanotoxins like microcystins can show phytotoxic effects on the growth of *Solanum tuberosum* seedlings and also cause a reduction in the chlorophyll content of the plant (Du et al. 2019). The growth inhibition percentage and reduction in chlorophyll contents determined by spectrophotometric analysis can be used as a good method for bioassay of these cyanotoxins.

Procedure

1. Prepare MS basal salt mixture with 1% agar as a solidifying agent, pour it into media bottles, autoclave it, and allow it to solidify.
2. Filter sterilize the extracted cyanotoxin sample and add it to the media bottles except in control.
3. Under sterile LAF conditions inoculate the media bottles with shoot segments by sub-culturing 1.5–2 cm long shoot segments. The shoot segments should have 1–2 nodal regions.
4. Incubate the cultures for 16 days under controlled conditions at 15 °C.
5. Observe the growth rate by measuring the shoot length and fresh weights of the plantlets.

Chlorophyll estimation of the cultured plantlets under toxin effects

1. Prepare plant extract by grinding the plantlet in a pestle and mortar using 0.5–1 mL 90% acetone under dark conditions.
2. Remove the debris by centrifuging the extract and collect the supernatant.
3. Using a spectrophotometer, measure the absorbance of the chlorophyll extract at 664 nm and 647 nm.
4. Calculate the amount of chlorophyll a and chlorophyll b by substituting the values obtained by spectrophotometer into the following equation (Geider and Osborne 1992).

$$\text{Chl a (mg/dm)} = 11.92 \times A_{664} - 1.93 \times A_{647}$$

$$\text{Chl b (mg/dm)} = 20.36 \times A_{647} - 5.5 \times A_{664}$$

A significant amount of growth inhibition is observed at microcystin concentrations of 0.005 µg/mL. Chlorophyll reduction can be observed at concentrations of 0.5 µg/mL to 5 µg/mL (McElhiney et al. 2001).

References

Armah A, Hiskia A, Kaloudis T, Chernoff N, Hill D, Antoniou MG, Dionysiou DD (2013) A review on cylindrospermopsin: the global occurrence, detection, toxicity and degradation of a potent cyanotoxin. Environ Sci Process Impacts 15(11):1979–2003

Bláha L, Cameán AM, Fessard V, Gutiérrez-Praena D, Jos Á, Marie B, Metcalf JS, Pichardo S, Puerto M, Törökné A, Vasas G, Žegura B (2016) Bioassay use in the field of toxic cyanobacteria. In: Handbook of cyanobacterial monitoring and cyanotoxin analysis, pp 272–279

Carmichael WW (1997) The cyanotoxins. In: Advances in botanical research, vol 27. Academic Press, pp 211–256

Chen G, Wang L, Wang M, Hu T (2021) Comprehensive insights into the occurrence and toxicological issues of nodularins. Mar Pollut Bull 162:111884

Chorus I (2001) Cyanotoxins: Occurrence, Causes, Consequences. Springer, Berlin. https://doi.org/10.1007/978-3-642-59514-1

Chorus I (ed) (2012) Cyanotoxins: occurrence, causes, consequences. Springer Science & Business Media

Chorus I, Bartram J (1999) Toxic cyanobacteria in water—a guide to their public health consequences. Monitoring and Management, Routledge, London and New York

Dawson RM (1998) The toxicology of microcystins. Toxicon 36(7):953–962

Díez-Quijada L, Puerto M, Gutiérrez-Praena D, Llana-Ruiz-Cabello M, Jos A, Cameán AM (2019) Microcystin-RR: occurrence, content in water and food and toxicological studies. A Review. Environ Res 168:467–489

Du X, Liu H, Yuan L, Wang Y, Ma Y, Wang R, Zhang H (2019) The diversity of cyanobacterial toxins on structural characterization, distribution and identification: a systematic review. Toxins 11(9):530

Ferrão-Filho ADS, Kozlowsky-Suzuki B (2011) Cyanotoxins: bioaccumulation and effects on aquatic animals. Mar Drugs 9(12):2729–2772

Gehringer MM, Kewada V, Coates N, Downing TG (2003) The use of *Lepidium sativum* in a plant bioassay system for the detection of microcystin-LR. Toxicon 41(7):871–876

Geider RJ, Osborne BA (1992) Algal photosynthesis. Chapman & Hall, New York

Habig WH, Pabst MJ, Jakoby WB (1974) Glutathione S- transferases: the first step in mercapturic acid formation. J Biol Chem 249:7130–7139

Haque F, Banayan S, Yee J, Chiang YW (2017) Extraction and applications of cyanotoxins and other cyanobacterial secondary metabolites. Chemosphere 183:164–175

Imam TS, Tsagero MH, Ari HA, Aminu SB, Ugya AY (2020) Emerging approach to uncover cyanotoxins in aquatic environment: a concise review. Curr World Environ 15(1):100

Kokociński M, Cameán AM, Carmeli S, Guzmán-Guillén R, Jos A, Mankiewicz-Boczek J, Metcalf JS, Moreno IM, Prieto AI, Sukenik A (2017) Cylindrospermopsin and congeners. In: Meriluoto J, Spoof L, Codd GA (eds) Handbook of cyanobacterial monitoring and cyanotoxin analysis, 1st edn. John Wiley & Sons; Hoboken, NJ, USA, pp 127–137

Kos P, Gorzo G, Suranyi G, Borbely G (1995) Simple and efficient method for isolation and measurement of cyanobacterial hepatotoxins by plant tests (*Sinapis alba* L.). Anal Biochem 225(1):49–53

Máthé C, Vasas G, Borbély G, Erdődi F, Beyer D, Kiss A, Surányi G, Gonda S, Jámbrik K, M-Hamvas M (2013) Histological, cytological and biochemical alterations induced by microcystin-LR and cylindrospermopsin in white mustard (Sinapis alba L.) seedlings. Acta Biol Hung 64(1):71–85

McElhiney J, Lawton LA, Leifert C (2001) Investigations into the inhibitory effects of microcystins on plant growth, and the toxicity of plant tissues following exposure. Toxicon 39(9):1411–1420

Metcalf JS, Codd GA (2012) Cyanotoxins. In: Ecology of cyanobacteria II: their diversity in space and time. Springer, Dordrecht, pp 651–675

Metcalf JS, Barakate A, Codd GA (2004) Inhibition of plant protein synthesis by the cyanobacterial hepatotoxin, cylindrospermopsin. FEMS Microbiol Lett 235(1):125–129

Ramya M, Umamaheswari A, Elumalai S (2020) Global health concern of cyanotoxins in surface water and its various detection methods. Curr Bot 11:65–74

Saqrane S, Oudra B (2009) Cyano HAB occurrence and water irrigation cyanotoxin contamination: ecological impacts and potential health risks. Toxins 1(2):113–122

Shi L, Du X, Liu H, Chen X, Ma Y, Wang R, Zhang H (2021) Update on the adverse effects of microcystins on the liver. Environ Res 195:110890

Sivonen K, Jones G (2009) Cyanobacterial toxins. Encyclopedia of microbiology, vol 290, p 307

Sotton B, Domaizon I, Anneville O, Cattanéo F, Guillard J (2015) Nodularin and cylindrospermopsin: a review of their effects on fish. Rev Fish Biol Fish 25(1):1–19

Šrédlová K, Šilhavecká S, Linhartová L, Semerád J, Michalíková K, Pivokonský M, Cajthaml T (2021) The sensitivity of multiple ecotoxicological assays for evaluating *Microcystis aeruginosa* cellular algal organic matter and contribution of cyanotoxins to the toxicity. Toxicon 195:69–77

Sutradhar M (2022) The current scenario and future aspects of cyanotoxins: a review study. J Mater Environ Sci 13(7):768–777

Ujvárosi AZ, Riba M, Garda T, Gyémánt G, Vereb G, M-Hamvas M, Vasas G, Máthé C (2019) Attack of Microcystis aeruginosa bloom on a Ceratophyllum submersum field: Ecotoxicological measurements in real environment with real microcystin exposure. Sci Total Environ 662: 735–745

Vasas G, Gáspár A, Surányi G, Batta G, Gyémánt G, M-Hamvas M, Máthé C, Grigorszky I, Molnár E, Borbély G (2002) Capillary electrophoretic assay and purification of cylindrospermopsin, a cyanobacterial toxin from Aphanizomenon ovalisporum, by plant test (blue-green Sinapis test). Anal Biochem 302(1):95–103

Wendel A (1981) Glutathione peroxidase. Methods Enzymol 77:325–332

In Vitro Assay for Determining Cyanotoxin Using Serological Methods

34

Sakshi R. Varshney, V. Z. Misbah Rehman, and Lokesh Ravi

Abstract

Cyanobacteria are free living organisms and the probability of finding them in water body is very high. These are proven to be harmful for humans in the form of hepatotoxins, neurotoxins, or alkaloids (forms of cyanotoxins). Thus, detection of these harmful agents becomes a necessity for a healthy environment at all forms of life. This chapter mainly focuses on the serologic methods of detecting cyanotoxins in blood sera using enzyme-linked immunosorbent assay. Although several other methods under analytical tests like chromatographic techniques can be employed to know the accurate concentration, ELISA on the other hand also furnishes good quality results. The following content will describe the protocol of performing experiments to detect cyanotoxins in their classified form.

Keywords

Serum preparation · Cyanotoxin extraction · Microcystin · Nodularin · Serological analysis · ELIZA · Limitations

S. R. Varshney · V. Z. Misbah Rehman
Department of Chemical Engineering, CPDMED, TBI, Hoyasan Labs Pvt. Ltd., Indian Institute of Science, Bengaluru, Karnataka, India

L. Ravi (✉)
Department of Food Technology, Faculty of Life and Allied Health Sciences, MS Ramaiah University of Applied Sciences, Bengaluru, India

N. Thajuddin et al. (eds.), *Protocols for Cyanobacteria Sampling and Detection of Cyanotoxin*, https://doi.org/10.1007/978-981-99-4514-6_34

34.1 Introduction

Surprisingly, global climate change has its effects deep down onto minute organisms like cyanobacteria as well. The increase in climate change leads to production of toxins released by cyanobacteria, and therefore this leads to contamination of waterbodies and finally affecting us humans because of the quality of water we drink. Molecular nature of these toxins includes cyclic peptides, alkaloids, and organophosphate moiety (Skulberg et al. 1993). There are two ways to detect cyanotoxins present in a sample, analytical and bio-analytical method. Bio-analytical methods include ELISA, Quantitative real-time PCR, colorimetric PPIA, the mouse bioassay, etc. Among the various types of ELISA methods, direct and indirect are mostly employed as illustrated in Fig. 34.1. These methods have low equipment requirements to operate and are therefore easy to operate, produce rapid and effective results, and allow sensitive detection of toxin (Koreivienėźand Belous 2012). Serological method ELISA comes in the form of kits as well, and the easy availability of these kits make the test handier because of the convenience and zero labor offered. Other analytical methods which help in cyanotoxin determination are more sophisticated. Other methods are high performance liquid chromatography (HPLC), liquid chromatography/mass spectrometry (LC/MS), and thin layer chromatography (TLC).

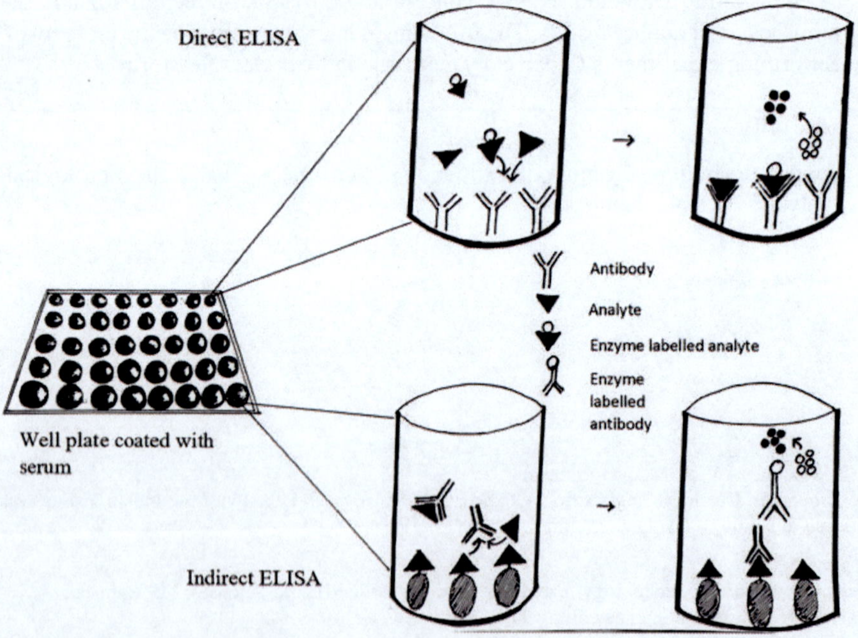

Fig. 34.1 Mechanism of different types of ELISA, direct and indirect

Microcystins are the most common and widely associated toxins of cyanobacteria. These are either freely floating or are bound in human serum. Some common freshwater cyanobacteria are *Anabaena flos-aquae, Microcystis aeruginosa, Oscillatoria agardhii*, and *Nodularia spumigena* (Sivonen et al. 1990). These cyanobacteria lead to production of harmful toxins, microcystins, and nodularin which have adverse effects on both human and animal life.

Serum samples are analyzed using ELISA—enzyme-linked immunosorbent assay. Free microcystins are directly analyzed using ELISA and an indirect method is also employed for determination of protein bound microcystins using GC/MS to find the amount of MMPB (2-methyl-3-methoxy-4-phenylbutyric acid) derived from microcystins as a result of chemical oxidation. This represents the total amount of microcystins present (Hilborn et al. 2007). Microcystins are categorized under hepatotoxins and liver being their target organ. Microcystin-leucine arginine is a common cyanobacterial toxin released by cyanobacteria which results in harmful effects on the reproductive organs leading to reproductive toxicity.

34.2 Protocol

34.2.1 Using a Commercially Available ELISA KIT from Enviro Logix, Inc. Portland, ME, USA

1. Take the human blood samples that need to be identified as intoxicated by microcystins.
2. Separate the serum from the blood cells.
3. Freeze the separated serum at −70 °C.
4. The serum sample is thawed.

Sample Preparation

5. In a 15 mL corex tube made of glass, add 1 mL aliquot of the serum sample followed by 10 mL of methanol. Multiple such tubes can be set up.
6. These are then centrifuged for 30 min at 9000 rpm.
7. In a 20 mL scintillation vial decant the supernatant.
8. Treat the pellet with 5 mL of methanol and allow it to suspend.
9. This is once again centrifuged at 9000 rpm for about 30 min and added to a scintillation tube.
10. Add 5 mL of hexane and vortex it.
11. Discard the hexane layer only and retain the methanol layer.
12. Wash the methanol layer with 5 mL hexane and repeat this for at least three times.
13. In a Speed VAC concentrator, dry the samples under vacuum at 40 °C, this is then taken up in 2 mL of 5% HOAc.
14. The solution should be passed through the Oasis HLB solid phase extraction cartridge.

15. The cartridge should be conditioned with 1:1 column volume of methanol and water, respectively.
16. The cartridge is then washed with 5 mL of methanol in water 30% (v/v).
17. In order to elute the microcystin, the samples are treated with 5 mL of methanol followed by drying.
18. Resuspend the samples in 10% (v/v) of methanol in water in a volume of 1 mL.
19. Centrifuge this for 3.5 h at 10,200 rpm through a YM-10 molecular weight cut off filter.
20. Divide these treated serum samples into wells of the ELISA KIT- EnviroLogix, Inc.Portland, ME, USA.
21. Preferred purchase of standard toxins is extractions from cultures of 98% *Microcystis aeruginosa.* (Wright State University).

ELISA procedure

22. Free microcystins are detected by direct competitive ELISA tests. These contain a rabbit antibody that is conjugated to bovine serum albumin against microcystin-LR. The analyte competes for the antibody-binding site against the microcystin-LR peroxidase (microcystin enzyme conjugate). On the addition of a substrate, there is an induction of color change which is read at a wavelength of 450 nm by ELISA plate reader.
23. Lighter coloration indicates higher concentration of the toxin and darker coloration shows lower concentration of the toxin.
24. All the reagents and standards for performing the assay are included in the kit.
25. Run the serum samples in triplicate.
26. Add 125 μL of assay diluent to each of the wells.
27. 20 μL of the negative control and 20 μL of calibrator is added to their respective vials.
28. And the serum samples of 20 μL should be added to their respective wells.
29. Mix the contents of the wells for a good 20–30 s.
30. Cover the well plate with parafilm.
31. Incubate the setup for 30 min at room temperature.
32. Now add about 100 μL of the microcystin-enzyme conjugate to each of the wells in the well plate.
33. Now cover the plates with parafilm and again incubate it for 30 min at room temperature after mixing the contents prior to incubation.
34. For washing, flood the contents of each well with the PBS solution for about 4 cycles (four times) followed by shaking to empty the well each time using a multi-reagent plate washer.
35. Now add about 100 μL of the substrate provided.
36. Mix the contents in the wells and follow it up with parafilm covering.
37. Incubate the plate at room temperature for 30 min.
38. Add 100 μL of stop solution provided in the kit, and thoroughly mix the contents in each well.

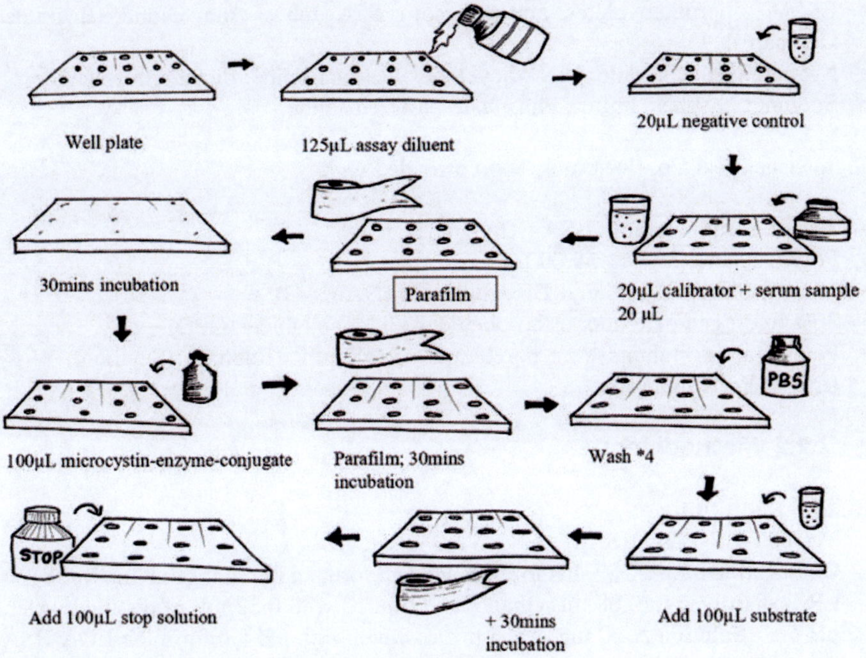

Fig. 34.2 Commercial ELISA kit protocol flowchart

39. Now read the plate at 450 nm wavelength (Hilborn et al. 2005). Figure 34.2 shows a detailed diagrammatic representation of the above-described procedure.

34.2.1.1 Observation

In order to estimate the amount or concentration of the cyanotoxin—microcystin in the tested serum sample, the plate with serum sample must be read for the intensity of color through a colorimetric method at a wavelength of 450 nm. This value should be estimated on the basis of the standard concentration of microcystin used whose concentration is already known. In general, the serum sample set up that has higher concentrations of the toxin will show lighter coloration whereas the serum sample having lower concentrations of the toxin will show deeper coloration.

34.2.2 Direct Competitive ELISA and Indirect ELISA

34.2.2.1 Materials

- Standard microcystins MCYST-LR, MCYST-RR, MCYST leucine-alanine (MCYST-LA), MCYST leucine-tyrosine (MCYST-LY).
- Ethylenediamine (EDA)-modified BSA.

- ELISA microtiter plates and minisorp RIA tubes from Nunc (Roskilde, Denmark).
- N-hydroxysuccinimide; N,N′-dicyclohexylcarbodiimide; dimethylformamide.
- Sodium phosphate buffer; Phosphate-buffered saline.

Sigma Chemical Co. (St. Louis, Mo.) provided with

- Bovine serum albumin BSA (RIA grade).
- Poly-L-lysine (MW = 58,000).
- N-acetylmuramyl-L-alanyl-Disoglutamine (NAMAG).
- 30% hydrogen peroxide; ortho-phenylenediamine tablets (OPD)
- Female albino rabbits were purchased from Smith's Rabbitry (Seymour, Wis.) (free of Pasteurella spp.).

34.2.2.2 Methodology

Building Immunity

- MCYST-LR-EDA-BSA:
  Conjugation of MCYST-LR to EDA-BSA to form an immunogen 5 mg MCYST-LR is dissolved in 0.08 mL ethanol and diluted with 0.32 mL of distilled water, pH 5 → Solution A 20 mg in 0.4 mL of water with pH 5.0 form the EDA-BSA solution → Solution B.

  Solution A and B with a few drops of 0.2 N NaOH added dropwise to Solution C. Solution C, EDPC solution is made by dissolving 1.32 g of EDPC in 1.5 mL of water with pH 5. The entire reaction is carried out at room temperature for 5 h, with a constant maintenance of pH 5. Post completion the entire solution undergoes dialysis 0.1 M NaCl (2 L) at 6 °C for 24 h and later lyophilized.
- MCYST-LR-PLL for indirect ELISA:
  1.8 mg of MCYST-LR is added to a solution, this solution contains 0.28 mg of N-hydroxysuccinimide dissolved in 0.28 mL of dimethylformamide and 0.5 mg of N,N′-dicyclohexylcarbodiimide is again dissolved in 0.5 mL of dimethylformamide solvent. The final solution is incubated at room temperature for 30 min. Activated succinimide ester is added dropwise to PLL solution (2.7 mg in 2 mL of 0.13 M freshly prepared bicarbonate) and stirred for an hour. The reaction mixture later undergoes dialyzation against 0.01 M sodium phosphate buffer (2 L) with pH 7.5 for 3 days, and in the end lyophilized (Chu et al. 1989).

Antisera Extraction

Once the toxins are prepared, they are induced into the host (rabbit) and antisera from rabbits are extracted and converted to a saturated form. The extraction of antisera is chronologically shown in Fig. 34.3.

34.2.2.3 Performing ELISA

Direct competitive ELISA:

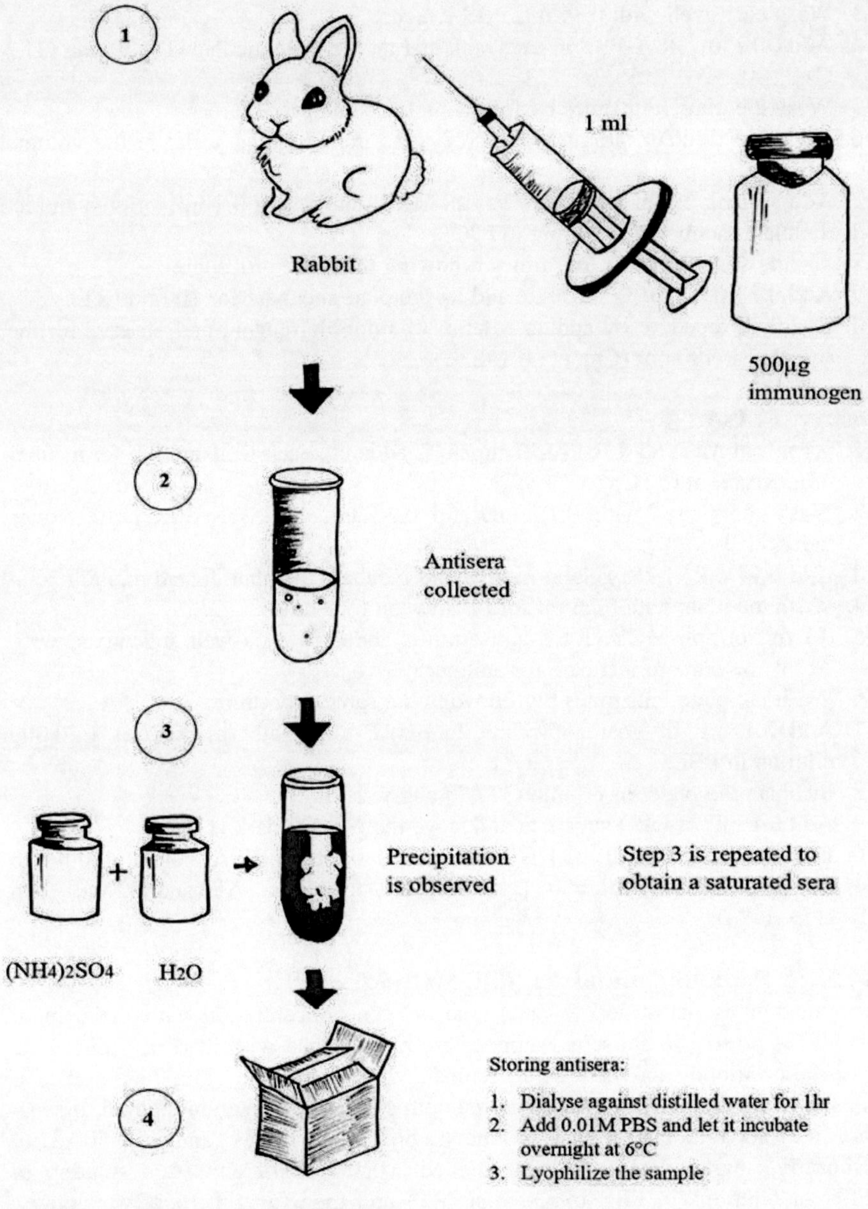

Fig. 34.3 Extraction of antisera from rabbit

1. ELISA plate used in this experiment is provided from Dynatech Immunlon I, or equivalent.
2. 0.1 mL of antiserum coating is spread in each well after diluting antibody with 0.01 M PBS and left for overnight incubation (6 °C).

3. Wash each well with 0.35 mL PBS-tween.
4. Add 0.17 mL BSA-PBS in each well and let the plate incubate for 30 min (37 ° C).
5. Wash the plate four times by following the same procedure.
6. Now add diluted variants of microcystins in individual wells in the volume 0.05 mL.
7. Add a blank buffer with MCYST-LR-HRP (marker) of 0.1 mL volume and let the plate incubate for 60 min (37 °C).
8. Again, wash the plate four times following the same procedure.
9. Add 0.1 mL of OPD substrate and let the plate incubate for 10 min (RT).
10. End the procedure by adding 0.1 mL of stooping reagent and observe for the samples at 490 nm (Chu et al. 1987).

Indirect ELISA:

1. 0.1 mL of MCYST-LR-PLL (antigen) is added to each well and left for incubation overnight (6 °C).
2. Wash the plates with 0.16 mL and 0.35 mL of PBS-tween buffer twice individually.
3. Add 0.17 mL of 1% gelatin in PBS and incubate the plate for 30 min (37 °C).
4. Wash the plate with 0.35 mL of PBS-tween four times.
5. 0.1 mL of anti-MCYST-LR-antiserum is added in each well and mixed well. Allow the plate to incubate for an hour (37 °C).
6. Wash the plate four times by following the same procedure.
7. Add 0.1 mL of goat anti-rabbit IgG-HRP conjugate (marker) at 1:20,000 dilution in PBS.
8. Incubate the plate for 60 min (37 °C) and wash it.
9. Add 0.1 mL of OPD substrate and leave the plate in dark at RT.
10. In the end add 0.1 mL of 1 N HCl and observe the ELISA plate at 490 nm in automatic reader (MR 600; Dynatech Industries, Inc., Alexandria, Va.) (Fan et al. 1984).

34.2.2.4 Radioimmunoassay (RIA) Method

This method is performed to determine antigen concentration with the help of antibodies present in antisera. Antigens are radiolabeled with markers which bind to specific antibody and increase the radioactivity thereby indicating the presence of antigen. Radioimmunoassay uses ammonium sulfate precipitation method to separate free (not radiolabeled) and bound (radiolabeled) toxins (antigen). 50 μL of radioactive marker ligand is incubated with 0.15 mL of antiserum solution of different dilutions in PBS for 30 min (RT) and then for 1 h (6 °C) or longer. Separation of bound and free ligand takes place by ammonium sulfate precipitation method. This method provided a clear conclusion on the free and bound toxins present in antiserum. Antibody titer is the reciprocal of the total amount of antiserum (mL) required to give 50% binding of tritiated microcystin (reduced MCYST-LR). In the experiment performed, total amount of antiserum used is 50 μL and marker ligand used in antibody titer is reduced MCYST-LR. Radioactivity was determined using liquid scintillation spectrometer (model LS-5801; Beckman Instruments, Inc., Fullerton, Calif.) in 4.5 mL volume of Ecolume (ICN Biochemicals, Inc., Irvine, Calif.).

MCYST-LR (0.5 mg) → reduced MCYST-LR

MCYST-LR was treated with below mentioned reagents in sequential order:

3.4 μmol of NaBH4 in 1mL redistilled 2-propanol; 0.85mL ethanol; 0.15mL of diluted acetic acid (0.1% glacial acetic acid in ethanol).

This reduced toxin is used in RIA as a marker for detection of specificity and determining antibody titers.

34.3 Conclusion

RIA titer values of MCYST-LR-EDA-BSA were the highest, 300–2500. Therefore, rabbits immunized with MCYST-LR-EDA-BSA were further studied (3 rabbits) and others were not.

The three rabbits ELISA and RIA reports showed nearly similar titer values. Production of antibody is observed as early as 4 weeks of time.

Direct ELISA is more versatile as it is less time consuming than indirect ELISA method (Msagati et al. 2006).

34.4 Advantages and Disadvantages of ELISA

Advantages	Disadvantages
Do not require radioactive compounds	Variable cross-reactivities hence may underestimate concentration of cyanotoxins
Evaluates only total value of cyanotoxin in a sample	Does not identify individual isoforms
Less equipment requirements, rapid and effective results	Does not assess the toxicity

- ELISA method helps in the determination of only free microcystins. Microcystins usually target the liver and mostly stay bound to the proteins. Therefore, the total count of microcystins can be detected only by Gas Chromatography/ Mass Spectroscopy.
- Limit of detection of ELISA for polyclonal antibody is 2.5 mg/L, and for monoclonal antibody—0.1 mg/L (Singh et al. 2012).

References

Chu FS, Fan TS, Zhang GS, Xu YC, Faust S, Mcmahon PL (1987) Improved enzyme-linked immunosorbent assay for aflatoxin Bj in agricultural commodities. J Assoc Off Anal Chem 70(5):854–857

Chu FS, Huang XUAN, Wei RD, Carmichael WW (1989) Production and characterization of antibodies against microcystins. Appl Environ Microbiol 55(8):1928–1933
Fan TS, Zhang GS, Chu FS (1984) An indirect enzyme-linked immunosorbent assay for T-2 toxin in biological fluids. J Food Prot 47(12):964–967
Hilborn ED, Carmichael WW, Yuan M, Azevedo SM (2005) A simple colorimetric method to detect biological evidence of human exposure to microcystins. Toxicon 46(2):218–221
Hilborn ED, Carmichael WW, Soares RM, Yuan M, Servaites JC, Barton HA, Azevedo SMFO (2007) Serologic evaluation of human microcystin exposure. Environ Toxicol 22(5):459–463
Koreivienė J, Belous O (2012) Methods for cyanotoxins detection. Bot Lith 18(1):58–65
Msagati TA, Siame BA, Shushu DD (2006) Evaluation of methods for the isolation, detection and quantification of cyanobacterial hepatotoxins. Aquat Toxicol 78(4):382–397
Singh S, Srivastava A, Oh HM, Ahn CY, Choi GG, Asthana RK (2012) Recent trends in development of biosensors for detection of microcystin. Toxicon 60(5):878–894
Sivonen K, Carmichael WW, Namikoshi M, Rinehart KL, Dahlem AM, Niemelä SI (1990) Isolation and characterization of hepatotoxic microcystin homologs from the filamentous freshwater cyanobacterium Nostoc sp. strain 152. Appl Environ Microbiol 56(9):2650–2657
Skulberg OM, Carmichael WW, Codd GA, Skulberg R (1993) Taxonomy of toxic Cyanophyceae (cyanobacteria). In: Algal toxins in seafood and drinking water, pp 145–164

In Vitro Assay for Determining Cyanotoxin Using Cell Line Method: Hepatotoxicity (Cell Lines—HepG2, Caco-2, and V79)

35

H. A. S. N. Abeysiri, K. T. Dilrukshi, S. A. Kulasooriya, and P. M. Manage

Abstract

Harmful cyanobacterial blooms are increasing and becoming a worldwide concern as many bloom-forming cyanobacterial species can produce toxic metabolites named cyanotoxins. These include microcystins, saxitoxins, anatoxins, nodularins, and cylindrospermopsins, which can adversely affect humans, animals, and the environment. Cyanotoxins, microcystin-LR (MC-LR), nodularin (NOD), and cylindrospermopsin (CYN), represent hazardous waterborne contaminants and potent human hepatotoxins. The application of in vitro models in evaluating toxicity has enhanced in recent decades. Many different in vitro models have been in use over the years, in which cell lines are the best model for the assessment of toxicity study of toxic compounds. Immortalized cell lines and primary isolated liver cells are currently the most widely used in vitro models for liver toxicity testing. Higher cytotoxic activity was gained in the 3-(4,5-dimethylthiazol-2-yl)-2,5-diphenyltetrazoliumbromide (MTT) assay compared with the PrestoBlue assay, with IC_{50} values after 24 h exposure, respectively. In the single-cell gel, electrophoresis assay also induces oxidative DNA damage and generated Reactive Oxygen Species (ROS), which is concentration-dependent correlated with the depletion of mitochondrial membrane potential and apoptosis.

H. A. S. N. Abeysiri · K. T. Dilrukshi · P. M. Manage
Centre for Water Quality and Algae Research, Department of Zoology, University of Sri Jayewardenepura, Nugegoda, Sri Lanka

Faculty of Graduate Studies, University of Sri Jayewardenepura, Nugegoda, Sri Lanka

S. A. Kulasooriya (✉)
National Institute of Fundamental Studies, Kandy, Sri Lanka
e-mail: ananda@kulasooriya.com

N. Thajuddin et al. (eds.), *Protocols for Cyanobacteria Sampling and Detection of Cyanotoxin*, https://doi.org/10.1007/978-981-99-4514-6_35

Keywords

Cyanotoxins · In vitro assays · Hepatotoxicity · Cell lines · Toxicity assessment

35.1 Introduction

Nowadays, a proliferation of cyanobacterial species can be seen globally because of water eutrophication and climate change (Diez-Quijada et al. 2019). Cyanotoxins are toxic secondary metabolites produced by various species of cyanobacteria, which involved an ample variety of compounds with different structural and physicochemical properties (Codd et al. 2016). Humans may be exposed to cyanotoxins via different routes, but oral exposure by means of contaminated water and foods (fish, crops, vegetables, and food supplements) is by far the most important (Roy-Lachapelle et al. 2017).

Microcystins (MCs), nodularin (NOD), and cylindrospermopsins (CYN) are among the most frequently investigated cyanotoxins due to their toxicity and extensive distribution. Microcystin-LR (MC-LR) and NOD are cyclic peptides with a molecular weight of 994 and 824, respectively. Their toxicities resulted on the inhibition of catalytic units of serine/threonine protein phosphatases types 1 (PP-1), 2A (PP-2A), and 3 (PP-3) (Dawson 1998). It has also been shown that oxidative damage might be involved in the hepatotoxicity of microcystins and nodularin (Lankoff et al. 2002). In cultured rat hepatocytes, cyanobacteria extract or microcystins are able to enhance intracellular production of Reactive Oxygen Species (ROS), to deplete intracellular glutathione (GSH), and to induce cell injury and lipid peroxidation (Ding et al. 2001). CYNs are guanidine alkaloid hepatotoxins with five known analogues. CYN is highly water soluble and chemically stable at high temperatures and a wide range of pH (Chiswell et al. 1999). For these reasons, humans can be more likely exposed to CYN than to other cyanotoxins as up to 90% of total CYN is presented in surrounding waters. The main mechanisms of CYN toxicity are the irreversible inhibition of protein synthesis and glutathione depletion related to the oxidative stress induced by CYN. Moreover, the bioactivation of CYN by cytochrome P-450 plays an important role in its mechanism of toxicity (Norris et al. 2002).

They have different chemical structures and mechanisms of action, thus interaction phenomena such as synergism, antagonism, or toxicity potentiation must be considered. Cell lines are extensively used for assessment of liver toxicity because they display similar genotypic and phenotypic characteristics of normal liver cells with functional enzymes responsible for phase I and phase II metabolism of xenobiotics (Strom et al. 2010). Liver cell lines are the best choice for toxicological and pharmacological testing for detection of toxic chemicals and evaluating their cellular mechanism of toxicity. Liver toxicity implies chemical-derived damage that led to several acute and chronic liver diseases. Many hepatic in vitro models are used in understanding of toxicity. Liver cell lines HepG2, Caco-2, and V79 are commonly immortalized liver-derived cell lines used for in vitro assessment of liver toxicity study (Guguen-Guillouzo and Guillouzo 2010).

35.2 Materials

Acrylamide (AA), dimethyl sulfoxide (DMSO), 3-(4,5-dimethylthiazol-2-yl)-2,5-diphenyltetrazolium bromide (MTT), Dulbecco's Modified Eagle's Medium (DMEM), ethylenediaminetetraacetic acid (EDTA), hydrogen peroxide (H_2O_2), low melting point (LMP) agarose, NaCl, NaOH, normal melting point (NMP) agarose, propidium iodide (PI), phosphate-buffered saline (PBS), streptomycin and penicillin, Triton X-100, Tris, Foetal Bovine Albumin (BSA), trypan blue, Foetal Bovine Serum (FBS), 4-(2-hydroxyethyl)-1-piperazineethanesulfonic acid (HEPES), and GlutaMAX™.

35.3 Methods

35.3.1 Cell Culture (HepG2, Caco-2, and V79) (Nowak et al. 2017a)

- Culture the cells in DMEM supplemented with 10% FBS, 4 mMGlutaMAX™, 25 mM HEPES, 100 μg/mL streptomycin, and 100 IU/mL penicillin.
- Incubate the cells in a humidifier at 37 °C with 5% CO_2 for 7–10 days to achieve 80% confluence.
- Wash the cells every 3 days with 0.1 M PBS (pH 7.2) and renew the medium.
- Detach the confluent cells from the culture with TrypLE™ Express (37 °C, 10 min) and suspend in sterile PBS, and gently aspirate.
- Centrifuge the cell suspension (182 × *g*, 5 min) and decant, and then re-suspend the pellet in fresh DMEM. After performing a cell count by hemocytometer and determining cell viability by trypan blue exclusion, it is ready to use the Caco-2 cells.

35.3.2 MTT and PrestoBlue Assays (Nowak et al. 2017a)

- Seed 1 × 10^4 Caco-2 cells in each well of a 96-well plate in complete culture medium.
- Incubate the cells overnight at 37 °C under 5% CO_2.
- Aspirate the medium and add AA in DMEM to each well to achieve the final tested concentrations [mM]: 0.2, 0.4, 0.8, 1.6, 3.2, 6.4, 12.5, 25, and 50.
- Add DMEM without AA to the negative control.
- Incubate the cells at 37 °C under 5% CO_2 for 24–72 h. After incubation, remove the medium with AA from each well and add MTT (0.5 mg/mL) and incubate at 37 °C under 5% CO_2 for 3 h.

- After that time, remove the MTT and solubilize the formazan precipitates by adding DMSO.
- Measure the absorbance at 550 nm with a reference filter of 620 nm, using a microplate reader.
- Perform the PrestoBlue assay according to the MTT assay procedure. After incubation, remove the medium with AA from each well and add the PrestoBlue reagent (10% solution in PBS) to each well and incubate at 37 °C under 5% CO_2 for 2 h.
- Measure the fluorescence at λ_{ex} 560 nm and λ_{em} 590 nm using the microplate reader.

35.3.3 Cytotoxicity and Half Maximal Inhibitory Concentration (IC_{50}) (Nowak et al. 2017a; Organization for Economic Cooperation and Development (OECD) protocol 2019)

- The absorbance/fluorescence of the control sample (untreated cells) represent the 100% cell viability.
- Calculate the cell viability (%) as follows:

$$[\text{Sample OD or fluorescence}/\text{control OD or fluorescence}] \times 100\%$$

- And determine the cytotoxicity (%) as 100 of cell viability (%).
- Present the results as mean ± standard deviation (SD)/standard error of the mean (SEM).
- Determine the IC_{50} value from curves according to OECD protocol, 2019.

35.3.4 Genotoxicity Testing (Comet Assay) (Nowak et al. 2017b)

- Perform the incubation of suspended Caco-2 cells (10^5 cells/mL) with AA final concentrations (0.2, 0.8, 3.2, 6.4, and 12.5 mM) and without AA as a negative control (1 h at 37 °C).
- The positive control contains 50 μM H_2O_2.
- Incubate all samples in DMEM without supplements.
- Set the final amount of each sample to 1 mL.
- Perform the comet assay under alkaline conditions (pH $>$ 13).
- After incubation, centrifuge the aliquots of suspended cells (182× *g*, 15 min, 4 ° C); decant; suspend in 0.75% LMP agarose; distribute onto slides, precoat with 0.5% NMP agarose; and immerse in lysing solution consisting of 2.5 M NaCl, 1% Triton X-100, 100 mM EDTA, and 10 mM Tris, at pH 10 (4 °C, 1 h).
- After lysis, allow DNA to unwind for 20 min in a solution consisting of 300 mM NaOH and 1 mM EDTA.
- Then, subject the slides to horizontal gel electrophoresis in an electrophoretic solution containing 300 mM NaOH and 1 mM EDTA, pH $>$ 13.

- Conduct the electrophoresis at 4 °C for 30 min at an electric field strength of 0.73 V/cm (300 mA).
- Then, neutralize the slides with distilled water.
- The following day, stain the slides with 1 mg/mL PI and visualize at 200× magnification with a fluorescence microscope.
- Randomly select 100 images from each sample and measure the percentage of DNA in the comet tail.
- Present the results as mean ± standard error of the mean.

References

Chiswell RK, Shaw GR, Eaglesham G, Smith MJ, Norris RL, Seawright AA, Moore MR (1999) Stability of cylindrospermopsin, the toxin from the cyanobacterium, Cylindrospermopsis raciborskii: effect of pH, temperature, and sunlight on decomposition. Environ Toxicol Int J 14:155–161

Codd GA, Meriluoto J, Metcalf JS (2016) Introduction: cyanobacteria, cyanotoxins, their human impact, and risk management. In: Handbook of cyanobacterial monitoring and cyanotoxin analysis, pp 1–8

Dawson RM (1998) The toxicology of microcystins. Toxicon 36:953–962

Diez-Quijada L, Prieto AI, Guzman-Guillen R, Jos A, Camean AM (2019) Occurrence and toxicity of microcystin congeners other than MC-LR and MC-RR: a review. Food Chem Toxicol 125: 106–132

Ding WX, Shen HM, Ong CN (2001) Critical role of reactive oxygen species formation in microcystin-induced cytoskeleton disruption in primary cultured hepatocytes. J Toxicol Environ Health A 64:507–519

Guguen-Guillouzo C, Guillouzo A (2010) General review on in vitro hepatocyte models and their applications. In: Hepatocytes: methods and protocols, pp 1–40

Lankoff A, Banasik A, Nowak M (2002) Protective effect of melatonin against nodularin-induced oxidative stress. Arch Toxicol 76:158–165

Norris R, Seawright A, Shaw G, Senogles P, Eaglesham G, Smith M, Chiswell R, Moore M (2002) Hepatic xenobiotic metabolism of cylindrospermopsin in vivo in the mouse. Toxicon 40:471–476

Nowak, A., Sójka, M. and Klewicka, E., 2017a. Lipinska, L.; Klewicki, R.; Kołodziejczyk, K. Ellagitannins from Rubusidaeus L. exert geno-and cytotoxic effects against human colon adenocarcinoma cell line Caco-2. J Agric Food Chem, 65, 2947–2955

Nowak A, Bakuła T, Matusiak K, Gałęcki R, Borowski S, Gutarowska B (2017b) Odorous compounds from poultry manure induce DNA damage, nuclear changes, and decrease cell membrane integrity in chicken liver hepatocellular carcinoma cells. Int J Environ Res Public Health 14(8):933

OECD (2019) Guidelines for the testing of chemicals, section 4. Test no. 442D: in vitro skin sensitisation ARE-Nrf2 luciferase test method. https://ntp.niehs.nih.gov/iccvam/suppdocs/feddocs/oecd/oecd-tg442d-508.pdf. Accessed 19 Nov 2019

Roy-Lachapelle A, Solliec M, Bouchard MF, Sauvé S (2017) Detection of cyanotoxins in algae dietary supplements. Toxins 9:76

Strom SC, Davila J, Grompe M (2010) Chimeric mice with humanized liver: tools for the study of drug metabolism, excretion, and toxicity. In: Hepatocytes: Methods and Protocols, pp 491–509

36 In Vitro Assay for Determining Cyanotoxin Using Cell Line Method: Neurotoxicity (Neuro-2a Neuroblastoma Cell)

M. Haripriyaa and K. Suthindhiran

Abstract

Cyanobacteria are phytoplankton prevalent in freshwater surface waters that can aggregate on the surface as blue-green "scums" or concentrate in surface water supplies as blooms. These secondary metabolites produce various harmful toxins known as cyanotoxins that impair treated water sources and pose a risk to human health and aquatic life by exploiting the food chain. They are of concern because they can produce a variety of toxins that are harmful to both human and animal health and exhibit hepatotoxic, neurotoxic, cytotoxic, and dermatotoxic behavior patterns. With regard to neurotoxicity, the brain can be a crucial target for various cyanotoxins escalating several neurodegenerative diseases. These toxins can block the voltage-gated sodium channels hindering to perform basic neurological functions. Also, the precise target/interacting proteins, the signaling cascades initiating the cell responses, and the downstream pathways of toxicity and cell injury still need to be discovered. Therefore, this chapter summarizes the detection of neurotoxicity using in vitro conditions.

Keywords

Phytoplankton · Neurotoxic · Cyanotoxin · In vitro

M. Haripriyaa · K. Suthindhiran (✉)
Marine Biotechnology and Bioproducts lab, Department of Biomedical Sciences, School of Biosciences and Technology, Vellore Institute of Technology, Vellore, India
e-mail: ksuthindhiran@vit.ac.in

N. Thajuddin et al. (eds.), *Protocols for Cyanobacteria Sampling and Detection of Cyanotoxin*, https://doi.org/10.1007/978-981-99-4514-6_36

36.1 Introduction

The severe toxicity of cyanotoxins for mammalian cells, including brain cells, has been revealed through intense research. They are known to bioaccumulate in the food chain and resist chemical and biological degradation. With relevance to their mechanism of action, they can be divided into three main categories: hepatotoxins (microcystins and cylindrospermopsin), neurotoxins (anatoxins), and dermatotoxins (Sazdova et al. 2022; Takser et al. 2016). Certain cyanotoxins are known to impair an organism's ability to operate quickly, with the nervous system as the prime target. Rapid neural communication depends on the production of two different types of electrical signals, with dendrites responding to neurotransmitters released by neighboring neurons by slowly depolarizing junctional potentials that evoke action potentials. When this reaches the end of the axon, the neurotransmitter is released and the whole process starts again (Florczyk et al. 2014; Metcalf et al. 2021).

Cyanotoxins have been categorized using various methods, the two most popular being their target of action or chemical composition (Kubickova et al. 2019). Among the harmful cyantotoxins, microcystin-LR (MC-LR) is thought to primarily exhibit its tumor-promoting effects by reorganizing the control of phosphorylated-dependent proteins. The pleiotropic downstream mechanisms connect tumor promotion and neoplastic transformation by inducing cell growth, producing reactive oxygen species (ROS), oxidative stress, damaging mitochondrial DNA, and altering cell phenotype to MC-LR-dependent inhibition of eucaryotic protein phosphatases (PPs) PP1, PP2A, phospho-PP4, and phospho-PP5 (Br'ozman et al. 2020).

For toxicological studies, cell cultures are very practical because they exhibit the mechanisms of cytotoxic effects, the affected tissues, intracellular targets, and strategies to reduce cytotoxicity. It is quick and efficient to evaluate the harmful impacts on humans and also aids in determining the most vulnerable tissues by using human cell lines in toxicological investigations. Therefore, cell line investigations can provide a scaffold for clinical trials in animal subjects by significantly aiming the target of interest.

For the identification of marine biotoxins in seafood products, the neuroblastoma cell-based assay (CBA-N2a) is frequently utilized (Viallon et al. 2020). Due to the large variety of congeners present in minute quantities in polluted biological matrices, detection and quantification of these classes of marine toxins continue to be extremely difficult (Caillaud et al. 2010). Therefore, we aim to reveal the effect of cyanotoxin on human cell lines to assess the neurotoxicity in Nero-2A neuroblastoma cells and identify the intracellular targets of the cyanotoxins.

36.2 Materials

- HEPES
- Fetal Bovine Serum
- RPMI Medium
- Streptomycin

- Penicillin
- Amphotericin B
- Phosphate-Buffered Saline
- Trypsin-EDTA
- Sterile Distilled Water
- Ouabain Octahydrate
- MeOH
- MTT
- DMSO
- Veratridine

36.3 Methods

36.3.1 Characterization of N2a Cell Growth and Initial Viability (Caillaud et al. 2010)

1. To resume the process, the growth curve of the N2a cell line should be assessed in 96-well microplates over a period of 4 days. For more convenience, growth experiment can be conducted in 25 cm^2 culture flasks. Therefore, twelve culture flasks are seeded with 335,000 cells resuspended in 11 mL 10% FBS culture medium solution, and left to incubate for 22 h. Cell densities can be monitored by sacrificing one flask every 6 h starting from 22 to 76 h.
2. The two last flasks should be sacrificed at 85 and 100 h, respectively. Cell enumeration will be achieved by manual counts ($n = 10$/flask) using KOVA Glasstic slides (Hycor), and the results can be plotted against time.
3. Cell density data should be normalized to a surface of 0.32 cm^2 and used as a proxy of cell densities in 96-well microplates. Using this rationale, the initial cell density in a 96-well microplate can be estimated at 4288 cells per well.
4. Proceeding this, the initial viability of N2a cells is to be examined under two different % FBS (5% and 10% in culture medium) and six different MTT incubation times (15–65 min). Cell numbers before seeding should be determined from $n = 5$ counts performed on an aliquot of the cell stock suspension and applied for all further experiments. To this end, twelve microplates (six microplates per % FBS) are seeded with 200 μL of ten different cell seeding densities ranging from 10,000 to 100,000 cells/well.
5. After 26 h growth, assess the cell viability via the MTT assay by sacrificing one microplate at 15 min and the remaining every 10 min.
6. For this, after removing the culture medium from the microplate, fill the 60 inner wells with 60 μL of MTT solution at 0.83 mg/mL prepared in 2% FBS culture medium. After MTT incubation at 37 °C in a 5% CO_2 incubator, the wells should be emptied and fill the 60 inner wells and 12 outer wells with 100 μL of DMSO. Following cell layer lysis with DMSO and manual homogenization, absorbance data should be measured at 570 nm. All viability data will be expressed in absorbance data.

36.3.2 Cytotoxicity Assays

36.3.2.1 Detection of Microcystins (MCs) and Cylindrospermopsin (CYN) (Viallon et al. 2020)

1. For the cytotoxicity assays, seed the N2a cells at a density of 5×10^5 cells/mL in 96-well culture plates and incubated at 37 °C and 5% CO_2 for 24 h before the exposure. From the stock solution of 4000 μg/mL Microcystins (MCs) and Cylindrospermopsin (CYN) {MC-LR}, serial dilutions are prepared from 0 to 200 μg/mL MC-LR. In the case of CYN, from the stock solution of 1000 μg/mL CYN, serial dilutions are prepared from 0 to 25 μg/mL CYN.

2. The concentrations are selected in order to obtain mean effective concentration (EC_{50}) values. The solvent control (MeOH) for MC-LR and negative control (non-treated cells) were also included. All the dilutions should be prepared in a medium without serum.

3. These cells are treated with the exposure solutions for 4, 24 and 48 h at 37 °C. The cytotoxicity biomarkers are assayed for total protein content (TP) that can be evaluated by Bradford method and identification of reduction of the tetrazolium salt MTS.

36.3.3 Detection of Saxitoxin (Diez-Quijada et al. 2022; Constante et al. 2022)

1. The strain used in the experiment is *Raphidiopsis raciborskii* (T-3), formerly belonging to the genus Cylindrospermopsis. Saxitoxins extracts are obtained from 150 mL cultures of *Raphidiopsis raciborskii* cells from the Culture Collection of the Laboratory of Cyanobacteria and Phycotoxins (LCF).
2. N2a cells are maintained as a monolayer in 75 cm^2 flasks in Dulbecco's Modified Eagle's Medium, supplemented with 10% heat-inactivated fetal bovine serum and antibiotics (1% of penicillin/streptomycin solution and 0.1% of gentamycin solution) and kept in a humidified 5% CO_2 atmosphere at a temperature of 37 °C until confluence is reached. SH-SY5Y cells are seeded and cultured for 24 h.
3. CBMN-Cyt assay is carried out by treating the incubated cells with different concentrations (1.25–10 μg/L) of Raphidiopsis extracts containing saxitoxin variants as well as negative control (0.05 M HCL) and positive control (bleomycin 3 μg/mL) for 24 h of treatment.
4. After treatment, SH-SY5Y cells are washed twice in Dulbecco's phosphate saline buffered and cytochalasin B was added to a final concentration of 3 μg/mL in complete fresh medium. After two cell cycles, Cyt B was removed and the cells were collected and harvested by cytocentrifugation. Transfer 160 μL of cell suspension to cytocentrifuge cups and centrifuge for 5 min at 700 rpm to produce 1 spot per slide. After staining, slides are air-dried and examined under 400× magnifications using a light microscope.

Acknowledgments The authors are thankful to Vellore Institute of Technology, Vellore, for providing the necessary facilities to carry out this research.

References

Br'ozman O, Kubickova B, Babica P, Laboha P (2020) Microcystin-LR does not alter cell survival and intracellular signaling in human bronchial epithelial cells. Toxins 12(3):165

Caillaud A, de la Iglesia P, Darius HT, Pauillac S, Aligizaki K, Fraga S, Chinain M, Diogène J (2010) Update on methodologies available for ciguatoxin determination: perspectives to confront the onset of ciguatera fish poisoning in Europe. Mar Drugs 8:1838

Constante JS, Al Khateeb JE, de Souza AP, Conter FU, Lehmann M, Yunes JS, Dihl RR (2022) In vitro and in silico assessment of cytotoxicity and chromosome instability induced by saxitoxin in human derived neural cell line. An Acad Bras Cienc 94:1–13

Diez-Quijada L, Puerto M, Gutiérrez-Praena D, Turkina MV, Campos A, Vasconcelos V, Cameán AM, Jos Á (2022) In vitro toxicity evaluation of cyanotoxins Cylindrospermopsin and microcystin-LR on human kidney HEK293 cells. Toxins 14(7):1–20

Florczyk M, Łakomiak A, Woźny M, Brzuzan P (2014) Neurotoxicity of cyanobacterial toxins. Environ Biotechnol 10:26–43

Kubickova B, Babica P, Hilscherova K (2019) Effects of cyanobacterial toxins on the human gastrointestinal tract and the mucosal innate immune system. Environ Sci Eur 31(1):1–27

Metcalf JS, Tischbein M, Cox PA, Stommel EW (2021) Cyanotoxins and the nervous system. *Toxins* 13(9):660

Sazdova I, Keremidarska-Markova M, Chichova M, Uzunov B, Nikolaev G, Mladenov M, Schubert R, Stoyneva-Gärtner M, Gagov HS (2022) Review of Cyanotoxicity studies based on cell cultures. J Toxicol 2022:5647178

Takser L, Benachour N, Husk B, Cabana H, Gris D (2016) Cyanotoxins at low doses induce apoptosis and inflammatory effects in murine brain cells: potential implications for neurodegenerative diseases. Toxicol Rep 3:180

Viallon J, Chinain M, Darius HT (2020) Revisiting the neuroblastoma cell-based assay (CBA-N2a) for the improved detection of marine toxins active on voltage gated sodium channels (VGSCs). Toxins 12(5):281

Genotoxicity of Cyanotoxins 37

H. D. D. Sadeepa, K. R. V. Bandara, G. Y. Liyanage, P. M. Manage, and S. A. Kulasooriya

Abstract

According to the stability and propensity for biomagnification along food webs, the cyanotoxins produced by more than 40 species of cyanobacteria have a detrimental effect on the aquatic food web. A more plausible exposure scenario involves a combination of cyanotoxins, and interaction phenomena cannot be completely ruled out. Therefore, the combined effects of multiple cyanobacterial toxins can potentially be more dangerous than just one of them. In brackish and freshwater blooms, the cyclic heptapeptides microcystins (MCs) and nodularin (NODs) are the most ubiquitous and common cyanotoxins. The primary route through which humans are exposed to cyanotoxins is oral ingestion. Their hepatotoxic, dermatoxic, neurotoxic, cytotoxic, and genotoxic properties pose some risks to the general public's health with increasing attention. Well-known cyanotoxins with genotoxicity include microcystins (MCs), nodularin (NODs), and cylindrospermopsin (CYN), which are commonly found in freshwater environments. Several in vitro and in vivo assays may be used to examine the

H. D. D. Sadeepa · K. R. V. Bandara · P. M. Manage
Department of Zoology, Centre for Water Quality and Algae Research, University of Sri Jayewardenepura, Nugegoda, Sri Lanka

Faculty of Graduate Studies, University of Sri Jayewardenepura, Nugegoda, Sri Lanka

G. Y. Liyanage
Department of Zoology, Centre for Water Quality and Algae Research, University of Sri Jayewardenepura, Nugegoda, Sri Lanka

Faculty of Urban and Aquatic Resources Management, University of Sri Jayewardenepura, Nugegoda, Sri Lanka

S. A. Kulasooriya (✉)
National Institute of Fundamental Studies, Kandy, Sri Lanka
e-mail: ananda@kulasooriya.com

N. Thajuddin et al. (eds.), *Protocols for Cyanobacteria Sampling and Detection of Cyanotoxin*, https://doi.org/10.1007/978-981-99-4514-6_37

potential genotoxicity of cyanotoxins. Thus, the in vitro and in vivo assay protocols for determining the genotoxicity of cyanotoxins produced by cyanobacteria are covered in this chapter.

Keywords

Cyanobacteria · Cyanotoxins · Genotoxicity · In vitro and in vivo assays · Assay protocols

37.1 Introduction

Eutrophication of waterbodies and global warming is the long-term effect of environmental pollution, and they cause lots of damage to the flora and fauna diversity of the ecosystem. Consequently, the manifestation of cyanobacterial blooms in the aquatic environment has surprisingly increased in past decades. More than 40 cyanobacteria genera produce cyanotoxins that adversely affect the biome (Žegura et al. 2011). Toxin-producing cyanobacteria blooms have been recorded from terrestrial, freshwater bodies, viz., ponds, rivers, lakes, and brackish water throughout the world (Žegura et al. 2011; Yilmaz et al. 2022).

Cyanotoxins are precarious for humans, other animals, and aquatic organisms. In humans, intoxication with cyanobacteria-contaminated water causes symptoms such as nausea/vomiting, weakness, skin irritation, and illnesses ranging from gastroenteritis and pneumonia to hepatoenteritis (Zanchett and Oliveira-Filho 2013). Oral ingestion is the main method by which humans are exposed to cyanotoxins and carry some dangers for the public's health with increased exposure due to their hepatotoxic, dermatoxic, neurotoxic, cytotoxic, and genotoxic effects of them (Table 37.1). Cases of poisoning in humans and animals due to consuming contaminated water with cyanotoxins were reported for the first time in the late 1800s. The mechanisms behind the toxic effects of cyanotoxins are different because they are structurally diverse chemicals (Žegura et al. 2011; Zanchett and Oliveira-Filho 2013).

Microcystins, nodularin, and cylindrospermopsin are well-documented cyanotoxins with genotoxicity, and these cyanotoxins are commonly found in freshwater environments. The genotoxicity effect of the cyanotoxin can be analyzed using a set of in vitro and in vivo assays (Žegura et al. 2011). The in vitro assays can be performed using bacteria systems, viz., *Salmonella typhimurium* TA1537, *Escherichia coli* PQ37, *Bacillus subtilis* 168, animal cell lines, viz., rat hepatocytes, hamster ovary cell lines, and human cell lines such as human hepatoma cell line, human lymphocytes, and human glioblastoma cell lines (Žegura et al. 2011; Tejs 2008; Guo et al. 2020). Further, the genotoxicity of cyanobacteria can be assessed using animal models such as Swiss albino mice, Kunming mice, Fischer 344 rats, and Sprague–Dawley rats as test organisms and those studies have been categorized as in vivo assays. Thus, this chapter discusses the in vitro and in vivo assay protocols for detecting the genotoxicity of cyanotoxins produced by cyanobacteria.

Table 37.1 Types of cyanotoxins, toxin-producing cyanobacteria and effect on human health

Type of cyanotoxin	Toxin-producing cyanobacteria	Effect of human health	References
Microcystins (MCs)	*Microcystis aeruginosa, Cyanobium bacillare*	Neurotoxins, Genotoxin	Žegura et al. (2011), Yilmaz et al. (2022), Zanchett and Oliveira-Filho (2013)
Nodularin (NOD)	*Nodularia spumigena*	Hepatotoxins, Genotoxins	Žegura et al. (2011), Yilmaz et al. (2022)
Cylindrospermopsin (CYN)	*Anabaena lapponica, Cylindrospermopsis raciborskii, Lyngbia wollei*	Hepatotoxins, Genotoxins	Žegura et al. (2011), Yilmaz et al. (2022)
Anatoxin-a (Antx-a)	*Anabaena planctonica, Anabaena spiroides, Microcystis* sp.	Hepatotoxins, Neurotoxins	Žegura et al. (2011), Yilmaz et al. (2022), Zanchett and Oliveira-Filho (2013)
Anatoxin a-(s) (Antx-a(s))	*Anabaena flos-aquae and Anabaena lemmermannii*	Neurotoxins	Žegura et al. (2011), Zanchett and Oliveira-Filho (2013)
Saxitoxins (STX	*Anabaena circinalis, Cylindrospermopsis raciborskii*	Neurotoxins	Žegura et al. (2011), Zanchett and Oliveira-Filho (2013)
Microviridin J	*Microcystis* spp.		Žegura et al. (2011), Zanchett and Oliveira-Filho (2013)

37.2 Materials

- Cyanotoxin standards
- Bacteria culture media
- Bacteria strains
- Cell lines
- Cell culture media
- Cell culture treated culture plates
- Trypsin-EDTA
- Cell staining solutions

37.3 Methods

37.3.1 In Vitro Studies

37.3.1.1 Bacteria System

Bacterial Reverse Mutation Test (Ames test)

The Ames Salmonella/microsome mutagenicity assay (Salmonella test; Ames test) is a short-term bacterial reverse mutation assay specifically designed to detect a wide range of chemical substances that can produce genetic damage that leads to gene mutations (Tejs 2008).

1. Steps are taken prior to experimenting:
   (a) Inoculate Salmonella cultures 12 h prior to performing the experiment.
   (b) Label an appropriate number of minimal glucose (GM) agar plates and sterile test tubes for each test chemical.
   (c) Prepare metabolic activation system and keep on ice until use.
   (d) Prepare cyanotoxin dilutions.
   (e) Melt top agar supplemented with 0.05 mM histidine and biotin and maintained at 43 °C to 48 °C.
2. To the 100 mL sterile tubes maintained at 43 °C, add in the following order with mixing after each addition:
   (a) 0.01 mL of the test cyanotoxin dilution.
   (b) 2.5 mL of molten top agar with overnight culture of the *Salmonella* strain.
   (c) 0.5 mL of metabolic activation (S-9) mix.
3. The contents of the test tubes are then mixed and poured onto the surface of GM agar plates.
4. When the top agar has hardened (2-3 min), the plates are inverted and placed in a 37 °C incubator for 48 h.
5. The colonies are then counted, and the results are expressed as the number of revertant colonies per plate.

37.3.1.2 Cell Lines

Micronucleus (MN) Assay Using Human Hepatic Cell Line HepaRG

The micronucleus (MN) assay is a core test used to evaluate the genotoxic potential of xenobiotics. The traditional in vitro MN assay is usually conducted in cells lacking metabolic competency or by supplementing cultures with an exogenous rat S9 metabolic system. This creates a significant assay limitation for detecting genotoxic metabolites (Guo et al. 2020; Díez-Quijada et al. 2019).

1. Prepare HepaRG human hepatoma cell line.
   (a) Seed proliferative HepaRG cells at passages 13–19 at 1.3×10^4 cells cm^2 in a 100-mm tissue culture dish and culture in 10 mL William's E medium supplemented with 1% GlutaMax, 1% Pen-Strep solution, and growth

additives (Lonza, Walkersville, MD) at 37 °C for 14 days in a humidified atmosphere with 5% CO_2.
   (b) Differentiate cells in William's E medium supplemented with differentiation additives (Lonza) for an additional 14 days.
   (c) Detach the fully differentiated HepaRG cells by adding trypsin—EDTA and reseed in a 96-well plate at a density of 5×10^4 cells/well in William's E differentiation medium and cultured for 3 days prior to cyanotoxin treatment.
   (d) Prepare toxin stock solutions by dissolving each cyanotoxin standard in DMSO.
   (e) Serially dilute working solutions in 100 μL differentiation medium, and the final concentration of DMSO in the medium never exceed 1%.
   (f) Freshly prepared Cisplatin, ENU, and hydroquinone (HQ) solutions before each experiment. Following 24-h treatment, remove media and maintain HepaRG cells in fresh differentiation medium supplemented with 100 $ngml^{-1}$hEGF for an additional 72 h prior to the MN assay.
2. Perform cytokinesis-block micronucleus (CBMN) assay.
   (a) Detach the cells by trypsinization and adhere to Goldseal microscope slides (ThermoFisher) by centrifuging 200 to 300 μL cell suspension at 850 rpm for 4 min in a CytoFuge 2 (Statspin).
   (b) Briefly air dry the slides and fix in absolute methanol for 10 min, and store at ambient temperature. Stain the HepaRG cell slides by applying 25 μL 250 $ng\mu l^{-1}$propidium iodide in 1× DPBS, covering with a 25 mm^2 cover glass and incubating for 5 min at ambient temperature.
   (c) Wick away excess PI staining solution from the slides with a Kimwipe and mount the cells in 25 μL SlowFade Gold Antifade with DAPI.
   (d) Examine slides with a DAPI-Texas Red-FITC triple band pass filter at 400× and MN verify using a DAPI single band pass filter at 1000 × magnification on an Eclipse 80i microscope.
   (e) Evaluate the cytotoxic and cytostatic effect of each treatment using the cytokinesis-block proliferation index (CBPI), by counting 500 cells that were either single-nucleated, BN or multi-nucleated.

37.3.2 In Vivo Studies

Although the genotoxicity of cyanotoxin has been extensively studied in vitro, limited data are available on its in vivo genotoxicity effect of cyanotoxins (Díez-Quijada et al. 2019).

37.3.2.1 Comet Assay Using Wister Rats

1. Obtain ethical clearance for the animal study.
2. Purchase male Wistar rats and weigh the initial body mass of each animal.
3. Acclimates the animals to the environmental conditions.

   (a) 12-h dark/light cycle, control temperature (23 ± 1 °C), relative humidity (55 ± 10%) for 1 week before the experiment.
   (b) During this time, feed the animals with a standard laboratory diet and water.
   (c) Use 5 animals per group and the 3 animals for positive controls.
4. Treat with different concentrations of cyanotoxins.
   (a) Weigh the body mass of each animal in order to ensure that weight variation did not exceed ±20% and randomly divide into groups.
   (b) Prepare the cyanotoxin concentration series in water in a final volume of 1 mL.
   (c) Feed the animals by gavage using an enteral feeding tube at 0; 24 and 45 h.
   (d) The animals were sacrificed 3 h after the final dose administration comet.
   (e) Record the clinical signs and body weight of each animal during the treatment period.
5. Sample collection.
   (a) Remove and dissect the elected tissues, then rinse with cold saline solution and weigh.
   (b) After weight, process stomach and liver quickly for the comet assay. Collect blood samples and maintain in Vacutainer® sodium heparin tubes.
   (c) For the histopathological study, collect a portion of stomach and liver samples from each individual separately and store.
6. Prepare samples for comet assay.
   (a) Prepare single cell suspensions from the stomach and liver.
   (b) Wash tissues with Merchant's buffer (MB) (0.14 M NaCl, 1.47 nM $KH_2 PO_4$, 2.7 mMKCl, 8.1 mM $Na_2 HPO_4$, 10 mM Na_2 EDTA, pH 7.4) and discard.
   (c) Homogenize a portion of each tissue in cold.
   (d) Centrifuge the homogenates and mix with 5 mL buffer after filtration until slide preparation.
   (e) Mix heparinized blood samples v/v (1/1) with phosphate-buffered saline (PBS) solution and isolate the lymphocytes with Histopaque® (Sigma-Aldrich, Madrid, Spain) and centrifuge (400 G, 30 min).
   (f) Wash the cells twice with PBS and re-suspend in PBS at a concentration of 2×10^5 cells/mL.
7. Perform comet assay.
   (a) For blood samples, mix 30 μL of cells suspension with 140 μL of 0.5% low-melting point agarose.
   (b) Place 5 μL aliquots on a microscope slide.
   (c) For the stomach and liver, mix the cells suspensions with 1% low-melting point agarose and place the mixtures on a microscope slide similar to blood samples.
   (d) Perform standard and modified comet assays.
   (e) Briefly, after lysis, wash slides three times for 5 min with enzyme buffer (40 mM HEPES; 0.1 M KCl; 0.5 mM EDTA; 0.2 mg/mL bovine serum albumin; pH 8).

(f) Expose two gels in each slide to 30 μL lysis solution; enzyme buffer alone (buffer F); buffer F containing Fpg or Endo III for 30 min in a metal box at 37 °C.
(g) Denature the nuclei and perform electrophoresis for 20 min, 0.81 V/cmup to 400 mA.
(h) Neutralize the DNA in PBS, wash with water and fix with 70% and absolute ethanol before staining.
(i) Analyze at least 150 randomly selected nuclei per animal with the image analysis software Comet Assay IV.

References

Díez-Quijada L, Llana-Ruiz-Cabello M, Cătunescu GM, Puerto M, Moyano R, Jos A, Cameán AM (2019) In vivo genotoxicity evaluation of cylindrospermopsin in rats using a combined micronucleus and comet assay. Food Chem Toxicol 132:110664

Guo X, Seo JE, Petibone D, Tryndyak V, Lee UJ, Zhou T, Robison TW, Mei N (2020) Performance of HepaRG and HepG2 cells in the high-throughput micronucleus assay for in vitro genotoxicity assessment. J Toxic Environ Health A 83(21–22):702–717

Tejs S (2008) The Ames test: a methodological short review. Environ Biotechnol 4:7–14

Yilmaz S, Ülger TG, Göktaş B, Öztürk Ş, Karataş DÖ, Beyzi E (2022) Cyanotoxin genotoxicity: a review. Toxin Rev 41(2):699–712

Zanchett G, Oliveira-Filho EC (2013) Cyanobacteria and cyanotoxins: from impacts on aquatic ecosystems and human health to anticarcinogenic effects. Toxins 5(10):1896–1917

Žegura B, Štraser A, Filipič M (2011) Genotoxicity and potential carcinogenicity of cyanobacterial toxins–a review. Mutat Res 727(1–2):16–41

38 In Silico Analysis of Cyanotoxin Using Computational Tools

Santhiya Panchalingam, Govindaraju Kasivelu, and Raja Muthuramalingam Thangavelu

Abstract

Danger lurks in the waters: Cyanobacteria, or blue-green algae, can produce toxic cyanotoxins that pose serious risks to human health, as well as to plants, aquatic organisms, and animals. These toxins, which include Microcystins, Cylindrospermopsin, Anatoxins, and Saxitoxins, can bind to host proteases and cause infection. In this study, we used computational methods like molecular docking and molecular dynamics simulation to uncover the binding site and structure of these toxins, and to understand how they interact with their hosts. Our findings highlight the harmful effects of cyanotoxins and the need for new approaches to combat them. Dive into the research and discover the power of in silico predictions to protect against these aquatic dangers.

Keywords

In silico · Cyanotoxin · Microcystins · Cylindrospermopsin · Anatoxins · Saxitoxins

S. Panchalingam · G. Kasivelu (✉)
Centre for Ocean Research (DST-FIST Sponsored Centre), Sathyabama Institute of Science and Technology, Chennai, India

R. M. Thangavelu
Department of Plant Pathology and Ecology, The Connecticut Agricultural Experiment Station, New Haven, CT, USA

N. Thajuddin et al. (eds.), *Protocols for Cyanobacteria Sampling and Detection of Cyanotoxin*, https://doi.org/10.1007/978-981-99-4514-6_38

38.1 Introduction

Cyanobacteria are Gram-negative microorganisms that produce energy through photosynthesis (Haselkorn 2009). Despite being commonly referred to as blue-green algae, cyanobacteria are not classified as algae (Anon 2012). The blue pigment they use for photosynthesis is a by-product of their metabolism, and they are considered primitive photosynthetic organisms that have existed for approximately 3.5 billion years (Schopf 2022). Many organisms, including fungi, protozoans, plants, and sponges, have symbiotic relationships with cyanobacteria, which provide fixed nitrogen to help these organisms survive in low-nitrogen environments (Antonia and Enrique 2008). At least 12 different types of cyanobacteria are known to produce toxins that can cause illness in humans and wildlife, as well as negative environmental impacts (Okamoto and Fleming 2005). These toxins, called cyanotoxins, are not necessary for cyanobacteria to grow or reproduce and can be extremely potent, including strong cytotoxins, endotoxins, hepatotoxins, and neurotoxins. The most common acute illness caused by drinking water contaminated with cyanobacteria is gastroenteritis. Exposure to the cyanobacterial neurotoxin BMAA has been suggested as a possible environmental cause of neurodegenerative conditions such as ALS, Parkinson's disease, and Alzheimer's disease (Holtcamp 2012). The well-known hepatotoxins called microcystins are produced by certain cyanobacterial species (Percival and Williams 2014).

Hepatotoxins are particularly harmful to the liver because they are transported into liver cells by specific transport systems and damage the organ by disrupting the cytoskeletal structure of hepatocytes. Secondary metabolites, which are substances produced by plants and animals that are not essential for normal growth, development, or reproduction, often play a role in defense against phytopathogens and other interspecies interactions (Malik et al. 2020). In contrast, secondary metabolites produced by cyanobacteria are found in many different genera and geographical regions. Saxitoxins, a secondary metabolite of cyanobacteria, are listed as a Chemical Weapons Convention Schedule 1 substance (Yunes 2019).

Many environmental stress factors, including nutrient availability, photoperiod, temperature, and others, can influence the production of cyanotoxins by cyanobacteria. For example, conditions with limited nitrogen are often associated with higher levels of the cyanotoxin cylindrospermopsin (CYN), which is produced by cyanobacteria that fix nitrogen. The intensity of light and temperature are also important factors in cyanotoxin production (Kaebernick and Neilan 2001; Griffiths and Saker 2003). In addition, stressors like grazing pressure can affect cyanobacterial toxin production. It has been suggested that cyanotoxins may have evolved as a chemical defense against grazers, as CYN has been shown to be toxic to a wide range of aquatic organisms (Nogueira et al. 2004). It is also believed that cyanotoxins may play an allelopathic role in the environment, acting as a threat to certain other plant and algae species in competition for light, nutrients, and habitat (Holland and Kinnear 2013).

Cyanotoxins have a wide range of physicochemical properties and levels of toxicity. There are several ways to classify cyanotoxins, such as based on their

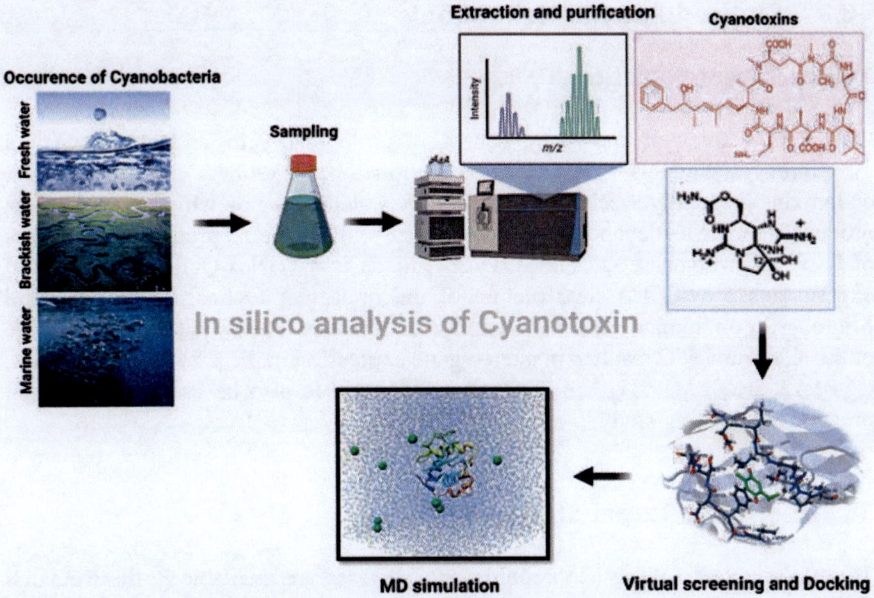

Fig. 38.1 Detection of cyanotoxin using in silico approach

chemical structures or the bacterial species that produces them or based on their toxicological mechanism and the primary organs affected in the body (Van Apeldoorn et al. 2007; Kaebernick and Neilan 2006). However, the actual physiological role and ecological regulatory oversight of cyanobacterial toxins are not well understood. This study aims to use computational tools to investigate the binding mechanism of cyanotoxins to host DNA or protein structures. Understanding these systems can help guide further research on cyanotoxins and is part of a new era of research that focuses on understanding environmental interactions in terms of their molecular factors (Fig. 38.1).

38.2 List of Computational Tools

- RCSB-Protein Data Bank (https://www.rcsb.org/)
- PubChem (https://pubchem.ncbi.nlm.nih.gov/)
- MarvinSketch (https://chemaxon.com/marvin)
- Open Bable (http://openbabel.org)
- CASTp (http://sts.bioe.uic.edu/castp/index.html?2cpk)
- AutoDock4 (https://autodock.scripps.edu/)
- LigPlot+ (https://www.ebi.ac.uk/thornton-srv/software/LigPlus/)
- Chimera (https://www.cgl.ucsf.edu/chimera/)
- GROMACS (https://www.gromacs.org/)
- Mathematica (https://www.wolfram.com/mathematica/online/)

38.3 Methodology of Using Tools

38.3.1 Target Selection

Cyanotoxins are well known for causing neurological, skin, and gastrointestinal symptoms in humans. They can be peptides (polyketides, cyclic peptides), endotoxins (Lipopoly saccharides), or alkaloids, depending on which cyanobacteria produces them. Most cyanotoxins target Serine and Threonine protein phosphatases of host cell (Maynes et al. 2006). Diez-Quijada et al. (Diez-Quijada et al. 2022) experiments provided a clear picture of the impact of Cylindrospermopsin and Microcystin on human cell protein expression changes. The simultaneous exposure of this CYN and MC resulted in altered gene expression in the CYP1A1, CDKN1A, CYP1A2, BCL2, GPX1, and TP53 genes. These can also be used as host target proteins for docking study.

38.3.2 Prepare Target Structure

The 3D structures of serine/threonine phosphatases are available in the Research Collaboratory for Structural Bioinformatics Protein Data Bank (RCSB PDB). For the purpose of this study on inhibition analysis, the PDB structure of human alpha thrombin in complex with an inhibitor (PDB ID: 1BHX) is used. The first step in docking using the AutoDock4 suite is to prepare the protein structure (Morris et al. 1998). HETATM or hetero atoms and water molecules should be removed from the protein structure as they may interfere with the binding of ligand molecules to the target. The protein structure is then assigned polar hydrogen bonds and kollman charges (electrostatic potential values for each amino acid). The target file should be saved in PDBQT (partial charge and atom type) format after adding charges. PDB format is not allowed for docking; extended format PDBQT files are required.

38.3.3 Ligand Selection and Preparation

You can obtain the 2D ligand structures of cyanotoxins such as microcystin, strophanthidin, and nodularin from the Zinc (Irwin et al. 2012) and PubChem (Kim et al. 2021) databases. Downloaded the ligand structure of strophanthidin (PubChem id: 6185) in sdf format and converted it into a format recognizable by AutoDock using the Open Bable tool (O'Boyle et al. 2011). During conversion, make sure to select the option for 3D coordinates; otherwise, the transformed PDB structure will be in a planar form. Once the ligand is selected for preparation, define the rigid root, which is the fixed component of the ligand from which various rotatable bonds emerge. You have the option to manually configure the root or use the automatic default selection option. Then, give the ligand molecule Gasteiger partial charges and save the file in PDBQT format.

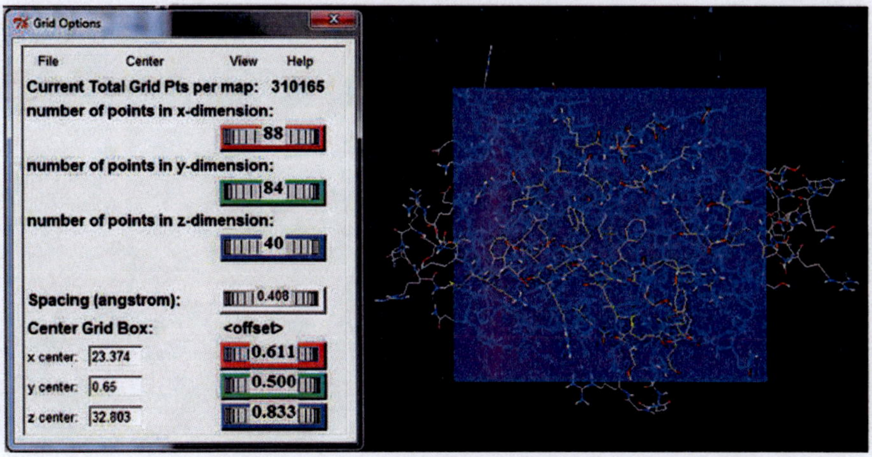

Fig. 38.2 AutoDock grid box generation with box dimension 88 Å 84 Å 40 Å and box center ($x = 23.374$, $y = 0.65$ and $z = 32.803$). The box represents the ligand binding site in the receptor

38.3.4 Grid Selection

Before docking the ligand to a specific protein site, you must identify the binding site. If you are not sure of the binding site, you can perform blind docking. To predict the binding site, you can use the CASTp (Tian et al. 2018) server. During blind docking, the docking algorithm searches for the best location to bind the ligand molecule. To define the grid box around the binding site, it is common to use a size of 126 × 126 × 126 with a spacing of 1 Å for the whole protein structure. The grid parameter file format (.gpf) is used to save the auto grid file, while the glg file is used to save the log file. The gpf file contains the grid map, grid points (in *x*, *y*, and *z* coordinates), and grid point spacing (Fig. 38.2).

38.3.5 Docking of Target and Ligand Molecule

Lamarckian, genetic algorithm, simulated annealing, and local search are the four search algorithms used by AutoDock; however, a Lamarckian genetic algorithm (LGA) is the most effective one for global optimization. The user can define the population size, and 50 is often thought to be a desirable value. For typical systems, AutoDock runs multiple times to provide a number of docked conformations. The optimum solution is then determined by combining an examination of the expected energy and the consistency of results (Fig. 38.3).

38.3.6 Docking Conformation Analysis

There are several methods for analyzing the results of docking using AutoDock tools, such as grouping the results based on conformational similarity, visualizing

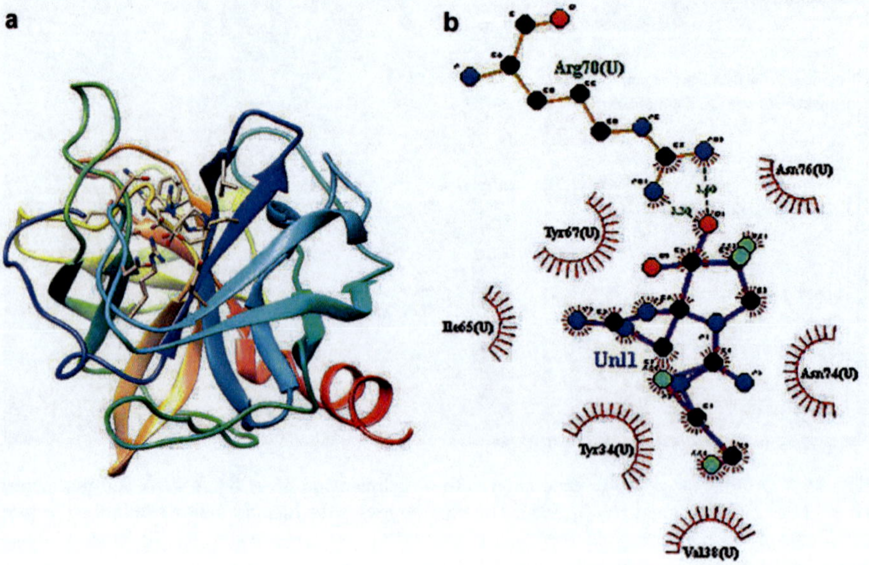

Fig. 38.3 (**a**) Docked complex of strophanthidin with human alpha thrombin. Stick conformation represents the ligand structure and cartoon represents the receptor structure. (**b**) The interaction between ligand and receptor and their corresponding H-bond is denoted as dotted line (green color) with distance

interactions between ligands and proteins, and displaying the affinity potentials generated by AutoGrid. The docked complex can be loaded and the various conformations, as well as information about hydrogen bonds and Pi-Pi bonds, can be viewed along with estimates of their binding energies. The best conformation with the highest number of hydrogen bonds and the lowest binding energy can be selected.

38.3.7 Molecular Dynamics Simulation of Docked Complex

Molecular dynamics (MD) simulations can be used to study the process of protein-ligand binding and unbinding by considering both energy and geometrical factors. In this computational method, the dynamic behavior of molecular systems, such as proteins, nucleic acids, and ligands, is modeled as a function of time, with all objects in the simulation treated as flexible. To create the necessary topological files for proteins and ligands, various tools such as PRODRG 2.5 (Schüttelkopf and van DMF 2004), CGenFF, Antechamber, acpype (Sousa Da Silva and Vranken 2012), ATB (Malde et al. 2011), Topol build, TopolGen, and LigParGen can be used, with the force field from these tools being applied to the ligand topology.

Using Gromacs, protein topology was created, structural minimization and MD simulations were then carried out using the CHARMM27 (Brooks et al. 1983) all atom force field in the Gromacs (Bekker et al. 1993).To create a neutral system, sodium counter-ions and water molecules were solvated into the protein-ligand complex. The TIP3P (includes the LJ interactions between oxygen and hydrogen) water solvent model was used to construct a water box with a 10Å thickness to completely encircle the protein-ligand complex. The simulation was run, using a periodic boundary condition. The particle-mesh Ewald method (Fischer et al. 2015) was used to handle long-range electrostatic interactions, and a cut-off value of 8 was used for non-bonding interactions. After 5000 steps of minimization using the steepest descent technique, (Kiss et al. 2014) a 1-ns NVT equilibration was performed, during for 150 ps, the equilibrations were carried out in an ensemble with constant normal temperature and pressure (NTP). The proteins were permitted to relax while the ligands were positionally constrained during equilibration. Then the minimum simulation run time of 100 ns is used to complete the production run for the protein-ligand complex.

38.4 Result Analysis

38.4.1 Docking Result Analysis

Protein-ligand docking entails several phases, including determining the active sites, the flexibility of the ligand, and the strength of the contact between the ligand and the protein. The docking poses generated by the docking algorithms can be imported into PyMOL using the plugin, or simpler analysis can be performed using Ligplot+ tool and 2d pose view server. Each tool has a specific cutoff for hydrogen bonds, hydrophobic interactions, Pi-Pi, Pi-cation, and van der Waals attraction. The docked ligands and their corresponding binding poses can be exported to PDB files or other file formats in Pymol. For example, docking findings may be exported in a CSV file format, which may be easily imported into applications like Excel. For 3D interaction analysis, the built-in Python Molecular Viewer (PMV) in AutoDock can be used. Ligplot + and 2d pose viewer accurately depicts hydrogen bonds, hydrophobic interactions, and other interactions with distance.

38.4.2 Simulation Result Analysis

Calculating the root-mean-square deviation (RMSD) is a more advanced method for analyzing conformational changes. This technique involves measuring the difference between the positions of atoms and a reference set of atom coordinates. RMSD graphs can be used to visualize the relative motion of entire proteins or domains during folding or ligand binding. RMSD calculations can also be used to compare computed docking configurations to known binding poses, to verify the accuracy of the docking method. Alternatively, the root-mean-square fluctuation (RMSF) calculation can be used to visualize the fluctuation of residues from their mean structure, allowing for the identification of flexible and stable regions of a protein in simulations. The compactness of a protein structure can be assessed using the radius of gyration (RoG). A protein may be more difficult to fold if it is excessively compact. RoG is typically plotted for a protein-ligand complex after molecular dynamics (MD) simulation, to provide insight into the stability of the complex in addition to the RMSD. RoG serves as a measure of how well-packed the regular secondary structures are in the three-dimensional structure of the protein.

38.4.3 Hydrogen Bond Analysis

A ligand-protein complex is usually formed through non-covalent interactions such as hydrogen bonds (Bissantz et al. 2010). These interactions are attractive because the partially positively charged hydrogen atom in the hydrogen bond donor points towards a partially negatively charged hydrogen bond acceptor atom, such as oxygen or nitrogen. Studying hydrogen bonds can provide valuable information about the stability of a ligand-protein complex and can be used to modify a lead molecule to increase its activity. The strength of a hydrogen bond can be calculated by measuring the angles and distances between the donor, hydrogen, and acceptor atoms. The Gromacsg_hbond command can be used to analyze the number of H-bonds formed between the ligand and protein molecule.

Acknowledgment The authors PS & KG thank Sathyabama Institute of Science and Technology, Chennai, for infrastructural support.

References

Anon (2012) Life history and ecology of cyanobacteria. University of California Museum of Paleontology. Archived from the original on 19 September 2012. Retrieved 17 July 2012

Antonia H, Enrique F (2008) The cyanobacteria: molecular biology, genomics, and evolution. Caister Academic Press. ISBN: 978-1-904455-15-8

Bekker H, Berendsen HJC, Dijkstra EJ, Achterop S, van Drunen R, van der Spoel D, Sijbers A, KeegstraH RB, Renardus MKR (1993) Gromacs: a parallel computer for molecular dynamics simulations. In: de Groot RA, Nadrchal J (eds) Physics computing 92. World Scientific, Singapore, pp 252–256

Bissantz C, Kuhn B, Stahl M (2010) A medicinal chemist's guide to molecular interactions. J Med Chem 53:5061–5084

Brooks BR, Bruccoleri RE, Olafson BD, States DJ, Swaminathan S, Karplus M (1983) CHARMM: a program for macromolecular energy, minimization, and dynamics calculations. J Comput Chem 4:187–217

Diez-Quijada L, Puerto M, Gutiérrez-Praena D, Turkina MV, Campos A, Vasconcelos V, Camean AM, Jos A (2022) *In vitro* toxicity evaluation of cyanotoxins cylindrospermopsin and microcystin-LR on human kidney HEK293 cells. Toxins 14:429

Fischer NM, van Maaren PJ, Ditz JC, Yildirim A, van der Spoel D (2015) Properties of organic liquids when simulated with long-range lennard-jones interactions. J Chem Theory Comput 11: 2938–2944

Griffiths DJ, Saker ML (2003) The Palm Island mystery disease 20 years on: a review of research on the cyanotoxin cylindrospermopsin. Environ Toxicol 18:78–93

Haselkorn R (2009) Cyanobacteria. Curr Biol 19:R277–R278

Holland A, Kinnear S (2013) Interpreting the possible ecological role(s) of cyanotoxins: compounds for competitive advantage and/or physiological aide? Mar Drugs 11:2239–2258

Holtcamp W (2012) The emerging science of BMAA: do cyanobacteria contribute to neurodegenerative disease? Environ Health Perspect 120:a110–a116

Irwin JJ, Sterling T, Mysinger MM, BolstadES CRG (2012) ZINC: a free tool to discover chemistry for biology. J Chem Inf Model 52(7):1757–1768

Kaebernick M, Neilan BA (2001) Ecological and molecular investigations of cyanotoxin production. FEMS Microbiol Ecol 35:1–9

Kaebernick M, Neilan BA (2006) Ecological and molecular investigations of cyanotoxin production. FEMS Microbiol Ecol 35:1–9

Kim S, Chen J, Cheng T, Gindulyte A, He J, He S, Li Q, Shoemaker BA, Thiessen PA, Yu B, Zaslavsky L, Zhang J, Bolton EE (2021) PubChem in 2021: new data content and improved web interfaces. Nucleic Acids Res 49(D1):D1388–D1395

Kiss PT, Sega M, Baranyai A (2014) Efficient handling of gaussian charge distributions: an application to polarizable molecular models. J Chem Theory Comput 10:5513–5519

Malde AK, Zuo L, Breeze M, Stroet M, Poger D, Nair PC, Oostenbrink C, Mark AE (2011) An automated force field topology builder (ATB) and repository: version 1.0. J Chem Theory Comput 7:4026–4037

Malik JK, Bharti VK, Rahal A, Kumar D, Gupta RC (2020) Chapter 31 - Cyanobacterial (blue-green algae) toxins. In: Gupta RC (ed) Handbook of toxicology of chemical warfare agents, 3rd edn, pp 467–478. https://doi.org/10.1016/B978-0-12-819090-6.00031-3

Maynes JT, Luu HA, Cherney MM, Andersen RJ, Williams D, Holmes CF, James MN (2006) Crystal structures of protein phosphatase-1 bound to motuporin and dihydromicrocystin-LA: elucidation of the mechanism of enzyme inhibition by cyanobacterial toxins. J Mol Biol 356: 111–120

Morris GM, Goodsell DS, Halliday RS, Huey R, Hart WE, Belew RK, Olson AJ (1998) Automated docking using a lamarckiangenetic algorithm and empirical binding free energy function. J Comput Chem 19:1639–1662

Nogueira ICG, Saker ML, Pflugmacher S, Wiegand C, Vasconcelos VM (2004) Toxicity of the cyanobacterium *Cylindrospermopsis raciborskii* to *Daphnia magna*. Environ Toxicol 19:453–459

O'Boyle NM, Banck M, James CA, Morley C, Vandermeersch T, Hutchison GR (2011) Open babel: an open chemical toolbox. J Cheminform 3:33

Okamoto K, Fleming LE (2005) Algae. In: Wexler P (ed) Encyclopedia of toxicology, 2nd edn, pp 68–76. ISBN: 9780080548005

Percival SL, Williams DW (2014) Chapter five - cyanobacteria. In: Percival SL, Yates MV, Williams DW, Chalmers RM, Gray NF (eds) Microbiology of waterborne diseases, 2nd edn. Academic Press, pp 79–88. https://doi.org/10.1016/B978-0-12-415846-7.00005-6

Schopf JW (2022) The fossil record: tracing the roots of the cyanobacterial lineage. In: Whitton B, Potts M (eds) *The ecology of cyanobacteria*. Kluwer Academic Publishers, Dordrecht, The Netherlands, pp 13–35

Schüttelkopf AW, van DMF A (2004) PRODRG: a tool for high-throughput crystallography of protein-ligand complexes. Acta Crystallogr D Biol Crystallogr 60:1355–1363

Sousa Da Silva AW, Vranken WF (2012) ACPYPE – Antechamber python parser interface. BMC Res Notes 5:367

Tian W, Chen C, Lei X, Zhao J, Liang J (2018) CASTp 3.0: computed atlas of surface topography of proteins. Nucleic Acids Res 46:W363–W367

Van Apeldoorn ME, Van Egmond HP, Speijers GJA, Bakker GJI (2007) Toxins of cyanobacteria. Mol Nutr Food Res 51:7–60

Yunes JS (2019) Chapter 22 - cyanobacterial toxins. In: Mishra AK, Tiwari DN, Rai AN (eds) Cyanobacteria, pp 443–458. https://doi.org/10.1016/B978-0-12-814667-5.00022-2

Part III

Cyanobacterial Toxin Extraction, Detection and Quantification

Extraction of Cyanotoxins Using Filtration and Other Methods

39

K. P. A. Imanthi, P. A. B. P. Abeyrathne, P. A. K. C. Wijerathna, G. Y. Liyanage, P. M. Manage, and S. A. Kulasooriya

Abstract

Certain types of cyanobacteria, commonly referred to as blue-green algae, which are present in aquatic habitats, produce chemical substances called cyanotoxins. Cyanobacterial blooms, which occur when these bacteria multiply rapidly in a body of water, are one of the major causes of cyanotoxin production. Nutrient pollution, hot temperatures, and low water levels are the most effective things that might result in blooms. To preserve public health, it is crucial to keep an eye out for cyanobacteria blooms in recreational and drinking water sources and to take action to reduce or prevent them. The health effects of these toxins, which include skin rashes, allergic responses, respiratory illnesses, gastrointestinal problems, and liver damage, can be hazardous to aquatic species and humans. In extreme circumstances, they may result in the death of children. Additionally, they may affect aquatic creatures, causing population declines or even extinction. Due to their low concentration and frequent presence in complicated combinations with

K. P. A. Imanthi · P. A. B. P. Abeyrathne · P. A. K. C. Wijerathna · P. M. Manage
Department of Zoology, Centre for Water Quality and Algae Research, University of Sri Jayewardenepura, Nugegoda, Sri Lanka

Present Address: Faculty of Graduate Studies, University of Sri Jayewardenepura, Nugegoda, Sri Lanka

G. Y. Liyanage
Department of Zoology, Centre for Water Quality and Algae Research, University of Sri Jayewardenepura, Nugegoda, Sri Lanka

Faculty of Urban and Aquatic Resources Management, University of Sri Jayewardenepura, Nugegoda, Sri Lanka

S. A. Kulasooriya (✉)
National Institute of Fundamental Studies, Kandy, Sri Lanka
e-mail: ananda@kulasooriya.com

N. Thajuddin et al. (eds.), *Protocols for Cyanobacteria Sampling and Detection of Cyanotoxin*, https://doi.org/10.1007/978-981-99-4514-6_39

other compounds, cyanotoxins can be challenging to extract from aquatic habitats. Filtration, liquid-liquid extraction, and solid-phase extraction are a few techniques developed for the extraction and detection of cyanotoxins. These techniques can be costly and time-consuming. Thus, there is a need for more effective ones.

Keywords

Cyanotoxins · Ecosystems · Extraction · Filtration · Liquid-liquid extraction · Solid phase extraction

39.1 Introduction

Cyanobacteria are photosynthetic and commonly found in freshwater systems. They are a group of microorganisms that have a long evolutionary history (4). As a result, they have developed a wide range of adaptations that allow them to thrive in various environments. They can colonize bodies of water with different salinity levels, including freshwater, brackish water, and saltwater (Aba et al. 2021). They can also survive in various temperatures, from extreme cold to extreme heat. In addition, cyanobacteria can adapt to changes in their environment, including climatic, geochemical, and anthropogenic changes. These adaptations allow cyanobacteria to thrive in a wide range of habitats and make them a common component of aquatic ecosystems (Aba et al. 2021; El-Nahhal et al. 2021). Factors that can influence the production of cyanotoxins include the presence of certain nutrients and environmental conditions such as temperature and pH. Many species of cyanobacteria including *Microcystis* sp., *Anabaena* sp., *Nodularia* sp., *Planktothrix* sp., *Aphanizomenon* sp., and *Lyngbya* sp. can produce cyanotoxins. These cyanobacteria are found in freshwaters such as lakes, streams, and reservoirs (Manubolu et al. 2018).

It is important to monitor cyanobacteria blooms. Cyanotoxins can impact humans, animals, and plants and are classified according to their toxicological target in hepatotoxins, cytotoxins, neurotoxins, and dermatoxins. The most common cyanotoxins are microcystins, hepatotoxins that can cause liver damage. Other cyanotoxins include anatoxins, which can affect the nervous system, and cylindrospermopsins, which can damage cells (Kulasooriya 2011). These can harm human health if they are ingested through contaminated drinking water or contact to the skin during recreational activities such as swimming. The symptoms of cyanotoxin poisoning can range from mild to severe, including headache, fever, respiratory paralysis, and in rare cases, death. They can also harm aquatic organisms, leading to reduced populations or even extinction (Kulasooriya 2011; Rao et al. 2002).

It is important to ensure that the levels of cyanotoxins in drinking water do not exceed the recommended limits set by the World Health Organization (WHO) and the Environmental Protection Agency (EPA). According to WHO standards, Microcystin concentration must be lower than 1 μg/L, whereas EPA recommends an acceptable level of ≤0.3 μg/L for children under 6 years of age and ≤ 1.6 μg/L for adults (de Magalhaes et al. 2001). Nevertheless, the concentration of other cyanotoxins in drinking water is not yet well-regulated in many countries, including the European Union (EU). There is a need to develop effective methods for extracting low concentrations of cyanotoxins from water to protect public health and test for the presence of cyanotoxins to ensure the safety of drinking and recreational water.

Several methods have been developed for the extraction of cyanotoxins from aquatic ecosystems. Several methods have been developed to extract cyanotoxins from aquatic samples, including filtration, liquid-liquid extraction, ultracentrifugation, and solid-phase extraction (de Magalhaes et al. 2001). Each method for extracting cyanotoxins from water has its own advantages and limitations, and the most appropriate method will depend on the application's specific needs. Some factors to consider when choosing a method for extracting cyanotoxins from water include the type of cyanotoxins present, the concentration of the cyanotoxins, and the desired level of accuracy and precision. It is also important to consider the cost and practicality of the method and any potential impacts on the environment. In some cases, it may be necessary to use a combination of methods to extract all of the cyanotoxins from a water sample effectively.

39.2 Materials

- Environmental water samples
- Whatman filter papers with a pore size of 0.45 μM or less
- Filter unit with filtration cup and vacuum pump
- Activated carbon filters
- Reverse osmosis filters
- Immunoaffinity chromatography filters
- Solid Phase Extraction (SPE) unit
- Centrifuge machine (≥100,000 rpm)
- Cool box
- DI water/ Distilled water
- 80% Methanol (AR grade)
- 70% Ethanol
- SPE extraction solution (Methanol: DI water 75:20 v/v)
- Sterilized glass bottles
- Sterilized centrifuge tubes (15 mL)
- Sterilized Petri dishes
- Sterilized measuring cylinders
- Sterilized volumetric flask
- Sterilized beakers
- Sterilized separation funnel
- Funnel holders
- Forceps

- Tissue papers
- Cotton wools

39.3 Methodology

39.3.1 Types of Extraction Methods and Mechanism of the Method

39.3.1.1 Filtration Methods

Filtration methods involve using filtration to remove cyanobacteria and cyanotoxins from water. This can be done using the most common method, filter papers with pore sizes small enough to remove cyanobacteria (Filters with a pore size of 0.45 μM or less are typically used for this filtration), or by using filters impregnated with substances that can bind to and remove cyanotoxins. Subsequently, Activated carbon filters, Reverse osmosis filters, and Immunoaffinity chromatography filters are also used widely (Sanseverino et al. 2017; Gijsbertsen-Abrahamse et al. 2006; Drogui et al. 2012; Vuori et al. 1997) (Fig. 39.1).

Activated Carbon Filters

These filters use activated carbon, a highly porous form of carbon with a large surface area. The cyanotoxins bind to the surface of the activated carbon and can then be eluted, or washed off, the filter and collected for analysis (Gijsbertsen-Abrahamse et al. 2006). Activated carbon filters effectively remove a wide range of contaminants, including cyanotoxins, from water. Activated carbon

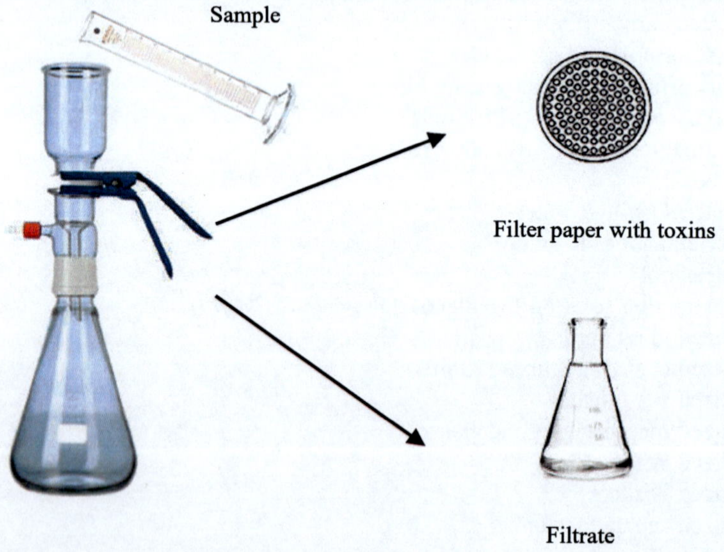

Fig. 39.1 Filter unit with filtration cup and filter paper

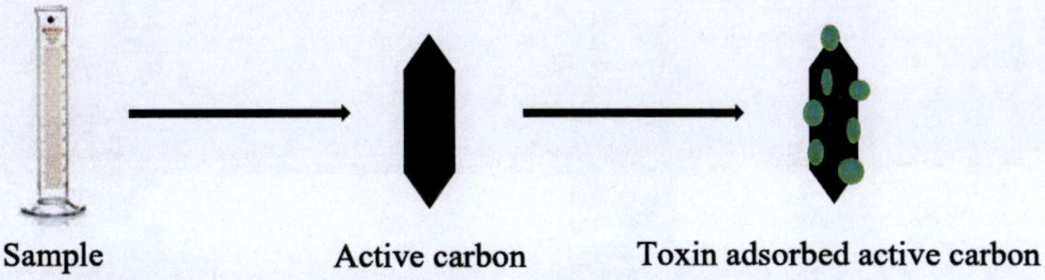

Fig. 39.2 Adsorption by Active Carbon Granules

Fig. 39.3 Mechanism of the reverse osmosis filtration

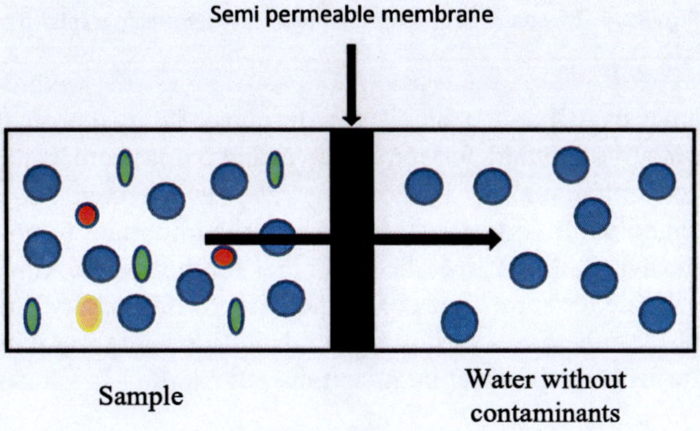

filters have a limited capacity for removing contaminants and will eventually become saturated and need to be replaced. They also need to be adequately maintained and replaced to ensure working efficiency. Additionally, activated carbon filters may not effectively remove all types of cyanotoxins, so it is important to test the water for the specific contaminants present before selecting a treatment method (Fig. 39.2).

Reverse Osmosis Filters

A semi-permeable membrane (SPM) filters out contaminants from water, including cyanotoxins. The water is forced through the membrane under pressure, and the contaminants are left behind and then eluted, washed off, the filter and collected for analysis. Reverse osmosis filters require a high-pressure pump to force the water through the membrane and may not be suitable for all water sources (Drogui et al. 2012; Vuori et al. 1997). They also have a limited capacity for removing contaminants and will eventually become saturated and need to be replaced (Fig. 39.3).

Immunoaffinity Chromatography Filters

These filters use antibodies specific for a particular cyanotoxin to extract the toxin from water. The water is passed through a column containing the antibodies, and the cyanotoxins bind to the antibodies while the other molecules pass through the column. The cyanotoxins can then be eluted from the column and collected for

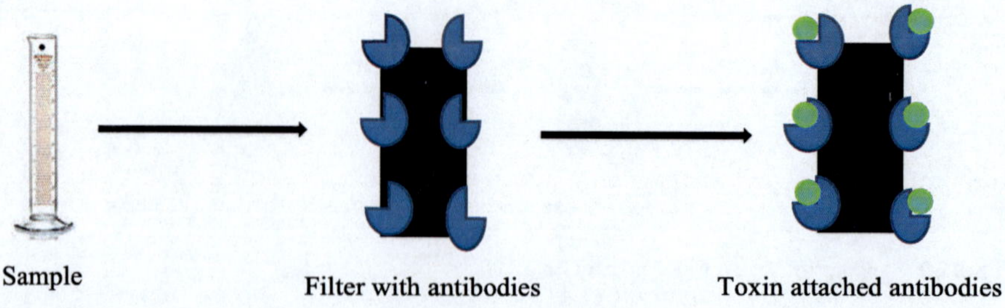

Fig. 39.4 Mechanism of the immunoaffinity chromatography filters

analysis (Chow et al. 1997). Immunoaffinity chromatography is a precise and sensitive method for removing cyanotoxins from water. It can remove very low concentrations of cyanotoxins and is effective at removing many different cyanotoxins (Fig. 39.4). However, it is important to note that this method requires the availability of specific antibodies for the cyanotoxins of interest and may only be suitable for some types of cyanotoxins. Additionally, immunoaffinity chromatography can be time-consuming and labor-intensive and may need to be more practical for large-scale water treatment operations (de Magalhaes et al. 2001).

1. Collect environmental sample into 1 L sterilized glass bottle.
2. Transfer it into the laboratory in a cool box (22 °C ± 2 °C).
3. Mix well and take 300 mL into a sterilized measuring cylinders.
4. Pour it into the filtration cup with filter paper (0.45 μM) or pass it through the filter columns with activated carbon, antibodies, or the semi-permeable membrane.
5. Remove the filter paper with a forceps and keep it in a petri dish/ wash the columns with 80% methanol or DI water into a volumetric flask.
6. Then store the filter paper, filtrate/washed column samples at 4 °C for further analysis.

39.3.1.2 Liquid-Liquid Extraction

It is a method to separate compounds based on their solubility in different solvents. In this method, the sample is mixed with a solvent in which the cyanotoxins are soluble, and the mixture is then separated into two layers. The cyanotoxins are extracted into the solvent, while the other molecules remain in the original sample. The solvent containing the cyanotoxins can then be separated and collected for analysis (El-Nahhal et al. 2021). This method can be used to extract cyanotoxins from water samples although it is generally less widely used than other methods, such as filtration or solid phase extraction. One advantage of liquid-liquid extraction is that it can be used to selectively extract specific types of cyanotoxins, depending on the solvents used. However, it is a more time-consuming and expensive method than filtration, and it can also be less efficient at recovering certain types of

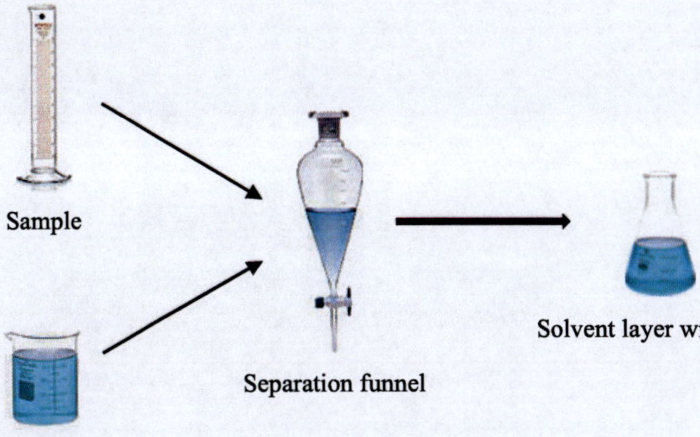

Fig. 39.5 Liquid-liquid extraction mechanism

cyanotoxins and may not be effective at removing very low concentrations of cyanotoxins.

1. Collect environmental sample into 1 L sterilized glass bottle.
2. Transfer it into the laboratory in a cool box (22 °C ± 2 °C).
3. Mix well and take 100 mL into a sterilized measuring cylinders.
4. Take a separation funnel and add 100 mL sample and 100 mL of 80% methanol.
5. Mix well and keep for the separation.
6. Remove the methanol layer and use it for further analysis and quantification (Fig. 39.5).

39.3.1.3 Solid Phase Extraction (SPE)

Solid phase extraction is a method for separating and purifying specific molecules, such as cyanotoxins, from a sample by adsorbing them onto a solid support. In this method, the sample is passed through a column containing the solid support, and the cyanotoxins are adsorbed onto the surface of the support (Kondo et al. 2000). The cyanotoxins can then be eluted, or washed off, the column and collected for analysis. SPE is a relatively simple and inexpensive method for extracting cyanotoxins from water. It is relatively easy to perform and does not require specialized equipment. However, it is important to choose appropriate solid support that can effectively adsorb the cyanotoxins from the sample and carefully control the extraction conditions to minimize losses. Additionally, SPE may only be suitable for some cyanotoxins and may not effectively remove deficient concentrations of cyanotoxins.

1. Collect environmental sample into 1 L sterilized glass bottle.
2. Transfer it into the laboratory in a cool box (22 °C ± 2 °C).
3. Mix well and take 100 mL into a sterilized measuring cylinders.

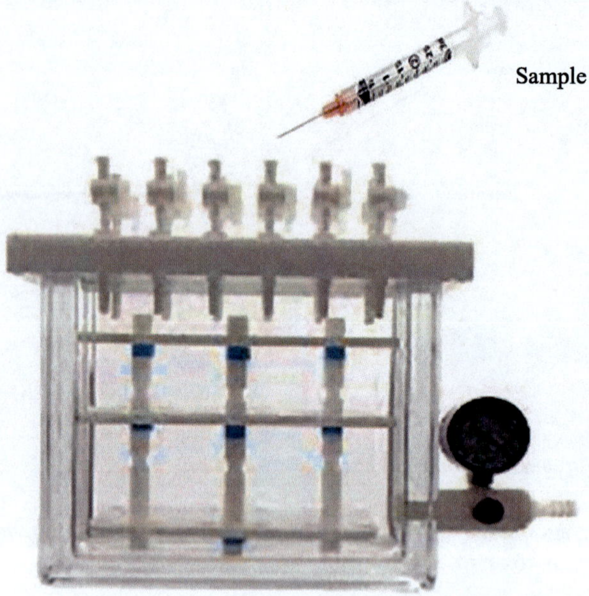

Fig. 39.6 Solid Phase Extraction (SPE) unit

4. Pass the sample through a column packed with solid support coated with C14 (Cyanotoxins in the sample are adsorbed onto the surface of the C14-coated solid support).
5. The cyanotoxins are then eluted using 80% methanol or DI water (relevant toxin solvent) and stored for analysis (Fig. 39.6).

39.3.1.4 Ultracentrifugation

Ultracentrifugation is a technique that separates particles based on their size, shape, and density. It involves using a high-speed centrifuge to generate very high centrifugal forces, which can be used to separate different types of molecules or particles from a mixture. In the case of cyanotoxins, ultracentrifugation can be used to extract and purify these toxins from water samples (Manubolu et al. 2018). This can be done by preparing a sample of the water and centrifuging it at high speeds (typically greater than 100,000 × *g*) to separate the cyanotoxins from other contaminants or impurities. The purified cyanotoxins can then be collected and need further purify the cyanotoxins using additional techniques, such as filtration.

1. Collect environmental sample into 1 L sterilized glass bottle.
2. Transfer it into the laboratory in a cool box (22 °C ± 2 °C).
3. Mix well, take 6 centrifuge tubes and add 10 mL of sample into each.
4. Place into the centrifuge and operate at 10,000 rpm for 10 min.
5. Separate the supernatant from pellet carefully.
6. Store the pellet for further purification and analysis.

As there are several extraction methods in use, the choice of extraction method always depends on the specific cyanotoxins present in the water sample and the desired concentration and purity of the extracted cyanotoxins.

Acknowledgment The authors acknowledge the Centre for Water Quality and Algae Research, Department of Zoology, University of Sri Jayewardenepura, Sri Lanka, for the immense support and the guidance.

References

Aba RP, Mugani R, Hejjaj A, Brugerolle de Fraissinette N, Oudra B, Ouazzani N, Mandi L (2021) First report on cyanotoxin (MC-LR) removal from surface water by multi-soil-layering (MSL) eco-technology: preliminary results. Water 13(10):1403

Chow CWK, Panglisch S, House J, Drikas M, Burch MD, Gimbel R (1997) A study of membrane filtration for the removal of cyanobacterial cells. J Water Supply Res Technol 46:324

Drogui P, Daghrir R, Simard MC, Sauvageau C, Blais JF (2012) Removal of microcystin-LR from spiked water using either activated carbon or anthracite as filter material. Environ Technol 33(4):381–391

El-Nahhal H, Yassin M, Alzaharna M, El-Nahhal I, El-Nahhal Y (2021) Extraction methods of cyanotoxins aqueous media and sediments. Am J Anal Chem 12(09):311

Gijsbertsen-Abrahamse AJ, Schmidt W, Chorus I, Heijman SGJ (2006) Removal of cyanotoxins by ultrafiltration and nanofiltration. J Membr Sci 276(1–2):252–259

Kondo F, Matsumoto H, Yamada S, Tsuji K, Ueno Y, Harada KI (2000) Immunoaffinity purification method for detection and quantification of microcystins in lake water. Toxicon 38(6): 813–823

Kulasooriya SA (2011) Cyanobacteria: pioneers of planet earth. Ceylon J Sci 40(2):71–88

de Magalhaes VF, Soares RM, Azevedo SM (2001) Microcystin contamination in fish from the Jacarepagua lagoon (Rio de Janeiro, Brazil): ecological implication and human health risk. Toxicon 39:1077–1085

Manubolu M, Lee J, Riedl KM, Kua ZX, Collart LP, Ludsin SA (2018) Optimization of extraction methods for quantification of microcystin-LR and microcystin-RR in fish, vegetable, and soil matrices using UPLC–MS/MS. Harmful Algae 76:47–57

Rao PV, Gupta N, Bhaskar AS, Jayaraj R (2002) Toxins and bioactive compounds from cyanobacteria and their implications on human health. J Environ Biol 23:215–224

Sanseverino I, António DC, Loos R, Lettieri T (2017) Cyanotoxins: methods and approaches for their analysis and detection. Joint Research Centre

Vuori E, Pelander A, Himberg K, Waris M, Niinivaara K (1997) Removal of nodularin from brackish water with reverse osmosis or vacuum distillation. Water Res 31(11):2922–2924

Extraction of Cyanotoxins by Filtration and Other Methods from Biological Materials (Animal Tissue: Fish and Mussels)

40

Milena Bruno, Valentina Messineo, Ilaria Di Marco Pisciottano, and Pasquale Gallo

Abstract

Anthropogenic activities like monocultural agriculture, industrial enterprises (pulp and papermills) and wastewater discharges from civilian settlements have caused a progressive increase in the trophic status of Italian waterbodies, often giving rise to cyanobacterial dominance and toxic blooms occurrence. Studies on different Italian lakes affected by cyanobacterial blooms have documented phenomena of accumulation of cyanotoxins in fish fauna. These natural events, through the emissaries, can even reach the seacoasts flanking the mouths. Freshwater fish production in Italy is an important economic field and is estimated at 320,000 tonnes/year (7% of total production in 2005). In order to avoid health risks to consumers, the most common cyanotoxins (microcystins, cylindrospermopsins, BMAA, anabaenopeptins) require to be detected with specific tissue research, using immunological and instrumental methods. Some of these main extraction methods are presented here.

Keywords

Toxin · Hepatotoxic compound · Tumor promoter · Algal bloom · Protein phosphatases PP1/PP2A · Oxidative damage

M. Bruno (✉) · V. Messineo
FAST, Core Facilities, National Institute of Health, Rome, Italy
e-mail: milena.bruno@iss.it; valentina.messineo@iss.it

I. D. M. Pisciottano · P. Gallo
Coordination of Chemistry Department, Experimental Zoo prophylactic Institute of Southern Italy, Naples, Italy

N. Thajuddin et al. (eds.), *Protocols for Cyanobacteria Sampling and Detection of Cyanotoxin*, https://doi.org/10.1007/978-981-99-4514-6_40

40.1 Introduction

The powerful toxins named microcystins (Fig. 40.1), a family of more than 100 toxic variants (Welker and von Döhren 2006) produced by many cyanobacterial species (Chorus and Bartram 1999), are PP1 and PP2A inhibitors, known to be hepatotoxic (Codd 1995; Dawson 1998), tumor promoter (Nishiwaki-Matushima et al. 1991) and possibly carcinogenic to humans (Grosse et al. 2006). When microcystins are released into the water during bloom decay, a wide range of aquatic organisms are directly exposed to the toxins in solution. Large-scale fish death outbreaks have been associated to massive occurrence of cyanobacteria in waterbodies (Zimba et al. 2006, 2001; Jewel et al. 2003). Studies on fish contaminations have shown species-specific sensitivities to microcystins; the uptake of these cyanotoxins in fish results primarily from oral ingestion, and to a minor extent from absorption via the gill epithelium (Ernst et al. 2006).

Microcystins can concentrate in various fish tissues (Xie et al. 2005): liver disease, nephropathy, and gill disease are present in fish exposed to cyanotoxins, due to the specific inhibition of protein phosphatases and other downstream effects, such as an increase of liver enzyme values in serum.

Microcystin-LR, the most studied variant of this group, is a very effective tumor promoter (classified as 2B in the carcinogenic scale of the International Agency for Research on Cancer, IARC) (IARC 2010) whose action, mediated by inhibition of the protein phosphatases PP1/PP2A, induces oxidative damage to DNA (Zegura et al. 2003), the activation of the proto-oncogenes c-jun, c-fos, c-myc (Li et al. 2009), and of the nuclear factor Nrf2 (erythroid-related factor 2 nuclear factor 2) (Gan et al. 2014), repression of the tumor suppressor genes baxa, gadd45α, and the proto-oncogene junba (Fonseca et al. 2018).

Studies of polluted lakes have highlighted the risk of human exposure to microcystins through both direct (drinking water, recreational activities) and indirect (fish consumption) routes (Zhang et al. 2009; Chen et al. 2009).

In the past, the risk to human consumers of gutted fish was traditionally considered low because microcystins were thought to accumulate primarily in fish liver.

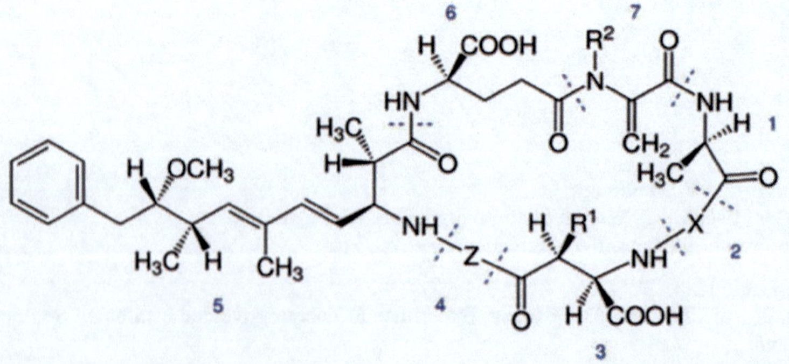

Fig. 40.1 Microcystin structure

Recent studies, however, have found concentrations of microcystins up to 337.3 μg/kg (*Tilapia rendalli*, (Magalhães et al. 2001)), 102 μg/kg (*Oreochromis niloticus*, (Mohamed et al. 2003)), 96.5 μg/kg (*Hypophthalmichthys molitrix*, (Chen et al. 2006)), and 28 μg/kg (*Oncorhynchus mykiss*, (Wood et al. 2006)) in wild or farmed fish muscle tissue, indicating that fish muscle tissue consumption could also constitute a threat to human health. Furthermore, microcystins can be transferred along the food chain (Smith and Haney 2006) representing a potential threat to human consumers not only due to their presence in raw fish, but also due to their stability and concentration during cooking (Bruno et al. 2009) and for the increase of free microcystins in fish tissue during a specific cooking, such as boiling (Zhang et al. 2013). This potential risk has led to the development of different extraction techniques and analytical methods for the detection of microcystins in fish tissues.

Other cyanotoxins that can be detected in fish and mussel tissues are:

- BMAA (see chapter "Extraction and quantification of BMAA from animal tissues").
- Cylindrospermopsin (CYN), that has been shown to have environmental toxicity to amphibians (Vasas et al. 2002). Recent evidence has shown that plant growth and metabolism (Metcalf et al. 2008) are also inhibited by CYN, with negative implications for current spray irrigation practices.

There are to date few studies on the accumulation of CYN in fish fauna: CYN has been proven to contaminate shrimp (*Cherax quadricarinatus*, up to 4.3 μg/g) and fish (*Melanotaenia eachamensis*, up to 1.2 μg/g g) from a small Australian aquaculture pond (Saker and Eaglesham 1999), and freshwater molluscs (*Anodonta cygnea*, up to 2.52 μg/day) in a 16-day exposure study (Saker et al. 2003). The presence of cylindrospermopsin was detected for the first time in Italy in 2004 (Manti et al. 2005).

In the last 20 years, many other groups of bioactive peptides have been discovered in cyanobacteria: among others aeruginosins, microginins, cyanopeptolins, anabaenopeptins, anabaenopeptilides, microviridines, and nostophycines. To date, more than 600 cyanobacterial peptides have been described (for a review, see (Welker and von Döhren 2006)). Despite the growing scientific production, the publication of studies on cyanopeptides (other than microcystins) represents less than 10% of scientific studies. The main challenge for a systematic risk assessment is the great diversity of cyanopeptides.

Cyanopeptides have been detected in the tissue of exposed fish, frogs, snails, and mussels, but the toxicokinetics of internal distribution and bioaccumulation are not fully understood (Gkelis 2006). Some toxicological studies on microcystins offer insights into the potential toxicity of other cyanopeptides, in cases where toxic effects exceeding those expected for microcystins alone are observed, or in cases where symptoms cannot be associated with microcystins.

Anabaenopeptins (Fig. 40.2) are unique cyclic peptides that have the common part of the cyclic peptide linked to tyrosine, arginine, lysine, and phenylalanine, through a urea bond (Fujii et al. 1996, 2002). Anabaenopeptilides are 19-component

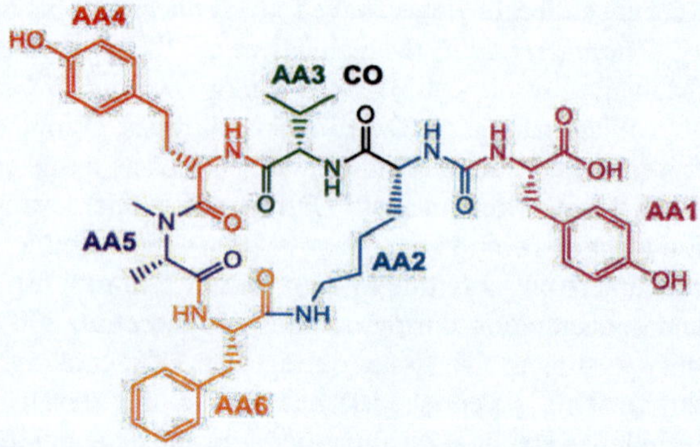

Fig. 40.2 Structure of Anabaenopeptins; an unique cyclic peptides

cyclic depsipeptides containing a single bonded residue, 3-amino-6-hydroxy-2-piperidone (Ahp) (Fujii et al. 1996, 2002). These peptides exhibit different bioactivities such as inhibition of serine, provapsin, or chymotrypsin protease (Welker and von Döhren 2006; Elkobi-Peer and Carmeli 2015; Ferranti et al. 2011).

40.2 Materials

- Volumetric Rotavapor flasks (100 mL)
- Volumetric flasks (5 mL)
- Paper filters
- Eppendorf vials
- Gilson micropipette
- Bidistilled H_2O
- 100% MeOH (HPLC grade)
- Homogenizer Ultra-Turrax T8, IKA Werke, Staufen, Germany
- Rotavapor Buchi
- Ultracentrifuge 11,000 rpm, Beckman LT-55
- Oasis HLB 3 cc Cartridge (Waters Corporation)
- 50 mL polypropylene tubes
- Methanol/Milli-Q water (2/1, v/v) 0.1% of acetic acid
- 20 mM $NaHCO_3$ buffer
- ISOLUTE C18 (EC) SPE cartridges (6 mL reservoir and 1 g of sorbent)
- 20 mM $NaHCO_3$ buffer/methanol (95/5, v/v)
- N_2 flow

40.3 Methods

40.3.1 Tissue Extraction for Microcystin Detection (Elisa Analysis)

1. Immediately homogenize tissues for 15′ with 10 mL of 100% MeOH (HPLC grade).
2. Centrifuge it at 5000 g for 5′.
3. Filter the supernatant on a paper filter and collect it in a Rotavapor flask.
4. Repeat extraction on the solid sample (point 1)
5. Filter the sample directly on a paper filter and collect it in the same Rotavapor flask of the first extraction.
6. Wash the paper filter with 5 mL of 100% MeOH for three times (tot. 15 mL)
7. Bring the methanolic extract to a little volume of 1–2 mL by drying it in Rotavapor.
8. Put the 1–2 mL in a 5 mL volumetric flask by washing the Rotavapor flask with MeOH to reach the total amount (sample + MeOH) of 5 mL.
9. Pick up 1 mL of the extract.
10. Dilute it in 1 mL of bidistilled water.
11. Condition the cartridge as follows:
12. Apply 1 mL of 100% MeOH
13. Apply 1 mL of bidistilled H_2O
14. Then transfer the sample on the head of the cartridge
15. Elute the sample with 1 mL of a 5% MeOH/H_2O solution
16. Collect the sample with 1 mL of 100% MeOH
17. Dry this fraction in Rotavapor
18. Reconstitute the sample washing the Rotavapor flask with 2 mL of bidistilled H_2O
19. The sample is ready to be tested with ELISA assay

40.3.2 Tissue Extraction for Cylindrospermopsin Detection (ELISA Analysis)

1. Homogenize approximately 5 g of tissue in 10 mL 100% MeOH for 15 min
2. Sonicate it 5 min at 30–40 °C in an ultrasonic bath (at 25 kHz) to disrupt the cells
3. Centrifuge the homogenate for 5 min at 5000 × g.
4. Decant the supernatant and filter it on a paper filter.
5. Repeat the extraction on the pellet (point 3 and 4) using the same filter previously used.
6. Wash the filter and funnel three times with small volumes of MeOH.
7. Collect the two extracts and the washings.
8. Dry the extracts with Rotavapor at 40 °C.
9. Resuspend the residue in 2 mL of distilled water.
10. Store it at −30 °C until analysis.

40.3.3 Tissue Extraction for Anabaenopeptins (A, B, and B1) Detection (Liquid Chromatography/High Resolution Mass Spectrometry [LC-HRMS] Analysis)

1. Homogenize tissue by knife mill.
2. Weight 5 g of the sample in a 50 mL polypropylene tube.
3. Add 10 mL of methanol/Milli-Q water (2/1, v/v) 0.1% of acetic acid.
4. Mix the sample by vortex for 1′.
5. Sonicate it in an ultrasonic bath for 10′.
6. Centrifuge the sample at 4000 rpm and 4 °C for 15′.
7. Transfer the supernatant in a clean 50 mL polypropylene tube.
8. Add the sample with further 10 mL of methanol/Milli-Q water (2/1, v/v) 0.1% of acetic acid.
9. Mix again the sample by vortex.
10. Sonicate and centrifuge as described before (point 5 and 6).
11. Collect the supernatant.
12. Add it to the first extract and mix by vortex.
13. Dilute 10 mL of the total extract with 10 of a 20 mM $NaHCO_3$ buffer (pH 6.3–6.5).
14. Mix it by vortex.
15. Load it on ISOLUTE C18 (EC) SPE cartridges (6 mL reservoir and 1 g of sorbent) previously conditioned with 5 mL of methanol and 5 mL of Milli-Q water.
16. Wash the cartridges with 5 mL of 20 mM $NaHCO_3$ buffer/methanol (95/5, v/v).
17. Dry applying vacuum for 2 min.
18. Elute the sample with 5 mL of methanol 0.1% acetic acid.
19. Evaporate the eluate to dryness under a gentle N_2 flow.
20. Dissolve the residue with 1 mL of mobile phase (A/B, 90/10, v/v) and 1 mL of methanol.
21. Mix it by vortex.
22. Centrifuge it at 12,000 rpm for 5′.
23. Transfer the supernatant in a glass vial for the LC-HRMS analysis.

References

Bruno M, Melchiorre S, Messineo V, Volpi F, Di Corcia A, Aragona I (2009) Microcystin detection in contaminated fish from Italian lakes using ELISA immunoassays and LC-MS/MS analysis. In: Gault PM, Marler HJ (eds) Handbook on cyanobacteria: biochemistry, biotechnology and applications. Nova Science Publishers, New York, NY, pp 191–210

Chen J, Xie P, Zhang D, Ke Z, Yang H (2006) In situ studies on the bioaccumulation of microcystins in the phytoplanktivorous silver carp stocked in Lake Taihu with dense toxic *Microcystis* blooms. Aquaculture 261:1026–1028

Chen J, Zhang D, Xie P, Wang Q, Ma Z (2009) Simultaneous determination of microcystin contaminations in various vertebrates (fish, turtle, duck and water bird) from a large eutrophic Chinese Lake, Lake Taihu, with toxic *Microcystis* blooms. Sci Total Environ 407:3317–3322

Chorus I, Bartram J (1999) Toxic cyanobacteria in water: a guide to their public health consequences, monitoring and management. CRC Press, Boca Raton, FL

Codd GA (1995) Cyanobacterial toxins: occurrence, properties and biological significance. Water Sci Technol 32:149–156

Dawson RM (1998) The toxicology of microcystins. Toxicon 36:953–962

Elkobi-Peer S, Carmeli S (2015) New prenylated aeruginosin, microphycin, anabaenopeptin and micropeptin analogues from a *Microcystis* bloom material collected in kibbutz Kfar Blum. Israel Mar Drugs 13:2347–2375

Ernst B, Hoeger SJ, O'Brien E, Dietrich DR (2006) Oral toxicity of the microcystin- containing cyanobacterium *Planktothrix rubescens* in European whitefish (*Coregonus lavaretus*). Aquat Toxicol 79:31–40

Ferranti P, Nasi A, Bruno M, Basile A, Serpe L, Gallo P (2011) A peptidomic approach for monitoring and characterising peptide cyanotoxins produced in Italian lakes by matrix-assisted laser desorption/ionisation and quadrupole time-of-flight mass spectrometry. Rapid Commun Mass Spectrom 25(9):1173–1183

Fonseca VB, Sopezki MS, Zanette J (2018) Effect of a toxic Microcystis aeruginosa lysate on the mRNA expression of proto-oncogenes and tumor suppressor genes in zebrafish. Ecotoxicol Environ Saf 161:729–734

Fujii K, Harada KI, Suzuki M, Kondo F, Ikai Y, Oka H, Carmichael WW, Sivonen K (1996) Occurrence of novel cyclic peptides together with microcystins from toxic cyanobacteria, *Anabaena* species. In: Yasumoto T, Oshima Y, Fukuyo Y (eds) Harmful and toxic algal bloom. Intergovernmental Oceanographic Commission of UNESCO, Paris, pp 559–562

Fujii K, Sivonen K, Nakano T, Harada K (2002) Structural elucidation of cyanobacterial peptides encoded by peptide synthetase gene in *Anabaena* species. Tetrahedron 58:6863–6871

Gan F, Zhang S, Rockwell NC, Martin SS, Lagarias C, Bryant DA (2014) Extensive remodeling of a cyanobacterial photosynthetic apparatus in far-red light. Science 345:1312–1317

Gkelis S (2006) Characterisation of planktic cyanobacteria and their bioactive peptides. Aristotle University of Thessaloniki, Thessaloniki

Grosse Y, Baan R, Secretan B, Straif K, El Ghissassi F, Cogliano V (2006) Carcinogenicity of nitrate, nitrite, and cyanobacterial peptide. Toxins 7:628–629

IARC (2010) Monograph. p 94. https://monographs.iarc.who.int/list-of-classifications/

Jewel MA, Affan MA, Khan S (2003) Fish mortality due to cyanobacterial bloom in an aquaculture pond in Bangladesh. Pak J Biol Sci 6:1046–1050

Li Z, Guo JS, Fang F, Chen J, Zhang C, Tian G (2009) Seasonal variation of nitrogen in Xiaojiang backwater area, three gorges reservoir. Environ Sci 30:1588–1594

Magalhães VF, Soares R, Azevedo S (2001) Microcystin contamination in fish from the Jacarepagua Lagoon (Rio de Janeiro, Brazil): ecological implication and human health risk. Toxicon 39:1077–1085

Manti G, Mattei D, Messineo V, Melchiorre S, Bogialli S, Sechi N, Casiddu P, LugliÉ A, Di Brizio M, Bruno M (2005) First report of *Cylindrospermopsis raciborskii* in Italy. Harmful Algae News 28:8–9

Metcalf JS, Banack SA, Lindsay J, Morrison LF, Cox PA, Codd GA (2008) Co-occurrence of beta-N-methylamino-L-alanine, a neurotoxic amino acid with other cyanobacterial toxins in British waterbodies, 1990–2004. Environ Microbiol 10:702–708

Mohamed ZA, Carmichael WW, Hussein AA (2003) Estimation of microcystins in the freshwater fish *Oreochromis niloticus* in an Egyptian fish farm containing a *Microcystis* bloom. Environ Toxicol 18(2):137–141

Nishiwaki-Matushima R, Nishiwaki S, Ohta T, Yoszawa S, Suganuma M, Harada K, Watanabe MF, Fujiki H (1991) Structure-function relationships of microcystins, liver-tumor promoters, in interaction with protein phosphatase. Jpn J Cancer Res 82:993–996

Saker M, Eaglesham GK (1999) The accumulation of cylindrospermopsin from the cyanobacterium *Cylindrospermopsis raciborskii* in tissues of the Redclaw crayfish *Cherax quadricarinatus*. Toxicon 37:1065–1077

Saker M, Metcalf JS, Codd GA, Vasconcelos VM (2003) Accumulation and depuration of the cyanobacterial toxin cylindrospermopsin in the freshwater mussel *Anodonta cygnea*. Toxicon 43:185–194

Smith JL, Haney JF (2006) Foodweb transfer, accumulation and depuration of microcystins, a cyanobacterial toxin in pump-kinseed sunfish (*Lepomis gibbosus*). Toxicon 48:580589

Vasas G, Gaspar A, Suranyi G, Batta G, Gyemant G, M-Hamvas M, Mathé C, Grigorszky I, Molnar E, Borbely G (2002) Capillary electrophoretic assay and purification of Cylindrospermopsin, a cyanobacterial toxin from *Aphanizomenon ovalisporum*, by plant test (blue-green Sinapis test). Anal Biochem 302:95–103

Welker M, von Döhren H (2006) Cyanobacterial peptides–Nature's own combinatorial biosynthesis. FEMS Microbiol Rev 30:530–563

Wood SA, Briggs LR, Sprosen J, Ruck JG, Wear RG, Holland PT, Bloxham M (2006) Changes in concentration of microcystins in rainbow trout, freshwater mussels, and cyanobacteria in lakes Rotoiti and Rotoehu. Environ Toxicol 21:205–222

Xie L, Xie P, Guo L, Li L, Miyabara Y, Park HD (2005) Organ distribution and bioaccumulation of microcystins in freshwater fish at different trophic levels from the eutrophic Lake Chaohu, China. Environ Toxicol 20:293–300

Zegura B, Sedmak B, Filipič M (2003) Microcystin-LR induces oxidative DNA damage in human hepatoma cell line HepG2. Toxicon 41(1):41–48

Zhang DW, Xie P, Liu YQ, Qiu T (2009) Transfer, distribution and bioaccumulation of microcystins in the aquatic food web in Lake Taihu, China, with potential risks to human health. Sci Total Environ 407:2191–2199

Zhang DW, Deng X, Xie P, Chen J, Guo L (2013) Risk assessment of microcystins in silver carp (*Hypophthalmichthys molitrix*) from eight eutrophic lakes in China. Food Chem 140:17–21

Zimba PV, Khoo L, Gaunt PS, Bittain S, Carmichael WW (2001) Confirmation of catfish, *Ictalurus punctatus* (Rafinesque) mortality from *Microcystis* toxins. J Fish Dis 24:41–47

Zimba PV, Camus A, Allen EH, Burkholder JM (2006) Co-occurrence of white shrimp, *Litopenaeus vannamei*, mortalities and microcystin toxin in a southeastern USA shrimp facility. Aquaculture 261:1048–1055

Extraction and Quantification of Nodularins from Fish Samples by LC-MS/MS

41

Guhanraj Radhamanalan and D. Dhanasekaran

Abstract

Freshwater harmful algal blooms (HABs) and aquaculture practices have lately risen globally. Cyanotoxins are chemical compounds produced by cyanobacterial mats in aquatic settings. These might damage human health as well as aquatic creatures. Because of their ubiquity in aquatic environments, removing these poisons is usually difficult. Limited methods for quantifying cyanotoxins in complex matrices have hampered scientific understanding of HAB impacts on aquatic environments. HABs can create cyanotoxins, several of which are hepatotoxins. To rupture the cell wall, most extraction processes involve freeze-drying or freeze-thawing, and ultrasonication is widely used to improve the burgeoning field of study. Aquaculture is increasing the need for faster and more cost-effective nodularin detection. One method for detecting nodularin and dimethyl nodularin in fish has been developed using LC-MS/MS. Although genetic approaches exist to determine whether or not cyanobacteria have the toxin gene, quantitative analysis, such as LC/MS/MS for liquid chromatography-mass spectrometry is required to determine whether or not cyanobacteria produce cyanotoxins. The goal of this study is to evaluate the fish sample extracted from nodularin and quantified using LC/MS/MS.

Keywords

Nodularin · Cyanotoxin · Fish · Aquaculture · LC/MS/MS

G. Radhamanalan (✉)
Department of Microbiology, School of Life Sciences, Bharathidasan University, Tiruchirappalli, Tamil Nadu, India

D. Dhanasekaran
Department of Microbiology, National Repository for Microalgae and Cyanobacteria - Freshwater (NRMC-F), Bharathidasan University, Tiruchirappalli, Tamil Nadu, India

N. Thajuddin et al. (eds.), *Protocols for Cyanobacteria Sampling and Detection of Cyanotoxin*, https://doi.org/10.1007/978-981-99-4514-6_41

41.1 Introduction

Cyanobacteria, commonly known as blue-green algae, are found in fresh, brackish, and marine habitats, as well as in soil and on moist surfaces. Cyclic heptapeptide hepatotoxins and microcystins (>70) identified from many freshwater taxa are the most frequently reported cyanobacterial toxins. *Planktothrix* (*Oscillatoria*), *Anabaena*, and *Nostoc*. The cyclic pentapeptide hepatotoxins Nodularins (<10) are found in the brackish water cyanobacterium *Nodularia* (Spoof et al. 2003). Cyanobacteria constitute the foundation of many terrestrial food systems and are finding increasing applications in biotechnology, from the production of fertilizers to the creation of potential pharmaceuticals. Cyanobacteria may be found in a variety of habitats in aquatic settings. In certain environments, dense populations of cyanobacteria can form blooms, scums, biofilms, or mats. NODs are cyclic pentapeptide hepatotoxins. In the often-occurring nodularin R, the second amino acid residue (Z) is lArg (Catherine et al. 2016). Several nonchromatographic cyanotoxin analysis approaches have been proposed. All of the methodologies described have limitations that restrict their applicability to large-scale occurrence surveys and human exposure investigations. In recent years, LC coupled to MS has emerged as the method of choice for detecting cyanotoxins in a wide range of samples. The multiple-reaction monitoring (MRM) data-gathering approach with LC-MS/MS provides excellent signal/noise enhancement, selectivity, and quantitation capabilities. Although the potential of LC-MS/MS for assessing cyanotoxins in water bodies has been thoroughly explored, only one technique based on LCMS/MS has been developed for detecting nodularin and dimethyl nodularin in fish and flounders (Bogialli et al. 2005). NODs are toxins generated by cyanobacteria in both fresh water and aqua (Butt et al. 2017). The purpose of this study was to look at the formation of NOD in fish samples. NOD was detected by liquid chromatography-mass spectrometry (LC-MS/MS).

41.2 Materials

- Nodularin's (NOD) test sample extracted from fish
- ATX-A
- MMPB
- CYN N15
- Acetonitrile
- Potassium permanganate
- LC-MS grade
- Formic acid
- Sodium (meta) periodate
- Sodium bisulfite
- Solid-phase extraction cartridges (250 mg, 3 cm^3)
- Oasis HLB and cartridges (60 mg, 3 cm^3)
- HPLC grade Water 18 MΩ cm and 3 ppb

- Syringe filters
- Methanol
- Dichloromethane
- Trifluoroacetic acid
- 2 mL sterile syringe
- Ultrasonic bath
- Analytical balance
- Microcentrifuge
- Pipettes
- Graduated cylinders

41.3 Methods

41.3.1 Extraction of Cyanotoxins (Greer et al. 2017)

1. Collect the fish from an aquaculture farm at Parangipettai, Cuddalore.
2. Transport the samples to Department of Microbiology, Bharathidasan University, where the temperature is kept at 22 °C ± 2 °C for the experimental method.
3. Homogenized lyophilized blank fish sample (100 mg) aliquots.
4. After centrifugation at 12,000 rpm for 15 min, 2 mL of 75% methanol (v/v) (0.5% acetic acid v/v) was added and immediately vortexed, mixed, and sonicated at room temperature for 45 min.
5. The supernatant was Take it out and dilute it five times. Before being put onto SPE cartridges, with samples placed onto the Oasis HLB cartridge
6. Collect the flow and afterward pass it through the ENVI cartridges
7. Wash the Oasis HLB cartridges with water and 20% methanol (v/v)
8. Toxins was eluted into clean glass tubes with 2 × 2 mL of 80% methanol (v/v) containing 0.1% trifluoroacetic acid (v/v) for the Oasis and 2 × 2 mL of methanol/dichloromethane (4:1 v/v) containing 5% formic acid (v/v).
9. Toxin was eluted into clean glass tubes using 80% methanol (v/v) containing 0.1% trifluoroacetic acid (v/v) for Oasis and 2 mL of (4:1 v/v) methanol/dichloromethane containing 5% formic acid (v/v) for ENVI.
10. Homogenize samples after lyophilization and store at −80 °C for further analysis.

41.3.2 Nodularin's Quantification

41.3.2.1 LC-MS/MS

The extracts of fish are examined using a liquid chromatography-mass spectrometer/mass spectrometry LC-MS/MS (Figs. 41.1 and 41.2) (Tables 41.1 and 41.2).

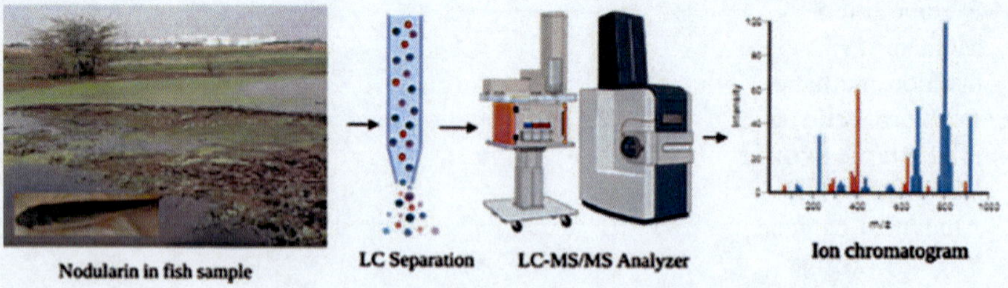

Fig. 41.1 Quantification of Nodularins from fish samples by LC-MS/MS

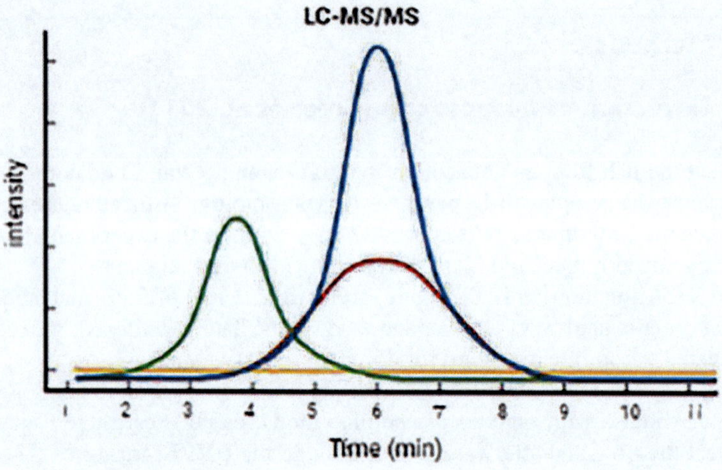

Fig. 41.2 Ion mass chromatogram

Table 41.1 Quantification of Nodularin (NOD)

S. No	Sample	LOD[1] (Fmol inj^{-1})	LOQ[2] (Fmol inj^{-1})	Calibration range (μgL^{-1})	Curve shape

41.3.2.2 Liquid Chromatography

Under varied experimental settings, the target chemicals can be separated using reversed-phase chromatography.

Table 41.2 Nodularin detection by LC-MS/MS

Sample	Precursor ion m/z	Production m/z	CE value

1. Column: C18 30 mm × 2.1 mm, 3 μm particle size.
2. Column temperature: 30 °C.
3. LC mobile phase: Solvent A: acetonitrile—0.5% HCOOH, Solvent B: water—0.5% HCOOH.
4. Gradient: 25% Solvent A (0–1 min) to 70% A (10.5 min) to 90% A (12.5–13.5 min) to initial conditions.
5. Flow rate: 0.2 mL min–1.
6. Injection volume: 20 μL.

41.3.2.3 Mass Spectrometry (Brown et al. 2018)

Calibration of mass spectrometry operating settings for positive ionization mode should be performed in accordance with the manufacturer's instructions and laboratory protocols. To identify and quantify target compounds, precursor-to-product ion transitions (SRMs) and chromatographic retention durations are employed.

1. Separate and isolate fish [0.100 + 0.005 g (d.w.)]
2. Extract the toxin and sonicate the sample in water bath 25 min and a 5 mL centrifuge tube containing 0.1 M acetic acid and 75% methanol at 12,000 rpm for 10 min.
3. The supernatant was retained, and the pellets were washed with 5 mL of 75% acidified methanol.
4. Separate methanolic supernatants 1 mL n-butanol washing.
5. Solid phase extraction, the supernatants were dried with N_2 at 60 °C and reconstituted (5 ml milli-Q3 water) (200 mg for methanolic extracts, 100 mg for butanol extracts).
6. After washing the columns with 5% MeOH (2 mL), they are eluted with 90% acetonitrile.
7. Elute was dried (N_2 at 60 °C), reconstituted (1 mL, 5% MeOH), and purified through 0.2 mm PVDF.
8. LC Separate sample components in a column (150 + 2.1 mm) with mobile phases A (2 mM formic acid and 3.6 mM ammonium formate in milli-Q3 water) and B (95% acetonitrile (v/v) in 2 mM formic acid and 3.6 mM ammonium formate). HPLC Water, acetonitrile, ammonium formate, and water (98%) (HPLC).

9. Solvent A maintained 70% for 5 min, 70–65% for 8 min, 65% for 2 min, 65–30% for 4 min, 30–70% for 2 min, and 70% for 3 min while flowing at 0.2 mL min + 1.
10. Chromatographic injections of 20 mL were performed for 24 min.

41.4 Observation

Acknowledgments The authors would like to thank the Department of Microbiology at Bharathidasan University. We gratefully acknowledge DST financial support for the research facility, which was provided through the DST Promotion of University Research and Scientific Excellence (PURSE) scheme-Phase II, Rashtriya Uchchatar Shiksha Abhiyan (RUSA)-2.0-Biological Sciences (TN RUSA: 311/RUSA (2.0) /2018 dt. 02/12/2020).

References

Bogialli S, Bruno M, Curini R, Di Corcia A, Laganá A, Mari B (2005) Simple assay for analyzing five microcystins and nodularin in fish muscle tissue: hot water extraction followed by liquid chromatography—tandem mass spectrometry. J Agric Food Chem 53(17):6586–6592

Brown A, Foss A, Miller MA, Gibson Q (2018) Detection of cyanotoxins (microcystins/nodularin) in livers from estuarine and coastal bottlenose dolphins (*Tursiops truncatus*) from Northeast Florida. Harmful Algae 76:22–34

Butt C, Quinn-Paquet A, Borton C (2017) Quantitation of microcystins and Nodularins in water samples using LC-MS/MS. Structure 1:2

Catherine A, Bernard C, Spoof L, Bruno M (2016) Microcystins and nodularins. In: Handbook of cyanobacterial monitoring and cyanotoxin analysis, pp 107–126

Greer B, Maul R, Campbell K, Elliott CT (2017) Detection of freshwater cyanotoxins and measurement of masked microcystins in tilapia from Southeast Asian aquaculture farms. Anal Bioanal Chem 409(16):4057–4069

Msagati TA, Siame BA, Shushu DD (2006) Evaluation of methods for the isolation, detection and quantification of cyanobacterial hepatotoxins. Aquat Toxicol 78(4):382–397

Spoof L, Vesterkvist P, Lindholm T, Meriluoto J (2003) Screening for cyanobacterial hepatotoxins, microcystins and nodularin in environmental water samples by reversed-phase liquid chromatography–electrospray ionisation mass spectrometry. J Chromatogr A 1020(1):105–119

Extraction and Quantification of Nodularins from Shrimp Samples by ELISA

42

Guhanraj Radhamanalan and D. Dhanasekaran

Abstract

Nodularins are potent hepatotoxic cyanotoxins produced by harmful algal blooms (HABs). Aquaculture is boosting the demand for faster and more cost-effective Nodularins detection systems. ELISAs, on the other hand, are efficient, sensitive, and affordable, and have the potential to replace or enhance existing procedures. This suggests that the ELISA approach can detect nodularin at concentrations as low as 1 ppb. The goal of this study is to evaluate a shrimp sample extracted from nodularin and quantified using ELISA.

Keywords

Nodularin · Cyanotoxin · Shrimp · Aquaculture · ELISA

42.1 Introduction

Cyanotoxins are cyanobacterial neurotoxins and hepatotoxins (blue-green algae). Identification is an important element of any natural substance toxicity assessment. It provides both a qualitative and quantitative approach to identifying and measuring toxins. There are two types of cyanotoxins: cytotoxins and biotoxins. Biological techniques include small animals (such as mice, fish, and invertebrates) and microbial cultures (Bacteria). This ELISA has a sensitivity of ppb and a detection limit of

G. Radhamanalan (✉)
Department of Microbiology, School of Life Sciences, Bharathidasan University, Tiruchirappalli, Tamil Nadu, India

D. Dhanasekaran
Department of Microbiology, National Repository for Microalgae and Cyanobacteria - Freshwater (NRMC-F), Bharathidasan University, Tiruchirappalli, Tamil Nadu, India

N. Thajuddin et al. (eds.), *Protocols for Cyanobacteria Sampling and Detection of Cyanotoxin*, https://doi.org/10.1007/978-981-99-4514-6_42

0.05–0.5 ppb. This indicates that the ELISA method may detect nodularin at levels as low as 1 ppb (Carmichael and An 1999). ELISA is a simple and effective approach for detecting algal toxins that are commonly used as a basic screening test, especially for samples that have been collected. One significant disadvantage of ELISA and other immunoassay methods is that the extraction methods necessary for solid materials, such as the tissues of fish, are not usually suitable for field use. To detect the Nod-R, a commercially available competitive ELISA was also employed. This ELISA utilizes an antibody directed against the unusual amino acid which is present only in MCs and Nod-R and nowhere else in nature. The ELISA is meant to detect microcystins and Nod-R exceedingly specifically, even in complex materials, as nothing more than a reason (Geis-Asteggiante et al. 2011). Nodularin (cyclic pentapeptide hepatotoxin) is a direct carcinogen produced by the filamentous brackish water cyanobacterium *Nodularia spumigena*. NOD may be transmitted to aquatic fish by food consumption. There is a growing demand for faster and more cost-effective nodularin detection methods (Zhou et al. 2011). The goal of this study was to examine Nodularin in shrimp utilizing an ELISA based on nodularin detection.

42.2 Materials

- 12 × 8 strips of microtiter plate coated with a nodularin-conjugated protein
- Standards (6) at variable volumes ppb
- Control 0.80 ± 0.185 ppb
- Sample diluent (Laboratory reagent blank LRB)
- Nodularin-Antibody solution
- Nodularin-Conjugate solution
- 5× wash solution
- Substrate (color)
- Solution (TMB)
- Stop solution
- Multichannel pipette
- Stepper pipette
- Micropipettes
- Disposable plastic tips
- Deionized water
- Container
- Graduated cylinder
- Paper towels
- Tissue paper
- Timer
- Tape or parafilm
- ELISA plate reader
- Microtiter plate washer
- Shrimp sample

42.3 Methods

42.3.1 Extraction of Nodularins from Shrimps (Msagati et al. 2006)

1. Collect the aquatic (Shrimp) samples from Sivagangai zone aquaculture farms.
2. Homogenize fresh shrimp samples in 100% methanol with 1% trifluoroacetic acid (v/v).
3. Ultrasonicate and spin the sample at 8000 rpm for 12 min.
4. Separate the supernatant and evaporate until dry at 40 °C.
5. Load the 5% acetic acid shrimp extracts into solid-phase extraction cartridges.
6. Mix an equal volume of methanol extract and hexane before discarding the hexane layer.
7. Filter the extract was using C18 SPE cartridges.

42.3.2 Nodularin's Quantification

42.3.2.1 Enzyme-Linked Immunosorbent Assay (ELISA)

Test Preparation

1. Before using the chemicals or samples, keep them at room temperature.
2. Open the resealable packet and remove the required amount of microtiter plate strips. Store the remaining strips in a bag next to the sterilizer (tightly sealed).
3. The standard, control, sample diluent (LRB), antibody, enzyme conjugate, substrate, and stop solutions are all ready to use and do not require further preparation.
4. Make the Wash Solution Diluted (5×) Concentrate in a 1:5 ratio with deionized (or) distilled water. If using the full bottle (100 mL), mix well with 400 mL of deionized (or) distilled water.

Assay Procedure

1. Pour 50 μL of stock solution, control, LRB, or samples into the test strip wells. It is advisable to conduct the analysis twice or three times.
2. Add 50 μL of the antibody solution to each well using a multichannel pipette.
3. Cover the wells with parafilm and mix the liquids for 30 s on the tabletop by turning the strip holder in a circular pattern.
4. At room temperature, incubate the strips for 90 min.
5. After removing the top, decant the liquid from the wells into a drain.
6. The strips should be washed three times with the 1× wash buffer solution.
7. Each well and washing step should contain at least 250 μL of wash buffer.
8. Wipe the plate dry with a pile of paper towels to remove any leftover buffer in the wells.
9. Add 100 μL of the enzyme conjugate liquid to every single well using a multichannel pipette. Cover the wells with parafilm or tape and mix the contents for 30 s on the workstation by turning the strip holder in a rotational movement.

Table 42.1 ELISA determined by shrimp samples

Concentration of NOD μg/ml^{-1}	Recovery (%)

Table 42.2 ELISA quantitative analysis

Aquatic samples	ELISA determination (μg l–1)	OD at 480nm

Make certain that the contents are not spilled. Incubate the strips at room temperature for 30 min.

10. Empty the contents of the wells by removing the top. The strips should be washed three times with the 1× washing buffer solution. For each well and washing step, please use at least 250 μL of wash buffer. Wipe the plate dry with a stack of paper towels to remove any residual buffer from the wells.
11. Fill each well with 100 μL of substrate (color) solution using a multichannel pipette. Cover the wells with cling film and mix the contents for 30 s on the tabletop by turning the strip holder horizontally. Make certain that the contents are not spilled. Incubate the strips for 20–30 min at room temperature. Remove the strips from direct sunlight.
12. Using a multichannel pipette, add 50 μL of stop solution in the same ratio as the substrate (color) solution to the wells.
13. Use an ELISA reader to determine the absorbance at 480 nm within 15 min of adding the stop solution (Tables 42.1 and 42.2).

42.4 Observation

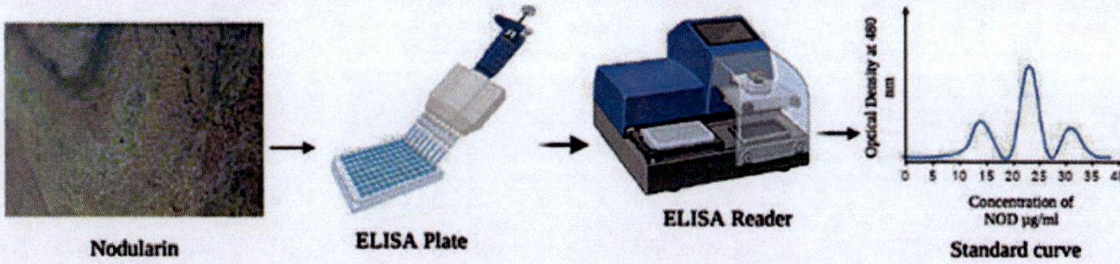

Acknowledgments The authors express their gratitude to the DST-FIST (Department of Science and Technology-Fund for Improving Science and Technology Infrastructure), New Delhi, for the instrumentation facilities (DST Sanction Order No.: SR/FIST/LSI-013/ 2012/Dt.13.08.2012). The Department of Science and Technology (DST) provided funding for the research facility through the DST Promotion of University Research and Scientific Excellence (PURSE) scheme-Phase II, Rashtriya Uchchatar Shiksha Abhiyan (RUSA)-2.0-Biological Sciences (TN RUSA: 311/RUSA (2.0) /2018 dt. 02/12/2020).

References

Carmichael WW, An J (1999) Using an enzyme-linked immunosorbent assay (ELISA) and a protein phosphatase inhibition assay (PPIA) for the detection of microcystins and nodularins. Nat Toxins 7(6):377–385

Geis-Asteggiante L, Lehotay SJ, Fortis LL, Paoli G, Wijey C, Heinzen H (2011) Development and validation of a rapid method for microcystins in fish and comparing LC-MS/MS results with ELISA. Anal Bioanal Chem 401(8):2617–2630

Msagati TA, Siame BA, Shushu DD (2006) Evaluation of methods for the isolation, detection and quantification of cyanobacterial hepatotoxins. Aquat Toxicol 78(4):382–397

Zhou Y, Li YS, Zhi BH, Lu SY, Ren HL, Zhang YY, Li ZH, Shen QF, Meng XM, Liu ZS, Zhang JH (2011) Detection of nodularin based on a monoclonal antibody in water and aquatic fish samples. Food Control 22(5):797–800

43 Extraction and Quantification of Cylindrospermopsins from Aquatic Samples by ELISA

Milena Bruno and Valentina Messineo

Abstract

Most of the world's population relies on surface freshwaters as its primary source for drinking water. The drinking water industry is constantly challenged with surface water contaminants that must be removed to protect human health. Toxic cyanobacterial blooms are an emerging issue worldwide due to increased source water nutrient pollution caused by eutrophication. Cylindrospermopsin is a toxin produced by several different strains of cyanobacteria (blue-green algae) and has been found in freshwater throughout the world. Certain strains of *Cylindrospermopsis raciborskii* (synonym of *Raphidiopsis raciborskii*) (Wołoszyńska) (found in Australia, Hungary, and the United States), *Umezakia natans* (found in Japan), and *Aphanizomenon ovalisporum* (synonym of *Chrysosporum ovalisporum*) (found in Australia and Israel) have been found to produce cylindrospermopsin.

Human exposure to cylindrospermopsin can occur through ingestion of contaminated water or food (fish) or during recreational activities in which water is swallowed. Dermal contact with cylindrospermopsin may occur during showering or bathing, or during recreational activities such as swimming or boating. These toxins mediate their toxicity by inhibiting liver function and are protein synthesis and glutathione potent inhibitors, leading to cell death.

Keywords

Cylindrospermopsin · Hepatotoxicity · Mutagenicity · Genotoxicity · Contaminated drinking water and food

---

M. Bruno (✉) · V. Messineo
FAST, Core Facilities, National Institute of Health, Rome, Italy
e-mail: milena.bruno@iss.it; valentina.messineo@iss.it

N. Thajuddin et al. (eds.), *Protocols for Cyanobacteria Sampling and Detection of Cyanotoxin*, https://doi.org/10.1007/978-981-99-4514-6_43

43.1 Introduction

Acute poisoning of humans and animals constitutes the most obvious problem from toxic cyanobacterial blooms and, in several cases, has led to death. Cylindrospermopsin (CYN) is a tricyclic alkaloid with the following chemical formula: C15H21N5O7S (Fig. 43.1), produced by several cyanobacterial species. The toxin has been recognized as a nationwide threat due to the invasive nature of its main producer, *Cylindrospermopsis raciborskii* (Fig. 43.2) (Svirčev et al. 2019) and *Aphanizomenon ovalisporum* (Fig. 43.3). CYN has been reported in surface waters at concentrations up to 173 μg/L (Buratti et al. 2017) and also to co-occur with MCs (Díez-Quijada et al. 2018).

Cylindrospermopsins, the second class of cyanotoxins in terms of frequency of detection in Italy, have been the subject of particular scientific interest in the last decade due to their spread from tropical to temperate areas, in apparent agreement with the occurrence of phenomena related to global warming. Cylindrospermopsins

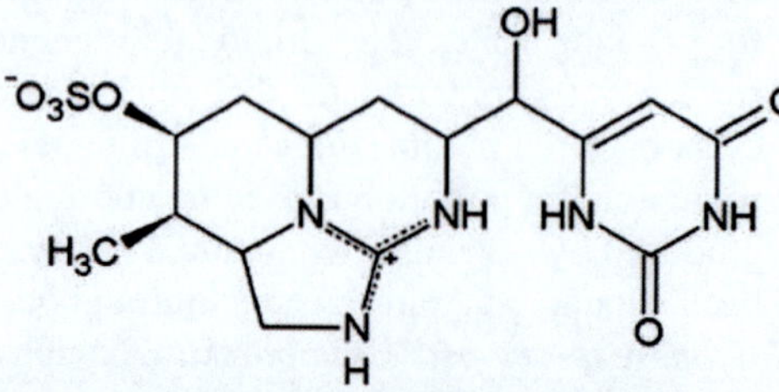

Fig. 43.1 Structure of cylindrospermopsin

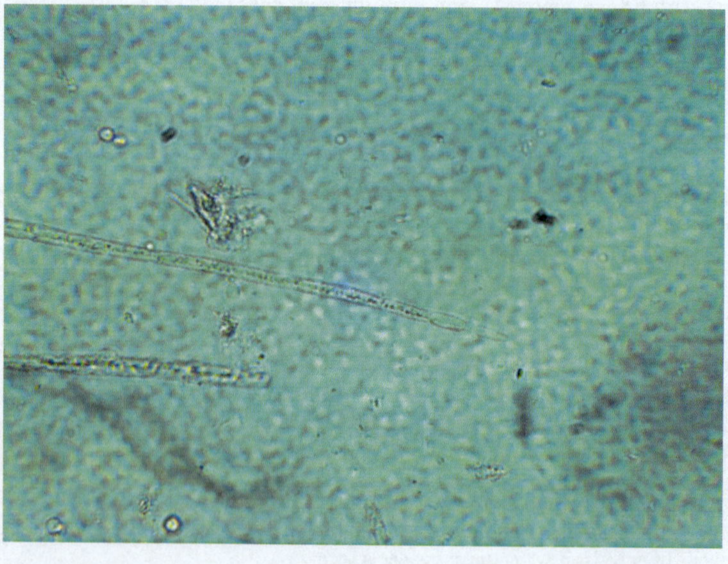

Fig. 43.2 *Cylindrospermopsis raciborskii,* synonym of *Raphidiopsis raciborskii* (Wołoszyńska 1912; Aguilera et al. 2018)

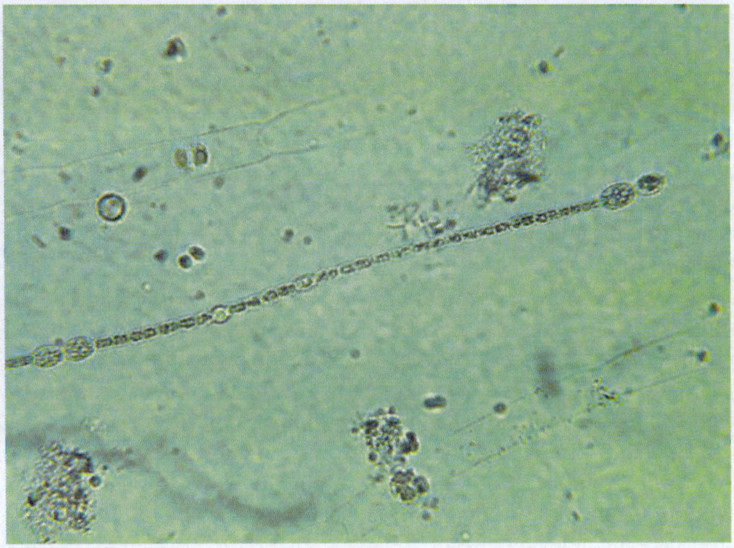

Fig. 43.3 *Aphanizomenon ovalisporum,* synonym of *Chrysosporum ovalisporum* (Forti 1911) Zapomelová et al. 2012

(Cylindrospermopsin, CYN) are a class consisting of five chemical analogs: CYN, 7-epi-CYN, 7-deoxy-CYN, 7-deoxydesulfo-CYN, and 7-deoxydesulfo-12-acetyl-CYN. Cylindrospermopsin was first isolated from the cyanobacterium *Cylindrospermopsis raciborskii* which was found in an artificial drinking water reservoir during a hepatoenteritis outbreak on Palm Island in northern Queensland, Australia, in 1979 (Griffiths and Saker 2003). The molecule (Fig. 43.1) is a nephrotoxic, thymotoxic, and hepatotoxic sulfate-guanidine alkaloid, with a 5-substituted-2,4-dioxypyrimidine radical (uracil), (molecular weight 415 g/mol) (Banker et al. 1997). It is a potent inhibitor of protein synthesis and can cause cell death, with the liver being the main target organ.

In addition to the liver, toxic effects on the kidneys, lungs, heart, and thymus have also been described following CYN poisoning (Terao et al. 1994). In mice, skin sensitization reactions produced by pure toxin or aqueous suspensions of *C. raciborskii* containing CYN have been demonstrated (Stewart et al. 2008) as well as fetal toxicity after exposure in late gestation (Rogers et al. 2007).

The mutagenicity of CYN (Žegura et al. 2011; Štraser et al. 2011; Puerto et al. 2018; Pichardo et al. 2017) has recently been demonstrated in vitro, and there is also strong evidence for its carcinogenicity in vivo: DNA fragmentation and modification (Shen et al. 2002; Shaw et al. 2000) have been observed in the livers of treated mice, while a whole range of cytogenetic abnormalities have been observed in human lymphoblastoid cells exposed to CYN, with the formation of centromere-negative micronuclei indicating double-stranded DNA breakage (Humpage et al. 2000).

Genotoxic activity is caused by the ability of the toxin to induce DNA helix breaks, with loss of entire chromosomes due to damage to the centromere/kinetochore function. Furthermore, CYN induces stress responses in cultured human fibroblasts and HepG2 cells, which cause the activation of the transcription factor p53 (Bain et al. 2007). At the molecular level, CYN may interact with the ribosomes and inhibit the synthesis of proteins in the cells (Froscio et al. 2003). At low doses, CYN suppresses glutathione-conjugated protein synthesis, probably inhibiting ribosomal translation through binding to a protein associated with the eukaryotic translation system (Froscio et al. 2008); but at higher concentrations a more rapidly toxic, metabolism-dependent process dominates: its acute toxicity appears to be mediated by cytochrome p 450-generated metabolites (Humpage et al. 2005). The production of cylindrospermopsin seems to be strain specific rather than species specific (Moreira et al. 2013). CYN has been proven to have environmental toxicity towards amphibians (Kinnear et al. 2007) and antibacterial activity (Rasmussen et al. 2008). Recent evidence has shown that plant growth and metabolism (Vasas et al. 2002) as well as pollen germination (Metcalf et al. 2004) are also inhibited by CYN, with negative implications for current jet irrigation practices.

To date, there are few studies on CYN accumulation in fish fauna: CYN has been proven to contaminate shrimp (*Cherax quadricarinatus*, up to 4.3 μg/g) and fish (*Melanotaenia eachamensis*, up to 1.2 μg/g) from a small Australian aquaculture pond (Saker and Eaglesham 1999) and freshwater molluscs (*Anodonta cygnea*, up to 2.52 μg/g) in a 16-day exposure study (Saker et al. 2003). The presence of cylindrospermopsin was first revealed in Europe in 2002 (Kiss et al. 2002) and in Italy in 2004 (Manti et al. 2005). To protect against adverse health effects, the U.S. Environmental Protection Agency (EPA) has established guidelines for cylindrospermopsin in drinking water (EPA 2022):

- For children pre-school age and younger (less than 6 years old), 0.7 μg/L (ppb)
- For school-age children and adults, 3.0 μg/L (ppb)

43.2 Materials

- Ultracentrifuge 11,000 rpm, BeckmanLT-55
- Eppendorf vials
- Gilson micropipette
- Bidistilled H_2O
- Vortex Vibra-Cell, Sonics & Materials Inc.
- Ultrasonic bath (Elgasonic Swiss made)
- ABRAXIS® Cylindrospermopsin, ELISA Assay, 96-test
- EZ Read 400 ELISA Reader

43.3 Methods

43.3.1 Toxin Extraction

Extract water samples as follows:

1. Obtain fresh phytoplankton aliquots (10–130 mg) by centrifugation of water samples.
2. Stock in a vial the centrifugation water.
3. Suspend the sample aliquots in 2 mL of sterile bidistilled water.
4. Stir the solution in Vortex.
5. Sonic it for 5 min at 30–40 °C in an ultrasonic bath (Elgasonic Swiss made) and after 2.5 min put the sample in Vortex for 20 s, then continue with the ultrasonic bath.
6. Centrifuge for 10 min at 11,000 r.p.m. (Beckman L5–55 Ultracentrifuge) to eliminate debris.
7. Collect the supernatant.
8. Resuspend the pellet.
9. Repeat the whole process twice.
10. Pool the two supernatants.
11. Analyze for intracellular toxin.
12. Determine extracellular CYN concentrations by testing with ABRAXIS® Cylindrospermopsin ELISA assay the centrifugated water.

43.3.2 Toxin Determination with ELISA Assay ABRAXIS®

The cylindrospermopsin ELISA is an immunoassay for the quantitative and sensitive detection of cylindrospermopsin in water samples. The test is a direct competitive ELISA for the detection of cylindrospermopsin: it is based on the recognition of cylindrospermopsin by specific antibodies. Cylindrospermopsin, when present in a sample, and a cylindrospermopsin-HRP analog compete for the binding sites of rabbit anti-cylindrospermopsin antibodies in solution. The anti-cylindrospermopsin antibodies are then bound by a second antibody (goat anti-rabbit) immobilized on the wells of the microtiter plate. After a washing step and addition of the substrate solution, a color signal is generated. The intensity of the blue color is inversely proportional to the concentration of cylindrospermopsin present in the sample. The color reaction is stopped after a specified time and the color is evaluated using an ELISA reader. The concentrations of the samples are determined by interpolation using the standard curve constructed with each run.

Reagents and Materials Provided

1. Microtiter plate (12 × 8 strips) coated with a second antibody (goat anti-rabbit).
2. Standards: 0, 0.05, 0.10, 0.25, 0.50, 1.0, 2.0 ppb; 1 mL each.

3. Control: 0.75 ± 0.15 ppb, 1 mL, prepared from a secondary source, for use as a Quality Control Standard (QCS).
4. Sample diluent, 25 mL, for use as a Laboratory Reagent Blank (LRB) and for dilution of samples above the range of the standard curve.
5. Cylindrospermopsin-HRP Conjugate solution, 6 mL.
6. Antibody solution (rabbit anti-cylindrospermopsin), 6 mL.
7. Wash buffer (5×) Concentrate, 100 mL.
8. Substrate (Color) solution (TMB), 12 mL.
9. Stop solution, 12 mL.

Additional Materials

1. Micropipettes with disposable plastic tips (20–200 μL)
2. Multi-channel pipette (10–300 μL), stepper pipette (10–300 μL), or electronic repeating pipette with disposable plastic tips
3. Deionized or distilled water
4. Container with 500 mL capacity (for diluted 1× Wash Buffer)
5. Graduated cylinder
6. Paper towels or equivalent absorbent material
7. Timer
8. Tape or parafilm
9. Microtiter plate reader (wavelength 450)
10. Microtiter plate washer (optional)

Test Preparation

1. Allow the reagents and samples to reach ambient temperature before use.
2. Remove the number of microtiter plate strips required from the resealable pouch. Store the remaining strips in the pouch with the desiccant (tightly closed).
3. The standards, control, sample diluent (LRB), antibody, enzyme conjugate, substrate, and stop solutions are ready to use and do not require any further dilutions.
4. To prepare the wash buffer, dilute the Wash Buffer (5×). Concentrate at a ratio of 1:5 with deionized or distilled water. If using the entire bottle (100 mL), add to 400 mL of deionized or distilled water and mix thoroughly.
5. Handle with care the stop solution as it contains diluted H_2SO_4.

The microtiter plate consists of 12 strips of 8 wells, which can be used individually for the test. The standards must be run with each test. Never use the values of standards which have been determined in a test performed previously.

Std0–Std6: Standards
Contr.: Control (QCS)
LRB: Laboratory Reagent Blank
Samp1, Samp2, etc.: Samples

Assay Procedure

1. Add 50 μL of the standards, control (QCS), LRB, or samples into the wells of the test strips according to the working scheme given. Analysis must be run in triplicate.
2. Vortex enzyme conjugate solution.
3. Add 50 μL of the enzyme conjugate solution to the individual wells successively using a multi-channel, stepping, or electronic repeating pipette.
4. Add 50 μL of the antibody solution to the individual wells successively using a multi-channel, stepping, or electronic repeating pipette.
5. Cover the wells with parafilm or tape and mix the contents by moving the strip holder in a circular motion on the benchtop for 30 s. Be careful not to spill the contents.
6. Incubate the strips for 45 min at room temperature.
7. Remove the covering, decant the contents of the wells into a sink, and blot the inverted plate on a stack of paper towels.
8. Wash the strips four times using the diluted wash buffer. Use at least a volume of 250 μL of prepared wash buffer for each well and each washing step. Blot the inverted plate after each wash step on a stack of paper towels. After the last wash/blot, check the wells for any remaining buffer in the wells, and if necessary, remove by additional blotting.
9. Add 100 μL of substrate (color) solution to the individual wells successively using a multi-channel, stepping, or electronic repeating pipette.
10. Cover the wells with parafilm or tape and mix the contents by moving the strip holder in a circular motion on the benchtop for 30 s. Be careful not to spill the contents.
11. Incubate the strips for 30–45 min at room temperature. Protect the strips from sunlight.
12. Add 100 μL of stop solution to the wells in the same sequence as for the substrate (color) solution using a multichannel, stepping, or electronic repeating pipette.
13. Read the absorbance at 450 nm using a microplate ELISA photometer within 15 min after the addition of the stopping solution.

43.3.2.1 Evaluation

The evaluation of the ELISA can be performed using commercial ELISA evaluation programs such as 4-Parameter (preferred) or Logit/Log. For a manual evaluation, the mean absorbance value for each of the standards has to be calculated.

The %B/B_0 has to be calculated for each standard by dividing the mean absorbance value for each standard by the Zero Standard (Standard 0) mean absorbance. A standard curve can be constructed by plotting the %B/B_0 for each standard on the vertical linear (y) axis versus the corresponding cylindrospermopsin concentration on the horizontal logarithmic (x) axis on graph paper.

%B/B_0 for the control (QCS), LRB, and samples will then yield levels in ppb of cylindrospermopsin by interpolation using the standard curve. Results can also be determined by using a spreadsheet macro available from Eurofins Abraxis upon request.

The concentrations of the samples are determined using the standard curve run with each test. Samples showing a lower concentration of cylindrospermopsin than standard 1 (0.05 ppb) should be reported as containing <0.05 ppb of cylindrospermopsin. Samples showing a higher concentration than standard 6 (2.0 ppb) have to be diluted to obtain accurate results. The concentration of the positive control (QCS) provided should be 0.75 ± 0.15 ppb.

Semi-quantitative results can be derived by simple comparison of the sample absorbances to the absorbances of the standards. Samples with lower absorbances than a standard will have concentrations of cylindrospermopsin greater than that standard. Samples which have higher absorbances than a standard will have concentrations of cylindrospermopsin less than that standard.

Performance Data

Test sensitivity: The detection limit for this assay is 0.040 ppb (μg/L).

Test reproducibility: Coefficients of variation (CVs) for standards: <10%; for samples: <15%.

Specificity: This ELISA recognizes cylindrospermopsin and related compounds with varying degrees:

Cylindrospermopsin 100%

Deoxy-Cylindrospermopsin 112%

7-Epi-Cylindrospermopsin 157%

Numerous organic and inorganic compounds commonly found in water samples have been tested and found not to interfere with this test. However, due to the high variability of compounds that may be found in water samples, test interferences caused by matrix effects cannot be completely excluded.

Seawater samples must also be diluted to a concentration ≤20% to avoid matrix effects.

Mistakes in handling the test can cause errors. Possible sources for such errors include: inadequate storage conditions of the test kit, incorrect pipetting sequence or inaccurate volumes of the reagents, too long or too short incubation times during the immune and/or substrate reaction, and extreme temperatures during the test performance (lower than 10 °C or higher than 30 °C).

Direct and indirect Elisa methods require a pH environment usually between pH 6.0 and 9.0, or even more neutral (around 7.3), to allow antibody binding (Researchgate 2015) and the ending of the reaction in algal toxins Elisa is generally performed just by adding a stop solution made by diluted sulfuric acid.

References

Bain P, Shaw G, Patel B (2007) Induction of p53-regulated gene expression in human cell lines exposed to the cyanobacterial toxin Cylindrospermopsin. J Toxicol Environ Health: Part A 70(19):1687–1693

Banker R, Carmeli S, Hadas O, Teltsch B, Porat R, Sukenik A (1997) Identification of cylindrospermopsin in the cyanobacterium *Aphanizomenon ovalisporum* (Cyanophyceae) isolated from lake Kinneret, Israel. J Phycol 33:613–616

Buratti FM, Manganelli M, Vichi S, Stefanelli M, Scardala S, Testai E, Funari E (2017) Cyanotoxins: producing organisms, occurrence, toxicity, mechanism of action and human health toxicological risk evaluation. Arch Toxicol 91:1049–1130

Díez-Quijada L, Guzmán-Guillén R, Ortega AP, Llana-Ruíz-Cabello M, Campos A, Vasconcelos V, Jos Á, Cameán AM (2018) New method for simultaneous determination of microcystins and Cylindrospermopsin in vegetable matrices by SPE-UPLC-MS/MS. Toxins 10:406

EPA (2022) EPA drinking water health advisories for cyanotoxins. https://www.epa.gov/cyanohabs/epa-drinking-water-health-advisories-cyanotoxins

Froscio SM, Humpage AR, Burcham PC, Falconer IR (2003) Cylindrospermopsin-induced protein synthesis inhibition and its dissociation from acute toxicity in mouse hepatocytes. Environ Toxicol 18:243–251

Froscio SM, Humpage AR, Wickramasinghe W, Shaw G, Falconer I (2008) Interaction of the cyanobacterial toxin cylindrospermopsin with the eukaryotic protein synthesis system. Toxicon 51:191–198

Griffiths DJ, Saker ML (2003) The Palm Island mystery disease 20 years on: a review of research on the cyanotoxin cylindrospermopsin. Environ Toxicol 18(2):78–93

Humpage AR, Fenech M, Thomas P, Falconer IR (2000) Micronucleus induction and chromosome loss in transformed human white cells indicate clastogenic and aneugenic action of the cyanobacterial toxin, cylindrospermopsin. Mutat Res 472(1–2):155–161

Humpage AR, Fontaine F, Froscio S, Burcham P, Falconer IR (2005) Cylindrospermopsin genotoxicity and cytotoxicity: role of cytochrome P-450 and oxidative stress. J Toxic Environ Health A 68:739–753

Kinnear SHW, Fabbro LD, Duivenvoorden LJ, Hibberd E (2007) Multiple-organ toxicity resulting from cylindrospermopsin exposure in tadpoles of the cane toad (*Bufo marinus*). Environ Toxicol 22:550–558

Kiss T, Vehovszky A, Hiripi L, Kovacs A, Voros L (2002) Membrane effects of toxins isolated from a cyanobacterium, *Cylindrospermopsis raciborskii*, on identified molluscan neurones. Comp Biochem Physiol C Toxicol Pharmacol 131:167–176

Manti G, Mattei D, Messineo V, Melchiorre S, Bogialli S, Sechi N, Casiddu P, Luglìè A, Di Brizio M, Bruno M (2005) First report of *Cylindrospermopsis raciborskii* in Italy. Harmful Algae News 28:8–9

Metcalf J, Barakate A, Codd GA (2004) Inhibition of plant protein synthesis by the cyanobacterial hepatotoxin, cylindrospermopsin. FEMS Microbiol Lett 235:125–129

Moreira C, Azevedo J, Antunes A, Vasconcelos V (2013) Cylindrospermopsin: occurrence, methods of detection and toxicology. J Appl Microbiol 114:605–620. https://doi.org/10.1111/jam.12048

Pichardo S, Cameán AM, Jos A (2017) In vitro toxicological assessment of cylindrospermopsin: a review. Toxins 9:402

Puerto M, Prieto AI, Maisanaba S, Gutiérrez-Praena D, Mellado-García P, Jos Á, Cameán AM (2018) Mutagenic and genotoxic potential of pure Cylindrospermopsin by a battery of in vitro tests. Food Chem Toxicol 121:413–422

Rasmussen JP, Barbez PH, Burgoyne LA, Saint CP (2008) Rapid preparation of cyanobacterial DNA for real-time PCR analysis. Lett Appl Microbiol 46:14. https://doi.org/10.1111/j.1472-765X.2007.02252.x

Researchgate (2015). https://www.researchgate.net/post/Any_tips_for_ELISA_at_multiple_pHs#:~:text=pH%20affects%20the%20reactions%20in,before%20carrying%20out%20the%20ELISA.&text=The%20reaction%20between%20antibody%20and,fragment%20will%20bind%20your%20protein

Rogers EH, Zehr RD, Gage MI, Humpage AR, Falconer IR, Marr M, Chernoff N (2007) The cyanobacterial toxin, cylindrospermopsin, induces fetal toxicity in the mouse after exposure late in gestation. Toxicon 49(6):855–864

Saker M, Eaglesham GK (1999) The accumulation of cylindrospermopsin from the cyanobacterium *Cylindrospermopsis raciborskii* in tissues of the Redclaw crayfish *Cherax quadricarinatus*. Toxicon 37:1065–1077

Saker M, Metcalf JS, Codd GA, Vasconcelos VM (2003) Accumulation and depuration of the cyanobacterial toxin cylindrospermopsin in the freshwater mussel *Anodonta cygnea*. Toxicon 43:185–194

Shaw RG, Seawright AA, Moore MR, Lam PK (2000) Cylindrospermopsin, a cyanobacterial alkaloid: evaluation of its toxicological activity. Ther Drug Monit 22(1):89–92

Shen X, Lam PK, Shaw GR, Wickramasinghe W (2002) Genotoxicity investigation of a cyanobacterial toxin, cylindrospermopsin. Toxicon 40(10):499–501

Stewart I, Seawright AA, Shaw GR (2008) Cyanobacterial poisoning in livestock, wild mammals and birds—an overview. Cyanobacterial harmful algal blooms: state of the science and research needs. Springer, New York, pp 613–637

Štraser A, Filipiˇc M, Žegura B (2011) Genotoxic effects of the cyanobacterial hepatotoxin cylindrospermopsin in the HepG2 cell line. Arch Toxicol 85:1617–1626

Svirčev Z, Lalić D, Bojadžija Savić G, Tokodi N, Drobac Backović D, Chen L, Meriluoto J, Codd GA (2019) Global geographical and historical overview of cyanotoxin distribution and cyanobacterial poisonings. Arch Toxicol 93:2429–2248

Terao K, Ohmori S, Igarishi K, Ohtani I, Watanabe MF, Harada KI, Ito E, Watanabe M (1994) Electron microscope studies on experimental poisoning in mice induced by cylindrospermopsin isolated from blue-green alga *Umezekia natans*. Toxicon 32(7):833–843

Vasas G, Gaspar A, Suranyi G, Batta G, Gyemant G, Márta M, Mathé C, Grigorszky I, Molnar E, Borbely G (2002) Capillary electrophoretic assay and purification of Cylindrospermopsin, a cyanobacterial toxin from *Aphanizomenon ovalisporum*, by plant test (blue-green Sinapis test). Anal Biochem 302:95–103

Žegura B, Štraser A, Filipiˇc M (2011) Genotoxicity and potential carcinogenicity of cyanobacterial toxins—a review. Mutat Res Rev Mutat Res 727:16–41

44 Extraction and Quantification of Anatoxins from Aquatic Samples by HPLC with Fluorescence Detection

Gayathri Ganesan, Gomathi Selvam, Ambikapathy Varatharaju, and Panneerselvam Annamalai

Abstract

Prokaryotic organisms such as cyanobacteria are a large, diversified species that can both photosynthesis and breathe through heterotrophic routes. A significant problem with water quality that has an international impact on surface water management is cyanobacterial blooms. There is potential for a wide range of hazardous secondary metabolites to be produced by numerous aquatic cyanobacterial blooms. One example of a neurotoxic made by cyanobacteria is anatoxin. It has already been noted that some species of the genera *Anabaena* and *Oscillatoria* are capable of producing anatoxin. Anatoxin is a powerful cyanotoxin-like alkaloid that blocks neuromuscular transmission by depolarizing the post-synaptic membrane potential. In this protocol, we practiced about extraction and quantification of anatoxin from aquatic sample.

Keywords

Cyanobacteria · Aquatic sample · HPLC · Anatoxin

44.1 Introduction

Prokaryotic organisms known as cyanobacteria are a large, varied species that can both produce oxygen and breathe through heterotrophic routes. It has been demonstrated that many aquatic cyanobacteria in freshwater and marine habitats create compounds that may be hazardous to people and other species (Huisman et al. 2005). Harmful algal blooms (HABs) are a significant problem for almost every

G. Ganesan (✉) · G. Selvam · A. Varatharaju · P. Annamalai
PG and Research Department of Botany, A.V.V.M Sri Pushpam College (Autonomous), Poondi, (Affiliated to Bharathidasan University, Tiruchirappalli), Thanjavur, Tamilnadu, India

N. Thajuddin et al. (eds.), *Protocols for Cyanobacteria Sampling and Detection of Cyanotoxin*, https://doi.org/10.1007/978-981-99-4514-6_44

continent and nation, occurring in a variety of freshwater, brackish, and marine water systems worldwide (Paul Bertani 2021). A significant problem with water quality that has an international impact on surface water management is cyanobacterial blooms. These blooms have the potential to produce a variety of toxic secondary metabolites or cyanotoxins, including hepatotoxins, neurotoxins, and cytotoxins of significant health, veterinary, and agricultural importance, such as Cylindrospermopsin (CYN), in addition to affecting the palatability of water (Chorus and Bartram 1999; Codd 1999; John et al. 2019).

The first cyanobacterial toxin whose structure was clarified was an alkaloid called anatoxin-a (AN), which is 2-acetyl-9-azabicyclo 4.2.1 non-2-ene (Devlin et al. 1977). The genera *Anabaena flos-aquae, Oscillatoria, Anabaena circinalis, Aphanizomenon flos-aquae,* and *Cylindrospermum* sp. are responsible for producing AN. Anatoxin-a was first detected in *Anabaena flos-aquae* (Gorham 1964) and structurally elucidated by X-rays (Devlin et al. 1977). In addition to acting as a post-synaptic, depolarizing, and neuromuscular blocking agent, anatoxin is a powerful nicotinic agonist (Carmichael 1989).

A very small number of studies have focused on the techniques required to accurately identify cyanotoxins in environmental samples, and none have been published that can identify all of these cyanotoxins at once in a single sample. A variety of environmental samples, such as water, fish, and aquatic plant samples, include cyanotoxins (Al-Sammak et al. 2013). The nicotinic acetylcholine receptor is where anatoxin method of action involves binding, where it functions as an acetylcholine analogue. Due to the low demand for acrylonitrile, which is mostly utilized in the plastic and automotive industries, a global acetonitrile scarcity began at the end of 2008 (Azevedo et al. 2011).

44.2 Materials

- Glasswares
- Chemicals and standards
- Distilled water
- Water samples
- Turbo vap LV evaporator
- Centrifuge
- Vortex mixer
- Sonic bath

44.3 Methods (James et al. 1998; Rellan et al. 2007)

44.3.1 Instrumentation

1. The HPLC system consists of pump, column oven, and fluorescence detector with autosampler.

2. The analytical HPLC column used prodigy C18, 5 μm, 250 × 3.2 mm, Ultremex C18, 5 μm, 250 × 3.2 mm and Vydac.
3. Weak cation-exchange cartridges, C18 Mega bond elut and Whatman filter paper.
4. Chromatographies data handling performed using chromatography data station and data transfer Microsoft excel graphical presentation.

44.3.2 Sample Collection

1. Collect the aquatic samples from different sites.
2. Transfer the collected samples to laboratory.

44.3.3 Extraction of Anatoxin

1. Prior to solid-phase extraction (SPE), filter water sample (10 mL) adjust the pH to 7 using a mild cation-exchange medium, WCX, 3 mL cartridge.
2. Add water and 6 mL of methanol in SPE cartridge.
3. The sample in cartridge should be rinsed with methanol-water (1:1), air dry, and then transferred.
4. Extract the anatoxins extracted using a TurboVap LV evaporator at 500 °C under nitrogen.
5. Elute 10 mL of methanol containing 0.2% trifluoroacetic acid.
6. Samples reconstitute in methanol, transfer to 2-mL amber vials and evaporate.

44.3.4 Derivatization of Anatoxins

1. Sample of anatoxin (up to 100 mg) reconstitute 2-mL amber vial with sodium borate (0.1 M).
2. To add 50 mL of NBD-F (1 mg mL^{-1} in acetonitrile) left to stand in dark room temperature.
3. To add hydrochloric acid (1 M, 50 μL) stop the reaction.
4. HPLC-FL performed directly on the products (20 μL injection) using Prodigy C_{18} columns at 35 °C.
5. Detect the fluorometric mobile phase flow rate of 0.5 mL/min (λ = 5470 nm, λ = 530 nm).

44.4 Observation

Detection and quantification limits were based on the signal-to-noise ratio of anatoxin present in taken sample (aquatic sample) through the HPLC with fluorescence detection.

Acknowledgment The authors are thankful to PG and Research Department of Botany, A.V.V.M Sri Pushpam College (Autonomous), Poondi, Thanjavur, and also to director and staff members of Indian Biotrack Research Institute, Thanjavur for providing the lab facilities.

References

Al-Sammak MA, Hoagland KD, Snowc DD, Cassada D (2013) Methods for simultaneous detection of the cyanotoxins BMAA, DABA, and anatoxin-a in environmental samples. Toxicon 76:316–325. https://doi.org/10.1016/j.toxicon.2013.10.015

Azevedo J, Osswald J, Guilhermino L, Vasconcelos V (2011) Development and validation of an SPE-HPLC-FL method for the determination of Anatoxin-a in water and trout (*Oncorhynchus mykiss*). Anal Lett 44(8):1431–1441. https://doi.org/10.1080/00032719.2010.512682

Bertani P, Lu W (2021) Cyanobacterial toxin biosensor for environmental monitoring and protection. In: Medicine in Novel Technology and Devices, vol 10. Elsevier, Amsterdam, p 100059. https://doi.org/10.1016/j.medntd.2021.100059

Carmichael WW (1989) Natural toxins: characterization. In: Ownby CL, Odell GV (eds) Pharmacology and Therapeutics, Pergamon Press, Oxford, p 3

Chorus I, Bartram J (1999) Toxic cyanobacteria in water: a guide to their public health consequences, monitoring and management. E and FN Spon, London and New York

Codd GA (1999) Cyanobacterial toxins, exposure routes and human health. Eur J Phycol 34:405–415

Devlin JP, Edwards OE, Gorham PR, Hunter NR, Pike RK, Starvick B (1977) Anatoxin-a a toxic alkaloid from Anabaena flos-aquae NCR-44h. Can J Chem 55:1367–1371

Gorham PR (1964) Toxic algae. In: Jackson DF (ed) Algae, man, and the environment, vol 307–336. Plenum Press, New York, p 434

Huisman J, Matthijs H, Visser P (2005) Aquatic ecology series: *harmful cyanobacteria 3*. Springer, Cham, p 241

James KJ, Furey A, Sherlock IR, Stack MA, Twohig M, Barry Caudwell F, Skulberg OM (1998) Sensitive determination of anatoxin-a, homoanatoxin-a and their degradation products by liquid chromatography with fluorometric detection. J Chromatogr A 798:147–157

John N, Baker L, Ansell BRE, Newham S, Crosbie ND, Jex AR (2019) First report of anatoxin-a producing cyanobacteria in Australia illustrates need to regularly up-date monitoring strategies in a shifting global distribution. Sci Rep 9:10894. https://doi.org/10.1038/s41598-019-46945-8

Rellan S, Osswald J, Vasconcelos V, Gago-Martinez A (2007) Analysis of anatoxin-a in biological samples using liquid chromatography with fluorescence detection after solid phase extraction and solid microextraction. J Chromatogr A 1156:134–140

45 Extraction and Quantification of Saxitoxins from Aquatic Samples by LC-MS/MS

Sharon Beaula, Manigundan Kaari, Ramesh Kumar, Kishore Kumar Annamalai, Radhakrishnan Manikkam, Gopikrishnan Venugopal, and Sakthivel Krishnan

Abstract

The most well-known paralytic shellfish toxin and strong neurotoxin are called saxitoxin. Paralytic shellfish poisoning is an ailment caused by people ingesting saxitoxin, typically by consumption of shellfish tainted by toxic algal blooms. Saxitoxin has a significant negative influence on the ecology and the economy since it regularly causes the harvesting of commercial and recreational shellfish in high temperate coastal water across the world. Bivalve shellfish includes mussels, clams, oysters, scallops, etc. Liquid chromatography with tandem mass spectrometry-mass spectrometry (LC-MS/MS) is a powerful analytical technique utilized for the extraction and identification of small organic compounds. Here, this technique discussed to utilize for confirmatory identification and quantification of saxitoxin from aquatic samples with a high degree of specificity and sensitivity.

Keywords

Saxitoxin · LC-MS/MS · Paralytic shellfish poisoning

S. Beaula · R. Kumar
Department of Biotechnology, Sathyabama Institute of Science and Technology, Chennai, Tamil Nadu, India

M. Kaari (✉) · K. K. Annamalai · R. Manikkam · G. Venugopal
Natural Product Discovery Laboratory, Centre for Drug Discovery and Development, Col. Dr. Jeppiaar Research Park, Sathyabama Institute of Science and Technology, Chennai, Tamil Nadu, India
e-mail: manigundan.cddd@sathyabama.ac.in

S. Krishnan
Crop Protection Section, ICAR-Indian Institute of Oilseed Research, Hyderabad, Telangana, India

N. Thajuddin et al. (eds.), *Protocols for Cyanobacteria Sampling and Detection of Cyanotoxin*, https://doi.org/10.1007/978-981-99-4514-6_45

45.1 Introduction

Saxitoxins (SXT) and their derivatives are neurotoxic alkaloids with an affinity for Na^+ and Ca^{2+} channels that have an impact on ion transport in bacteria as well as the creation and transmission of nerve impulses in metazoan muscle and nerve cells. Cyanobacteria and Dinoflagellates, two kinds of evolutionarily remote creatures from separate domains, produce these toxins, which are members of the paralytic shellfish toxin group (Patria et al. 2020). Saxitoxins, which concentrate poisons as a result of ingesting poisonous organisms, cause intoxication in marine habitats when fish and, particularly, filter-feeding mollusks are consumed. As they are among the most neurotoxic compounds known to cause significant sickness with the possibility of deadly intoxication consequences when swallowed in larger concentrations, it is crucial to keep an eye out for the accumulation of these toxins in edible bivalves (Yue et al. 2020).

LC-MS/MS is a strong analytical technique used for aquatic toxin separation and identification. In LC-MS/MS, mixtures are separated based on their physical and chemical characteristics, and the components within each peak are then identified and detected using the mass spectra of those components. A separation technique called liquid chromatography (LC) is used to separate the various parts of a mixture (Harju et al. 2015). An instrument that gauges the mass-charge ratio of charged particles is a mass spectrometer (MS). An atmospheric pressure ionization source, typically an ESI source or an atmospheric pressure chemical ionization (APCI) source, is one of the components of an LC-MS/MS instrument. This source is connected to a first mass-filtering device by an ion-inlet and focusing component, which provides both transitions from atmospheric pressure to vacuum and ion-focusing (Yun et al. 2018) Using this instrument, individual compounds can be identified and quantified. Also, multiple compounds can be identified and quantified in one run. The LC-MS/MS technique was validated in terms of specificity, linearity, sensitivity, accuracy, precision, matrix effects, and recovery (Grebe and Singh 2011; Simone and Maria 2017). It is used to calculate the masses of particles, identify the elements present in a sample or molecule, and clarify the chemical structures of molecules like peptides and other chemical compounds.

45.2 Materials

- Marine sample
- Tap water
- Cutter
- Grinder
- PBS solution
- 20 mL centrifuge tubes
- Glass container
- 30 and 60% acetone water solution
- Tris-HCl buffer solution

- STX monoclonal antibody solution
- Sepharose
- NAHCO and $NaCl_3$
- Formic acid
- Ammonium formate
- 95% acetonitrile
- TSK gel amide 80
- Standard solution

45.3 Methods

45.3.1 Sample Preparation (Patria et al. 2020)

1. 1 kg of a fresh marine sample (oyster, mussels, etc.) is taken and cleaned with tap water.
2. Open the sample by cutting the adductor muscle and remove the tissue from the cells.
3. Then the sample can be drained for 5 min and homogenized and utilized for further extraction procedure.

45.3.2 Extraction (Yue et al. 2020)

1. Take 1.0 g of the prepared sample and put it in a 20 mL plastic centrifuge tube.
2. Add 5.0 mL of PBS solution to the above tube.
3. Boil the homogenate for 5 min and cool it to room temperature.
4. Ultrasonic extraction is done for 10 min.
5. Then, centrifuge it at 10,000 rpm for 10 min.
6. Collect the supernatant and repeatedly extract the precipitate.
7. Combine the supernatant obtained two times and centrifuge it to obtain the extracting solution.
8. The extracting solution is then taken and made to pass through a super fiber filter paper.
9. Take the filtrate and pass it through an STX immunoaffinity column.
10. Control the flow rate to be 1–2 drops.
11. Then the liquid is drained.
12. Clean the filtrate 1 time by using 10 mL of distilled water.
13. Then, control the flow rate to 2–3 drops/s and completely drain the liquid using air.
14. Add 2 mL of 1% acetic acid-methanol solution for elution.
15. Then, collect the eluent and dissolve the eluent by using an initial mobile phase solution.

16. Fix the volume to 1 mL and filter by using a 0.22 μm membrane to obtain a sample detection solution.
17. The preparation method of the STX immune affinity column comprises the following steps:
    - Take unactivated Sepharose 4B and suspend the sepharose in the HCl solution.
    - Swell and wash the Sepharose 4B with HCl solution to obtain wet Sepharose 4B gel.
    - Then wash the Sepharose 4B wet glue with 30% acetone water solution and 60% acetone water solution.
    - Then add PBS solution and stir uniformly to get the treated gel.
    - Take the STX monoclonal antibody solution and add NAHCO containing $NaCI_3$.
    - Oscillate this gel and the treated gel in a glass container with a plug at 4 °C overnight.
    - Wash and perform suction filtration to obtain coupled STX-Sepharose 4B.
    - Add tris-HCl buffer solution on the Sepharose 4B with a volume of two to three times.
    - Leave it overnight at 4 °C.
    - Then wash the gel to obtain the STX immunoaffinity column.

45.3.3 Quantification (Yue et al. 2020)

1. Perform liquid chromatography-tandem mass spectrometry on the sample detection solution.
2. Conditions for liquid chromatography compromise
    - Mobile phase—compromises of component A and component B
    - Component A—50 mmol/L formic acid aqueous solution containing 2 mmol/L ammonium formate
    - Component B—95% acetonitrile aqueous solution containing 2 mmol/L ammonium formate and 50 mmol/L formic acid
    - The flow rate of the mobile phase—is 0.4 mL/min
    - Column temperature—40 °C
    - Sample injection amount—20 μL
    - Chromatographic column—TSK—GEL Amide—80 (2.0 × 250 mm, 5 μm)
3. Conditions for tandem mass spectroscopy compromise
    - Dry gas temperature—350 °C
    - Dry gas flow—10 L/min
    - Atomic pressure—38 psi
    - Capillary voltage—4000 V
    - Positive electricity sprays the ion source
4. Prepare a standard curve

   - Inject the STX standard solutions with different concentrations prepared by using the initial mobile phase solution into a liquid chromatogram-tandem mass spectrometer.
   - Draw a standard curve by taking the mass concentration of STX as a horizontal coordinate and the corresponding peak area as a vertical coordinate.
5. Inject the sample solution into a liquid chromatography-tandem mass spectrometer for measurements to obtain the peak area of STX in the sample solution.
6. Obtain the mass concentration of the STX in the sample solution from the standard curve.
7. The deviation between the retention time of the STX in the sample solution and the STX time in the standard solution should be within ±2.5%, and the relative abundance of the detected STX qualitative ion should be consistent with the relative abundance of the STX qualitative ion in the standard solution having a close mass concentration.
8. Then the presence of the test compound in the sample can be determined.

45.4 Observation

Sample name	Sample source	STX(μg/kg)	Detected /not detected

A spike in the concentration of the sample determines the presence of saxitoxin.

References

Grebe SK, Singh RJ (2011) LC-MS/MS in the clinical laboratory—where to from here? Clin Biochem Rev 32(1):5–31

Harju K, Rapinoja ML, Avondet MA, Arnold W, Schar M, Burrell S, Luginbuhl W, Vanninen P (2015) Optimization of sample preparation for the identification and quantification of saxitoxin in proficiency test mussel sample using liquid chromatography-tandem mass spectrometry. Toxins 25:4868–4880

Patria FP, Pekar H, Zuberovic-Muratovic A (2020) Multi-toxin quantitative analysis of paralytic shellfish toxins and Tetrodotoxins in bivalve mollusks with ultra-performance hydrophilic interaction LC-MS/MS-an in-house validation study. Toxins 12:452

Simone P, Maria CC (2017) Influence of environmental variables on saxitoxin yields by *Cylindrospermopsis raciborskii* in a mesotrophic subtropical reservoir. J Water Health 15(4): 509–518

Yue Y, Zhu B, Lun L, Xu N (2020) Quantifications of saxitoxin concentrations in bivalves by high performance liquid chromatography-tandem mass spectrometry with the purification of immunoaffinity column. J Chromatogr B Anal Technol Biomed Life Sci 1147:122133

Yun SS, Bentson-Royal A, Collins K, Book S, Figlar-Barnes R (2018) Development and application of LC/MS based analysis for marine algal toxins in Hood Canal Salish Sea Ecosystem Conference. 110

Extraction and Quantification of Guanitoxin from Aquatic Samples by HPLC-MS

46

Angel Alisha, Manigundan Kaari, Masilamani Selvam, Kishore Kumar Annamalai, Radhakrishnan Manikkam, Gopikrishnan Venugopal, and Sakthivel Krishnan

Abstract

Guanitoxin is a neurotoxin formed by certain cyanobacterial strains such as *Dolichospermum* and *Sphaerospermopsis*. Due to its permanent suppression of acetylcholinesterase, Guanitoxin is the only naturally occurring organophosphate that may kill the animals. Even though GNT is very dangerous, it is still very hard to find it in biological samples. Guanitoxin's primary mode of action is the irreversible inhibition of acetylcholinesterase's active site, which results in an excess of acetylcholine in the parasympathetic and peripheral nervous systems and causes poisoning via activation of nicotinic and muscarinic cholinergic receptors. For the identification and confirmation of this toxin's stability, high performance liquid chromatography combined with mass spectrometry (HPLC-MS) is used. Guanitoxin may progressively degrade at ambient temperature (>23 °C) over a period of 96 h; however, it is often stable in acidic media (pH = 3.0). As a consequence, the findings presented here provide useful information for investigations intending to extract and quantify guanitoxin from aquatic materials.

A. Alisha · M. Selvam
Department of Biotechnology, Sathyabama Institute of Science and Technology, Chennai, Tamil Nadu, India

M. Kaari (✉) · K. K. Annamalai · R. Manikkam · G. Venugopal
Natural Product Discovery Laboratory, Centre for Drug Discovery and Development, Col. Dr. Jeppiaar Research Park, Sathyabama Institute of Science and Technology, Chennai, Tamil Nadu, India
e-mail: manigundan.cddd@sathyabama.ac.in

S. Krishnan
Crop Protection Section, ICAR-Indian Institute of Oilseed Research, Hyderabad, Telangana, India

N. Thajuddin et al. (eds.), *Protocols for Cyanobacteria Sampling and Detection of Cyanotoxin*, https://doi.org/10.1007/978-981-99-4514-6_46

Keywords

Guanitoxin · Cyanobacteria · Anatoxin · Acetylcholinesterase · High performance liquid chromatography coupled with mass spectrometry

46.1 Introduction

The most powerful naturally occurring neurotoxic produced by freshwater cyanobacteria is anatoxin-a(S). In 1989, the structure of guanitoxin was first determined. It is made up of a cyclic N-hydroxyguanine organophosphate with a phosphate ester moiety (Walls et al. 2018). Guanitoxin's primary mode of action is the irreversible inhibition of acetylcholinesterase's active site, which results in an excess of acetylcholine in the parasympathetic and peripheral nervous systems and causes poisoning through activation of nicotinic and muscarinic cholinergic receptors. Excessive salivation, lacrimation, urine incontinence, chromodacryorrhoea (in rats), muscle twitching, muscular weakness, convulsion, respiratory difficulty, and/or failure are the major symptoms of high levels of guanitoxin exposure (Kaloudis et al. 2013).

The majority of Guanitoxin is labile. Although it decomposes quickly in basic solutions, it typically remains quite stable in neutral or acidic solutions (pH 3.0). It slowly undergoes hydrolysis when kept at −20 °C, producing (5S)-5-[(dimethylamino)methyl]-2-imino-1-imidazolidinol and monomethyl-phosphate, as well as the more slowly formed (S)-1-(2-iminoimidazolidin-4-yl)-N, N-dimethyl methanamine. In addition, considerable hydrolysis of guanitoxin to (5S)-5-[(dimethylamino)methyl]-2-imino-1-imidazolidinol was produced by air evaporation of the toxin.

Toxin monitoring and confirmation have recently been proposed using high performance liquid chromatography coupled to mass spectrometry (HPLC-MS) techniques that investigate the most prevalent guanitoxin loss fragments in positive mode with electrospray ionization, using quadrupole ion trap, triple quadrupole, and quadrupole time-of-flight mass analysers (Zervou et al. 2017; Wang et al. 2012). By using LC-MS/MS in aqueous samples with various pH values (acid-neutral base) connected to the variables of temperature and exposure period, the stability of guanitoxin was examined.

46.2 Materials

1. Aquatic samples
2. Tap water
3. Formic acid (FA)
4. Ammonium hydroxide solution (NH_4OH)
5. HPLC grade methanol (MeOH)
6. Acetonitrile (ACN)

7. Ultrapure water (H_2O)
8. MC-RR, MC-LR, MC-YR, and MC-LW solutions

46.3 Methods

46.3.1 Extraction of Guanitoxin (Palagama et al. 2017)

1. Mix ethanol and water (80:20 v/v) with 1.0 M acetic acid.
2. Filtration on gel resin with 0.1 M acetic acid.
3. Samples were subjected to solid-phase extraction followed by another SPE with weak cation exchange.
4. The SPE process elution's were performed with water and different concentrations of acetic acid.

46.3.2 HPLC-MS/MS Analysis (Merlo et al. 2020)

1. HPLC apparatus Agilent 1260 Infinity is coupled with an Agilent 6460 MS spectrometer HESI-MS/MS for quantitative analysis.
2. Use Zorbax Eclipse Plus C18 column preceding by a pre-column C18 and thermostat both at 30 °C, for chromatographic separation.
3. Perform elution by (A) H_2O and (B) ACN and both should contain 0.5% FA, which should be added to improve the ionization of the compounds.
4. Elution program is started at 95% (A) (maintain for 3 min), decreased to 80% A in 1 min (maintain for 2 min), to 65% A in 1 min (maintain for 7 min), to 30% A in 14 min and to 10% in 1 min, and then to 2% A in 1 min (maintain for 10 min).
5. Re-establish the initial conditions by a 10-min equilibration time.
6. The observed flow rate is 0.2 mL/min and the sample injection volume is 10 μL.
7. Tune up the MS spectrometer by direct injection of individual analytic solutions (0.5 mg/L in MeOH) and the ionization of the analytes is carried out in positive mode.
8. Operating parameters of the MS detector are: drying gas (N_2) and sheath gas temperature 350 °C and 355 °C, respectively; drying gas and sheath gas flow 12 L/min; nebulizer 50 psi; capillary voltage 5000 V positive, 0 V negative; nozzle voltage 1000 V positive, 0 V negative; electron multiplier voltage (EMV) 0 V both positive and negative; cell accelerated voltage 4 V positive, 1 V negative.
9. The Multiple Reaction Mode (MRM) is adopted for the quantification of the target toxins, using the highest and characteristic precursor/product ion transitions of each compound obtaining from MS spectra.

46.4 Observations

Mean Recoveries (%)					
Lake Water		River Water		Sea Water	

Mean recovery (%) in the water samples.

Retention Time(min)	*m/z* Precursor Ion	*m/z* and Hypothesized Formula for Production*	Dwell Time (ms)	Fragmentor (V)	Collision Energy (V)

Instrumental conditions

References

Kaloudis T, Zervou SK, Tsimeli K, Triantis TM, Fotiou T, Hiskia A (2013) Determination of microcystins and nodularin (cyanobacterial toxins) in water by LC–MS/MS. monitoring of Lake Marathonas, a water reservoir of Athens, Greece. J Hazard Mater 263:105–115

Merlo F, Maraschi F, Piparo D, Profumo A, Speltini A (2020) Simultaneous pre-concentration and HPLC-MS/MS quantification of phycotoxins and cyanotoxins in inland and coastal waters. Int J Environ Res Public Health 17(13):4782

Palagama DS, West RE III, Isailovic D (2017) Improved solid-phase extraction protocol and sensitive quantification of six microcystins in water using an HPLC-orbitrap mass spectrometry system. Anal Methods 9(13):2021–2030

Walls JT, Wyatt KH, Doll JC, Rubenstein EM, Rober AR (2018) Hot and toxic: temperature regulates microcystin release from cyanobacteria. Sci Total Environ 610:786–795

Wang Z, Maucher-Fuquay J, Fire SE, Mikulski CM, Haynes B, Doucette GJ, Ramsdell JS (2012) Optimization of solid-phase extraction and liquid chromatography–tandem mass spectrometry for the determination of domoic acid in seawater, phytoplankton, and mammalian fluids and tissues. Anal Chim Acta 715:71–79

Zervou SK, Christophoridis C, Kaloudis T, Triantis TM, Hiskia A (2017) New SPE-LC-MS/MS method for simultaneous determination of multi-class cyanobacterial and algal toxins. J Hazard Mater 323:56–66

Detection of Anatoxins from Aquatic Samples by LC-MS

47

A. Swedha, G. Kanimozhi, and A. Panneerselvam

Abstract

Aquatic ecosystems connect people, land, and wildlife through water, wetlands, rivers, lakes, and coastal estuaries are all aquatic ecosystems. Wetlands connect land and water, serving as natural filters, reducing pollution, controlling flood, and acting as nurseries for many aquatic species. Rivers, lakes, and estuaries serve as important recreation and wildlife hubs. Anatoxin-a is a neurotoxic alkaloid, produced by several freshwater planktonic and benthic cyanobacteria. Aquatic water contamination occurs due to anatoxins. Anatoxin-a occurs naturally and may be present when a harmful algal bloom has formed on a lake, pond, or stream. Not all algal blooms contain the cyanobacteria that produce anatoxin-a. However, there is no clear way to know if an algal bloom contains harmful toxins. *Chrysosporum, ovalisporum, Cuspidothrix, Cylindrospermopsis, Cylindrospermum, Dolichospermum, Microcystis, Oscillatoria, Planktothrix, Phormidium, Anabaena flos-aquae,* etc. produce anatoxins. This chapter summarizes the detection of anatoxins from aquatic sample by LC-MS analysis methodology. LC-MS involves separating mixtures in accordance with their physical and chemical properties, then identifying the components with each peak and detecting based on their mass spectrum. This method used to quantitation of algal toxins.

A. Swedha (✉) · G. Kanimozhi
PG and Research Department of Microbiology, A.V.V.M. Sri Pushpam College (Autonomous), Affiliated to Bharathidasan University, Tiruchirappalli-24, Thanjavur, Tamilnadu, India

A. Panneerselvam
PG and Research Department of Botany, A.V.V.M Sri Pushpam College (Autonomous), Poondi, (Affiliated to Bharathidasan University, Tiruchirappalli), Thanjavur, Tamilnadu, India

N. Thajuddin et al. (eds.), *Protocols for Cyanobacteria Sampling and Detection of Cyanotoxin*, https://doi.org/10.1007/978-981-99-4514-6_47

Keywords

Algal bloom · Anatoxin-a · *Anabaena* spp. · Liquid chromatography-mass spectrometry

47.1 Introduction

Cyanobacteria survive extreme environments such as desiccation, high UV radiation, low light, low nutrient levels, and high salt concentrations. The poisonings associated of cyanobacteria are the primary reason for harmful algal blooms. Anatoxin-a(s) is the only naturally occurring organophosphate and has been isolated from *Anabaena flos-aquae* and *Anabaena lemmermannii* (Kaushik and Balasubramanian 2013). Cyanotoxins are classified based on functional properties, hepatotoxins, neurotoxins, dermatotoxins, and cytotoxins (Mohamed 2022). The chemical structure of cyanotoxins is mainly divided into cyclic peptides, alkaloids, and lipopeptides. Basic organic compounds of alkaloids containing nitrogen. They vary in type and structure, so the properties of different alkaloids can greatly vary. Alkaloids have a complex cyclic structure. The following alkaloids are cyanotoxins cylindrospermopsins, saxitoxins, anatoxin-a, anatoxin-a (s), lyngbyatoxins, and aplysiatoxins (Du et al. 2019). The alkaloids produced blue green algae of cyantoxins known as cylindrospermopsin and anatoxin-a. Anatoxin-a is a neurotoxin that produces very rapid toxic effects, including loss of coordination, twitching, convulsions, and rapid death by respiratory paralysis (Jaramillo and O'Shea 2019). Alkaloids, in general, are a broad group on heterocyclic nitrogenous compounds usually of low to moderate molecular weight. The bloom-forming freshwater cyanobacterial genera including *Anabaena, Aphanizomenon, Oscillatoria,* and *Cylindrospermum* produce the neurotoxin, anatoxin a, an alkaloid with a high toxicity to animals. Anatoxin-a is a commonly encountered toxin that has been found throughout much of Europe, as well as Canada and Japan (Hikmet et al. 2004). Anatoxin-a is one of the predominant detected cyanotoxins in freshwater. Anatoxin-a is a natural, toxic alkaloid produced by several species of cyanobacteria from different genera, including Dolichospermum (Anabaena), Aphanizomenon, Oscillatoria, Planktothrix, and Cylindrosperum. Humans can be exposed to anatoxin-a not only by drinking contaminated water but also by consuming algae-containing dietary supplements, or by eating fish products and bivalve mollusks (Cevallos-Cedeño et al. 2022). Cyanobacteria-producing cyanotoxins contain a global distribution and most of them are able to bloom in aquatic environments, posing human health and environmental risks. The appearance and duration cyanobacterial bloom seasons mainly depends on the climatic and environmental conditions of the region, such as the air temperature, concentration of nutrients, pH, salinity, and sunlight. The sensitive LC-MS/MS method for ATX detection is developed in order to identify and separate different toxin analogs. The total amount of toxins is quantified and compared by this method (Sanchez et al. 2014). Cyanobacteria are able to produce a large number of bioactive secondary metabolites. Amongst these metabolites different toxin types have been identified, comprising neurotoxins, hepatotoxins, cytotoxins, and lipopolysaccharide endotoxins all grouped under the term cyanotoxins. LC-MS methods have been developed for the simultaneous detection of different cyanotoxins (Gaget et al. 2017). The following genera of cyanobacteria that produce anatoxin-a are *Anabaena* (Dolichospermum), *Aphanizomenon, Cylindrospermopsis, Cylindrospermum, Lyngbya, Microcystis, Nostoc, Oscillatoria, Phormidium, Planktothrixx, Raphidiopsis, Tychonema,* and *Woronichinia.*

47.2 Materials

- Algal bloom
- Lugol's solution
- Microscope
- Ammonium acetate
- Methanol
- Acetic acid
- Ammonium hydroxide
- Distilled water

47.3 Methods

47.3.1 LC-MS Analysis of Anatoxins (Harada et al. 1993)

1. Collect heavy algal bloom for toxicological analyses. Store the samples in ice chests and transport to the laboratory.
2. Preserve the samples in 1% Lugol's solution and observe under microscope for phytoplankton identification.
3. Anatoxin-a has been successfully extracted efficiently from cyanobacterial cells using acidified solvents, either just water or methanol or a mixture of both.
4. The following strain and bloom samples were using for analysis: *Anabaena flos-aquae*, LCMS was carried out under the following conditions: LC: pump, Shimadzu (Kyoto. Japan) LC-9A; detector, Shimazdu SPD-6A; Column, Cosmosil 5PH (150 × 4.6 mm); mobile phase. Methanol: O.1 M ammonium acetate (adjusted pH 5 with TFA) = 1486; flow rate, 1.2 mL/min; detection, UV (227 nm). MS: mass spectrometer, Shimadzu LC/MS-QP 1000; SIM, m/z 166.182 and 184; interface. Shimadzu TSP-100; control temperature, 155–160 ° C; tip temperature, 230–238 °C; block temperature, 250–270 °C; vapor temperature, 240–255 °C; tip heater temperature, 250–265 °C.
5. Lyophilized cells (100 mg) were extract three times with 10 mL of 0.05 M acetic acid for 30 min while stirring. The extract allowing for centrifugation at 4000 rpm for 15 min and the supernatant adjusted to pH 10 with 7% ammonium hydroxide. It applied to a preconditioned ODS silica gel cartridge (Baker, Phillipsburg, NJ,

USA), which was then washed with 10 mL of water, followed by 10 mL of water methanol (9, 1). The toxic fraction was eluted with 20 mL of methanol and evaporated to dryness. The residue was dissolved in 10 mL of water and was subjected to cation exchange on organosilane bonded to silica gel (Baker 10 COOH). The cartridge was rinsed with 10 mL of water, followed by 10 mL of water and the desired toxic fraction was eluted with 20 mL of 0.01% TFA: methanol. The fraction was evaporated and the resulting residue was dissolved in methanol. A sample solution with an appropriate concentration was injected into HPLC or LC/MS. The LC-MS analyses confirmed the presence of anatoxin.

6. Correct data processing is a fundamental step in generating good quality quantitative data.
   Most modern software packages contain automated algorithms for integrating peaks and these are preferred over manual integration.
   Most modern quantitation software allows direct export of results into word processing packages or spreadsheets. LIMS systems are also designed to integrate with most LC-MS instruments.

Acknowledgment The authors are grateful to the Secretary and Correspondent, Principal and Head, Department of Microbiology, A.V.V.M. Sri Pushpam College, Poondi, Thanjavur.

References

Cevallos-Cedeño RE, Quiñones-Reyes G, Agulló C, Abad-Somovilla A, Abad-Fuentes A, Mercader JV (2022) Rapid immunochemical methods for Anatoxin-a monitoring in environmental water samples. Anal Chem 94(30):10857–10864

Du X, Liu H, Yuan L, Wang Y, Ma Y, Wang R, Zhang H (2019) The diversity of cyanobacterial toxins on structural characterization, distribution and identification: a systematic review. Toxins 11(9):530

Gaget V, Lau M, Sendall B, Froscio S, Humpage AR (2017) Cyanotoxins: which detection technique for an optimum risk assessment? Water Res 118:227–238

Harada KI, Nagai H, Kimura Y, Suzuki M, Park HD, Watanabe MF, Carmichael WW (1993) Liquid chromatography/mass spectrometric detection of anatoxin-a, a neurotoxin from cyanobacteria. Tetrahedron 49(41):9251–9260

Hikmet K, Beril SA, Tahir A (2004) Microalgal toxin (s): characteristics and importance. Afr J Biotechnol 3(12):667–674

Jaramillo M, O'Shea KE (2019) Analytical methods for assessment of cyanotoxin contamination in drinking water sources. Curr Opin Environ Sci Health 7:45–51

Kaushik R, Balasubramanian R (2013) Methods and approaches used for detection of cyanotoxins in environmental samples: a review. Crit Rev Environ Sci Technol 43(13):1349–1383. https://doi.org/10.1080/10643389.2011.644224

Mohamed Mohamed (2022) Anatoxin-a. https://www.researchgate.net/publication/365780272_Anatoxin-a

Sanchez JA, Otero P, Alfonso A, Ramos V, Vasconcelos V, Aráoz R, Botana LM (2014) Detection of anatoxin-a and three analogs in *anabaena* spp. cultures: new fluorescence polarization assay and toxin profile by LC-MS/MS. Toxins 6(2):402–415

48 Detection of Guanitoxin from Aquatic Samples by HPLC/MS

Stany B, Shatakshi Mishra, and Lokesh Ravi

Abstract

A potent natural neurotoxin produced by freshwater cyanobacteria that can kill both domestic and wild animals is called Guanitoxin (formerly known as anatoxin). Due to the unintentional consumption of cyanobacterial cells, Guanitoxin has been linked to animal deaths in domestic and wild environments. The subject of aquatic pathology can benefit significantly from analytical methods that interpret the specifics of Guanitoxins employing high-performance liquid chromatography as a mode. One of the most popular techniques for determining the purity of medications is HPLC. The cost may go up, but the accuracy and specificity gained from the trade-off are higher. HPLC may offer advantages over conventional purification techniques. Since the pharmaceutical sector started using his HPLC, this technology has demonstrated its efficacy. This protocol summarizes the origin, toxicity, and analysis of Guanitoxins using high-performance liquid chromatography. Also, it includes a brief discussion about the advantages, disadvantages, and applications linked to HPLC.

Keywords

HPLC · Guanitoxin · Cyanobacteria · Algal toxin · Mobile phase · Stationary phase

S. B · S. Mishra
Department of Biotechnology, School of Bio Sciences and Technology, Vellore Institute of Technology, Vellore, India

L. Ravi (✉)
Department of Food Technology, Faculty of Life and Allied Health Sciences, MS Ramaiah University of Applied Sciences, Bengaluru, India

N. Thajuddin et al. (eds.), *Protocols for Cyanobacteria Sampling and Detection of Cyanotoxin*, https://doi.org/10.1007/978-981-99-4514-6_48

48.1 Introduction

Guanitoxin (formerly known as anatoxin) is a potent, naturally occurring neurotoxin produced by freshwater cyanobacteria that can kill wild and domestic animals (Fernandes et al. 2020). Guanitoxin has been associated with death in domestic and wild animals that inadvertently ingested cyanobacterial cells. These animals' most common clinical signs were excessive salivation, muscle tremors, convulsions, and respiratory failure (Matsunaga et al. 1989; Mahmood and Carmichael 1986; Cook et al. 1989). Guanitoxin has no known variants. However, species of Dolichospermum and Sphaerospermopsis are the leading producers of this cyanotoxin (Fiore et al. 2020; Werner et al. 2012). Acetylcholinesterase was used as a biomarker to detect the presence of toxins in aqueous samples. Therefore, the LC-MS analytical method is more recommended as it offers high specificity and sensitivity for identifying guanitoxins. Although this toxin is less common than other cyanotoxins such as microcystin, very high concentrations of Guanitoxin have been detected in water samples. Monitoring Guanitoxin in waterbodies for consumption has yet to be mandated, and the World Health Organization still needs to set detection limits for guanitoxin. The main factors limiting mandatory monitoring of Guanitoxin in the aquatic environment are the lack of consistent toxicological data and analytical standards to quantify Guanitoxins. Environmental factors such as toxin instability at elevated temperatures and slightly alkaline pH suggest a lack of data on Guanitoxin occurrence in aquatic environments. However, there are conflicting results about the timing of molecular degradation, and it is unclear whether guanitoxin is resistant to other chemicals.

48.2 Principle

This standard operating procedure uses HPLC-coupled mass spectrometry to measure and quantify Guanitoxin in whole water or filter samples (LC-MS). The method described here is a modification of the technique used to detect particulate guanitoxin in water samples from various sources, and it does not employ solid-phase extraction. Designed to use heptafluorobutyric acid as an extraction solvent protocol (Quilliam et al. 1989). Unsuitable for tissue or other samples with complex matrix turbulence sediment or soil. In HPLC, a small amount of liquid sample is injected into a tube filled with tiny particles (3–5 μm in diameter, called the stationary phase), where the individual components of the sample are mixed with a liquid (the mobile phase) and are moved through the column and down the packing tube by the high pressure supplied by the pump. The column packing method is used for component separation. It involves a series of chemical and physical interactions between molecules and packing particles. A detector that measures the number of separated components detects them at the column outlet. The "liquid chromatogram" is the output of this detector. LC-MS technology uses HPLC to separate the individual components of a mixture, followed by ionization and separation of ions based on their mass-to-charge ratio as shown in the Fig. 48.1. The ions are then

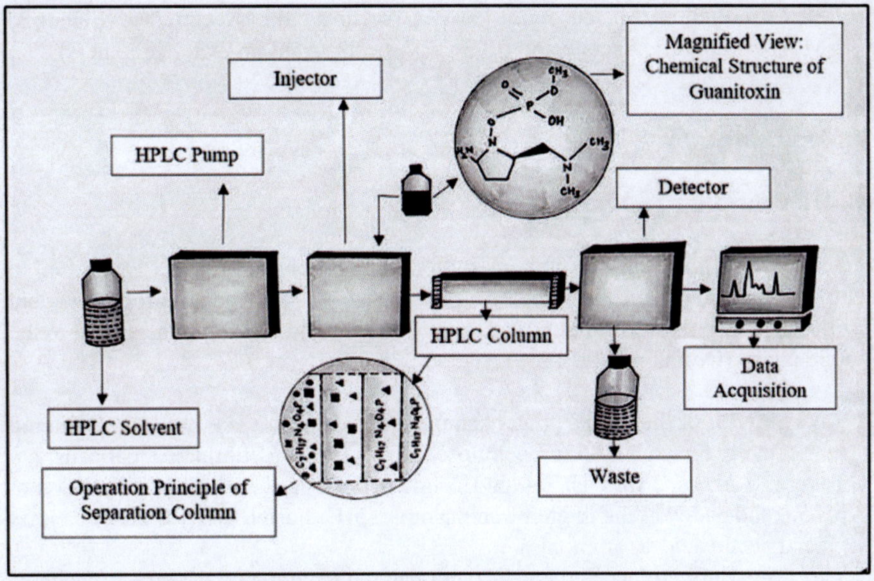

Fig. 48.1 High-performance liquid chromatography: instrumentation

separated, leading to an electron-multiplied tubular photodetector, which identifies and quantifies each ion. The ion source is an essential component of any MS analysis as it aids in efficiently generating ions for analysis. Ion sources for intact ionizing molecules include APCI (atmospheric pressure chemical ionization), ESI (electron spray ionization), and others (Lawton et al. 1994). The chemical nature of the analyte, polar or nonpolar, also affects the choice of the ion source. Due to the molecular level of analysis, this technology's key benefits include sensitivity, specificity, and accuracy. Additionally, the analyte's structural features can be understood.

48.3 Materials

- Heptafluorobutyric Acid (HFBA)
- 1 M Acetate Buffer
- Glacial Acetic Acid or Ammonia
- Acetone
- Methanol
- Deionized Water
- Acidified Methanol
- Trifluoroacetic Acid (TFA)
- 2 Mm HFBA In Water, Ph 3.5 (With Ammonia, Eluent A)

- 2 Mm Ammonium Acetate and 0.1% V/V Acetic Acid in Acetonitrile (Eluent B)
- 2 Mm Ammonium Acetate and 0.1% V/V Acetic Acid in Water (Eluent C)

48.4 Methods

48.4.1 Sample Preparation

Guanitoxin is a polar cation compound; therefore, an ion-pair reagent is required to neutralize the charge on the sample (Lawton et al. 1994). Heptafluorobutyric acid (HFBA) is chosen as a compound for sample enrichment and ion-pair reagent in the mobile phase (Quilliam et al. 1989).

1. Take 500 mL of the aquatic water sample is collected in glass bottles and filtered immediately through 0.45 μm cellulose nitrate filters to eliminate algal cells.
2. Take 130 μL of 2 mM HFBA and 2 mL of 1 M acetate buffer maintained at pH 5.0 and added to the filtered water sample (pH adjusted to 5 ± 0.05 with either glacial acetic acid or ammonia).
3. The 1.5 g Bulk Isolate Sorbent C_{18} end-capped (International Sorbent Technology, Mid-Glamorgan, UK) material is used for solid phase extraction (SPE) and is twice treated with acetone, methanol, and water.
4. The passage of water samples through the SPE material.
5. The tubes are dried in a nitrogen stream.
6. Water sample is eluted with 3 mL of acidified methanol (0.1% v/v TFA).
7. Eluates are re-suspended in a solution of mobile phase eluent A (80 μL) and acetonitrile after evaporating to dryness in a moderate stream of nitrogen (20 μL).

48.4.2 Preparation of Mobile Phase

1. Eluent A—2 mM HFBA in pure deionized water, pH 3.5 (adjusted with ammonia)
2. Eluent B—2 mM ammonium acetate and 0.1% v/v acetic acid in acetonitrile
3. Eluent C—2 mM ammonium acetate and 0.1% v/v acetic acid in deionized water

The linear gradation conditions of the eluent in HPLC are mentioned in Table 48.1.

48.4.3 Ensuring Desirable Settings

1. Make sure the waste line is in the trash can.
2. Set the mobile phase flow rate to 0.2 mL/min.
3. Set the minimum and maximum pressure to 250 psi and 4000 psi, respectively.
4. To set the detector to the pure mobile phase, press "zero" on the front panel.

Table 48.1 Linear gradient conditions in HPLC of Guanitoxin

Time (min)	Eluent A (%)	Eluent B (%)	Eluent C (%)	Flow (mL min^{-1})
0.00	96	4	0	0.2
3.00	80	20	0	0.2
12.00	0	68	32	0.2
15.00	0	68	32	0.2
16.00	0	95	5	0.2
20.00	0	95	5	0.2
30.00	96	4	0	0.2
45.00	96	4	0	0.2

48.4.4 Injection of the Sample and Collection of the Data

1. Inject 10 μL of the sample solution/effluent through the septal opening with the injector handle in the loaded position.
2. To allow three peaks to elute across the detector, set the data acquisition program to collect data for 300 s.
3. In the data acquisition (provides a signal to save the data file after 300 s) program on the computer, start the test by clicking "Start Test" immediately after rotating the injector handle to the injection position.
4. Tracks the number of seconds it takes to compile each attempt to identify each component.
5. After removing the syringe from the septum, complete the rest of the working standard procedure.
6. The effluent is entirely transported into the MS after being initially observed by UV at 235 nm.

48.4.5 MS Detection

1. Ionization in the electrospray ion source is set at 5200 V (positive mode), with nebulizer and curtain gas settings of 60 and 40 psi, respectively.
2. The heater's gas flow is set to 400 °C at 75 psi.
3. At 2 s $scan^{-1}$, full scan spectra are obtained for the mass range m/z 100–1200.
4. Single ion mode (SIM) and product ion scan are used (For compound identification).
5. The quantification is conducted using the multiple reaction monitoring (MRM) method.
6. The "quantitative optimization" program of the MS software produces the ion-dependent MS conditions for the individual chemicals.

Schematic representation of the HPLC protocol is shown in Fig. 48.2 as a flowchart.

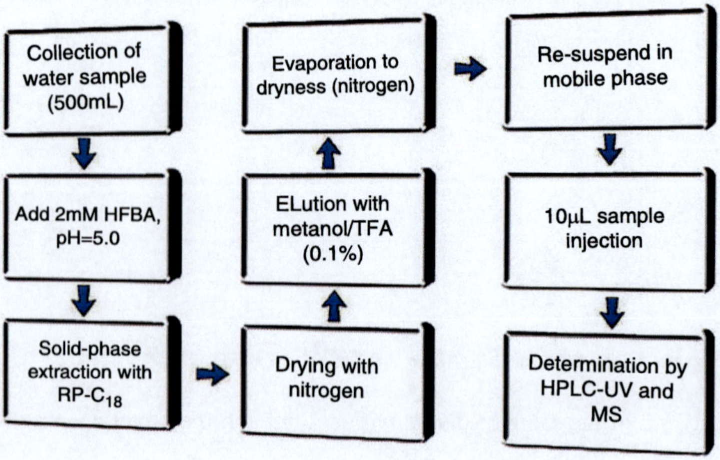

Fig. 48.2 Schematic representation of the HPLC experimentation

48.5 Observation

Determine the concentration of all components in the sample solution by optimizing the pH and gradient of the mobile phase. A graph shows the time (min) on the X-axis and intensity on the Y-axis. The signal intensity of the neurotoxin is tenfolds higher, resulting in a lower detection limit (TDL) of 17 fg for Anatoxin-a, which was conducted by (Pietsch et al. 2001), the quantification of the fragment ions m/z for anatoxin-a was found to be 149.2 and the collision energy (V) is found to be between 23 and 25. According to the data collected from the experiment conducted by (Pietsch et al. 2001), the standards were obtained and mentioned in Table 48.2.

An overview of the statistical calibration parameters derived in accordance with a German standard technique (DIN 32645) and the algal toxin recovery data is discussed in Table 48.3 (Lawton et al. 1994; Powell 1997).

48.6 Advantages of HPLC/MS

1. There is no doubt that HPLC is critical in environmental analysis.
2. It helps identify many compounds that are difficult to assess by GC.
3. Trace measurements in environmental samples using HPLC combined with atomic absorption spectrometry and other spectroscopy methods is significant for distinguishing between various organic and inorganic forms of toxic elements.
4. Headspace HPLC has the potential to be a fast and effective sample purification method for volatile organic and inorganic compounds.
5. HPLC with fluorescence detection is used to detect aflatoxins in food.

Table 48.2 Parameters for detection of algal toxins by ESI-MS (Pietsch et al. 2001)

	CAD (au)	CE (V)	DP (V)	$[M + H]^+$ (m/z)	Production (m/z)	TDL (fg)
Microcystin-LR	3	90	135	995.6	135.1	330
Microcystin-YR	3	90	150	1045.6	135.1	710
Microcystin-RR	3	45	25	520.1[a]	135.1	130
Microcystin-LA	3	75	75	910.6	135.1	420
Nodularin	3	80	115	825.4	135.1	260
Anatoxin-a	3	25	20	166.1	149.2	17
Saxitoxin	3	25	30	300.2	204.1	38

CAD collision activated dissociation, *CE* collision energy, *DP* declustering potential, $[M + H]^+$ protonated molecule, *TDL* total detection limit
S/N = 3;
[a]Doubly charged $[M + 2H]^{2+}$)

Table 48.3 Statistical parameters of calibration and recovery data

	Slope	RSD (%)	Correlation coefficient	LOD (ng L^{-1})	LOQ (ng L^{-1})	Recovery (%)
Microcystin-LR	119	1.7	0.9998	27	40	93 ± 3
Microcystin-YR	87.4	2.1	0.9997	33	49	94 ± 3
Microcystin-RR	183	1.7	0.9998	28	41	79 ± 6
Microcystin-LA	107	1.6	0.9998	29	44	96 ± 3
Nodularin	252	1.8	0.9998	29	44	96 ± 3
Anatoxin-a	2458	1.9	0.9998	30	44	50 ± 4
Saxitoxin	2.89	5.5	0.9971	425	634	3.2 ± 0.8

RSD relative standard deviation, *LOD* limit of detection, *LOQ* limit of quantification

6. HPLC method is used for quantification.
7. Very suitable for measuring polycyclic aromatic hydrocarbons.
8. A common approach for analyzing various chemicals, including organic and inorganic compounds, across multiple sample types is high-performance liquid chromatography (HPLC). It is used for forensic examination, environmental issues, and analysis of clinical trials.
9. High-resolution mass spectrometry is used when analyzing complex chemicals that cannot be accurately identified at low resolution.

48.7 Limitations of HPLC/MS

1. The lack of a universal detector is an issue though UV-Vis detector and refractive index detection can be used for chromatographic compounds.
2. The detection of the analyte with chromophores that absorb in the 190–600 nm range and sensitivity can be affected due to the wavelength selected for the analysis and the mobile phase properties for the former, and low sensitivity and incompatibility with gradient elution for the latter.
3. The introduction of UHPLC has extended (P_c) to 400–1000 varieties in a period of ~60 min.
4. 2D-LC can also increase P_c for the complete evaluation of very complicated samples in proteomics and metabonomics.
5. The complex instrumentation of HPLC makes it difficult for a beginner.

48.8 Applications of HPLC/MS

1. Catecholamines, including dopamine and epinephrine, are essential for various biological processes. By analyzing their precursors and metabolites, diseases such as Parkinson's, heart, and muscular dystrophy can be diagnosed.
2. HPLC is ideally suited for diagnostic purposes as it offers greater molecular separation and comparability than other techniques.
3. Reversed-phase HPLC (RP-HPLC) is one of the most used techniques due to its efficiency, column stability, and the ability to separate various chemicals. Retention time measurements on HPLC are used for compound identification.
4. The sympathetic nervous system is home to a potentially fatal tumor known as pheochromocytoma. Because they are formed from neural crest tissue, they are thought to secrete catecholamines. This makes HPLC ideal for diagnostics, but results vary depending on the source of the sample.
5. Both peripheral and central nervous system metabolites are present in urine specimens. HPLC is one of the most widely used methods worldwide for checking the purity of pharmaceuticals.

48.9 Conclusion

Guanitoxin (formerly known as anatoxin) is a potent, naturally occurring neurotoxin produced by freshwater cyanobacteria that can kill wild and domestic animals. Guanitoxin has been associated with death in domestic and wild animals that inadvertently ingested cyanobacterial cells. An analytical technique to interpret the details of Guanitoxin using HPLC as a mode would significantly contribute to the field of aquatic pathology. HPLC is one of the most widely used methods worldwide for checking the purity of pharmaceuticals. This may increase cost, but this sacrifice yields high accuracy and specificity. HPLC may be more beneficial than other methods to ensure purity. This method has shown its usefulness since the beginning of his HPLC use in the pharmaceutical industry. The knowledge of the same would also enable one to work on the solutions that would eradicate the toxic trait or nature from the cyanobacterial forms.

References

Cook WO, Dellinger JA, Singh SS, Dahlem AM, Carmichael WW, Beasley VR (1989) Regional brain cholinesterase activity in rats injected intraperitoneally with anatoxin-a(s) or paraoxon. Toxicol Lett 49:29–34

Fernandes KA, Pinto E, Ferraz HG, Vereau F (2020) Availability of guanitoxin in water samples containing sphaerospermopsis torques-reginae cells submitted to dissolution tests. Pharmaceuticals 13:1–12

Fiore MF, de Lima ST, Carmichael WW, McKinnie SMK, Chekan JR, Moore BS (2020) Guanitoxin, re-naming a cyanobacterial organophosphate toxin. Harmful Algae 92:101737

Lawton LA, Edwards C, Codd GA (1994) Extraction and high-performance liquid chromatographic method for the determination of microcystins in raw and treated waters. Analyst 119:1525

Mahmood NA, Carmichael WW (1986) The pharmacology of anatoxin-a(s), a neurotoxin produced by the freshwater cyanobacterium Anabaena flos-aquae NRC 525-17. Toxicon 24:425–434

Matsunaga S, Moore RE, Niemczura WP, Carmichael WW (1989) Anatoxin-a(s), a potent anticholinesterase from Anabaena flos-aquae. J Am Chem Soc 111:8021–8023

Pietsch J, Fichtner S, Imhof L, Schmidt W, Brauch HJ (2001) Simultaneous determination of cyanobacterial hepato-and neurotoxins in water samples by ion-pair supported enrichment and HPLC-ESI-MS-MS. Chromatographia 54:339–344

Powell M (1997) Analysis of anatoxin-a in aqueous samples. Chromatographia 45:25–28

Quilliam M, Thomas B, Scott G, Sui KW (1989) Ion-spray mass spectrometry of marine neurotoxins. Rapid Commun Mass Spectrom 3:145–150

Werner VR, Laughinghouse HD IV, Fiore MF, Sant Anna CL, Hoff C, de Souza Santos KR, Neuhaus EB, Molica RJ, Honda RY, Echenique RO (2012) Morphological and molecular studies of Sphaerospermopsis torques-reginae (cyanobacteria, Nostocales) from South American water blooms. Phycologia 51:228–238

Detection of Cyanotoxin in Algal Food and Feed Supplements Using UHPLC-MS/MS

49

Ajith Kumar K, Shree Kumari G R, and Lokesh Ravi

Abstract

Over the years, global demand for algal food is growing immensely, and it is consumed for functional benefits beyond the traditional considerations of nutrition and health. Increased frequency and longevity of Cyanobacteria over time, Cyanobacteria has the capability of producing secondary metabolites called Cyanotoxins. These toxins are said to be present in algal food, consuming cyanotoxins in food supplements might result in consumer health hazards such gastrointestinal neurological, hepatic, or cutaneous toxicity. Thus, for the investigation of Cyanotoxins a targeted ultra-high-performance liquid chromatographic-tandem mass spectrometric (UHPLC-MS/MS) method is used in this chapter to detect and validate cyanotoxins in algal food and feed supplements like tablets/powders, capsules, liquids/syrup. UHPLC-MS/MS is becoming a popular approach because of its remarkable specificity and sensitivity for both their detection and quantification of toxins and in this particular chapter the following species were analysed Aphanizomenon on Flos-aquae, arthrospira platensis, spirulina, and Chlorella. According to the results provided here, a variety of algal supplement tablet powder, water, and Cyanobacteria can all be detected and quantified using the optimized LC-MS/MS method for cyanotoxins.

A. K. K.
Department of Life Sciences, Kristu Jayanti College (Autonomous), Bengaluru, Karnataka, India

S. K. G. R.
Department of Biomedical Sciences, School of Bio Sciences and Technology, Vellore Institute of Technology, Vellore, India

L. Ravi (✉)
Department of Food Technology, Faculty of Life and Allied Health Sciences, MS Ramaiah University of Applied Sciences, Bengaluru, India
e-mail: lokesh.ravi@sju.edu.in

N. Thajuddin et al. (eds.), *Protocols for Cyanobacteria Sampling and Detection of Cyanotoxin*, https://doi.org/10.1007/978-981-99-4514-6_49

Keywords

Cyanotoxins · Cyanobacteria · Ultra-high performance liquid chromatography · Tandem mass spectrometry · Algal food · Detection · Feed supplements

49.1 Introduction

According to the archaeological evidence from 14,000 years ago in Chile algae have been a part of the human diet for thousands of years (Dillehay et al. 2008). Towards the trend of increasing the nutritional demand for algal products and a larger usage of food additives on a global scale (Wells et al. 2017). Nowadays, as a result of anthropogenic activity and climate change, cyanobacteria blooms are now increasing globally (Díez-Quijada Jiménez et al. 2020). Cyanobacteria, commonly known as blue green algae, are Gram-negative, photosynthetic, prokaryotes that can be found in a range of habitats, including fresh, brackish, marine, hypersaline, volcanic ash, desert sand, rocks, and terrestrial environment (Reddy and Mastan 2011). Cyanobacteria have both advantageous and harmful characteristics (Masango 2008). In freshwater systems, cyanobacteria typically contribute to HABs (Harmful Algal Blooms) and have the ability to create cyanotoxins, which are extremely poisonous (Sundaravadivelu et al. 2022).

In this chapter, we examine various cyanobacterial algal food and dietary supplement products to screen for the presence of different cyanotoxins: anatoxin-a (ANA-a), cylindrospermopsin (CYN), saxitoxin (STX), and β-methylamino-L-alanine (BMAA). The Structure, Molecular weight (g/mol), and PubChem CID of the Cyanotoxins are gathered from the PubChem server (https://pubchem.ncbi.nlm.nih.gov/) and it is shown in Table 49.1.

High-resolution mass spectrometry (HRMS) detection in combination with other analytical techniques allows us to confirm the presence of several cyanotoxins that could not be analysed in a single analytical run due to their particular structure (Roy-Lachapelle et al. 2017). For the investigation of cyanotoxins, ultra-high-performance liquid chromatography coupled to tandem mass spectrometry (UHPLC-MS/MS) has emerged as the preferred method due to its exceptional specificity and sensitivity for both their detection and quantification (Greer et al. 2016). The instrumentation of UHPLC-MS/MS is shown in Fig. 49.1. The widespread availability, safe product labels, and even health benefits, dietary supplements could pose a risk to the general public's health if they contain unwanted and hazardous chemicals. We seek to evaluate the harmful potential of these supplements by screening a wide range of cyanotoxins (Roy-Lachapelle et al. 2017).

Table 49.1 Showing the structure, molecular weight (g/mol), PubChem CID of the cyanotoxins

Cyanotoxin	Structure	Molecular weight (g/mol)	PubChem CID
Saxitoxin (STX)	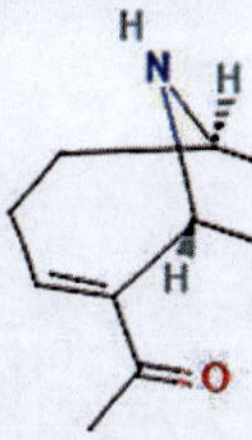	165.23	3,034,748
Cylindrospermopsin (CYN)	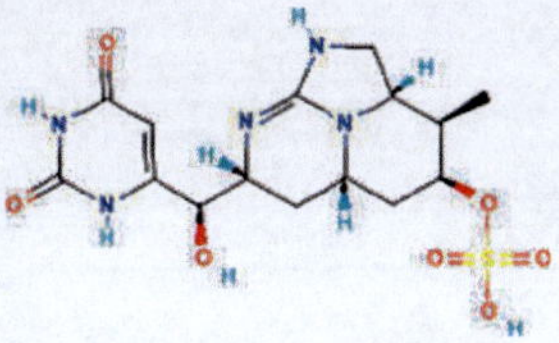	415.4	135,565,888
Anatoxin-A (ANA-a)	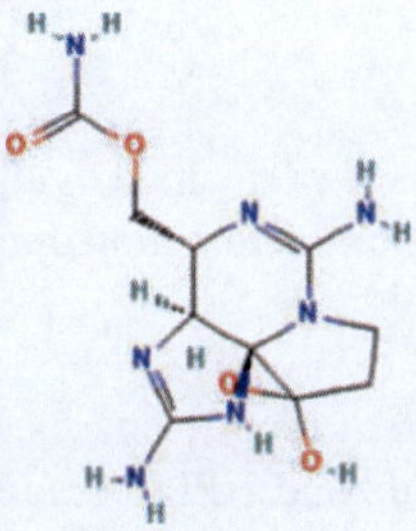	299.29	56,947,150
β-Methylamino-L-alanine	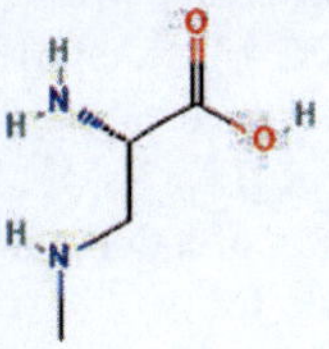	118.13	135,062,216

49.2 Materials (Turner et al. 2018)

49.2.1 Reference Standards

- 28.0 mL 50% aqueous MeOH
- 0.1% acetic acid
- Centrifuge (4500 g, 10 min)
- RM-BGA (280 mg)
- Calibration range—0.33 ng/mL to 327 ng/m per toxin
- Stock solution (Solid powders dissolved in 50% aqueous methanol)
- A mixed stock solution of seven-level suite of working calibration standards

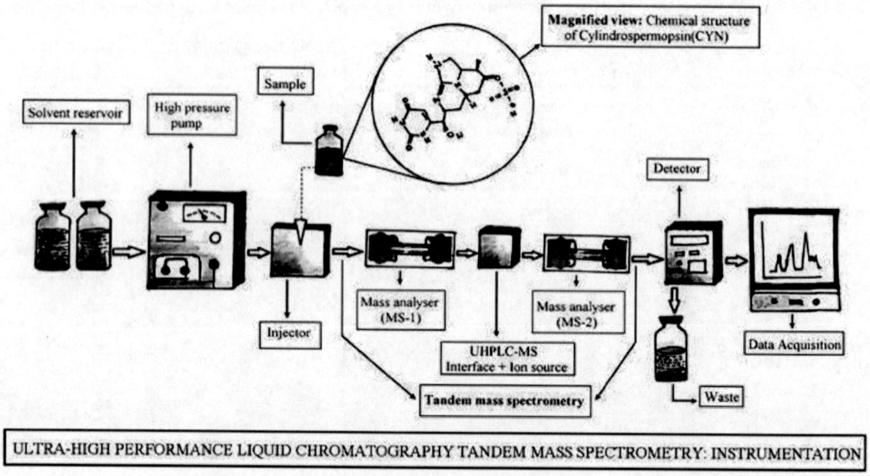

Fig. 49.1 Showing the instrumentation of the ultra-high-performance liquid chromatography-Tandem mass spectrometry (UHPLC-MS/MS)

(NOTE: external calibration of cyanotoxins is performed in all samples; seven-point calibration standards are utilized, and dilution factors are adjusted based on the type of extraction used)

- Instrument solvents used for preparation of mobile phases, chemicals and sample of preparation reagents are LC-MS-grade and HPLC grade.
- Standard references for toxins (MC-RR, MC-LA, MC-LY, MC-LF, MC-LW, MC-YR, MC-WR, MC-HilR, MC-HtyR, MC-LR, [Asp3] MC-LR, and Nod).
- Freeze-dried matrix reference material of blue-green algae that contains a range of MC concentrations and a verified standard of [Dha7]-MC-LR.
- Raw components are procured for the creation of a toxin-free blank material, and supplements were bought in tablet or capsule form.
- Ingredients of the tablets consisted of Aphanizomenon flos-aquae, arthrospira platensis, spirulina, and Chlorella.
- Tablets consisted either of 100% blue-green algae or contained small proportions of the additive rice maltodextrin and the anti-adherent magnesium stearate.
- Solid tablets were ground into a fine powder using a pestle and mortar and capsule is opened to decant the internal powders.
- Ground powder from the sources of dried algae is stored refrigerated in sealed vials until use.

49.3 Methods

49.3.1 UHPLC Conditions

1. The UHPLC-MS/MS (abbreviated further to LC-MS/MS) system is as follows.
2. Tandem quadrupole mass spectrometer (MS/MS) coupled to Waters Acquity UHPLC system used for LC-MS/MS analysis.
3. Chromatography conducted using 1.7 μm, 2.1 × 50 mm Waters Acquity UPLC BEH C18 column in conjunction with WatersVanGuard BEH C18 1.7 μm 2.1 × 5 mm guard cartridge.
4. Columns are held at +60 °C, with samples held in sample manager at +10 °C.
5. Take 5 μL of sample inject it and then mobile phase flow rate is 0.6 mL/min.
6. Mobile phase A—water +0.025% formic acid
   Mobile phase B—acetonitrile (MeCN) + 0.025% formic acid
7. UHPLC gradient is: 2% B initial conditions rising to 25%
   - B1 at 0.5 min holding until 1.5 min, rising to 40%
   - B at 3.0 min, increasing further to 50%
   - B at 4 min, quick rise to 95%
   - B held until 4.5 min until dropping back to 2%
   - B at 5 min (total run time 5.5 min)
8. Each instrumental sequence started with series of instrumental blanks, followed by toxin calibration standards, and the RM-BGA extract is used as an IQC and a matrix-based retention time marker.
9. Water and MeCN flushes are used to conclude the instrumental sequences, the first at 60 °C, and the second at 30 °C.
10. New columns are conditioned, injections of individual toxin solution are performed to determine retention times and confirm there is no significant cross-over between determinants.
11. Optimizing the UHPLC conditions of the cyanotoxins anatoxin-a (ATX), cylindrospermopsin (CYN), and β-N-methylamino-L-alanine (BMAA).

The conditions for UHPLC experimentation are graphically represented in Fig. 49.2.

49.3.2 MS/MS Conditions

1. The following is the Waters Xevo TQ tuning parameters: Source temperature of 150 °C, desolvation temperature of 600 °C, desolvation gas flow rate of 600 L/h, and collision gas flow rate of 0.15 mL/min. Capillary voltage at 1.0 kV.
2. For each toxin, the MS/MS approach incorporates Selected Reaction Monitoring (SRM) transitions employing positive mode acquisition.
3. Pure standards are injected into the mass spectrometer in the mobile phase together with parent and daughter ions, cone and collision voltages, and other variables.

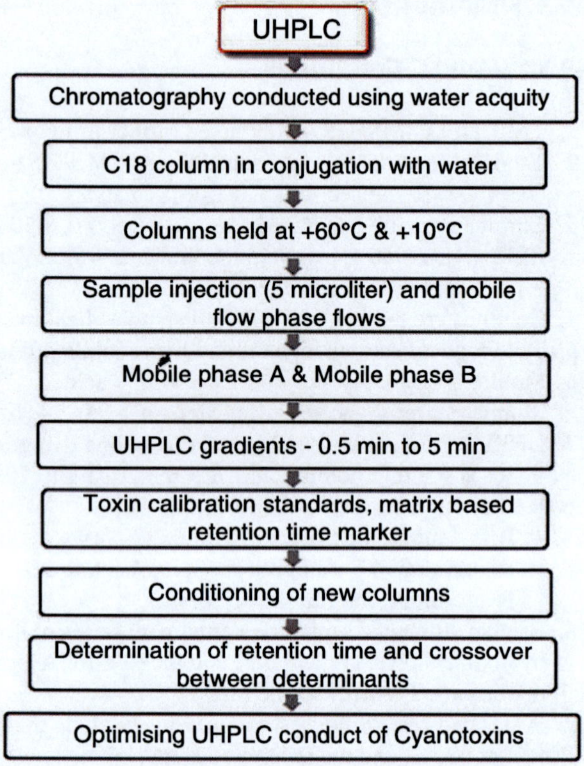

Fig. 49.2 Experimental conditions of UHPLC

4. A good separation of cyanotoxins is achieved throughout the course of the 5 min run period because the bulk of the toxins had distinctive SRM transitions and chromatographic retention durations.
5. The cyanotoxin toxins are directly quantified using the LC-MS/MS to certified reference standards that is available as working standards.
6. Utilizing external calibration, quantification is carried out, and results are calculated in terms of g/L of cultures.

49.3.3 Validation of the Cyanotoxin LC-MS/MS Method

1. In this study, the quantitative analysis of several Cyanotoxins analogues is subjected to validation.
2. Checking the analytical method's suitability for use over a suitable range of toxin concentrations in each of the four matrices is the goal (Specificity, Linearity, Sensitivity, and Recovery).

49.4 Observation

To avoid hazardous consequences, many institutes have put forward various TDI (Tolerable Daily Intake) recommendations for cyanotoxin intake. The WHO developed a TDI recommendation for MCs at 0.04 g/kg body weight as the most widely accepted figure. This figure translates to a daily intake of 2.4 g of MCs for a 60 kg adult (10). ANA-a (6 g), CYN (1.8 g), MCs (3 g), and STX (3 g) produced TDI values. Our findings suggest that consuming these "health goods" may have health concerns due to the presence of cyanotoxins in a number of cyanobacteria products (Roy-Lachapelle et al. 2017). Therefore, our take aside was to create an easy procedure that would enable accurate quantitation while requiring little in the way of analyst input and overall user expense (Turner et al. 2018).

49.5 Conclusion

Supplements containing cyanobacteria are widely marketed, accessible, and promoted as being good for your health. However, there is a limited quality control when these products are supplied, which could result in harmful dangers because of cyanotoxins contamination (Roy-Lachapelle et al. 2017). Due to the numerous confirmation criteria, including HRMS and MS/MS ions, the developed method in this chapter can be suggested for both environmental and food analysis (Filatova et al. 2020). Finally, we don't want to overstate the health dangers connected to the specific supplement we studied. We emphasized the significance of increased oversight of all dietary supplements derived from algae by the relevant authorities, inevitably lead to the implementation of maximum intake standards based on health to guarantee the safety of these items.

49.5.1 Advantages (Turner et al. 2018)

- The technique offers a useful early warning tool for the quick, routine extraction and analysis of natural waters, cyanobacterial blooms, algal powders, food supplements, and shellfish tissues. This enables monitoring labs to supplement conventional microscopy techniques and report toxicity results quickly after receiving sample material.
- This approach is straightforward, rapid, adaptable to a variety of matrices, and well-suited for use as a routine regulatory monitoring tool with a high degree of thoroughness and a short turnaround time.

49.5.2 Disadvantages (Lawrence et al. 2001)

- The UHPLC narrow peaks that demand a quick duty cycle, which is only possible on the most recent generations of MS equipment. As a result, some analysers (such as QqQ or TOF/MS) are more easily compatible with UHPLC than others (e.g., ion trap or FT-MS).

- Extra-column band widening may become severe as a result of the decreased column volume, which will lower the kinetic performance of the UHPLC-MS system.
- The electrospray ionization source must be able to give high sensitivity at flow rates since the mobile phase linear velocity is higher in UHPLC.

49.5.3 Applications (Rathod et al. 2019; Qin et al. 2020)

It is used in various type of research like

- Pharmacokinetics
- Toxicology screening and confirmation
- Forensic research
- Veterinary drugs in animal matrices and feed
- Pesticides in vegetables and fruits
- Toxins in vegetables, fruits, and cereals
- Marine toxins in fish and shellfish

Acknowledgement The authors acknowledge the Department of Botany, St Joseph's University, Bengaluru, India, for the infrastructural support.

References

Díez-Quijada Jiménez L, Guzmán-Guillén R, Cătunescu GM, Campos A, Vasconcelos V, Jos Á, Cameán AM (2020) A new method for the simultaneous determination of cyanotoxins (microcystins and Cylindrospermopsin) in mussels using SPE-UPLC-MS/MS. Environ Res 185:109284

Dillehay TD, Ramírez C, Pino M, Collins MB, Rossen J, Pino-Navarro JD (2008) Monte Verde: seaweed, food, medicine, and the peopling of South America. Science 320:784–786

Filatova D, Núñez O, Farré M (2020) Ultra-trace analysis of cyanotoxins by liquid chromatography coupled to high-resolution mass spectrometry. Toxins (Basel) 12:247. https://doi.org/10.3390/toxins12040247

Greer B, McNamee SE, Boots B et al (2016) A validated UPLC-MS/MS method for the surveillance of ten aquatic biotoxins in European brackish and freshwater systems. Harmful Algae 55: 31–40

Lawrence JF, Niedzwiadek B, Menard C, Lau BP, Lewis D, Kuper-Goodman T, Carbone S, Holmes C (2001) Comparison of liquid chromatography/mass spectrometry, ELISA, and phosphatase assay for the determination of microcystins in blue-green algae products. J AOAC Int 84:1035–1044

Masango MG (2008) A comparative analysis of the cytotoxicity of cyanotoxins using in vitro (cell culture) and in vivo (mouse) assays. University of Pretoria, Pretoria

Qin N, Shen M, Xiang P, Wen D, Shen B, Deng H, Qiang H, Song F, Shi Y (2020) Determination of 37 fentanyl analogues and novel synthetic opioids in hair by UHPLC-MS/MS and its application to authentic cases. Sci Rep 10:11569. https://doi.org/10.1038/s41598-020-68348-w

Rathod RH, Chaudhari SR, Patil AS, Shirkhedkar AA (2019) Ultra-high performance liquid chromatography-MS/MS (UHPLC-MS/MS) in practice: analysis of drugs and pharmaceutical formulations. Futur J Pharm Sci 5:1–26. https://doi.org/10.1186/s43094-019-0007-8

Reddy MRK, Mastan SA (2015) Algal toxins and their impact on human health. Biomed Pharmacol J 4(1):129–134

Roy-Lachapelle A, Solliec M, Bouchard MF, Sauvé S (2017) Detection of cyanotoxins in algae dietary supplements. Toxins 9:76

Sundaravadivelu D, Sanan TT, Venkatapathy R, Mash H, Tettenhorst D, Danglada L, Frey S, Tatters AO, Lazorchak J (2022) Determination of cyanotoxins and prymnesins in water, fish tissue, and other matrices: a review. Toxins 14:213

Turner AD, Waack J, Lewis A, Edwards C, Lawton L (2018) Development and single-laboratory validation of a UHPLC-MS/MS method for quantitation of microcystins and nodularin in natural water, cyanobacteria, shellfish and algal supplement tablet powders. J Chromatogr B Anal Technol Biomed Life Sci 1074–1075:111–123

Wells ML, Potin P, Craigie JS, Raven JA, Merchant SS, Helliwell KE, Smith AG, Camire ME, Brawley SH (2017) Algae as nutritional and functional food sources: revisiting our understanding. J Appl Phycol 29:949

Rapid Detection Method of Microcystin in Water

50

G. Y. Liyanage, S. P. Manage, H. A. S. N. Abeysiri, P. M. Manage, and S. A. Kulasooriya

Abstract

Microcystins (MCs) are secondary metabolites produced by some cyanobacteria, a class of cyclic heptapeptide toxins that are stable in the environment. Microcystins can create various adverse health effects in humans, animals, and plants through contaminated water. Global occurrence and concern about microcystin contamination in water have prompted the development of a range of detection methods for their identification and quantification. However, most protocols are relatively time-consuming and expensive and require laboratory expertise. The development of the inhibition assays and biosensor protocols offered rapid and accurate detection, high reproducibility, and portability. This chapter incorporates the protocols of Protein Phosphatase Inhibition Assay

G. Y. Liyanage
Centre for Water Quality and Algae Research, Department of Zoology, University of Sri Jayewardenepura, Nugegoda, Sri Lanka

Department of Aquatic Bio resources, Faculty of Urban and Aquatic Bioresources, University of Sri Jayewardenepura, Nugegoda, Sri Lanka

S. P. Manage
Centre for Water Quality and Algae Research, Department of Zoology, University of Sri Jayewardenepura, Nugegoda, Sri Lanka

H. A. S. N. Abeysiri · P. M. Manage
Centre for Water Quality and Algae Research, Department of Zoology, University of Sri Jayewardenepura, Nugegoda, Sri Lanka

Present Address: Faculty of Graduate Studies, University of Sri Jayewardenepura, Nugegoda, Sri Lanka

S. A. Kulasooriya (✉)
National Institute of Fundamental Studies, Kandy, Sri Lanka
e-mail: anada@kulasooriya.com

N. Thajuddin et al. (eds.), *Protocols for Cyanobacteria Sampling and Detection of Cyanotoxin*, https://doi.org/10.1007/978-981-99-4514-6_50

(PPIA), Surface Plasmon Resonance (SPR) immunosensors, Fluorescent immunosensors, Electrochemical Biosensors, and Lateral flow dipstick (Immunostrip) as methods used for MCs detection in terms of their novel development and usage.

Keywords

Microcystins · Protein Phosphatase Inhibition Assay (PPIA) · Surface Plasmon Resonance (SPR) immunosensors · Fluorescent immunosensors · Electrochemical biosensors

50.1 Introduction

Cyanobacterial harmful algal blooms (Cyano HABs) are globally on the increase in both frequency and intensity as a result of eutrophication and climate change (Lawton and Edwards 2008). These algal outbreaks secrete toxic secondary metabolites and cyanobacterial toxins, of which microcystins (MCs) are the most widely distributed and toxic type of cyanobacterial toxins in cyanobacterial bloom pollution (Manage et al. 2009).

MCs cause considerable concern due to both their acute (Carmichael 2001) and chronic effects, including the potential development of cancer in exposed populations (Falconer et al. 1994; Ueno et al. 1996; Zhou et al. 2002). This has led to the publication of a guideline threshold for microcystin-LR in drinking water (WHO 1998) and the urgent requirement for sensitive, rapid, and reliable detection methods to identify and quantify MCs in natural and treated waters (McElhiney and Lawton 2005; Lawton and Edwards 2008).

Table 50.1 Illustrates the development of MC detection techniques. Chemical methods, such as HPLC and LC-MS, could identify various variants more sensitively, but they are expensive, the equipment is large in size, and expertise is needed to use them correctly, making it difficult to promote their application in actual water

Table 50.1 Currently available Microcystins (MCs) detection techniques

Types	Methods
Biochemical method	Protein phosphatase inhibition assay (PPIA)
	Enzyme-linked immunosorbent assay (ELISA)
Chemical method	High-performance liquid chromatography (HPLC)
	Liquid chromatography-mass spectrometry (LC-MS)
	High-performance capillary electrophoresis (HPCE)
	Gas chromatography (GC)
Biosensor	Surface plasmon resonance (SPR) immunosensors
	Luminescent immunosensors
	Fluorescent immunosensors
	SPR-DNA biosensor
	Immunostrips

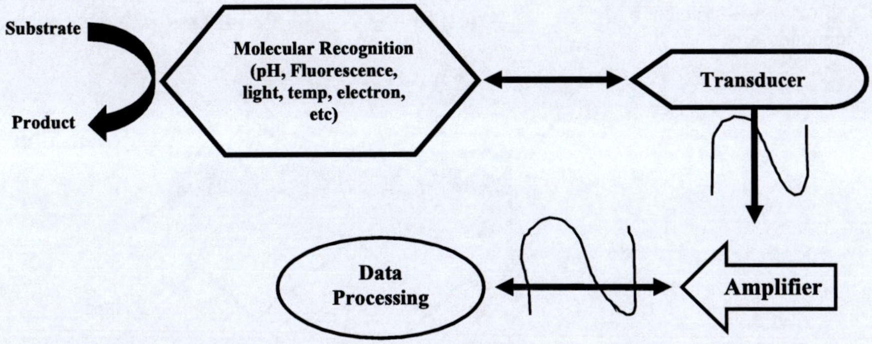

Fig. 50.1 The primary process of the biosensing principle

monitoring and not accessible for all parts of the world. The Protein Phosphatase Inhibition Assays (PPIAs), Enzyme-Linked Immunosorbent Assay (ELISAs), and other biological techniques are simpler and more specific. In terms of portability, the advantages of biosensors are more prominent. Emerging methods, such as biosensors, which are easy to detect in practice because of their rapidity and portability, provide new ideas for detecting MCs.

A Protein Phosphatase Inhibition Assay (PPIA), which usually needs to be performed on a 96-well microtiter plate (Bittencourt-Oliveira et al. 2011), is a simple, convenient, economical, and rapid method for MC detection with high sensitivity and is suitable for testing a large number of samples as a routine detection technique.

A biosensor is an analytical tool made up of a biological recognition element termed a bioreceptor in direct contact with a transducer. This analytical tool can be classified either by its biological recognition element or signal transduction methods. The principle of the biosensor is illustrated in Fig. 50.1. A bioreceptor in the biosensor is often combined with a suitable transduction method to generate a signal following interaction with the target molecule of interest (Singh et al. 2012; Yang et al. 2018; Qin et al. 2019; Pang et al. 2020). It is worthwhile noting that various natural and artificial biological elements, including whole cells, enzymes, antibodies, molecularly imprinted polymers (MIPs), and nucleic acids are employed in biosensors.

At present, enzyme-based biosensors (including optical and electrochemical biosensors), immunosensors (including electrochemical, piezoelectric, NMR-based and optical immunosensors such as Surface Plasmon Resonance (SPR), Evanescent Wave Fiber-optic immunosensors, Luminescent immunosensors, Fluorescent immunosensors, and Immunoarray biosensors), and Nucleic acid biosensors (including electrochemical DNA and SPR-DNA biosensor) have been developed for successful MCs detection (Fig. 50.2).

Biosensors can achieve low detection limits of MC-LR in dietary supplements and various aquatic settings such as drinking water, lakes, and reservoir waters due

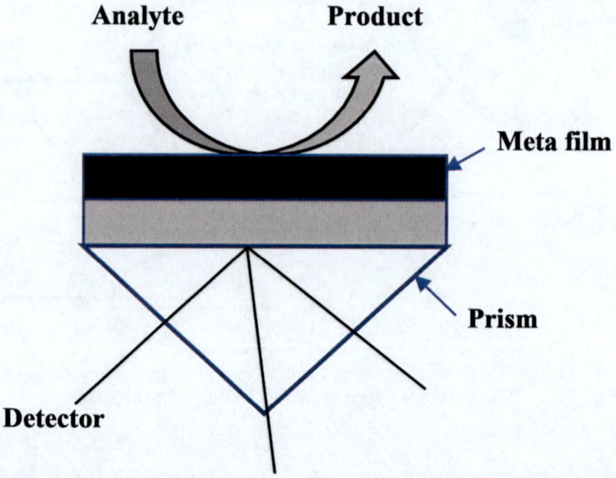

Fig. 50.2 Basic structure of an immunosensor

to the selective binding or reaction of the biological recognition element to the target analyte. The technique can also demonstrate good recovery, precision, and accuracy through the evaluation of the spiked water samples and can be readily extended toward the on-site real-time sensitive detection of other targets in the field of environment, food, and medical diagnosis (Strachan et al. 2002).

Lateral flow dipstick (Immunostrip) is another biosensor method which successfully used for a wide range of applications including the detection of drugs of abuse, pregnancy, and many environmental contaminants. The isolation of recombinant antibodies to microcystins (McElhiney et al. 2000; Strachan et al. 2002) allows almost limitless production of single-chain antibodies in a bacterial host, greatly reducing the production cost and eliminates the restriction on the amount of antibody used in assays. These recombinant antibodies have been exploited to develop a low-cost lateral flow dipstick (immunostrip) (Strachan et al. 2002).

However, to effectively manage and control MCs and prevent or minimize their health risks, sensitive, fast, and reliable screening methods capable of detecting these toxins are urgently required. Thus, the chapter aims to summarize the protocols of Protein Phosphatase Inhibition Assay (PPIA), Surface Plasmon Resonance (SPR) immunosensors, Fluorescent immunosensors, Electrochemical Biosensors, Lateral flow dipstick (Immunostrip), commercially available ELISA kits as methods used for MCs detection in terms of their novel development and usage. The chapter also puts forward some directions for future research toward method application and improvement.

50.2 Methods

50.2.1 Protein Phosphatase Inhibition Assay (PPIA)

50.2.1.1 Materials

40 mM Tris HCl
20 mMKCl
30 mM $MgCl_2$
3 mM DTT (dithiothreitol)
Bovine serum albumin
MC-LR standard

50.2.1.2 Preparation of Solutions

1. Prepare a stock buffer using 40 mMTrisHCl, 20 mMKCl, and 30 mM MgCl2 at pH 8.6.
2. Dilute A, B, C reagents using prepared stock solution [A: 3 mM DTT (dithiothreitol); B: 1.5 mg mlÿ1 BSA (bovine serum albumin); C: 1.5 mM MnCl2 (make fresh daily)].
3. Make reaction solution D using A:B: C in a 1:1:1 ratio.
4. Prepare substrate solution E using 40 mMpNPP in reaction solution D

(Note: It must be prepared fresh before use).

50.2.1.3 Performing the Assay

1. Dilute PP1 with reaction solution D to achieve a final concentration of 1 mU mL^{-1} (suppliers may recommend different concentrations in inhibition assays)
2. Add to each well of a 96-well microtiter plate: 10 mL of an unknown sample, 10 mL of distilled H_2O (control wells), MC-LR standards at 30, 10, 3, 1, and 0.3 mg L^{-1}.
3. To all wells, add 40 mL of diluted PP1 solution.
4. Incubate at 37 °C for 5 min.
5. Add 50 ml of substrate solution E to each well to start the enzymatic reaction.
6. Monitor the color production of each sample and standard at 405 nm, taking a reading at 5 min intervals for 40 min (a total of 8 readings).
7. Set the plate reader software to read a single wavelength (405 nm) and generate the inhibition curve using the linear (kinetic) mode.
8. Calculate the MC concentration in the unknown samples using the standard curve generated in each run from the MC-LR standards.

50.2.2 Surface Plasmon Resonance (SPR) Immunosensors

50.2.2.1 Materials

MC-LR with >95% purity (HPLC-grade)
The (MC-LR)–BSA bioconjugate

Anti-(MC-LR) monoclonal antibody (mAb)
N-hydroxysuccinimide (NHS)
1-ethyl-3-(3-dimethylaminopropyl)-carbodiimide hydrochloride (EDC)
Ethanolamine
Tween-20
Deionized water

50.2.2.2 Instrumentation

BIAcoreTM 3000 instrument
Matrix-assisted laser desorption/ionization time-of-flight mass spectrometry (MALDI-TOF-MS) instrument

50.2.2.3 Immobilization Protocol

1. Determine the molecular weight (MW) of BSA and (MC-LR)–BSA to be 66,354 and 68,708, respectively.
2. Function the CM5 sensor chip by the EDC–NHS amine coupling as follows: the sensor chip is activated twice by mixing equal volumes of NHS (0.1 M) and EDC (0.4 M) and injecting the mixture over the sensor chip for 10 min at a flow rate of 10 L min^{-1}.
3. Dissolve (MC-LR)–BSA (100 g mL^{-1}) in sodium acetate buffer (10 mM, pH 4.3) and inject over the sensor chip surface at a flow rate of 5 L min^{-1}.
4. Inject NaOH (25 mM) to wash the surface, and capping of the unreacted sites was achieved by injection of ethanolamine (1 M, pH 8.0) for 7 min.

50.2.2.4 SPR Immunoassay

1. MC-LR standards or samples (40 μL) were incubated with an equal volume of mAb for 30 min at room temperature.
2. The equilibrated mixture was injected over the functional surface at a flow rate of 5 L min^{-1}; the remaining free mAb was captured by (MC-LR)–BSA on the surface.
3. Real-time monitoring will be displayed in a sensor gram as the optical response (RU) versus time (in seconds).
4. To convert the response back to baseline, the sensor chip was regenerated by injecting NaOH (40 mM) at a flow rate of 20 L min^{-1}.
5. The next sample was analyzed randomly.

50.2.3 Fluorescent Immunosensors

50.2.3.1 Materials

Fluorescent dye Cy5.5 N-hydroxysuccinimide (NHS) ester
Anti-MC-LR-antibody
Hapten conjugates
1 mg/L MC-LR stock solutions

Methanol
Phosphate-buffered saline (10 mM PBS, pH 7.4)

50.2.3.2 Fluorescent Immunoassay

1. The MC-LR-BSA conjugate is covalently immobilized on the chip surface by a similar procedure described by Long et al. (2009).
2. When performing the test cycle, 0.8 mL of sample solution and 0.2 mL of 0.3 μg/mL Cy5.5-labelled antibody solution (in 10 mM PBS containing 5.0 mg/mL BSA and 0.1 mg/mL thiomersal) is firstly transferred to the preincubation loop for 5 min to make the antibody-binding sites occupied with the analyte.
3. Subsequently, the mixture is delivered into the flow cell.
4. Antibodies with free binding sites remaining interact with the coated antigen immobilized on the biochip for 10 min.
5. To reduce the effect of free antibody in solution and its nonspecific adsorption on the detection result, the fluorescence signal is detected after the mixture is washed with PBS solution.
6. The amount of antibody coupled on the chip surface is inversely correlated to the concentrations of MC-LR in samples, which the fluorescence signal can reflect.
7. The signal intensities were fitted to a four-parameter logistic equation.

$$A = \frac{A_1 - A_2}{1 + (x/x_0)^p} + A_2$$

Where A is fluorescence intensity, x is the MC-LR concentration; A1 is the upper asymptote and A2 is the lower asymptote (background signal) to the titration curve; x 0 is the analyte concentration at inflexion and p is the slope at the inflexion point.

50.2.4 Electrochemical Biosensors

50.2.4.1 Materials and Reagents

Potassium ferricyanide (K3Fe (CN)6) (≥99.5%)
Potassium ferrocyanide (K4Fe (CN)6) (≥99.5%)
Potassium chloride (KCl) (≥99.5%)
Methanol (chromatography grade)
Ethanol (99.7%, analytical grade)
Sulfuric acid (95–98%, analytic grade)
Ultra-pure water (18 MΩ/cm)
Calf thymus DNA (ctDNA)
1.0 g/L stock solution of ctDNA
Microcystin-LR (MC-LR)

50.2.4.2 Fabrication of Biosensors

1. Carry out electrochemical experiments on an electrochemical workstation (RST5200F, Suzhou, China) at room temperature (25 °C), equipped with a three-electrode system comprising a gold working electrode or biosensor (3 mm diameter), a platinum wire as the counter electrode, and a calomel electrode as reference (GaossUnion, Shanghai, China).
2. To fabricate the biosensor, a bare gold electrode is first polished with 0.05 mm α-Al2O3 powders on a polishing micro cloth.
3. Sonicate in ethanol and ultra-pure water for 5 min, respectively, followed by washing with ultra-pure water (three times with a total volume of about 12 mL).
4. Activate through cyclic voltammetric scanning between −0.2 and + 1.4 V in 0.5 M H_2SO_4 solution until steady and ideal cyclic voltammogram is observed.
5. Wash the electrode with ultra-pure water and re-test in 0.1 M KCl containing 5.0 mM K3Fe(CN)6/K4Fe(CN)6 (1:1 molar ratio) to get the control result.
6. Carefully drop 5 μL of ctDNA solution (1 g/L) onto the pre-treated gold electrode and incubate at 37 °C for 3 h to obtain a ctDNA-immobilized electrode.
7. Wash the excess ctDNA with ultra-pure water (three times with a total volume of about 12 mL)
8. Under the optimal conditions, the amount of ctDNA immobilized on the electrode is evaluated by collecting and concentrating all washing solutions, followed by quantitation of ctDNA using Nano drop.

50.2.4.3 Electrochemical Detection Assay

1. Perform measurements on MC-LR Cyclic voltammetry (CV), electrochemical impedance spectroscopic (EIS), and differential pulse voltammetry (DPV) in RST electrochemical workstation (RST5200F, Suzhou, China).
2. The electrolyte solution is 0.1 M KCl containing 5.0 mMK3Fe(CN)6/K4Fe(CN) 6.
3. Add MC-LR standards to the detection solution; the solution is stirred for 3 min and settled for 4 min.
4. Record the CV data at the scan rate of 0.05 V/s.
5. Measure the EIS data within the frequency range from 1 Hz to 0.1 MHz.
6. Perform the DPV measurements between - 0.4 and 0.7 V with a step of 4 mV, pulse height of 0.05 V, the width of 40 ms, and period of 0.1 s and sampling width of 0.01 s.

50.2.5 Lateral Flow Dipstick (Immunostrip)

50.2.5.1 Materials

Glass fiber membrane
m PBS
1% BSA
0.05% Tween 20
MC-LR standards

50.2.5.2 Preparation of Immunoassay Test Strips

1. Pre-treat the sample pad (glass fiber membrane) with 0.01 M PBS buffer containing 1% BSA and 0.05% Tween 20.
2. Dry for 1 h at 50 °C to prevent nonspecific binding.
3. Cut the nitrocellulose membrane into sections (5 mm × 20 mm).
4. Dispense separately 1 μg of secondary antibody (control line) and 1.75 μg of MCLR-BSA conjugate (test line) at one end of the nitrocellulose membrane near absorption pad side.
5. The distance between these two lines is 5 mm.
6. Block the nitrocellulose membrane by 1% BSA for 30 min at room temperature in order to prevent nonspecific protein binding.
7. Dispense gold-antibody conjugate onto the conjugated pad.
8. Establish the absorption pad at the end of the nitrocellulose membrane.
9. The nominal detection limit is 5–10 μg L^{-1}.

50.2.5.3 Application of Immunostrip

1. Filter a small amount of lake water (~10 mL) through a 25-mm polycarbonate filter with a nominal pore size of 2 μm.
2. The filters are placed in a screw cap microfuge tube and boiled for 5 min to release the microcystins.
3. Microcystin immunostrips are also used to confirm the toxicity of the 50% methanol extracts prepared in the lab.
4. These extracts are highly concentrated.
5. Therefore, dilute with distilled water to achieve an apparent concentration factor of 50-fold over raw lake water (detection limit 0.2 μg L^{-1}).
6. This also diluted the extracts' methanol concentration to below 5%.

50.2.6 Immunological Assays Using ELISA

1. Freeze 25 mL water samples and thaw three times to lyse the cells and liberate the intracellular toxin.
2. Use commercially available ELISA test kits (BEACON, USA) to quantify intracellular and extracellular microcystin-LR.
3. Analyze each sample in triplicate, and follow the procedure exactly as directed by the manufacturer.
4. Measure the absorbance in a plate reader at 450 nm using spectrophotometer.

Acknowledgment The authors wish to thank the University of Sri Jayewardenepura.

References

Bittencourt-Oliveira MC, Oliveira MC, Pinto E (2011) Diversidade de genótiposprodutores de microcistinasemlinhagensbrasileiras de microcystis (Cyanobacteria). Braz J Biol 71:209–216

Carmichael WW (2001) Health effects of toxin-producing cyanobacteria: "the CyanoHABs". Hum Ecol Risk Assess Int J 7(5):1393–1407
Falconer IR, Burch MD, Steffensen DA, Choice M, Coverdale OR (1994) Toxicity of the blue-green alga (cyanobacterium) *Microcystis aeruginosa* in drinking water to growing pigs as an animal model for human injury and risk assessment. Environ Toxicol Water Qual 9(2):131–139
Lawton LA, Edwards C (2008) Conventional laboratory methods for cyanotoxins. In: Cyanobacterial harmful algal blooms: state of the science and research needs. Springer, New York, NY, pp 513–537
McElhiney J, Lawton LA (2005) Detection of the cyanobacterial hepatotoxins microcystins. Toxicol Appl Pharmacol 203(3):219–230
McElhiney J, Lawton LA, Pôrter AJ (2000) Detection and quantification of microcystins (cyanobacterial hepatotoxins) with recombinant antibody fragments isolated from a naive human phage display library. FEMS Microbiol Lett 193(1):83–88
Pang P, Lai Y, Zhang Y, Wang H, Conlan XA, Barrow CJ, Yang W (2020) Recent advancement of biosensor technology for the detection of microcystin-LR. Bull Chem Soc Jpn 93(5):637–646
Qin J, Sun X, Li D, Yan G (2019) Phosphorescent immunosensor for simple and sensitive detection of microcystin-LR in water. RSC Adv 9(22):12747–12754
Singh S, Srivastava A, Oh HM, Ahn CY, Choi GG, Asthana RK (2012) Recent trends in the development of biosensors for the detection of microcystin. Toxicon 60(5):878–894
Strachan IB, Pattey E, Boisvert JB (2002) Impact of nitrogen and environmental conditions on corn as detected by hyperspectral reflectance. Remote Sens Environ 80(2):213–224
Ueno, Y., Nagata, S., Tsutsumi, T., Hasegawa, A., Watanabe, M. F., Park, H. D., et al (1996) Detection of microcystins, a blue-green algal hepatotoxin, in drinking water sampled in Haimen and Fusui, endemic areas of primary liver cancer in China, by highly sensitive immunoassay
WHO (1998) Guidelines for drinking water quality addendum to Vol. 2. Health criteria and other supporting information, 2nd edn. World Health Organisation, Geneva, pp 95–110
Yang R, Song D, Fang S, Liu Y, Zhou X, Long F, Zhu A (2018) Development of novel portable and reusable fibre optical chemiluminescent biosensor and its application for sensitive detection of microcystin-LR. Biosens Bioelectron 121:27–33
Zhou L, Yu H, Chen K (2002) Relationship between microcystin drinking water and colorectal cancer. Biomed Environ Sci 15:166–171

Part IV
Advanced Methods in Cyanotoxins

Screening of Polyketide Synthase Genes in the Anatoxin-a-Producing Cyanobacteria

51

G. Y. Liyanage, S. A. Kulasooriya, S. P. Manage, D. M. P. I. Dissanayake, K. R. V. Bandara, Budhdhika Prabhath, and P. M. Manage

Abstract

Cyanobacterial blooms are a common occurrence in both freshwater and brackish environments. The cyanobacterial species in these habitats frequently produce toxins, making the water unsafe for human consumption. Various varieties of cyanobacteria, including *Anabaena, Oscillatoria*, and *Aphanizomenon* species, make the toxin anatoxin-a. Anatoxin production genes and the biosynthesis of these toxins in the *Oscillatoria* sp. strain were recently characterized. This chapter covered the anatoxin synthetase gene cluster (*anaA* to *anaG* and orf1; 29 kb) found in the *Anabaena* sp. variety. Molecular detection methods for potential anatoxin-a producers from the genera *Anabaena* sp., *Aphanizomenon* sp., and *Oscillatoria* sp. were developed by designing primers that recognize the *ana*C gene. Several species of cyanobacteria would have *ana*C genes from *Anabaena*

G. Y. Liyanage
Department of Zoology, Centre for Water Quality and Algae Research, University of Sri Jayewardenepura, Nugegoda, Sri Lanka

Faculty of Urban and Aquatic Resources Management, University of Sri Jayewardenepura, Nugegoda, Sri Lanka

S. A. Kulasooriya (✉)
National Institute of Fundamental Studies, Kandy, Sri Lanka
e-mail: ananda@kulasooriya.com

S. P. Manage · D. M. P. I. Dissanayake
Centre for Water Quality and Algae Research, Department of Zoology, University of Sri Jayewardenepura, Nugegoda, Sri Lanka

K. R. V. Bandara · B. Prabhath · P. M. Manage
Department of Zoology, Centre for Water Quality and Algae Research, University of Sri Jayewardenepura, Nugegoda, Sri Lanka

Faculty of Graduate Studies, University of Sri Jayewardenepura, Nugegoda, Sri Lanka

N. Thajuddin et al. (eds.), *Protocols for Cyanobacteria Sampling and Detection of Cyanotoxin*, https://doi.org/10.1007/978-981-99-4514-6_51

sp. and *Oscillatoria* sp. isolated from them using polymerase chain reaction. *ana*C amplicon RFLP analysis allowed simultaneous identification of *Anabaena, Oscillatoria*, and *Aphanizomenon* species as the three primary genera. The molecular techniques outlined in this chapter have effectively identified *Anabaena* sp. and *Oscillatoria* sp., two genera suspected of producing anatoxin, in freshwater and marine water. They could be used to conduct similar surveys for other potential neurotoxin producers in aquatic environments.

Keywords

Anatoxin-a · *ana*C · *Anabaena* sp. · *Oscillatoria* sp.

51.1 Introduction

Anatoxin-aand homoanatoxin-a are two potent cyanobacteria neurotoxins (Fig. 51.1) (Aráoz et al. 2010). Animal deaths have been linked to exposure to cyanobacteria that produce these alkaloids in a wide range of conditions worldwide (Christensen and Khan 2020; Colas et al. 2021). Therefore, releasing such cyanobacterial toxins into water sources and bodies is a significant cause of public health and environmental anxiety.

Recent studies have shown that the anatoxin-a biosynthetic gene cluster (*ana*) is present on the chromosomes of *Oscillatoria* sp. PCC 6506 and *Anabaena* sp. 37 (Cullen et al. 2019). Eight *ana* genes are known, with three of them encoding polyketide synthases (*ana*E, *ana*F, and *ana*G). A biosynthesis pathway for the two anatoxins has also been suggested (Kust et al. 2020). The adenylation enzyme *ana* C first converts proline into its active form and then loads it onto the acyl transport protein *ana*D (Cullen et al. 2019; Kust et al. 2020).

A gene responsible for the biosynthesis of the anatoxin-a and homoanatoxin-a toxins-producing cyanobacteria was identified as a reliable genetic marker. Hemscheidt et al. 1995 (Dittmann et al. 2013) proposed a biosynthetic pathway involving a polyketide synthase (PKS) in elongating the glutamic semialdehyde starter. The *Oscillatoria* sp. strain PCC 6506 was selected for this research because it produces anatoxin-a and homoanatoxin-a, multiplies under standard conditions and has lost its benthic characteristics after multiple subcultures since its isolation and used the mutated DKF and DKR primers developed by Moffitt and Neilan (Cadel-Six et al. 2009) to amplify PKS genes from the PCC 6506 genome.

Fig. 51.1 Structure of anatoxin-a

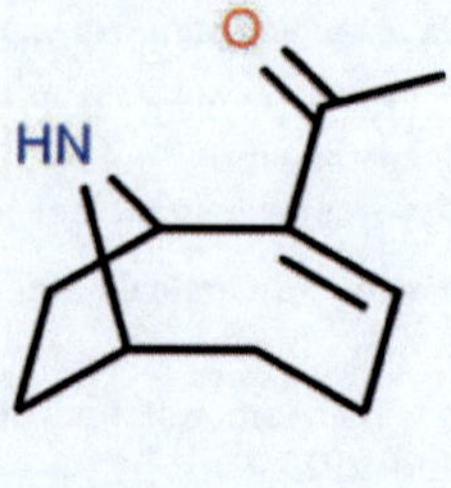

Genus-specific primers were used to differentiate between *Anabaena* and *Oscillatoria ana* C gene variants, while general primers were used to simultaneously detect all three producer species (Rantala-Ylinen et al. 2011). Additionally, restriction fragment length polymorphism (RFLP) analysis of the general PCR products (Rantala-Ylinen et al. 2011; Kasai et al. 2015) allowed for the simultaneous identification of the three producer species under study. Cyanobacterial strains and environmental DNA samples were used to develop molecular tools for detecting these potential anatoxin producers.

51.2 Methods

51.2.1 Cyanobacterial strains and growth conditions (Rantala-Ylinen et al. 2011)

- Prepare axenic cultures of *Anabaena* sp. and *Oscillatoria* sp. using Z8 and BG11 medium.
- Maintain cultures at approximately 23 °C with continuous illumination of 5 µmol photons s^{-1} m^{-2}.

51.2.2 DNA Extraction Using a Commercial Kit (Rantala-Ylinen et al. 2011; Pipan et al. 2018)

1. Disrupt harvested cells (approximately 150 mg) of *Anabeana* sp. with a Fast Prep FP120 bead beater for the 30S at a speed of 5 ms^{-1}.
2. Extract DNA from the prepared sample using the E.Z.N.A SP plant DNA Extraction Kit (Omega Bio-Tek) according to the manufacturer's instructions.
3. Store the purified DNA at −20 °C until the PCR analysis.
4. Harvest 10 cells of *Oscillatoria* stains from 40 mL cultures.
5. Freeze in liquid nitrogen before being lyophilized.
6. Extract DNA from the prepared sample using the NucleoBond AXG columns (macherey-Nagel, France) according to the manufacturer's instructions.
7. Store the purified DNA at −20 °C until the PCR analysis.

51.2.3 Sequencing of the anaC Gene (Rantala-Ylinen et al. 2011)

1. Amplify all the samples with primer pair anagen *ana* C (861 bp) or *ana* C (813 bp).
2. Amplicons either sequence directly with gene-specific primers or clone into the pCR2.1-TOPO vector according to the manufacturer's instructions.
3. Sequence inserts with vector (M13 and T7) or gene-specific primers on ABI 310 sequencer (Applied Biosystems).
4. Analyze sequences with the BioEdit sequence alignment editor.

51.2.4 PCR (Polymerase Chase Reaction) and RFLP (Restriction Fragment Length Polymorphism) (Rantala-Ylinen et al. 2011; Moreira et al. 2017)

1. Add1× DyNAzyme PCR buffer, 0.2 mM deoxynucleoside triphosphates (dNTPs), 0.5 μM primers, and 0.5 U DyNAzyme II polymerase (Finnzymes) in a total volume of 20 μL.
2. Use purified DNA (2–25 ng) as the template.
3. Adjust the PCR program to 94 °C for 2 min; 25–35 cycles of 94 °C for 30 s, 50–60 °C for 30 s, and 72 °C for 30 s; and 72 °C for 5 min.
4. Annealing temperatures for each primer are shown in Table 51.1.
5. For cloning, reaction mixtures are cycled 25 times, whereas, for specificity and restriction fragment length polymorphism (RFLP) experiments,
6. The success of amplification is assessed by running the PCRs in 1.5% agarose gels stained with ethidium bromide.

51.2.5 RFLP Analysis (Rantala-Ylinen et al. 2011; Moreira et al. 2017)

1. The 366-bp *ana* C amplicons from four parallel PCRs are combined and purified PCR products with an Amicon Ultra-0.5 kit (Millipore).
2. Check the length of the purified amplicon was by an agarose gel run (1.5% agarose in 0.5× Tris-acetate-EDTA [TAE]).
3. Measure the concentration with a NanoDrop ND-1000 spectrophotometer (NanoDrop Technologies, Inc.).
4. In single-strain reactions, 200 ng of the purified PCR product is digested with 1 μL HhaI or HinfIFastDigest enzyme (Fermentas) in 0.67× FastDigest Green Buffer (Fermentas) in a total volume of 30 μL.
5. In digestions of cyanobacterial strain mixtures, the total amount of amplified DNA is 200 ng for two-, 300 ng for three-, and 400 ng for four-strain mixtures.
6. Incubate reaction mixtures at 37 °C for 15 min and inactivate at 65 °C for 20 min.
7. Separate the restriction fragments in ethidium bromide-stained 3% MetaPhor (FMC BioProducts) or 3% TopVision (Fermentas) agarose gels in 1× Tris-borate-EDTA (TBE) buffer.
8. Gels are documented with a Kodak DC290 camera and the Kodak 1D imaging program, version 3.5.0.

Table 51.1 Annealing temperatures for the primers

Target	Primer	Primer sequence 5–3	Annealing temperature (°C)
ana C(813 bp)	anxC-F anxC-R	TGAGGGAACAAGTGAGTT ATCATCTCCGATCCCAATCC	52
ana C(861 bp)	anxgen-F anxgen-R	ATGGTCAGAGGTTTTACAAG CGACTCTTAATCATGCGATC	52

9. Estimate the sizes of fragments by comparing fragments of the size marker (O'GeneRuler low-range DNA ladder; Fermentas).

References

Aráoz R, Molgó J, Tandeau de Marsac N (2010) Neurotoxic cyanobacterial toxins. Toxicon 56: 813–828

Cadel-Six S, Iteman I, Peyraud-Thomas C, Mann S, Ploux O, Mejean A (2009) Identification of a polyketide synthase coding sequence specific for anatoxin-a-producing Oscillatoria cyanobacteria. Appl Environ Microbiol 75(14):4909–4912

Christensen VG, Khan E (2020) Freshwater neurotoxins and concerns for human, animal, and ecosystem health: a review of anatoxin-a and saxitoxin. Sci Total Environ 736:139515

Colas S, Marie B, Lance E, Quiblier C, Tricoire-Leignel H, Mattei C (2021) Anatoxin-a: overview of a harmful cyanobacterial neurotoxin from the environmental scale to the molecular target. Environ Res 193:110590

Cullen A, Pearson LA, Mazmouz R, Liu T, Soeriyadi AH, Ongley SE, Neilan BA (2019) Heterologous expression and biochemical characterisation of cyanotoxin biosynthesis pathways. Nat Prod Rep 36(8):1117–1136

Dittmann E, Fewer DP, Neilan BA (2013) Cyanobacterial toxins: biosynthetic routes and evolutionary roots. FEMS Microbiol Rev 37(1):23–43

Kasai C, Sugimoto K, Moritani I, Tanaka J, Oya Y, Inoue H, Tameda M, Shiraki K, Ito M, Takei Y, Takase K (2015) Comparison of the gut microbiota composition between obese and non-obese individuals in a Japanese population, as analyzed by terminal restriction fragment length polymorphism and next-generation sequencing. BMC Gastroenterol 15:1–10

Kust A, Méjean A, Ploux O (2020) Biosynthesis of anatoxins in cyanobacteria: identification of the carboxy-anatoxins as the penultimate biosynthetic intermediates. J Nat Prod 83(1):142–151

Moreira C, Matos A, Mendes R, Antunes A (2017) Plant cyanotoxins: molecular methods and current applications. In: Plant Toxins. Springer, Dordrecht, pp 339–360

Pipan B, Zupančič M, Blatnik E, Dolničar P, Meglič V (2018) Comparison of six genomic DNA extraction methods for molecular downstream applications of apple tree (Malus X *domestica).* Cogent Food Agric 4(1):1540094

Rantala-Ylinen A, Känä S, Wang H, Rouhiainen L, Wahlsten M, Rizzi E, Berg K, Gugger M, Sivonen K (2011) Anatoxin-a synthetase gene cluster of the cyanobacterium Anabaena sp. strain 37 and molecular methods to detect potential producers. Appl Environ Microbiol 77(20):7271–7278

NRPS and PKS Gene Analysis in Cyanobacteria Using PCR

52

Nazar Reehana, Mohamed Yousuff Mohamed Imran, Mohamed Iqbal Mohamed Mahroop Raja, N. Thajuddin, and D. Dhanasekaran

Abstract

Cyanobacteria are ancient prokaryotes with the capacity to perform oxygenic photosynthesis and can be found in a great diversity of habitats, including extreme environments that are challenging to most of the life forms in Earth. Cyanobacteria are a prolific source of bioactive compounds with promising therapeutic applications. To further evaluate the potential of cyanobacterial isolates to produce bioactive compounds, a Polymerase Chain Reaction (PCR) screening for the presence of genes encoding non-ribosomal peptide synthetases (NRPS) and polyketide synthases (PKS), targeting the adenylation (A) and ketosynthase (KS) domains, respectively, was performed. In addition to validating the use of degenerate primers for the identification of PKS and NRPS genes in cyanobacteria, this study also defines numerous gene fragments

N. Reehana (✉) · M. I. M. M. Raja
PG and Research Department of Microbiology, Jamal Mohamed College (Autonomous) [Affiliated to Bharathidasan University], Tiruchirappalli, Tamil Nadu, India

M. Y. M. Imran
National Repository for Microalgae and Cyanobacteria-Freshwater (NRMC-F) (Sponsored by the DBT, Government of India), Department of Microbiology, Bharathidasan University, School of Life Sciences, Tiruchirappalli, Tamil Nadu, India

N. Thajuddin
B.S. Abdur Rahman Crescent Institute of Science & Technology (Deemed to be University), Chennai, Tamil Nadu, India

Department of Microbiology, National Repository for Microalgae and Cyanobacteria - Freshwater (NRMC-F), School of Life Sciences, Bharathidasan University, Tiruchirappalli, Tamil Nadu, India

D. Dhanasekaran
Department of Microbiology, National Repository for Microalgae and Cyanobacteria - Freshwater (NRMC-F), Bharathidasan University, Tiruchirappalli, Tamil Nadu, India

N. Thajuddin et al. (eds.), *Protocols for Cyanobacteria Sampling and Detection of Cyanotoxin*, https://doi.org/10.1007/978-981-99-4514-6_52

that will be useful as probes for future studies of the synthesis of natural products in cyanobacteria.

Keywords

Bioactive compounds · Cyanobacteria · Non-Ribosomal Peptide Synthetases (NRPS) · Polyketide Synthases (PKS) · Natural products

52.1 Introduction

Cyanobacteria are a valuable and prolific source of biologically active secondary metabolites (Ehrenreich et al. 2005; Nunnery et al. 2010), and some of these products can be toxic to higher animals including humans. These cyanotoxins can be classified into five functional groups according to their primary target organ or effects being designated as hepatotoxins, neurotoxins, cytotoxins, dermatotoxins, and irritant toxins (Wiegand and Pflugmacher 2005; Dittmann et al. 2013). However, cyanobacteria can also produce secondary metabolites with promising therapeutic properties such as anticancer, antibiotic, and anti-inflammatory (Burja et al. 2001; Singh et al. 2011).

In recent years, many microbiologists and biochemists have studied the distribution and functions of non-ribosomal peptide synthetases (NRPSs) and polyketide synthases (PKSs), two similar molecular systems that are known to be involved in natural product synthesis in many bacteria, fungi, and plants (Cane et al. 1998; Moffitt and Neilan 2003). Both NRPSs and PKSs are large (200–2000 kDa), multifunctional enzymes that possess modular organization (Cane and Walsh 1999). NRPSs use amino acid monomers as substrates for synthesizing complex oligopeptides, whereas PKSs use acyl coenzyme A monomers to form elaborate chemical structures along a ketide backbone. NRPSs and PKSs have been found to synthesize a diverse array of biologically active compounds, including antibiotics, toxins (Cane et al. 1998), siderophores (Crosa and Walsh 2002), and immune suppressants (Cane et al. 1998). Discrete modules in the NRPSs and PKSs mediate the synthesis of their products. Similar to an assembly line, each enzymatic module is responsible for the addition of a single monomer to an elongating chain. These modules, in turn, also possess modular organization, as they consist of a number of enzymatic active sites that can be partitioned into sequence domains at the level of primary structure.

The majority of the bioactive metabolites isolated from cyanobacteria are polyketides (PK), non-ribosomal peptides (NRP) or a hybrid of these two, and they are synthesized by modular multienzymatic complexes, namely type I polyketide synthases (PKS) and non-ribosomal peptide synthetases (NRPS). Type I PKS and NRPS pathways are characterized by a modular organization in which each module is responsible for the incorporation of a structural carboxylic acid or amino acid unit, respectively (Kehr et al. 2011; Barrios-Llerena et al. 2007; Wase and Wright 2008). Although the scientific community is generally aware of the prevalence of NRPSs and PKSs among the cyanobacteria (Christiansen et al. 2001;

Moffitt and Neilan 2001; Neilan et al. 1999), the distribution and diversity of these genes across the phylum have to be well characterized.

52.2 Materials

1. **STE Buffer (50×)**
   - 50 mM Tris (pH–8.0)
   - 5 mM EDTA (pH–8.0)
   - 50 mM NaCl
2. **Lysozyme**
   - Dissolve 10 mg of lysozyme in 10 mL of double distilled water.
3. **4 M Ammonium Acetate (50 mL)**
   - Dissolve 15.416 g ammonium acetate in 50 ml distilled water.
4. **TE Buffer (10×)—Stock Solution**
   - 1 M Tris-HCl—100 mL (pH 7.5)
   - 500 mM EDTA—20 mL (pH 8.0)
   - Distilled water—880 mL
     – For 1× TE buffer, add 450 mL of distilled water in 50 mL of 10× buffer
5. **TBE Buffer (5×)—Stock Solution**
   - Tris base—54 g
   - Borate—27.5 g
   - 0.5 M EDTA (shake vigorously before use)—2.93gm (or) 20 mL
     – Make the volume to 1 L using double distilled water and adjust up to 8.5 using KOH.
     – Prepare 1× TBE buffer by adding 20 mL of 5× TBE buffer in 80mL of distilled water.
6. **Agarose gel (0.8%)**
   - Add 0.4 g of agarose in 50 mL of 1× TBE buffer.
7. **Primers used for Amplification and Sequencing of NRPS gene**

Primer	Sequence(5′-3′)	Reference
MTF2	GCNGGYGGYGCNTAYGTNCC	(Neilan et al. 1999)
MTR	CCNCGDATYTTNACYTG	

8. **Primers used for Amplification and Sequencing of PKS gene**

Primer	Sequence(5′-3′)	Reference
DKF	GTGCCGGTNCCRTGNGYYTC	(Moffitt and Neilan 2001)
DKR	GCGATGGAYCCNCARCARYG	

9. **Composition of Prime Taq premix**
   - *Prime Taq*TM DNA Polymerase 1 unit/10 μL, 20 mM Tris-HCl, 80 mM KCl, 4 mM $MgCl_2$, enzyme stabilizer, sediment, loading dye, pH 9.0, 0.5 mM of each dATP, dCTP, dGTP, dTTP

52.3 Methods

52.3.1 Molecular Characterization of Cyanobacteria

52.3.1.1 Extraction of DNA from Cyanobacteria (Smoker and Barnum 1988)

1. Take 1 mL of overnight grown cyanobacterial culture and centrifuged at 10,000 rpm for 5 min and collect the pellet.
2. Wash the collected pellet using 500 μL of STE buffer and vortex it.
3. Add 20 μL of lysozyme (10 mg/mL) and incubate in water bath at 55 °C for 30 min.
4. After incubation, add 10 μL of proteinase K (10 mg/mL) and 20 μL of 10% SDS and incubate in water bath at 60 °C for 1–2 h.
5. Allow the mixture to cool in ice and add equal volume of Phenol: Chloroform: Isoamyl alcohol (25:24:1).
6. Centrifuge the above mixture at 12,000 rpm for 10 min and collect the supernatant with care.
7. To the supernatant add equal volume of 4 M Ammonium acetate and two volumes of Isopropanol.
8. Again centrifuge the mixture at 14,000 rpm for 15 min and discard the supernatant, collect the pellet and wash with 70% ethanol then dry it.
9. Finally, dissolve the DNA pellet in 100 μL of TE buffer and store at 20 °C until further use.
10. To each DNA samples, add the loading dye and transfer the samples to respective wells in 0.8% agarose gel to carry out the electrophoresis using 1X TBE buffer at a constant supply of 80 V for 1 h.
11. When the sample reaches the end of the gel, disconnect the supply and visualize the bands under Gel Documentation system.

52.3.2 NRPS and PKS Gene Amplification by Polymerase Chain Reaction (PCR)

52.3.2.1 Procedure

1. Carry out the PCR amplification for the respective purified DNA samples.
2. For NRPS and PKS gene amplification, carry out the PCR with the following holds and cycles: 94 °C for 5 min; 35 cycles of 94 °C for 1 min, 50 °C for 1 min for NRPS or 50.8 °C for 1 min for PKS, and 72 °C for 2 min for NRPS or 72 °C

for 1 min for PKS; and 72 °C for 7 min using the respective primers MTF2 and MTR for NRPS gene, DKF and DKR for PKS gene.

3. 40 μL of reaction mixture for amplification of NRPS and PKS contains

Prime Taq premix	20 μL
Forward primer (10 pmoles/μL)	1 μL
Reverse primer (10 pmoles/μL)	1 μL
DNA (50 ng)	1 μL
Distilled water	17 μL
	40 μL

6. Carry out by electrophoresis on 1.2% agarose gel for the amplified products using 1X TBE buffer at a constant supply of 100 V for 30 min.
7. End the electrophoresis when the sample reaches the end of the gel.

52.4 Observation

Observe the bands under Gel Documentation System.

Acknowledgment The authors, Nazar Reehana and Mohamed Iqbal Mohamed Mahroop Raja, acknowledges the Secretary and Correspondent, the Principal and PG & Research Department of Microbiology, Jamal Mohamed College (*Autonomous*), Tiruchirappalli, Tamilnadu, India for the inherent support in preparing this data.

References

Barrios-Llerena ME, Burja AM, Wright PC (2007) Genetic analysis of polyketide synthase and peptide synthetase genes in cyanobacteria as a mining tool for secondary metabolites. J Ind Microbiol Biotechnol 34:443–456

Burja AM, Banaigs B, Abou-Mansour E, Burgess JG, Wright PC (2001) Marine cyanobacteria–a prolific source of natural products. Tetrahedron 57:9347–9377

Cane DE, Walsh CT (1999) The parallel and convergent universes of polyketide synthases and nonribosomal peptide synthetases. Chem Biol 6:R319–R325

Cane DE, Walsh CT, Khosla C (1998) Harnessing the biosynthetic code: combinations, permutations, and mutations. Science 282:63–68

Christiansen G, Dittmann E, Via Ordorika L, Rippka R, Herdman M, Borner T (2001) Nonribosomal peptide synthetase genes occur in most cyanobacterial genera as evidenced by their distribution in axenic strains of the PCC. Arch Microbiol 176:452–458

Crosa JH, Walsh CT (2002) Genetics and assembly line enzymology of siderophore biosynthesis in bacteria. Microbiol Mol Biol Rev 66:223–249

Dittmann E, Fewer DP, Neilan BA (2013) Cyanobacterial toxins: biosynthetic routes and evolutionary roots. FEMS Microbiol Rev 37:23–43

Ehrenreich IM, Waterbury JB, Webb EA (2005) Distribution and diversity of natural product genes in marine and freshwater cyanobacterial cultures and genomes. Appl Environ Microbiol 71: 7401–7413

Kehr JC, Gatte Picchi D, Dittmann E (2011) Natural product biosyntheses in cyanobacteria: a treasure trove of unique enzymes. Beilstein J Org Chem 7:1622–1635

Moffitt MC, Neilan BA (2001) On the presence of peptide synthetase and polyketide synthase genes in the cyanobacterial genus Nodularia. FEMS Microbiol Lett 196:207–214

Moffitt MC, Neilan BA (2003) Evolutionary affiliations within the superfamily of ketosynthases reflect complex pathway associations. J Mol Evol 56:446–457

Neilan BA, Dittmann E, Rouhiainen L, Bass RA, Schaub V, Sivonen K, Borner T (1999) Nonribosomal peptide synthesis and toxigenicity of cyanobacteria. J Bacteriol 181:4089–4097

Nunnery JK, Mevers E, Gerwick WH (2010) Biologically active secondary metabolites from marine cyanobacteria. Curr Opin Biotechnol 21:787–793

Singh RK, Tiwari SP, Rai AK, Mohapatra TM (2011) Cyanobacteria: an emerging source for drug discovery. J Antibiot (Tokyo) 64:401–412

Smoker JA, Barnum SR (1988) Rapid small-scale DNA isolation from filamentous cyanobacteria. FEMS Microbiol Lett 56(1):119–122

Wase NV, Wright PC (2008) Systems biology of cyanobacterial secondary metabolite production and its role in drug discovery. Expert Opin Drug Discovery 3:903–929

Wiegand C, Pflugmacher S (2005) Ecotoxicological effects of selected cyanobacterial secondary metabolites: a short review. Toxicol Appl Pharmacol 203:201–218

Identification of Microcystin, Nodularin Synthatase Gene Clusters in Toxic Cyanobacteria Using AntiSMASH Pipeline

53

Fayaazuddin Thajuddin, Asraf Sithikka Rasheed, N. Thajuddin, and D. Dhanasekaran

Abstract

A wide variety of toxins produced by cyanobacteria can cause severe health problems or even death in both people and animals. These cyanotoxins include neurotoxins, hepatotoxins, and dermatotoxins produced by the recurrent mass development of harmful algal blooms (HABs). Hepatotoxin microcystin and nodularin synthetase are the most commonly reported cyanotoxins. Aquatic food chains are known to be significantly impacted by microcystin bioaccumulation in aquatic life forms. Additionally, cyanotoxin biodegradation is very slow. Both microcystin and nodularin inhibit protein phosphate, lead to liver failure and hepatic hemorrhage. Consequently, it is rapidly developing into environmental and public concern. Thus, analysis of cyanotoxin gene clusters plays a vital role in identification of toxin-producing cyanobacteria. We describe a technique for locating gene clusters involved in the biosynthesis of cyanotoxins in the whole genome sequence of cyanobacterial species using the antiSMASH pipeline.

F. Thajuddin · A. S. Rasheed (✉)
Present Address: School of Life Sciences, Bharathidasan University, Tiruchirappalli, Tamil Nadu, India

N. Thajuddin
B.S. Abdur Rahman Crescent Institute of Science & Technology (Deemed to be University), Chennai, Tamil Nadu, India

Department of Microbiology, National Repository for Microalgae and Cyanobacteria - Freshwater (NRMC-F), School of Life Sciences, Bharathidasan University, Tiruchirappalli, Tamil Nadu, India

D. Dhanasekaran
Department of Microbiology, National Repository for Microalgae and Cyanobacteria - Freshwater (NRMC-F), Bharathidasan University, Tiruchirappalli, Tamil Nadu, India

N. Thajuddin et al. (eds.), *Protocols for Cyanobacteria Sampling and Detection of Cyanotoxin*, https://doi.org/10.1007/978-981-99-4514-6_53

Keywords

Microcystin · Nodularin synthetase · Whole genome sequence · antiSMASH

53.1 Introduction

Cyanobacteria are the most abundant autotrophic organisms on the earth and found in diverse environment, especially in the marine and freshwater. Many genera of cyanobacteria undergo rapid multiplication and their substantial growth can cause significant problems in management of inland waters. They are known to produce compounds like geosmin, 2-methylisoborneol causing dreadful odor and taste in drinking water (Falconer 1999). These cyanotoxins include neurotoxins, hepatotoxins, and dermatotoxins. As they have been linked with several animal and human poisonings (Briand et al. 2003; Griffiths and Saker 2003; Chorus and Welker 2021; Pouria et al. 1998). Humans are potentially vulnerable to cyanotoxins via regular activities like consumption and bathing in contaminated water resources. Increasing algal blooms, particularly blooms with harmful properties are termed as harmful algal blooms (HABs), and it is considered a risk to global water security (Brooks et al. 2016; Huisman et al. 2018; Arulprakasam and Dharumadurai 2021).

Cyanotoxins are mostly endotoxins and its liberation into the surrounding is dependent on suitable conditions and bloom growth stage (Pearson et al. 2010). CyanoHABs are influenced by toxigenic cyanobacterial genera such as *Microcystis, Cylindrospermopsis, Planktothrix*, and *Anabaena*. They are characterized by gene sequences encoding the development of cyanotoxins (Hisbergues et al. 2003; Van Dolah 2000). Hepatoxin microcystin (MCY) is the most commonly reported cyanobacterial toxin. Microcystin's basic chemical structure is cyclo (D-Ala-X-D-MeAsp-Z-Adda-DGlu-Mdha) (Botes et al. 1984), and it is usually produced in freshwater cyanobacteria like *Microcystis, Anabaena*, and *Planktothrix (*Chorus and Welker 2021*)*. Production of microcystin is more under nitrogen depleted condition in *Nodularia*, *Anabaena*, and *Aphanizomenon*, while *Microcystis* and *Planktothrix* exhibit higher microcystin production under nitrogen replete condition (Van Apeldoorn et al. 2007). Aquatic food chains are known to be significantly impacted by microcystin bioaccumulation in aquatic life forms. Additionally, cyanotoxin biodegradation is very slow (Dittmann et al. 1997). Detection of microcystin synthetase (mcy) gene cluster in PCC 7806 (Tillett et al. 2000; Al-Tebrineh et al. 2010) facilitated the advancement of highly distinct and sensitive polymerase chain reaction (PCR)-based techniques for evaluating the toxigenicity of bloom specimens (Al-Tebrineh et al. 2011; Baker et al. 2013; Moffitt and Neilan 2004).These techniques are now widely utilized by scientists and water quality managements worldwide. Nodularins have a close relationship to microcystin's and typically have cyclo (D-MeAsp-L-Arg-Adda-D-Glu-Mdhb) structure. The biosynthetic pathway of nodularin gene clusters was characterized for *Nodularia spumigena* using genome walking method. The nodularin synthetase evolved from a microcystin synthetase through a deletion event and a change in substrate selectivity, as demonstrated by the elucidation of the nda gene cluster (William and Laurens 2010).

Presently, the leading innovations in the field of microbial natural product research was the complementation of classical isolation and analytical techniques with genome mining methods. Thus, it enables the identification and characterization of the biosynthetic pathways based on (meta-) genome data (Villebro et al. 2019). antiSMASH is released in 2010 and it is considered as one of most popular software used for finding gene clusters important for the biosynthesis of secondary metabolites. It is a rapid and reliable source for genome analysis (Medema et al. 2011). antiSMASH analysis also provides detailed description of the secondary metabolites predicted. Thus, the toxin-producing gene clusters in the whole genome sequence of the cyanobacteria are revealed (Mushtaq et al. 2018). Genome-based drug discovery approach is applied to exhibit gene clusters that produce bioactive compounds and to propose novel therapeutic drugs. At this consideration, tools like antiSMASH that determines biosynthetic gene clusters have attained substantial importance (Shunmugam et al. 2017; Gehringer et al. 2012).

53.2 Materials Required

1. antiSMASH bacterial version 6.0 tool.
2. Whole-genome sequences of cyanobacterial species.
3. National Center for Biotechnology Information (NCBI) database.

53.3 Procedure

1. Use the following URL to access NCBI: https://www.ncbi.nlm.nih.gov/

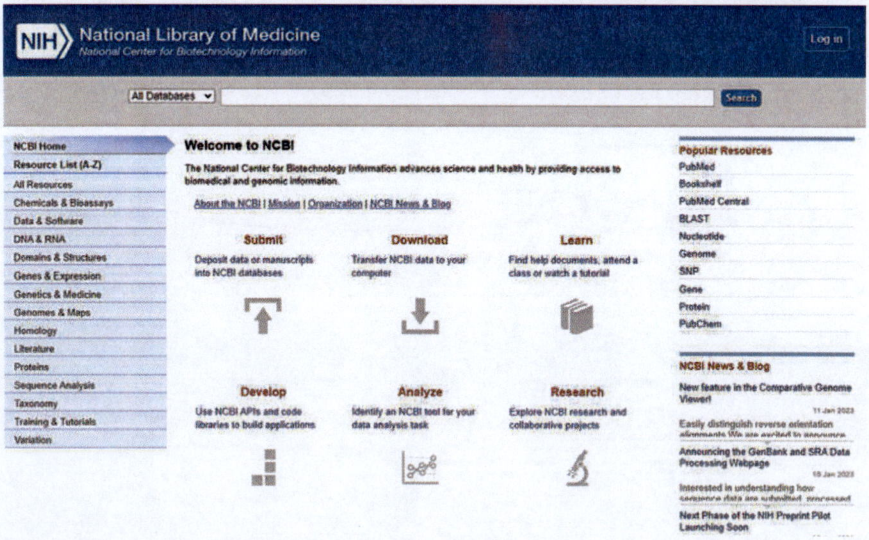

2. Utilize the nucleotide filter to look for the cyanobacteria *Microcystis aeruginosa*.
3. Copy the accession number or download the respective species whole genomic sequence.

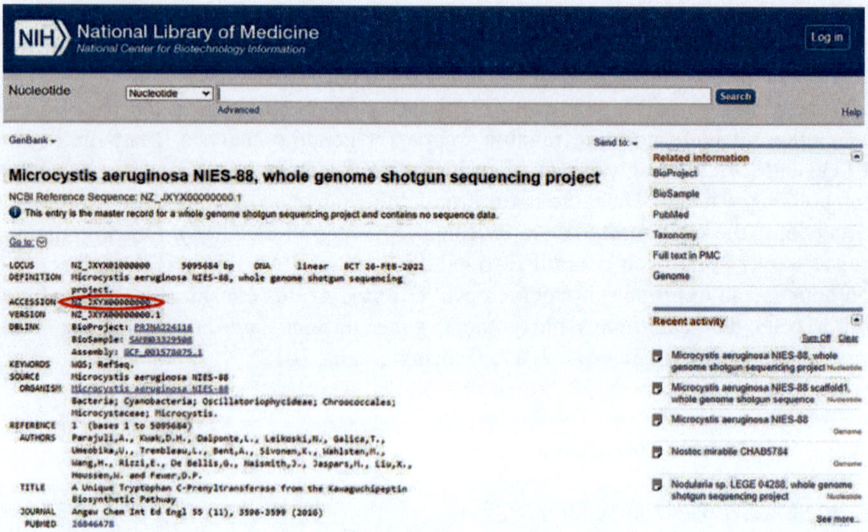

4. Use the following link to launch the antiSMASH bacterial version 6.0 tool: https://antismash.secondarymetabolites.org/#!/start
5. Input the gene accession number or upload the downloaded whole genome sequence in the data input box.

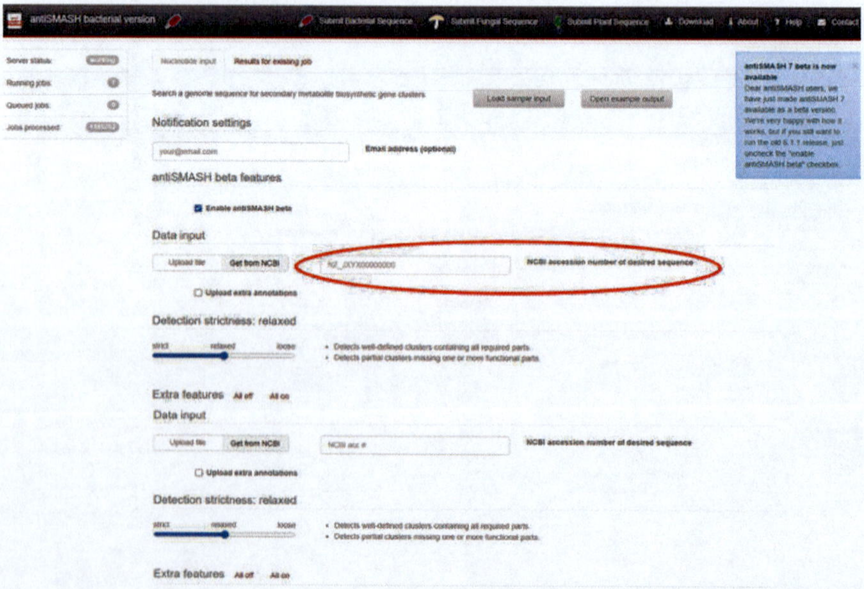

6. Select the submit option and wait for the antiSMASH tool to execute and evaluate.
7. Select the show results option when finished and make any necessary observations.

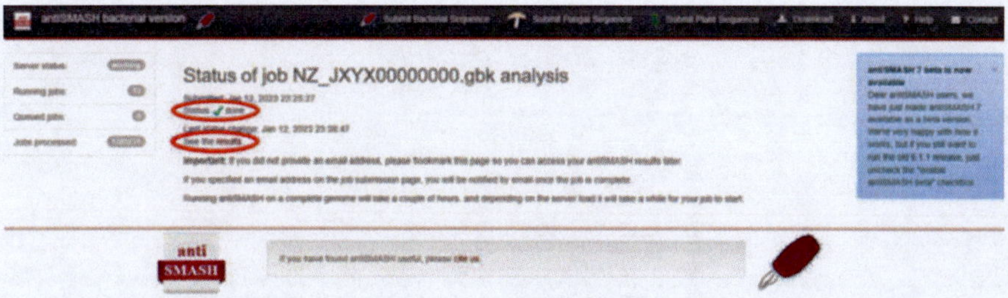

8. Study and interpret the cyanotoxin-producing biosynthetic gene clusters found in the whole genome sequence of *Microcystis* sp.

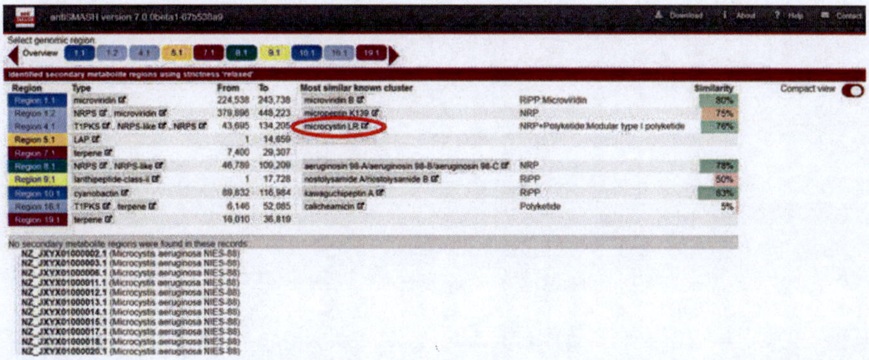

9. Same protocol is followed for whole genome sequence of other cyanobacteria species to check for the presence of microcystin and nodularin synthetase gene clusters using antiSMASH.

53.4 Conclusions

A total of 10 cyanobacteria species is selected to study the presence of cyanotoxin-producing microcystin and nodularin synthetase gene cluster in its whole genome sequence. The species selected are as follows: *Oscillatoria salina*, *Phormidium tenue, Lyngbya aestuarii, Microcystis aeruginosa, Microcystis flos-aquae, Nodularia spumigena, Nostoc linkia, Nostoc mirabile, Planktothrix agardhii* and *Planktothrix pseudoagardhii* (Table 53.1).

Out of these 10 cyanobacteria species, the presence of microcystin gene cluster is observed in *Microcystis aeruginosa, Microcystis flos-aquae, Nodularia spumigena,*

Table 53.1 Cyanotoxin gene clusters in whole genome sequence of cyanobacteria species

S. No	Microalgae and cyanobacteria species	Accession number	Cyanotoxin gene cluster	Location of gene cluster in WGS (nt)	Total no. of nucleotide (nt)
1	*Microcystis aeruginosa*	NZ_JXYX00000000	Microcystin	43,695–134,205	90,511
2	*Microcystis flos-aquae*	NZ_JACJSW000000000	Microcystin	4433–95,128	90,696
3	*Nodularia spumigena*	NZ_CP020114	Microcystin	5,060,300–5,143,361	830,692
4	*Nostoc linkia*	NZ_LAHB00000000	Microcystin	44,841–86,679	41,839
5	*Nostoc mirabile*	NZ_JAIVFS000000000	Microviridin	15,699–35,986	20,288
6	*Planktothrix agardhii*	NZ_LR882952	Microcystin	2,151,312–2,241,461	90,150
7	*Planktothrix pseudoagardhii*	NZ_LR882967	Microcystin	1,257,078–1,320,718	63,641
8	*Oscillatoria salina*	NZ_JAAHBQ000000000	Absent		
9	*Phormidium tenue*	NZ_MRCG00000000			
10	*Lyngbya aestuarii*	NZ_AUZM00000000			

Fig. 53.1 Structure of microcystin

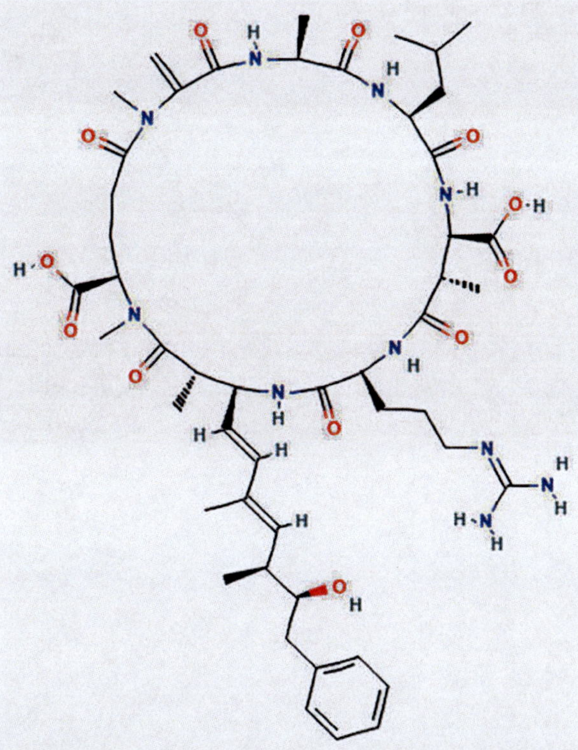

Nostoc linkia, Planktothrix agardhii and *Planktothrix pseudoagardhii.* While *Oscillatoria salina*, *Phormidium tenue*, and *Lyngbya aestuarii* lacked the presence of cyanotoxin gene clusters. *Nostoc mirabile* exhibited the presence of microviridin gene cluster. Nodularin synthetase gene clusters were not observed in any of the above selected species. The structure of microcystin and nodularin synthetase gene clusters are similar, but the size of both gene clusters are 55 kb and 48 kb, respectively (Figs. 53.1 and 53.2) [27]. Thus, utilization of antiSMASH pipeline is effective in locating cyanotoxin gene clusters in the whole genome sequence of cyanobacterial species.

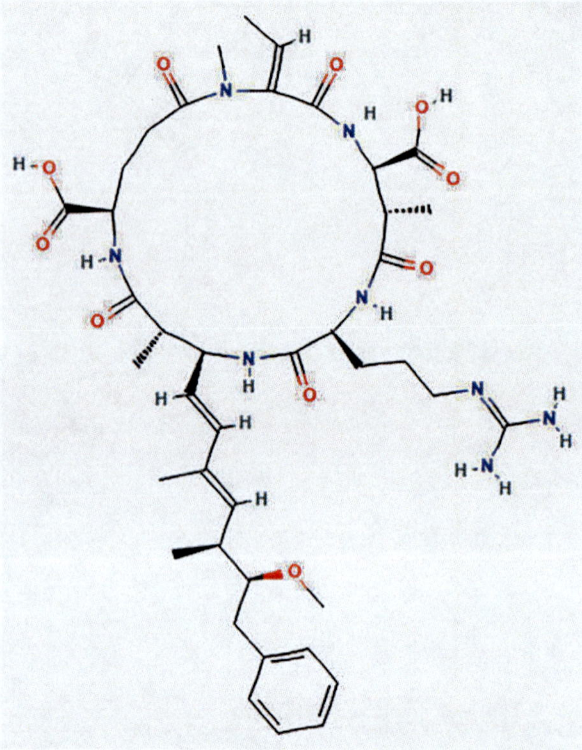

Fig. 53.2 Structure of nodularin

References

Al-Tebrineh J, Mihali TK, Pomati F, Neilan BA (2010) Detection of saxitoxin-producing cyanobacteria and Anabaena circinalis in environmental water blooms by quantitative PCR. Appl Environ Microbiol 76(23):7836–7842

Al-Tebrineh J, Gehringer MM, Akcaalan R, Neilan BA (2011) A new quantitative PCR assay for the detection of hepatotoxigenic cyanobacteria. Toxicon 57(4):546–554

Arulprakasam KR, Dharumadurai D (2021) Genome mining of biosynthetic gene clusters intended for secondary metabolites conservation in actinobacteria. Microb Pathog 161:105252

Baker L, Sendall BC, Gasser RB, Menjivar T, Neilan BA, Jex AR (2013) Rapid, multiplex-tandem PCR assay for automated detection and differentiation of toxigenic cyanobacterial blooms. Mol Cell Probes 27(5–6):208–214

Botes DP, Tuinman AA, Wessels PL, Viljoen CC, Kruger H, Williams DH, Santikarn S, Smith RJ, Hammond SJ (1984) The structure of cyanoginosin-LA, a cyclic heptapeptide toxin from the cyanobacterium Microcystis aeruginosa. J Chem Soc Perkin Trans 1:2311–2318

Briand JF, Jacquet S, Bernard C, Humbert JF (2003) Health hazards for terrestrial vertebrates from toxic cyanobacteria in surface water ecosystems. Vet Res 34(4):361–377

Brooks BW, Lazorchak JM, Howard MD, Johnson MVV, Morton SL, Perkins DA, Reavie ED, Scott GI, Smith SA, Steevens JA (2016) Are harmful algal blooms becoming the greatest inland water quality threat to public health and aquatic ecosystems? Environ Toxicol Chem 35(1):6–13

Chorus I, Welker M (2021) Toxic cyanobacteria in water: a guide to their public health consequences, monitoring and management. Taylor & Francis, Oxfordshire, London, p 858

Dittmann E, Neilan BA, Erhard M, Von Döhren H, Börner T (1997) Insertional mutagenesis of a peptide synthetase gene that is responsible for hepatotoxin production in the cyanobacterium Microcystis aeruginosa PCC 7806. Mol Microbiol 26(4):779–787

Falconer IR (1999) An overview of problems caused by toxic blue-green algae (cyanobacteria) in drinking and recreational water. Environ Toxicol 14(1):5–12

Gehringer M, Adler L, Roberts A et al (2012) Nodularin, a cyanobacterial toxin, is synthesized in planta by symbiotic Nostoc sp. ISME J 6:1834–1847. https://doi.org/10.1038/ismej.2012.25

Griffiths DJ, Saker ML (2003) The Palm Island mystery disease 20 years on: a review of research on the cyanotoxin cylindrospermopsin. Environ Toxicol 18(2):78–93

Hisbergues M, Christiansen G, Rouhiainen L, Sivonen K, Börner T (2003) PCR-based identification of microcystin-producing genotypes of different cyanobacterial genera. Arch Microbiol 180(6):402–410

Huisman J, Codd GA, Paerl HW, Ibelings BW, Verspagen JM, Visser PM (2018) Cyanobacterial blooms. Nat Rev Microbiol 16(8):471–483

Medema MH, Blin K, Cimermancic P, De Jager V, Zakrzewski P, Fischbach MA, Weber T, Takano E, Breitling R (2011) antiSMASH: rapid identification, annotation and analysis of secondary metabolite biosynthesis gene clusters in bacterial and fungal genome sequences. Nucleic Acids Res 39(suppl_2):W339–W346

Moffitt MC, Neilan BA (2004) Characterization of the nodularin synthetase gene cluster and proposed theory of the evolution of cyanobacterial hepatotoxins. Appl Environ Microbiol 70(11):6353–6362

Mushtaq S, Abbasi BH, Uzair B, Abbasi R (2018) Natural products as reservoirs of novel therapeutic agents. EXCLI J 17:420

Pearson L, Mihali T, Moffitt M, Kellmann R, Neilan B (2010) On the chemistry, toxicology and genetics of the cyanobacterial toxins, microcystin, nodularin, saxitoxin and cylindrospermopsin. Mar Drugs 8(5):1650–1680

Pouria S, de Andrade A, Barbosa J, Cavalcanti RL, Barreto VTS, Ward CJ, Preiser W, Poon GK, Neild GH, Codd GA (1998) Fatal microcystin intoxication in haemodialysis unit in Caruaru, Brazil. Lancet 352(9121):21–26

Shunmugam S, Gayathri M, Prasannabalaji N, Thajuddin N, Muralitharan G (2017) Unraveling the presence of multi-class toxins from *Trichodesmium* bloom in the Gulf of Mannar region of the Bay of Bengal. Toxicon 135:43–50

Tillett D, Dittmann E, Erhard M, Von Döhren H, Börner T, Neilan BA (2000) Structural organization of microcystin biosynthesis in Microcystis aeruginosa PCC7806: an integrated peptide–polyketide synthetase system. Chem Biol 7(10):753–764

Van Apeldoorn ME, Van Egmond HP, Speijers GJ, Bakker GJ (2007) Toxins of cyanobacteria. Mol Nutr Food Res 51(1):7–60

Van Dolah FM (2000) Marine algal toxins: origins, health effects, and their increased occurrence. Environ Health Perspect 108(suppl 1):133–141

Villebro R, Shaw S, Blin K, Weber T (2019) Sequence-based classification of type II polyketide synthase biosynthetic gene clusters for antiSMASH. J Ind Microbiol Biotechnol 46(3–4): 469–475

William PLIB, Laurens LML (2010) Microalgae as biodiesel and biomass feedstock: review and analysis of the biochemistry, energetic and economics. Energy Environ Sci 3:554–590

54 qPCR Assay in *sxtA* Gene in Saxitoxin-Producing Cyanobacteria

Zhongkun Zhou, Yunhao Ma, Yuqing Niu, Juan Lu, Lixue Tu, and Peng Chen

Abstract

Saxitoxin, as one type of paralytic shellfish toxin, can bind to sodium channels and modulate the flux of sodium in nerve and muscle cells. Saxitoxin-producing cyanobacteria cause serious threat to fishery and food safety, resulting in huge economic losses and thousands of deaths each year. Researches of natural product biosynthesis and genomics gradually elucidate the saxitoxin biosynthesis pathway. *sxtA* gene, encoding polyketide synthase (PKS)-like enzymes, initiates the biosynthesis, and provides an excellent opportunity for the detection of saxitoxin by molecular biology technology. Especially, quantitative PCR based on fluorescent dye, probe or multiple primers is accurate, fast, and cost-effective, thus providing an early warning system to detect saxitoxin-producing cyanobacteria before toxic metabolites are produced and released into the water. This chapter summarized the experimental principle, material preparation, operation process, and matters needing attention of SYBR Green-qPCR assay in *sxtA* gene detection of saxitoxin-producing cyanobacteria.

Keywords

qPCR · SYBR Green · sxtA · Saxitoxin · Cyanobacteria

Z. Zhou · Y. Ma · Y. Niu · J. Lu · L. Tu · P. Chen (✉)
School of Pharmacy, Lanzhou University, Lanzhou, People's Republic of China
e-mail: chenpeng@lzu.edu.cn

N. Thajuddin et al. (eds.), *Protocols for Cyanobacteria Sampling and Detection of Cyanotoxin*, https://doi.org/10.1007/978-981-99-4514-6_54

54.1 Introduction

Saxitoxinis a kind of paralytic shellfish toxin produced by freshwater cyanobacteria and marine dinoflagellates. It can block voltage-gated Na^+ channels and is acknowledged as one of the most potent neurotoxic alkaloids (Cusick and Sayler 2013). Saxitoxin causes about 2000 deaths per year and huge economic losses to fishery with the cost exceeding 895 million dollars (Wang et al. 2016; Orr et al. 2013). Therefore, monitoring saxitoxin-producing cyanobacteria or dinoflagellates is of great significance for food safety and aquaculture safety.

In the past decades, various methods have been developed for detecting saxitoxin, such as mouse bioassay, high-performance liquid chromatography (HPLC) (Watanabe et al. 2014), liquid chromatography-mass spectrometry (LC-MS) (Yue et al. 2020), enzyme-linked immunosorbent assay (ELISA) (Wharton et al. 2017), and LC–fluorescence detection (LC-FLD) method. With the advances in omics technology and bioinformatics, the biosynthetic pathway of saxitoxin has been clarified (Akbar et al. 2020). Especially, *sxtA* gene, as the starting point of biosynthesis, is evolutionarily conservative and provides an excellent opportunity to detect saxitoxin through identification of the core genes (Moustafa et al. 2009).

Being different from the detection of metabolites, emerging molecular methods can detect saxitoxin from cyanobacteria before they are produced and released into water. Recently, qPCR has been applied to detect toxic cyanobacteria and used as an early warning system. Additionally, this method is based on community DNA extracted from environmental samples, then it can provide an overview of the diversity of cyanobacteria (Pacheco et al. 2016). Quantification of genes involved in saxitoxin synthesis is cost-effective and can realize fast real-time dynamic detection. In the post COVID-19 era, related qPCR equipment, the training of professionals, and the production and transportation of raw materials have all been developed unprecedentedly. Thus, it is more convenient to monitor saxitoxin-producing cyanobacteria using qPCR.

In previous studies, traditional 16S rRNA gene, internal transcribed spacer (ITS) and phycocyanin loci were used to differentiate toxic strains, yet no general pattern was revealed. Until the characterization of saxitoxin biosynthetic gene cluster, specific primers targeting *sxtA* gene sequence were designed (Kellmann et al. 2008; Pearson and Neilan 2008). Using 16S rRNA gene sequence conserved in cyanobacteria as internal control, we can realize relative quantification of saxitoxin. Based on different technical principles and detection throughput, it can be classified as SYBR Green-qPCR (Al-Tebrineh et al. 2010), TaqMan probe-qPCR (Campo et al. 2013), and multiple qPCR (Al-Tebrineh et al. 2012). Taking the SYBR Green-qPCR as an example, briefly, Mix solution (SYBR Green I dye, DNA Polymerase, dNTP, Mg^{2+}), RNase-free ddH_2O, ROX reference dye, specific primers, and genomic DNA consist of the PCR reaction system. Under specific reaction conditions, *sxtA* gene in the genomic DNA is exponentially amplified. SYBR Green I is a dye that only combines with the small groove of double-stranded DNA (dsDNA) instead of single-stranded DNA strand. Moreover, it does not emit fluorescence in the free state, while emitting light when mixed into the DNA double strand. Thus, in the PCR

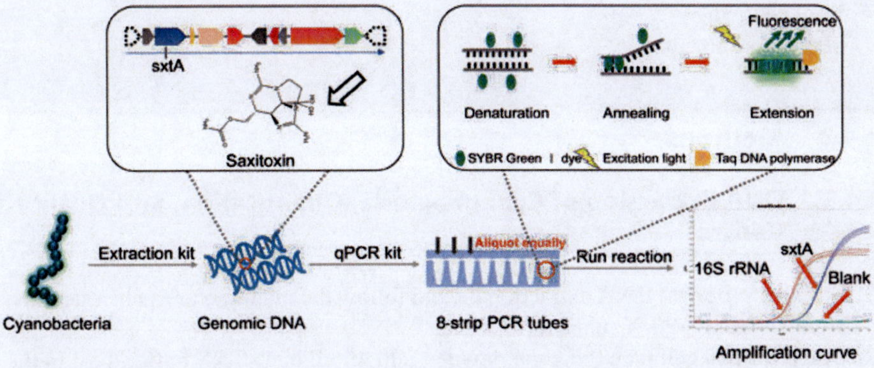

Fig. 54.1 Principle of qPCR assay for detecting *sxtA* gene of saxitoxin-producing cyanobacteria

reaction system, SYBR Green is mixed into dsDNA at the extension stage of each cycle. With the exponential amplification of specific PCR products, the fluorescence signal intensity is positively correlated with the number of PCR products (Fig. 54.1). When the fluorescence can be detected above the background signal, the cycle number is defined as cycle threshold (Ct) value. According to the $2^{-\Delta\Delta Ct}$ method, we can obtain the content of *sxtA* gene in the genome. The lower Ct value represents higher content of *sxtA* gene, while higher Ct value represents lower content of *sxtA* gene.

54.2 Materials

- dd H_2O
- Bacterial Genome Extraction Kit
- 6xDNA loading buffer
- DNA marker (e.g., 100 ~ 5000 bp)
- Agarose
- Microwave oven
- Nucleic acid dye (e.g., Gel-red)
- TAE buffer
- Gel electrophoresis apparatus
- Gel Imager
- Nanodrop
- qPCR kit (containing Mix solution, Rox dye)
- Forward and reverse primers of *sxtA* gene
- 96-well PCR tube box
- Ice bag or ice maker
- 0.2 mL PCR tubes
- 0.2 mL 8-strip PCR tubes and caps
- 1–10 μL, 10–100 μL, 20–200 μL pipettes

- Mini centrifuge
- qPCR instrument

54.3 Methods

54.3.1 DNA Extraction of Cyanobacteria, Quantitation, and Quality Control

1. Select appropriate DNA extraction kit and follow the manufacturer's instructions, store DNA at −20 °C until further use.
2. Prepare 1.5% gel: take 0.6 g gel powder, add 40 ml of 1×TAE buffer, heat in the microwave oven, then add 4 μL nucleic acid dye, mix well, and pour it into the glue making tank, then cool for 30 min.
3. Take a PCR tube, add 5 μL DNA and 1 μL loading buffer.
4. Add 3 μL DNA marker in the gel and then add 6 μL DNA (5 μL DNA and 1 μL loading buffer), run in the gel electrophoresis apparatus for about 30 min.
5. Put the gel in Gel Imager and scan image. Complete and bright DNA band is qualified.
6. Add 1 μL DNA in the Nanodrop to determine the concentration (ng/μL).

54.3.2 Preparation of the qPCR Reaction System and Testing

1. The reaction volume is 10 μL (Table 54.1); the concentration of both forward and reverse primers are 0.3 μM; the final concentration of both Mix solution and ROX dye are 1×.

2. Each sample and gene should set three duplicate wells. Meanwhile, negative control should be set each time. To reduce operating error, it is better to add DNA and other reagents separately, and mix the reagents needed for all wells and divided into aliquots. To reduce error caused by dispensing process, it is better to add more regents than theoretical required volume. All the operations should be carried out on ice.

Table 54.1 Reaction system of qPCR

Reagent	Volume (μL)
2×Mix solution	5.0
10 μM forward primer (F)	0.3
10 μM reverse primer (R)	0.3
DNA	1.0
50×ROX dye	0.2
RNase-free ddH_2O	3.2

3. For example, we have one DNA sample and use 16S rRNA gene as internal reference. We need three sample wells, three internal reference wells and three negative control wells. Take five PCR tubes and mark 1–5, respectively. Add 3.5 μL DNA sample to tube 1–2 and 3.5 μL ddH_2O to tube 3, respectively. Add 40 μL Mix solution+2.4 μL F + 2.4 μL R + 1.6 μL ROX + 25.6 μL H_2O to tube 4, add 12 μL Mix solution+0.9 μL F + 0.9 μL R + 0.8 μL ROX + 12.8 μL H_2O to tube 5. Mix well and take 31.5 μL solution from tube 4 and add it to tube 1 and tube 2, respectively. Take 31.5 μL solution from tube 5 and add it to tube 3. Finally, mix well and aliquot the solution of tube 1–3 into three wells of 8-strip PCR tubes (10 μL each well), respectively. Cover the 8-strip PCR tubes with caps.
4. Put the 8-strip PCR tubes into the Mini centrifuge to centrifuge for 3 s.
5. Put the 8-strip PCR tubes into the qPCR instrument, set the reaction conditions (e.g., 95 °C 15 min, 40 cycles for 95 °C 10 s, and 60 °C 30 s, finally 95 °C 60 s, 60 °C 60 s, 95 °C 60 s).

54.3.3 Calculation of Experimental Results

1. Calculate the mean Ct of each group of *sxtA* gene (mCt_{sxtA}) and internal reference genes (mCt_{16S}).
2. Calculate $\Delta Ct = mCt_{sxtA} - mCt_{16S}$, and ΔCt represents the relative content of *sxtA* gene in genomic DNA.

54.4 Observation/any Special Note (If Any)

1. From preparing sample solution to testing on the machine, it is better to operate on ice and keep away from light.
2. 10 μL reaction volume is more economic, while 20 μL reaction volume is more accurate.
3. The single peak of the melting curve indicates that the primer's specificity is good.
4. To determine the detection limit of this method, the plasmid containing the *sxtA* gene sequence should be constructed and diluted with ten times concentration gradient (at least 8 gradients) to establish the standard curve.
5. $E = 10^{(-1/k)} - 1$, E represents the amplification efficiency, k represents the slope of the standard curve. When the amplification efficiency is between 90 ~ 110%, the reaction system is qualified.
6. It is better to keep the Ct values between 15 and 35.
7. Select 0.1 mL or 0.2 mL 8-strip tubes and caps according to the experiment or instrument demand.

References

Akbar MA, Mohd Yusof NY, Tahir NI et al (2020) Biosynthesis of Saxitoxin in marine dinoflagellates: an omics perspective. Mar Drugs 18:103

Al-Tebrineh J, Mihali TK, Pomati F et al (2010) Detection of saxitoxin-producing cyanobacteria and Anabaena circinalis in environmental water blooms by quantitative PCR. Appl Environ Microbiol 76:7836–7842

Al-Tebrineh J, Pearson LA, Yasar SA et al (2012) A multiplex qPCR targeting hepato- and neurotoxigenic cyanobacteria of global significance. Harmful Algae 15:19–25

Campo E, Agha R, Cirés S et al (2013) First TaqMan assay to identify and quantify the cylindrospermopsin-producing cyanobacterium Aphanizomenon ovalisporum in water. Adv Micro 3:430–437

Cusick KD, Sayler GS (2013) An overview on the marine neurotoxin, saxitoxin: genetics, molecular targets, methods of detection and ecological functions. Mar Drugs 11:991–1018

Kellmann R, Mihali TK, Jeon YJ et al (2008) Biosynthetic intermediate analysis and functional homology reveal a saxitoxin gene cluster in cyanobacteria. Appl Environ Microbiol 74:4044–4053

Moustafa A, Loram JE, Hackett JD et al (2009) Origin of saxitoxin biosynthetic genes in cyanobacteria. PLoS One 4:e5758

Orr RJ, Stüken A, Murray SA et al (2013) Evolution and distribution of saxitoxin biosynthesis in dinoflagellates. Mar Drugs 11:2814–2828

Pacheco AB, Guedes IA, Azevedo SM (2016) Is qPCR a reliable indicator of cyanotoxin risk in freshwater? Toxins (Basel) 8:172

Pearson LA, Neilan BA (2008) The molecular genetics of cyanobacterial toxicity as a basis for monitoring water quality and public health risk. Curr Opin Biotechnol 19:281–288

Wang D, Zhang S, Zhang Y et al (2016) Paralytic shellfish toxin biosynthesis in cyanobacteria and dinoflagellates: a molecular overview. J Proteome 135:132–140

Watanabe R, Harada T, Matsushima R et al (2014) Indirect quantitation of saxitoxin by HPLC with post-column oxidation and fluorometric detection. Anal Sci 30:931–936

Wharton RE, Feyereisen MC, Gonzalez AL et al (2017) Quantification of saxitoxin in human blood by ELISA. Toxicon 133:110–115

Yue Y, Zhu B, Lun L et al (2020) Quantifications of saxitoxin concentrations in bivalves by high performance liquid chromatography-tandem mass spectrometry with the purification of immunoaffinity column. J Chromatogr B Anal Technol Biomed Life Sci 1147:122133

55 qPCR Assay for the Detection of Hepatotoxigenic Cyanobacteria

Megha J K and Lokesh Ravi

Abstract

Harmful algal blooms are contaminating fresh and marine waterbodies. The frequency and toxicity of these algal blooms are predicted to rise at an alarming rate. Cyanobacterial species present in the algal blooms can produce hepatotoxins such as microcystin, nodularin, and cylindrospermopsin. These toxins have detrimental effects on humans, domesticated animals, and wildlife. Hence, early detection of hepatotoxigenic cyanobacteria is crucial to prevent and protect the waterbodies from harmful algal blooms. Earlier PCR-based analytical techniques, such as competitive qPCR and most probable number-PCR, are ineffective, time-consuming, and resource-consuming. Hence, real-time quantitative PCR is the best analytical technique that is rapid, sensitive, reliable, and reproducible. The qPCR technique uses fluorescence-based detection to quantify the gene-copy number of hepatotoxin-synthesizing genes, thereby allowing us to determine the absolute number of potential toxin-producing cyanobacterial cells in the water body. This chapter discusses the principle and detailed protocol for absolute quantification of hepatotoxigenic cyanobacteria using SYBR-green dye qPCR.

Keywords

qPCR assay · Hepatotoxigenic cyanobacteria · Absolute quantification · SYBR-green dye

---

M. J. K
Department of Biotechnology, School of Bio Sciences and Technology, Vellore Institute of Technology, Vellore, Tamil Nadu, India

L. Ravi (✉)
Department of Food Technology, Faculty of Life and Allied Health Sciences, MS Ramaiah University of Applied Sciences, Bengaluru, India

N. Thajuddin et al. (eds.), *Protocols for Cyanobacteria Sampling and Detection of Cyanotoxin*, https://doi.org/10.1007/978-981-99-4514-6_55

55.1 Introduction

Cyanobacteria are omnipresent organisms known for producing a diverse range of secondary metabolites, some of which are toxic in nature (Martins and Vasconcelos 2011). They produce hepatotoxigenic compounds such as microcystins, nodularin, and cylindrospermopsins that cause liver injury by directly or indirectly damaging hepatocytes (Czaja 2004; Blaha et al. 2009). Microcystins and nodularins are cyclic heptapeptides and cyclic pentapeptide compounds, respectively. They cause liver damage by inhibition of protein phosphatase, disruption of conductance and membrane integrity, and are reported to upregulate tumour promotors. Whereas cylindrospermopsins can cause inhibition of protein synthesis, necrotic liver injury, and are genotoxic in nature (Blaha et al. 2009). Exposure of living organisms to these toxins in cyanobacterial blooms is hazardous, and it is crucial to detect potential toxin-producing cyanobacterial species in waterbodies to prevent such mishappenings (Falconer 1996; Briand et al. 2003). Hence, it is essential to develop a rapid, reliable, and applicable analytical technique. This chapter describes an accurate and rapid molecular biology-based analytical technique called real-time qPCR assay to detect hepatotoxigenic cyanobacteria from bloom samples.

The qPCR assay aims to detect the genes encoding the hepatotoxins biosynthesis and interpret the gene copy number as cell abundance. The biosynthesis of microcystin is encoded in *myc* gene cluster, first characterized in *Microcystis aeruginosa* PCC7806. The *myc* gene cluster is 55 kb in size and is made up of 10 genes. The genes are organized into two operons, *mycA-C* and *mycD-J* that are divergently transcribed. The genes of *mycD-J* operon, such as *myc D, E, and G* encode for polyketide synthase module and enzymes involved in non-ribosomal peptide synthase, genes *mycJ, F,* and *I* are tailoring genes, and *mycH* gene encodes for toxin transport. At last, the operon *mycA-C* codes three non-ribosomal peptide synthase enzymes (Tillett et al. 2000). The gene cluster *nda* encodes the biosynthesis of nodularin. It is 48 kb in size and is made up of nine open reading frames that is *ndaA-I.* It is transcribed bidirectionally from a regulatory promotor region (Moffitt and Neilan 2004). Cylindrospermopsin biosynthesis is encoded by *cyr (A-L)* gene cluster spanning 43 kb. It consists of 15 open reading frames, namely *cyrA* (amidinotransferase), *cyrB* (non-ribosomal peptide synthase and polyketide synthase module), *cyrC-F* (four polyketide synthase module), *cyrG-H* (two cytosine deaminase homologues), *cyrI* (hydroxylase), *cyrK* (efflux pump), and *cyrL* (Transposase) (Mihali et al. 2008).

Real-time qPCR, also known as quantitative PCR is an analytical technique used to quantify the number of gene copies in the sample, by using a fluorescence-based detection system to measure the amount of DNA or amplicons after each cycle during the exponential phase of the polymerase chain reaction.

55.1.1 Principle

The qPCR is based on the principle of conventional PCR, where a specific sequence of template DNA containing the gene of interest is exponentially amplified using precisely designed primers complementary to the gene of interest (Martins and Vasconcelos 2011). Unlike conventional PCR, which relies on endpoint analysis, qPCR monitors the amplification of DNA in real-time (Wittwer et al. 1997). Fluorescent markers such as SYBR-green dye is incorporated into the amplicons, and increased fluorescent signals would reflect the increase in the number of amplicons after each cycle at that specific time (Martins and Vasconcelos 2011; Heid et al. 1996). SYBR-green dye is fluorochromatic substance that emits fluorescence by binding to the minor groove of double-stranded DNA. The dye gets excited by the intercalation between the nitrogenous bases of double-stranded DNA and provides a significantly brighter fluorescence than the unbound dye (Fig. 55.1) (Life science technologies 2012).

The qPCR software tracks the incremental changes in fluorescence after every amplification cycle and generates an amplification curve. In the amplification curve, during the initial cycles, the fluorescence from DNA amplicons is indistinguishable from the background fluorescence emitted by the unbound dye (Kralik and Ricchi

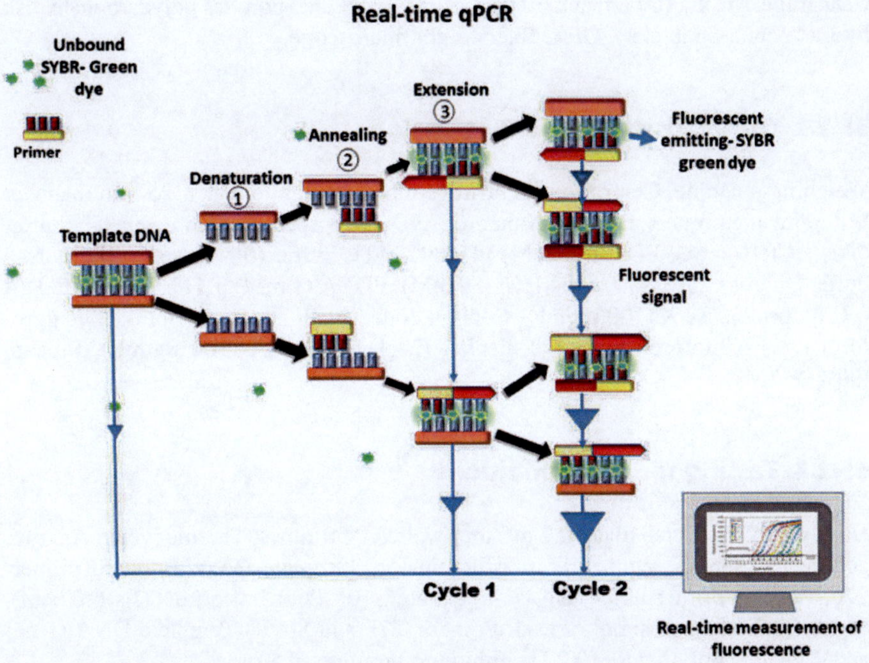

Fig. 55.1 Real-time qPCR—The figure illustrates the principle of qPCR detection by SYBR-green dye in real time. The fluorescence emitted by the SYBR-green dye bound to the double-stranded DNA is detected and analysed to generate the amplification curve

2017). However as the amplification proceeds, the intensity of inflorescence from the amplicon exceeds the background noise, the corresponding cycle number is called as Cycle threshold (Ct), and this phase is known as the exponential phase (Martins and Vasconcelos 2011; Heid et al. 1996). The number of the amplicons produced during the exponential phase is proportional to the initial number of template DNA molecules present in the sample, and the gene copy numbers of template DNA are quantified at these Ct values (Kralik and Ricchi 2017; Smith and Osborn 2009).

The quantification of the unknown concentration of the initial target sequence is calculated from the Ct values by using the absolute quantification method. In this method, a standard(/calibration) curve is constructed from the amplification of the target gene sequence, with initial template concentration ranging from 10^1 to 10^{10} and the Ct values are obtained for each template concentration. A simple linear regression of the obtained Ct values is plotted against the log of the initial copy numbers (Smith and Osborn 2009).

55.2 Materials

55.2.1 Sample Collection

Centrifuge, Mesh (100 μm and 50 μm), 3 μm—pore size poretics polycarbonate disk filter (47 mm, Diameter), GF/C filters, light microscope.

55.2.2 DNA Extraction and Purification

Weighing machine, Centrifuge, Centrifuge tubes, Millipore water, 25 mm diameter Millipore membrane, Beaker, Nanodrop ND-1000 spectrophotometer, TE buffer (Tris-HCl 10 mM), EDTA (1 mM) pH 8.0, 1×TE buffer (0.1% sarkosyl), Sucrose buffer (sucrose 25%, w/v), Tris-HCl (50 mM), EDTA (1 mM) pH 8, 100 mM EDTA (pH 8), proteinase K (100 μg/mL), Sodium dodecyl sulphate (2.5%), Phenol/chloroform (v/v), Chloroform-isoamyl alcohol (24:1), RNase A (100 μg/mL), Ethanol (90% & 70%).

55.2.3 Testing for PCR Inhibitions

Hand glove, PCR well plate, 0.2 mL thin-walled PCR tubes, Thermocycler, Agarose gel electrophoresis setup, UV transilluminator, Template DNA, Forward primer-27F, Reverse primer-809R, DNase-free water, 10x PCR buffer, dNTPs (10 mM), $MgCl_2$ (2.5 mM), 10 pmol of each Primers (27F and 809R) Template DNA (1 ng) and *Taq*DNA polymerase (0.2 U), Ethidium bromide (0.5 μg).

55.2.4 qPCR Assay

Well plate, UV Cabinets, Vortex, Centrifuge, Rotor gene instrument with Rotor gene analysis software 6.0.27, 1× platinum SYBR-Green SuperMix UDG (Invitrogen) master mix, Primers, Template DNA, PCR grade water, Optical adhesive seal.

55.3 Procedure

All the procedures of qPCR assay should be in compliance with MIQE guidelines. (*see* Note 1).

55.3.1 Sampling Collection

Sample collection is the most important step in the process of detection of hepatotoxin-producing cyanobacteria. This section provides general instruction to sample collection.

1. Clearly define the target population, select the sampling frame, and determine a suitable sampling technique and sample size (Hamed 2016) (*see* note 2).
2. After each session of water sample collection, transfer the water sample from the field to the laboratory and within 2 h, filter the sample by gently pouring it through the sieves with mesh sizes of 100 μm and 50 μm (*see* Note 3).
3. Identify the cyanobacterial species under light microscope.
4. Resuspend each colony size in 100 mL of lake water filtered through GF/C filters.
5. To collect the cells for DNA extraction, filter an aliquot of 25 mL through GF/C filters and weigh the fresh sample before subjecting it to DNA extraction (Kurmayer et al. 2003).
6. Alternatively, filter 100 mL of water sample through 3 μm—a pore size poretics polycarbonate disk filter with diameter size 47 mm. Store the cells with lysis buffer at −20 °C (Vaitomaa et al. 2003).

55.3.2 Cyanobacterial Cultures

Acquire cyanobacterial strains from microbial culture collections and culture them using suitable growth media, at optimum growth conditions.

55.3.3 DNA Extraction and Purification

A simple protocol for cyanobacterial DNA extraction was prescribed by Franche *et al.*, 1981 (Franche and Damerval 1988).

1. Weigh 5 g of cells and wash it twice with TE buffer. Initially, wash the cells with 20 mL of 1× TE buffer and for furthermore wash use 20 mL of 1× TE buffer solution containing 0.1% sarkosyl.
2. After washing, discard the buffer and resuspend the pellet in 20 mL of sucrose buffer and let it rest for 2 h at 4 °C.
3. Add 100 mM EDTA to the suspension and treat it with 5 mg/mL of lysozyme for 1 h at 37 °C to lyse the cell membrane.
4. Add 100 μg/mL of proteinase K to denature the interfering protein and 2.5% sodium dodecyl sulphate to the suspension and incubate at 37 °C until the solution becomes colourless.
5. Extract the lysate with an equal volume of phenol/chloroform (v/v) and centrifuge for 5 min at 13,000 rpm at room temperature.
6. Transfer the upper phase into a fresh centrifuge tube and repeat step 5 once again for double extraction.
7. Again transfer the DNA containing upper phase into a fresh tube.
8. Add chloroform-isoamyl alcohol at a volume ratio of 24:1 and centrifuge for 5 min at 13,000 rpm at room temperature.
9. Precipitate the extract with 90% (v/v) ethanol at −20 °C for 1 h.
10. Wash the precipitate with 70% (v/v) ethanol and resuspend the extracted DNA in 50 μL of Millipore water (Kurmayer et al. 2003).
11. To purify the extracted DNA, dialyse the DNA against TE buffer (Silhavy et al. 1984).
12. Take approximately 30–100 mL of 1× TE buffer in a beaker and immerse a 25 mm diameter Millipore membrane for approximately 5 min. Ensure that the membrane is completely wet and devoid of trapped air bubbles.
13. Carefully pipette the DNA extract onto the membrane (*see* Note 5).
14. Cover the beaker and allow the dialysis process for 1–4 h.
15. After completion, retrieve the DNA Droplet and transfer it into a microcentrifuge tube. Rinse the spot on the membrane with 50 μL of 1× TE buffer and transfer it to the microcentrifuge tube.
16. Finally, to prevent RNA contamination, add 100 μg/mL of RNase A and incubate for 3 h at 37 °C.
17. To determine the concentration of the DNA, measure the absorbance at 260 nm using a spectrophotometer (Nanodrop ND-1000 spectrophotometer) (Vaitomaa et al. 2003; Al et al. 2011).
18. Determine the purity of the DNA by calculating the ratio of the absorbance measured at 260 nm (A_{260}) to the absorbance at 280 nm (A_{280}) (Vaitomaa et al. 2003; Al et al. 2011).
19. Purity of DNA = A_{260}/A_{280}.

55.3.4 Testing for PCR Inhibitions

PCR was performed on the extracted DNA of bloom samples and the positive control (standard strain) using cyanobacterial-specific 16S rRNA primers 27F and

809R to detect PCR inhibitors in the DNA sample (Al et al. 2011; Lorenz 2012) (*see* Note 4).

1. Place the well plate into the ice bucket to hold the 0.2 mL thin-walled PCR tubes. The PCR tubes should be maintained at a cold temperature while the PCR reagents are added into the PCR tubes to prevent nuclease activity and nonspecific priming.
2. Pipette respective volumes of PCR reagents in the following order into Cold PCR tube: Q.S of Sterile water, 10× PCR buffer, dNTPs (10 mM), $MgCl_2$ (2.5 mM), 10 pmol of each Primers (27F and 809R), Template DNA (1 ng), and *Taq*DNA polymerase (0.2 U) (*see* Note 6).
3. Add 1 ng of Template DNA extracted from a standard strain of cyanobacteria as positive control and as the No-template control (NTP) add 1 ng of DNase water instead of template DNA.
4. Take caution to mix the reagents well with the pipette before adding them to the reaction mixture and avoid introducing bubbles while mixing them.
5. Close the caps of the PCR tubes and place them into the thermocycler. Firmly close the lid of the thermocycler and set up the cycling process as prescribed, initiate the cycling with 4 min denaturation at 94 °C, followed by 30 cycles of 10 s denaturation at 94 °C, 25 s of annealing at 55 °C, 45 s of extension at 72 °C and one cycle of final extension for 5 min at 72 °C.
6. Once the program is completed, the PCR tubes are removed and stored at 4 °C.
7. To observe any discrepancies caused by PCR inhibitors in the sample, the PCR products are stained with 0.5 μg of Ethidium bromide and electrophoresed through 1% agarose gel. After agarose gel electrophoresis, they are visualized under a UV transilluminator. (*see* Note 7).

55.3.5 Selection of Amplicon and Primer Selection

Selection of the appropriate primer is necessary to amplify the toxin-producing target gene. The universal primer sequences and their target gene to detect the abundance of the hepatotoxigenic cyanobacterial species in bloom sample are given in Table 55.1.

55.3.6 qPCR Assay

1. Create and print a plate template that matches the preferred qPCR plate, use the template to organize the placement of samples on the qPCR plate and to identify wells. Prepare each DNA template and the negative control reaction in triplicates. (*see* Note 4).
2. The qPCR Master Mix must be thawed at room temperature. Gently homogenize by vortexing and centrifuging (*see* Note 8).

Table 55.1 Universal primer sequences of hepatotoxin-coding genes and their annealing temperatures

Toxin	Target gene	Primer	Sequence (5′-3′)	Annealing temp (°C)	Reference
Microsystin	*mycA*	MAPF	CTAATGGCCGATTGGAAGAA	60 °C at 30 s	Martins and Vasconcelos (2011), Briand et al. (2003)
		MAPR	CAGACTATCCCGTTCCGTTG		
		M1r-F:	AGC GGT AGT CAT TGC ATC G	65 °C for 30 s	Martins and Vasconcelos (2011), Yoshida et al. (2007)
		M1r-R	GCC CTT TTT CTG AAG TCG CC		
		MSF	ATCCAGCAGTTGAGCAAGC	62 °C for 10 s	Martins and Vasconcelos (2011), Furukawa et al. (2006)
		MSR	TGCAGATAACTCCGCAGTTG		
		MSR-2R	GCCGATGTTTGGCTGTAAAT		
		MISYF	CGACCGAGGAATTTCAAGCT	55 °C for 30 s	Foulds et al. (2002), Pearson and Neilan (2008)
		MISYr	AGTATCCGACCAAGTTACCCAAAC		
		mcyA-Cd 1F	AAAATTAAAAGCCGTATCAAA	59 °C for 30 s	Hisbergues et al. (2003)
		mcyA-Cd 1R	AAAAGTGTTTTATTAGCGGCTCAT		
	mycB	MIf/Mir	GCAGCGAACTCTTGAAGGGTTTATG	55 °C for 2 min	Foulds et al. (2002), Pearson and Neilan (2008)
			GCGGATTCTGTGCAGCTTGTTCTTC		
		mcyB 2959F	TGGGAAGATGTTCTTCAGGTATCCAA	57 °C for 45 s	Martins and Vasconcelos (2011), Nonneman et al. (2002)
		mcyB 3278R	AGAGTGGAAACAATATGATAAGCTAC		
		30F	CCTACCGAGCGCTTGGG	60 °C for 15 s	Kurmayer et al. (2003)
		108R	GAAAATCCCCTAAAGATTCCTGAGT		
	mycD	mcyDF	GATCCGATTGAATTAGAAAG	56 °C for 30 s	Vaitomaa et al. (2003)
		mcyDR	GTATTCCCCAAGATTGCC		
		mcyD F2	GGTTCGCCTGGTCAAAGTAA	56 °C for 60 s	Kaebernick et al. (2000), Rinta-Kanto et al. (2005)
		mcyD R2	CCTCGCTAAAGAAGGGTTGA		

	mycE	mcyE-plaR3	CTCAATCTGAGGATAACGAT	57 °C for 30 s	Pearson and Neilan (2008), Rantala et al. (2006)
		mcyE-F2	GAAATTTGTGTAGAAGGTGC	56 °C for 30 s	
		mcyE-R4	AATTCTAAAGCCCAAAGACG		
		MicmcyE-R8	CAA TGG GAG CAT AAC GAG	58 °C for 5 s	Martins and Vasconcelos (2011), Vaitomaa et al. (2003), Nonneman et al. (2002)
		AnamcyE-12R	CAA TCT CGG TAT AGC GGC		
	mycF	MCY F1	TGGGAA GAT GTTCTT CAG GTATCC AA	57 °C for 45 s	Martins and Vasconcelos (2011), Nonneman et al. (2002)
		MCY R1	AGA GTG GAA ACA ATATGATAAGCT AC		
		MCY R2	GAG ATC CAT CTGTTGCAAGACATA G		
Nodularin	*ndaF*	ndaF8452	GTG ATT GAA TTT CTT GGT CG	61 °C for 30 s	Pearson and Neilan (2008), Koskenniemi et al. (2007)
		ndaF8640	GGA AAT TTC TAT GTC TGA CTC AG		
Microsystin and nodularin	*mcyE / ndaF*	DQmcyF	TTTA GAAC SGGV GATT TAGG	60 °C for 30 s	Al et al. (2011)
		DQmcyR	CGRB TVAD TTGR TATT CAAT TTCT		
Cylindrospermopsin	*aoaA*	A1f	GATATCCCAGGTGGCACAATG	60 °C for 25 s	Pearson and Neilan (2008), Rasmussen et al. (2008)
		A1r	TGTTCTGCCAGTCCAGTAAGTG		
	aoaB 5'	B1f	CGGACGGGCTGTCGCACATG		
		B1r	AACGGGAACATTGGCAAGCCATGA		
	aoaB 3'	B2f	TTGAGGGTATGAGTGGTGGCATCTTGA		
		B2r	CGAAGTTATCACTACCGGGGAGCAGAT		
	aoaC 5	C1f	TGTGATCTAGCACTAGCTGGTGGAGTC		
		C1r	GCTGTTGCATCCGAGAGCGTTTTAGA		
	aoaC 3'	C2f	CCGCCTCCGAGTTCAGACATACCA		
		C2r	CCTGTTCAACCATCCAACGAGCCA		

3. Prepare the qPCR Master Mix in a new sterile microtube—add the reagents using a sterile microtip for each transfer as follows: *Q.S* of PCR grade water, 1x platinum SYBR-Green SuperMix UDG (Invitrogen), 0.8 pmol of appropriate forward and reverse primer (*see* Note 9) (to a final volume of 25 μL).
4. Carefully transfer the 23 μL of reaction mixture into the well plates, in accordance with the template design.
5. Carefully add 2 μL template DNA to appropriate well plates.
6. Prepare at least two dilutions of the DNA extracted from bloom sample (Maza et al. 2020): (*see* Note 10).
7. To prepare dilution 1 (1/10): Pipette 0.2 μL from the stock DNA and add to 1.8 μL of PCR grade water in tube 1.
8. To prepare dilution 2 (1/20): pipette 0.2 μL from the stock DNA and add to 19.8 μL of PCR grade water to tube 2.
9. As a negative control, prepare No-template control (NTP), by adding 2 μL of PCR DNase-free water.
10. Apply an optical adhesive with caution to seal the plate. Seal the areas between the wells with a sterile pointed object. Before proceeding to the next step, ensure the wells are completely sealed.
11. Vortex the well plates for 5–10 s to ensure complete dissolution of the reagents and pellets. Centrifuge the well plate at low speed for 1 min and place the plate into the thermal cycler block.
12. Set up the cycling process as prescribed: Initiate the cycling with 2 min of initial heat activation at 50 °C, initial denaturation at 95 °C for 2 min followed by 45 quantification cycles, denaturation at 95 °C for 15 s and combined annealing and extension at appropriate temperature and time (*see* Note 11) for each cycle and melting curve analysis at 72–95 °C. The amplification of the target gene is measured at the end of each cycle at 60 °C, by quantification of fluorescence generated by PRC, through the Fam/SyBr channel (510 nm).
13. Determine the Ct value by second derivative maximum method using Rotor-Gene Analysis Software 6.0.27 (Al et al. 2011) (*see* Note 12).

55.3.6.1 Melting Curve Analysis

Melting curve analysis is conducted to ensure that there are no discrepancies in data due to primer dimer formation or RNA contamination, etc. To conduct melting curve analysis, program the qPCR instrument to include the melting profile as a continuation after the thermal cycling protocol. The instrument automatically gives the complete melting curve data. The presence of nonspecific amplification products would appear as an additional peak in the melting curve (Life science technologies 2012). Schematic representation of qPCR workflow for detection of hepatotoxigenic cyanobacteria is shown in Fig. 55.2.

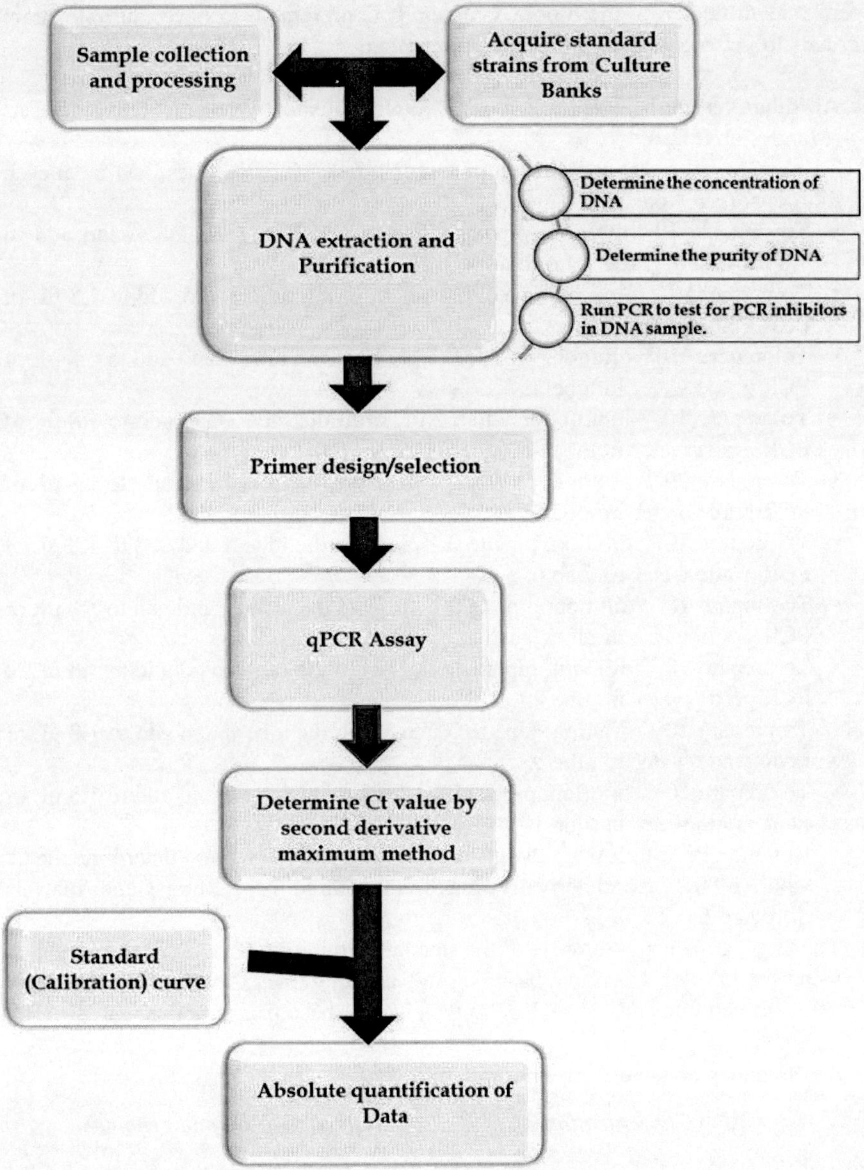

Fig. 55.2 The flowchart represents the overall workflow of qPCR assay

55.3.7 qPCR of Standard Strains to Generate the Standard Graph

To generate a standard graph, prepare tenfold serial dilutions of DNA extracted from the standard strains. The standard strains are the strain variants of cyanobacterial species identified from the bloom samples. It is preferable to have multiple strain variants to get accurate mean gene copy numbers.

1. All dilutions must be in triplicates to obtain statistically representative datasets (Maza et al. 2020).
2. To prepare a 2 μL of tenfold dilution series, take 10 PCR tubes, and label each dilutions: (*see* Note 10).
   - To prepare 10^{-1}dilution, pipette 0.2 μL from the stock DNA and add to 1.8μLof PCR grade water in tube 1.
   - To prepare10^{-2}dilution, pipette 0.2 μL from the tube 1 and add to 1.8 μL of PCR grade water in tube 2.
   - To prepare 10^{-3} dilution, pipette 0.2 μL from the tube 2 and add to 1.8 μL of PCR grade water in tube 3.
   - To prepare 10^{-4}dilution, pipette 0.2 μL from the tube 3 and add to 1.8 μL of PCR grade water in tube 4.
   - To prepare 10^{-5} dilution, pipette 0.2 μL from the tube 4 and add to 1.8 μL of PCR grade water in tube 5.
   - To prepare 10^{-6} dilution, pipette 0.2 μL from the tube 5 and add to 1.8 μL of PCR grade water in tube 6.
   - To prepare 10^{-7} dilution, pipette 0.2 μL from the tube 6, and add to 1.8 μL of PCR grade water in tube 7.
   - To prepare 10^{-8} dilution, pipette 0.2 μL from the tube 7 and add to 1.8 μL of PCR grade water in tube 8.
   - To prepare 10^{-9} dilution, pipette 0.2 μL from the tube 8 and add to 1.8 μL of PCR grade water in tube 9.
   - To prepare 10^{-10}dilution, pipette 0.2 μL from the tube 9 and add to 1.8 μL of PCR grade water in tube 10.
   - Run the qPCR following the instructions in Sect. 55.3.5 and determine the Ct values by the second derivative maximum method using Rotor-Gene Analysis Software 6.0.27.
3. The target gene copy numbers of the standard strains (per reaction) are calculated using the following formula based on two assumptions: There is only one target gene per genome and the molecular weight of 1 base pair is 660 g mol^{-1}.

$$\text{Number of gene copies per microlitre} = 6 \times 10^{23} \left(\mathit{Concentration\ of\ DNA} | \mathit{Molecular\ weight\ of\ one\ genome} \right)$$

   Where
   - 6×10^{23} = number of copies per mole.
   - The concentration of DNA is given in grams per microlitre.
   - The molecular weight of one genome is given in grams per mole.
4. Plot a standard graph using the Log of the initial copy number (calculated gene copy number per reaction) on the X-axis and the obtained Ct values of the standard strains on the Y-axis.

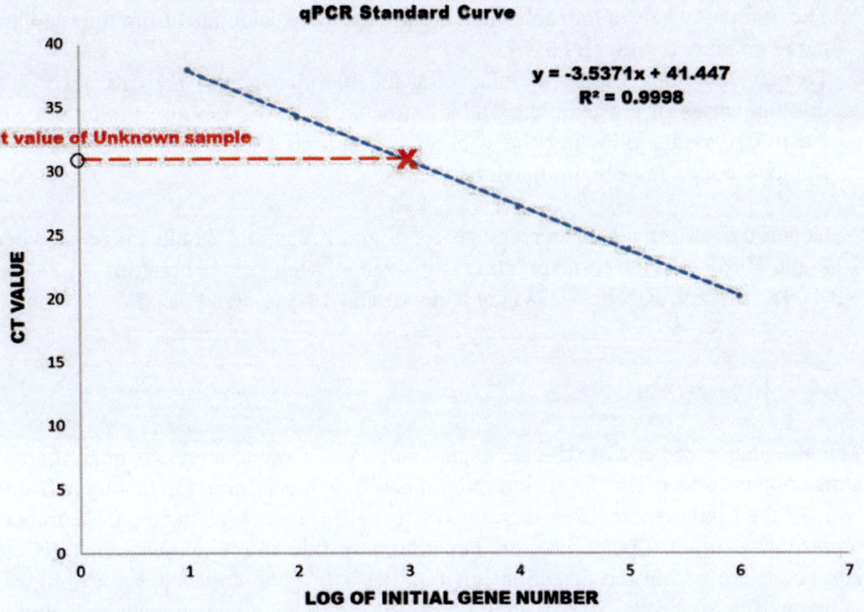

Fig. 55.3 Standard curves and the linear relation (regression) equation of Ct values obtained for serial dilutions of DNA samples containing target genes

5. Linear relation (regression) equation of the Ct values is obtained as a function of calculated gene copy numbers of standard strains.
6. To quantify the gene copy number of the unknown target template, compare the Ct values of the target template against the standard curve (Smith and Osborn 2009; Vaitomaa et al. 2003) (*see* Note 12). A sample standard curve interpretation for standard curves and the linear relation equation of Ct values is graphically represented in Fig. 55.3.

55.3.8 Determine the Amplification Efficiency

PCR efficiency represents the percentage of amplification efficiency, i.e. the PCR efficiency is 100%, when the template doubles in each consecutive cycle during the exponential phase of thermocycling. It is essential to calculate the PCR efficiency of the standard strains and the bloom samples while conducting the qPCR.

To determine the PCR efficiency by using the following equation:

$$E = 10^{\left(-\frac{1}{\text{slope}}\right)} - 1$$

where the slope value is derived from the standard curve.

1. The efficiency values for each standard strain can be calculated from the standard curve generated in Sect. 55.3.7.
2. To calculate the efficiency values for the bloom sample, prepare a six-fold dilution series of the template DNA extracted from the bloom sample and run the qPCR assay following the general instructions (Sect. 55.3.6). Construct a standard curve and obtain the slope values.

Generally, an efficiency percentage for a good reaction should range between 90% and 100%, and the corresponding slope values would range between −3.58 and −3.10 (Kralik and Ricchi 2017) (*see* Note 13 and 14).

55.4 Interpretation

The abundance of hepatotoxigenic cyanobacteria in a water sample is quantified as gene copy numbers. The data is interpreted based on the assumption that there is one copy of the toxin-synthesizing target gene per cell. Hence, high gene copy numbers represent a higher potential for hepatotoxin-producing cyanobacteria. If the amplicons are within the detection range of 10^1–10^6 gene copies per reaction, this indicates the presence of potential hepatotoxin-producing cyanobacterial strains (Al et al. 2011). Generally, Ct values are inversely proportional to the gene copy numbers. A lower Ct value would indicate the presence of a higher concentration of the target gene in the sample, hence would imply that there are more hepatotoxin-producing cyanobacteria in the sample (Smith and Osborn 2009).

55.5 Conclusion

Compared to previously reported molecular detection assays, qPCR is easier to use, more sensitive, and has the added benefits of being quantitative and adaptable to crude bloom samples. Even though the assay has its merits, it can lack specificity as the SYBR- green dye binds to all dsDNA, not only the target DNA. Nevertheless, post-PCR melting (dissociation) curve analysis may be used to evaluate the specificity of each result. The qPCR assay could also give an inaccurate interpretation of cell count, as there might be more than one gene copy per cell. Al-Tebrineh et al. (2011) reported 1.3 copies of gene per cell in *Microcystis (*Al et al. 2011*)*. Also, the presence of gene copy doesn't translate to toxins produced, as the genes might not be expressed due to mutations such as deletions. So qPCR assay gives data on "potentially" hepatotoxigenic cyanobacteria (Koskenniemi et al. 2007). Hence, water management authorities may quickly assess the danger of a bloom occurrence by using this assay to identify the possible toxin status. As a result, control measures might be put in place early on to prevent human exposure to cyanobacterial hepatotoxins before the toxin concentrations become detrimental.

Notes

1. The Minimum Information for Publication of Quantitative Real-time PCR Experiments (MIQE) guidelines provide the list of information required for qPCR assay publication. Kindly refer to Bustin et al. (2009) (Bustin et al. 2009)
2. Using the right kind of sampling technique is essential to obtain a representative sample for the analysis. There is no one-size-fits-all approach for the sampling collection, as it is highly dependent on the objective, sample location, and the target sample species.
3. Cyanobacteria are planktonic organisms; they are found floating on the surface as well as distributed vertically. The cyanobacterial population size also varies throughout the seasons. Hence, the sampling has to be conducted at regular intervals of time and at different depths of the water body to obtain a representative sample.
4. Wear hand gloves at all times during the procedure to prevent contamination. To prevent contamination, prepare the qPCR reaction mixtures in UV cabinets.
5. Carefully place a few μL of DNA droplets onto the centre of the membrane. If the droplet is not adsorbed on the membrane, discontinue the process and remove the organic solvents further before proceeding. If the droplet is adsorbed on the membrane, carefully pipette the remaining DNA extract onto the membrane (Silhavy et al. 1984).
6. *Quantum satis* (Q.S) is derived from Latin and it means required amount. PCR-grade water is used to make up the reaction mixture to the final volume. However, the water is added first into the PCR tube by pre-calculating the volumes of reagents before adding (Lorenz 2012).
7. To verify the specificity of PCR amplification, agarose gel electrophoresis is an easy and trustworthy technique that enables the identification of nonspecific amplicons or primer dimers. An optimum PCR reaction should show a single band when visualized under UV transilluminator (Maza et al. 2020).
8. Avoid foam formation and exposure to light. The SBYR-green dye must be wrapped in aluminium foil and stored at −20 °C) (Maza et al. 2020).
9. Prepare the stock solution of primers at a concentration of 100 Mm and store at −20 °C. To prepare the working solution, dilute the stock solution at a concentration of 0.8 pmol by using PCR-grade water. Avoid subjecting the stock solution to multiple freeze-thaw cycles (Maza et al. 2020)
10. Each primer has specific annealing temperature.
11. Precise pipetting is crucial in this step. To generate statistically representative data, prepare each dilution in triplicates.
12. To obtain an accurate Ct value, it is important to determine the baseline and the threshold. The baseline represents the background "noise" of the reaction. In real-time PCR, the baseline is established empirically for each reaction by manual or automatic examination of the amplification plot. To determine the threshold cycle (Ct) accurately as specified below, the baseline should be fixed carefully. The number of cycles needed to completely remove the background "noise" in the first few cycles of the amplification curve should be included in

the baseline determination; however, the cycles in which the amplification signal starts to rise beyond the background should not be included. The threshold of the real-time PCR reaction shows the statistically significant increase of fluorescent signal over the determined baseline signal. It is programmed to differentiate the significant amplification signal from the noise (Life science technologies 2012).

13. Each qPCR amplification is reported as a standard curve. To validate the data, for each standard curve, it is necessary to calculate the amplification efficiency, the y-intercept value (indicates the sensitivity of the reaction) and the linear regression coefficient (R^2) (Smith and Osborn 2009).
14. For every amplification curve of each cycle, the Ct value of NTC and its gene copy numbers should be reported (Smith and Osborn 2009).

References

Al-Tebrineh J, Gehringer MM, Akcaalan R, Neilan BA (2011) A new quantitative PCR assay for the detection of hepatotoxigenic cyanobacteria. Toxicon 57:546–554

Blaha L, Babica P, Marsalek B (2009) Toxins produced in cyanobacterial water blooms – toxicity and risks. Interdiscip Toxicol 2:36–41

Briand J-F, Jacquet S, Bernard C, Humbert J-F (2003) Health hazards for terrestrial vertebrates from toxic cyanobacteria in surface water ecosystems. Vet Res 34:361–377

Bustin SA, Benes V, Garson JA et al (2009) The MIQE guidelines: minimum information for publication of quantitative real-time PCR experiments. Clin Chem 55:611–622

Czaja MJ (2004) Hepatotoxins. In: Encyclopedia of Gastroenterology, pp 365–367

Falconer IR (1996) Potential impact on human health of toxic cyanobacteria. Phycologia 35:6–11

Foulds IV, Granacki A, Xiao C, Krull UJ, Castle A, Horgen PA (2002) Quantification of microcystin-producing cyanobacteria and E. coli in water by 5′-nuclease PCR. J Appl Microbiol 93:825–834

Franche C, Damerval T (1988) Tests on nil probes and DNA hybridizations. Methods Enzymal 167: 803–808

Furukawa K, Noda N, Tsuneda S, Saito T, Itayama T, Inamori Y (2006) Highly sensitive real-time PCR assay for quantification of toxic cyanobacteria based on microcystin synthetase a gene. J Biosci Bioeng 102:90–96

Hamed T (2016) Sampling methods in research methodology; how to choose a sampling technique for research. Int J Acad Res Manag 5:18–27

Heid CA, Stevens J, Livak KJ, Williams PM (1996) Real time quantitative PCR. Genome Res 6: 986–994

Hisbergues M, Christiansen G, Rouhiainen L, Sivonen K, Börner T (2003) PCR-based identification of microcystin-producing genotypes of different cyanobacterial genera. Arch Microbiol 180:402–410

Kaebernick M, Neilan BA, Börner T, Dittmann E (2000) Light and the transcriptional response of the microcystin biosynthesis gene cluster. Appl Environ Microbiol 66:3387–3392

Koskenniemi K, Lyra C, Rajaniemi-Wacklin P, Jokela J, Sivonen K (2007) Quantitative real-time PCR detection of toxic *Nodularia* cyanobacteria in the Baltic Sea. Appl Environ Microbiol 73: 2173–2179

Kralik P, Ricchi M (2017) A basic guide to real time PCR in microbial diagnostic: definitions, parameters, and everything. Front Microbiol 8:1–9

Kurmayer R, Christiansen G, Chorus I (2003) The abundance of microcystin-producing genotypes correlates positively with colony size in *Microcystis* sp. and determines its microcystin net production in Lake Wannsee. Appl Environ Microbiol 69:787–795

Life science technologies (2012) Melting curve analysis. In: Real-time PCR handbook, pp 15–16

Lorenz TC (2012) Polymerase chain reaction: basic protocol plus troubleshooting and optimization strategies. Journal of Visualizes Experiments 63:e3998

Martins A, Vasconcelos V (2011) Use of qPCR for the study of hepatotoxic cyanobacteria population dynamics. Arch Microbiol 193:615–627

Maza-Márquez P, Aranda E, González-López J, Rodelas B (2020) Evaluation of the abundance of fungi in wastewater treatment plants using quantitative PCR (qPCR). Quantitative Real-Time-PCR: Methods and Protocols 20:79–92

Mihali TK, Kellmann R, Muenchhoff J, Barrow KD, Neilan BA (2008) Characterization of the gene cluster responsible for Cylindrospermopsin biosynthesis. Appl Environ Microbiol 74:716–722

Moffitt MC, Neilan BA (2004) Characterization of the nodularin synthetase gene cluster and proposed theory of the evolution of cyanobacterial hepatotoxins. Appl Environ Microbiol 70: 6353–6362

Nonneman D, Zimba P, v. (2002) A PCR-based test to assess the potential for microcystin Occurrence in channel catfish production ponds 1, 2. J Phycol 38:230–233

Pearson LA, Neilan BA (2008) The molecular genetics of cyanobacterial toxicity as a basis for monitoring water quality and public health risk. Curr Opin Biotechnol 19:281–288

Rantala A, Rajaniemi-Wacklin P, Lyra C, Lepistö L, Rintala J, Mankiewicz-Boczek J, Sivonen K (2006) Detection of microcystin-producing cyanobacteria in Finnish lakes with genus-specific microcystin synthetase gene E (*mcyE*) PCR and associations with environmental factors. Appl Environ Microbiol 72:6101–6110

Rasmussen JP, Giglio S, Monis PT, Campbell RJ, Saint CP (2008) Development and field testing of a real-time PCR assay for cylindrospermopsin-producing cyanobacteria. J Appl Microbiol 104: 1503–1515

Rinta-Kanto JM, Ouellette AJA, Boyer GL, Twiss MR, Bridgeman TB, Wilhelm SW (2005) Quantification of toxic *Microcystis* spp. during the 2003 and 2004 blooms in Western Lake Erie using quantitative real-time PCR. Environ Sci Technol 39:4198–4205

Silhavy T, Berman M, Enquist L (1984) Experiments with gene fusions. Cold Spring Harbor Laboratory, New York, pp 291–298

Smith CJ, Osborn AM (2009) Advantages and limitations of quantitative PCR (Q-PCR)-based approaches in microbial ecology. FEMS Microbiol Ecol 67:6–20

Tillett D, Dittmann E, Erhard M, von Döhren H, Börner T, Neilan BA (2000) Structural organization of microcystin biosynthesis in Microcystis aeruginosa PCC 7806: an integrated peptide–polyketide synthetase system. Chem Biol 7:753–764

Vaitomaa J, Rantala A, Halinen K, Rouhiainen L, Tallberg P, Mokelke L, Sivonen K (2003) Quantitative real-time PCR for determination of microcystin Synthetase E copy numbers for Microcystis and anabaena in lakes. Appl Environ Microbiol 69:7289–7297

Wittwer CT, Herrmann MG, Moss AA, Rasmussen RP (1997) Continuous fluorescence monitoring of rapid cycle DNA amplification. BioTechniques 22:130–138

Yoshida M, Yoshida T, Takashima Y, Hosoda N, Hiroishi S (2007) Dynamics of microcystin-producing and non-microcystin-producing *Microcystis* populations is correlated with nitrate concentration in a Japanese lake. FEMS Microbiol Lett 266:49–53

Molecular Detection of *Mcy* Genes in Toxic Cyanobacteria: *Anabaena, Microcystis, Planktothrix*

56

H. A. S. N. Abeysiri, G. Y. Liyanage, K. Sirisena, S. A. Kulasooriya, and M. M. Pathmalal

Abstract

Toxic cyanobacteria pose a significant hazard to human health and the environment. The recent characterization of cyanotoxin synthetase gene clusters has resulted in an explosion of molecular detection methods for these organisms and their toxins. Conventional Polymerase Chain Reaction (PCR) tests targeting cyanotoxin biosynthesis genes provide a rapid and sensitive means for detecting potentially toxic populations of cyanobacteria in water supplies. Adapting these simple PCR tests into quantitative methods has additionally enabled the monitoring of dynamic bloom populations and the identification of particularly problematic species. Of the known toxins produced by cyanobacteria, microcystins are the most significant threat to human and animal health. Knock-out studies have confirmed that microcystins are produced non-ribosomally by a multi-enzyme complex consisting of peptide synthetases, polyketide synthases, and tailoring

H. A. S. N. Abeysiri · M. M. Pathmalal
Centre for Water Quality and Algae Research, Department of Zoology, University of Sri Jayewardenepura, Nugegoda, Sri Lanka

Present Address: Faculty of Graduate Studies, University of Sri Jayewardenepura, Nugegoda, Sri Lanka

G. Y. Liyanage
Department of Aquatic Bio resources, Faculty of Urban and Aquatic Bioresources, University of Sri Jayewardenepura, Nugegoda, Sri Lanka

K. Sirisena
Department of Environmental Technology, Faculty of Technology, University of Colombo, Colombo, Sri Lanka

S. A. Kulasooriya (✉)
National Institute of Fundamental Studies, Kandy, Sri Lanka
e-mail: ananda@kulasooriya.com

N. Thajuddin et al. (eds.), *Protocols for Cyanobacteria Sampling and Detection of Cyanotoxin*, https://doi.org/10.1007/978-981-99-4514-6_56

enzymes. Gene clusters for microcystin biosynthesis have been identified and sequenced in the distantly related cyanobacterial genera *Microcystis* spp., *Planktothrix* spp. and *Anabaena* spp. Subsequently, microcystin biosynthesis (mcy) genes have been used to establish molecular techniques for detecting toxigenic cyanobacteria in laboratory and field studies. *mcy* genes of unknown origin can be assigned to the producing species. Techniques are currently being developed for the quantification of *mcy* genes in field populations. These initial genetic investigations pave the way for molecular monitoring of microcystin-producing cyanobacteria and for studying the dynamics of toxic cyanobacteria in lakes. The experience gained on microcystin biosynthesis genes will be valuable for a risk assessment of microcystin in the environment and future water management and lake-restoration strategies.

Keywords

mcy genes · Toxic cyanobacteria · PCR · Microcystin biosynthesis

56.1 Introduction

Cyanobacterial mass occurrences are a frequent phenomenon worldwide. A survey of the blooms in freshwater has shown that, on average, 59% contain toxins, with hepatotoxic blooms being more common than neurotoxic blooms (Sivonen and Jones 1999). Toxic blooms expose water users to health risks and prevent the recreational use of water (Kuiper-Goodman et al. 1999).

Microcystins (MCs) are the most prevalent cyanobacterial hepatotoxins in freshwaters, where they are produced mainly by strains of the genera *Anabaena, Microcystis, Planktothrix*, and occasionally *Nostoc* (Sivonen and Jones 1999). The toxicity of MCs is due to the inhibition of eukaryotic protein phosphatases 1 and 2A (Mac Kintosh et al. 1990) in liver cells, where MCs enter via the bile acid transport system (Carmichael 1994). MCs are cyclic heptapeptides with a general structure of cyclo (-d-Ala-X-d-erythro-β-methylasparticacid-Z-Adda-d-Glu-N-methyldehydroalanine), where X and Z are various l-amino acids and Adda is 3-amino-9-methoxy-2,6,8-trimethyl-10-phenyldeca-4,6-dienoicacid. d-Glu and Adda form the part of the molecule that interacts with the protein phosphatases and thus are the crucial amino acids for the toxicity of MCs (Goldberg et al. 1995).

Non-ribosomal enzyme complexes produce MCs. Adda is synthesized and integrated into the MC molecule by the enzymes McyG, McyD, and McyE. McyE also incorporates d-Glu, the other crucial amino acid for toxicity. Microcystin synthetase (*mcy*) gene clusters that encode these biosynthetic enzymes have now been characterized from all the main MC-producing genera (Tillett et al. 2000). Biosynthetic genes have also been proven to be a prerequisite for MC production (Dittmann et al. 1997). Although intensively studied, only a few strains of *Microcystis* (Tillett et al. 2000) that contain mcy genes have yet to be shown to produce MCs. Recent results, however, indicate that this may be more frequent among *Planktothrix* strains (Kurmayer et al. 2004).

MC-producing genera include toxic strains (with the mcy genes) and nontoxic strains (without the *mcy* genes). Toxic and nontoxic strains, sometimes of more than one genus, can coexist even in the same bloom (Vezie et al. 1998). These strains cannot be separated from each other by microscope. The revelation of the genetic basis for MC production has enabled the development of molecular methods for the detection and identification of MC producers. Most of these methods are based on PCR using primers designed to recognize the mcy genes. Many studies have concentrated on solely detecting *Microcystis*, the most common and important MC producer worldwide (Yoshida et al. 2005) or *Planktothrix* (Mbedi et al. 2005). Often the target gene has been either mcy A or mcy B (Yoshida et al. 2005). Some studies have used a combination of several biosynthetic genes (Ouahid et al. 2005) but again concentrated only on *Microcystis* sp. Mbedi et al. (2005) used eight different genes (*mcy A, mcy B, mcy E, and mcy T*) and intergenic regions (*mcyCJ, mcyEG, mcyHA, and mcyTD*) to validate their usability in detecting MC-producing *Planktothrix* strains.

56.2 Materials

40 mM EDTA, 400 mM NaCl, 0.75 M sucrose, 50 mM Tris-HCl, lysozyme, Tris-EDTA

56.3 Methods

1. Collect the cells from 75 mL to 1000 mL of lake water for DNA extraction by filtration.
2. Depending on the concentration of humic or other substances clogging the filter, use a series of filters with various pore sizes (10 μm, 5 μm, 1 μm, and 0.2 μm) to maximize the sample volume.
3. Store the filters in 2 mL of lysis buffer (40 mM EDTA, 400 mM NaCl, 0.75 M sucrose, 50 mM Tris-HCl, pH 8.3) at −20 °C until use.

56.4 DNA Extraction and Purification

1. Extract the DNA from filters with a modified hot phenol method (Giovannoni et al. 1990).
2. For the lysis of cyanobacterial cells, use higher concentrations of lysozyme (1.25 mg/mL final concentration) and proteinase K (300 μg/mL final concentration).

3. After extraction of the DNA-containing aqueous phase with phenol-chloroform-isoamyl alcohol (25:24:1, vol/vol/vol) and chloroform-isoamyl alcohol (24:1, vol/vol), precipitate DNA with sodium acetate and ethanol at −20 °C overnight.
4. Wash precipitated DNA with 70% ethanol and resuspended it in Tris-EDTA (10: 1, vol/vol) buffer.
5. Further, purify the extracted DNA with NucleoTrap and QuickStep PCR purification kits according to the manufacturer's instructions.
6. For PCRs with *mcyE* gene-targeted primers, combine the DNAs extracted from different filters in equal amounts.
7. Measure the DNA concentration with a BioPhotometer.

56.5 Designing and Testing of a Planktothrix-Specific Reverse Primer

1. Designs a *Planktothrix-specific* reverse primer (mcyE-plaR3) based on the alignment of *mcyE* gene sequences from 30MC or nodularin-producing *Anabaena, Microcystis, Nostoc, Planktothrix,* and *Nodularia* strains (Rantala et al. 2004) for use with the mcyE—general forward primer mcyE-F2 (Vaitomaa et al. 2003).
2. Tests the specificity of the resulting primer pair with MC or nodularin-producing and non-MC or non-nodularin-producing Anabaena, Aphanizomenon, Hapalosiphon, Limnothrix, Microsystis, Nodularia, Nostoc, Phormidium, Planktothrix, and Synechococcus strains.
3. Performs the PCR in a total volume of 20 μl of 1 × DyNAzyme PCR buffer, including 1 μl of DNA, 250 μM each deoxynucleotide, a 0.5 μM concentration of both primers, and 0.5 U of DyNAzyme II DNA polymerase.
4. Consist of the PCR amplification for an initial denaturation for 3 min at 95 °C, 30 cycles of 30 s at 94 °C, 30 s at 57 °C, and 60 s at 72 °C, with a final extension of 10 min at 72 °C.
5. Load the whole PCR mixture into a 1.5% agarose gel dyed with ethidium bromide (0.15 μg/mL) to ensure the detection of even the faintest amplification products and analyze the images.

56.6 PCR with Water Samples

1. To exclude the possibility of PCR inhibition causing negative *mcyE* PCR results, perform a PCR with cyanobacterium-specific 16S rRNA gene-targeted primers (Nübel et al. 1997) with DNA extracted from different-cell-size fractions.
2. If there was no amplification, carry out an additional PCR with eubacterium-specific 16S rRNA primers in case the sample contained no cyanobacteria.
3. To identify potential MC producers in lake samples, combine the DNA extractions from fractions of different cell sizes before amplification to decrease the number of samples.

4. Perform the PCR with four primer pairs designed to amplify regions of the *mcyE* gene.
5. All reaction mixtures, use the same forward primer, mcyE-F2 (Vaitomaa et al. 2003).
6. Target all potential MC-producing genera with the use of a general reverse primer, mcyE-R4 (Rantala et al. 2004), and target the MC-producing *Anabaena*, *Microcystis*, and *Planktothrix* spp. with the genus-specific reverse primers mcyE-12R, mcyE-R8 (Vaitomaa et al. 2003), and mcyE-plaR3(5′--CTCAATCTGAGGATAACGAT-3′), respectively.
7. Prepare all reaction mixtures in a 20-μL total volume containing 30 ng of extracted lake DNA as a template, a 250 μM concentration of each deoxynucleotide, and a 0.5 μM concentration of both primers.
8. Perform the *Anabaena*- and *Microcystis*-specific PCRs with 1× DyNAzyme PCR buffer containing 1 U of DyNAzyme II DNA polymerase.
9. Took place the general and *Planktothrix*-specific PCRs in 1× Super Taq Plus PCR buffer with 1 U of Super Taq Plus polymerase and 1.25 μg/μL of bovine serum albumin.
10. Involve the PCR protocols in an initial denaturation for 3 min at 95 °C; 35 cycles of 30 s at 94 °C, 30 s at 56 °C (mcyE-R4), 57 °C (mcyE-plaR3), or 58 °C (mcyE-12R, mcyE-R8), and 60 s at 68 °C (Super Taq Plus) or72°C (DyNAzyme II); and a final extension of 10 min at 68 °C or 72 °C.
11. Load the whole PCR into a 1.5% agarose gel dyed with ethidium bromide (0.15 μg/mL) and analyze the images.

References

Carmichael WW (1994) The toxins of cyanobacteria. Sci Am 270:78–86

Dittmann E, Neilan BA, Erhard M, von Döhren H, Börner T (1997) Insertional mutagenesis of a peptide synthetase gene that is responsible for microcystin production in the cyanobacterium *Microcystis aeruginosa* PCC 7806. Mol Microbiol 26:779–787

Giovannoni SJ, DeLong EF, Schmidt TM, Pace NR (1990) Tangential flow filtration and preliminary phylogenetic analysis of marine picoplankton. Appl Environ Microbiol 56:2572–2575

Goldberg J, Huang H-B, Kwon Y-G, Greengard P, Nairn AC, Kuriyan J (1995) Three-dimensional structure of the catalytic subunit of protein serine/threonine phosphatase-1. Nature 376:745–753

Kuiper-Goodman T, Falconer I, Fitzgerald J (1999) Human health aspects. In: Chorus I, Bartram J (eds) Toxic cyanobacteria in water: a guide to their public health consequences, monitoring and management. E & FN Spon, London, pp 113–153

Kurmayer R, Christiansen G, Fastner J, Börner T (2004) Abundance of active and inactive microcystin genotypes in populations of the toxic cyanobacterium *Planktothrix* spp. Environ Microbiol 6:831–841

Mac Kintosh C, Beattie KA, Klumpp S, Cohen P, Codd GA (1990) Cyanobacterial microcystin-LR is a potent and specific inhibitor of protein phosphatases 1 and 2A from both mammals and higher plants. FEBS Lett 264:187–192

Mbedi S, Welker M, Fastner J, Wiedner C (2005) Variability of the microcystin synthetase gene cluster in the genus *Planktothrix* (Oscillatoriales, cyanobacteria). FEMS Microbiol Lett 245: 299–306

Nübel U, Garcia-Pichel F, Muyzer G (1997) PCR primers to amplify 16R rRNA genes from cyanobacteria. Appl Environ Microbiol 63:3327–3332

Ouahid Y, Pérez-Silva G, del Campo FF (2005) Identification of potentially toxic environmental *Microcystis* by individual and multiple PCR amplifications of specific microcystin synthetase gene regions. Environ Toxicol 20:235–242

Rantala A, Fewer DP, Hisbergues M, Rouhiainen L, Vaitomaa J, Börner T, Sivonen K (2004) Phylogenetic evidence for the early evolution of microcystin synthesis. Proc Natl Acad Sci 101: 568–573

Sivonen K, Jones G (1999) Cyanobacterial toxins. In: Chorus I, Bartram J (eds) Toxic cyanobacteria in water: a guide to their public health consequences, monitoring, and management. E & FN Spon, London, pp 41–111

Tillett D, Dittmann E, Erhard M, von Döhren H, Börner T, Neilan BA (2000) Structural organization of microcystin biosynthesis in *Microcystis aeruginosa* PCC7806: an integrated peptide-polyketide synthetase system. Chem Biol 7:753–764

Vaitomaa J, Rantala A, Halinen K, Rouhiainen L, Tallberg P, Mokelke L, Sivonen K (2003) Quantitative real-time PCR for determination of microcystin synthetase gene E copy numbers for *Microcystis* and *Anabaena* in lakes. Appl Environ Microbiol 69:7289–7297

Vezie C, Brient L, Sivonen K, Bertru G, Lefeuvre J-C, Salkinoja-Salonen M (1998) Variation of microcystin content of cyanobacterial blooms and isolated strains in Lake grand-lieu (France). Microb Ecol 35:126–135

Yoshida M, Yoshida T, Takashima Y, Kondo R, Hiroishi S (2005) Genetic diversity of the toxic cyanobacterium *Microcystis* in Lake Mikata. Environ Toxicol 20:229–234

Molecular Imprinting Polymer (MIP) in the Detection of Microcystin-LR

57

Yashaswini DM, Sathvik G, and Lokesh Ravi

Abstract

Algal blooms cause the release of numerous toxins into water resources which can have adverse effects on life forms in and around them. One such cyanotoxin is Microcystin-LR, produced by *Microcystis aeruginosa*, which is carcinogenic to humans and also to other aquatic food sources. So, for the removal of this toxin from the water bodies, molecular imprinted polymers can be used due to their high selectivity and target retrieval. This is done by the polymerization of suitable monomer units around a template molecule, thus formed MIP is later capable of capturing the same toxin when it is encountered again.

Keywords

Microcystin-LR · MIP · Molecular imprinted polymer · Cyanotoxin · Algal blooms · Polymerization · Monomer · Detection

D. M. Yashaswini and G. Sathvik contributed equally with all other contributors.

Y. D. M · S. G
Department of Botany, School of Life Sciences, St Joseph's University, Bengaluru, Karnataka, India

L. Ravi (✉)
Department of Food Technology, Faculty of Life and Allied Health Sciences, MS Ramaiah University of Applied Sciences, Bengaluru, India

N. Thajuddin et al. (eds.), *Protocols for Cyanobacteria Sampling and Detection of Cyanotoxin*, https://doi.org/10.1007/978-981-99-4514-6_57

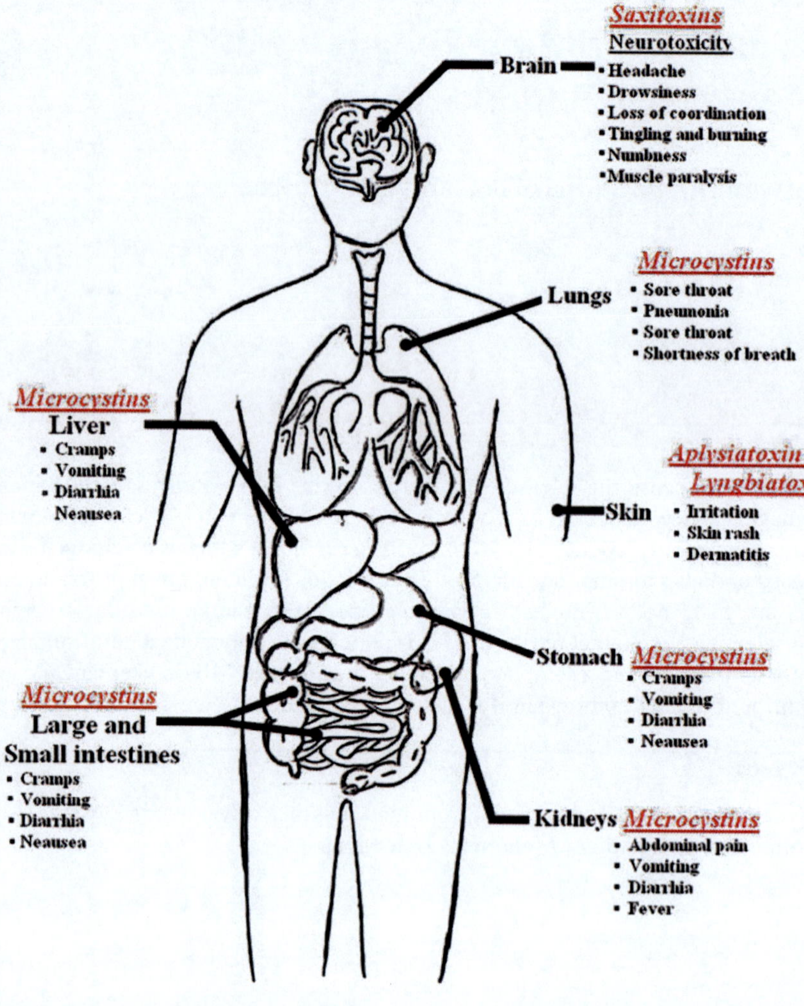

Fig. 57.1 Adverse effects of cyanotoxins and microcystin-LR in humans

57.1 Introduction

Cyanobacteria, commonly called blue-green algae are Gram-negative prokaryotes that are photosynthetic and possess the ability to colonize an entire aquatic and terrestrial ecosystem. This dominance over the other species is mainly because they thrive in much warmer conditions than other green algae meaning, their physiological function is also much quicker than others. Due to this dominant character of cyanobacteria, they are the primary cause of eutrophication (otherwise called "algal blooms") (Msagati et al. 2006). These cyanobacteria also can release harmful toxins as metabolites or during the lysing of their cell wall into the water. These can be broadly classified into three groups based on their toxicology as follows (shown in Fig. 57.1):

- Neurotoxins: Saxitoxin and neosaxitoxin or in other words, the tumor promoters
- Dermatotoxins: Lynbyatoxin A, apysiatoxins, and lipopolysaccharides.
- Heptatotoxins: Microcystins, nodularians, and cylindrospermopsins (McElhiney and Lawton 2005).

Since we are inclined toward the toxicological effects of Microcystins, we will look further into that, in this chapter.

57.1.1 Microcystin-LR

Microcystins generally are cyclic heptapeptides, containing seven peptide-linked amino acids (chemical structure shown in Fig. 57.2). The Microcystin group more or less has similar structures for all of the sub-variants, with only a few structural variations arising in these seven amino acids. Regardless, they all have three D-amino acids, N-methyldehydroalanine (Mdha), and C20B-amino acid-3-amino-9-methoxy-2, 6, 8 trimethyl-10-phenyldeca-4(E), 6(E)-dienoic acid (Adda) (Msagati et al. 2006; McElhiney and Lawton 2005; Van Apeldoorn et al. 2007) (Msagati et al. 2006; McElhiney and Lawton 2005). Microcystin-LR specifically speaking has a slight variation in the L-amino acids, which are located in positions 2 and 4 where they have Leucine (L) and Arginine (R) at these positions (McElhiney and Lawton 2005; Van Apeldoorn et al. 2007). The chemical structure of microcystin-LR is shown in Fig. 57.2 as retrieved from PubChem database (PubChem 2004).

57.1.2 Reason for Toxicity and Physiological Effects of MY-LR

The toxicity of cyanotoxins and microcystin-LR in particular is due to the inhibition of serine/threonine protein phosphatases 1 and 2A, which later causes hepatocyte necrosis, and hemorrhage, damage to the liver can lead to death eventually. However, low concentrations of Microcystins can also have adverse effects on the gastrointestinal tract, kidneys, heart, and even lungs. Many cases of MY-LR toxic poisoning have also been directly related to consuming contaminated water or aquatic food like fish, muscles, and prawns, that swim in these toxic bloom waters (McElhiney and Lawton 2005; Abdallah et al. 2021).

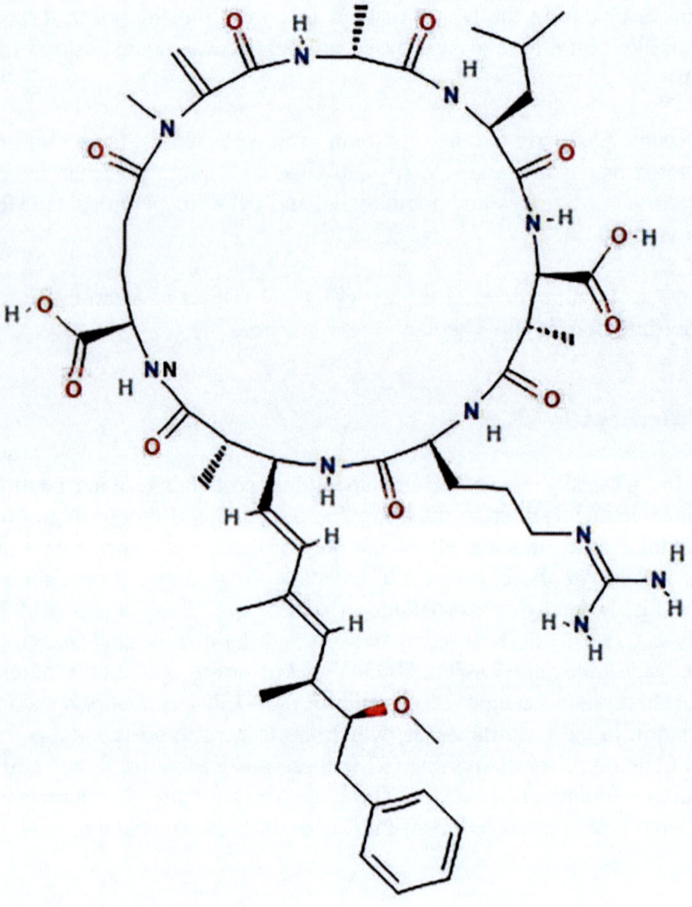

Fig. 57.2 Structure of microcystin-LR from PubChem

57.1.3 Molecular Imprinted Polymer

The molecular immunology antibody formation theory proposed by L Pauling paved the way for the development of molecular imprinting, it is the creation of an imprinted cavity that matches the shape of the template molecule (they form specific binding sites for the template molecule)in a polymer matrix, and it involves the polymerization of functional monomers by cross-linking (Xiao et al. 2018; Mosbach 1994; Zaidi 2016). Molecular imprinted polymers are highly stable and capable of target extraction, so their potential is exploited for drug delivery, chemical analysis, and organic synthesis (Ansari and Karimi 2017).

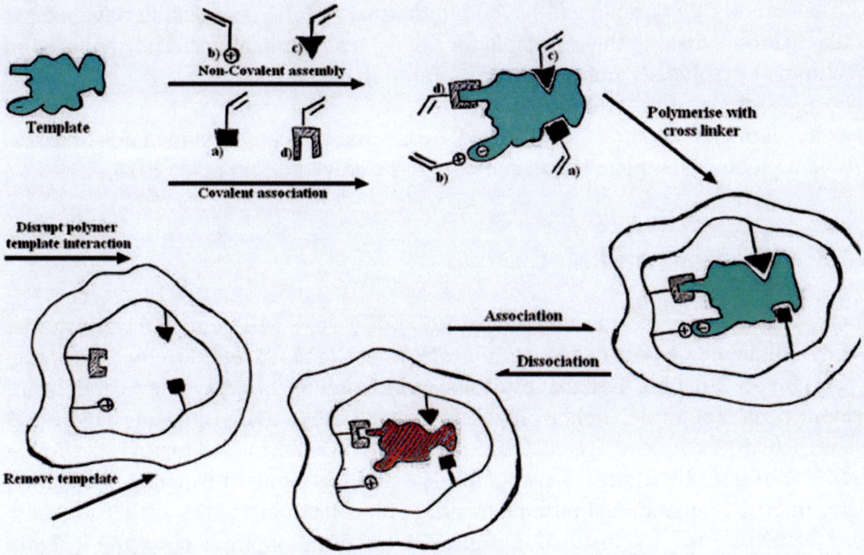

Fig. 57.3 A scheme for synthesis and imprinting process of MIP

Covalent and non-covalent approaches have been developed for the creation of MIPs (Mosbach 1994; Ansari and Karimi 2017). The template or the target molecule is introduced into a solvent with a cross-linker and monomer (which undergoes polymerization around the template molecule), this makes up the polymer matrix. Once the template is removed from the polymer matrix, a permanent three-dimensional cavity matching the shape and size of the target molecule is left behind which helps in target extraction by selectively rebinding with the target molecules during their detection (Zaidi 2016; Ansari and Karimi 2017). A graphical schematic representation of the MIP synthesis is shown in Fig. 57.3.

57.1.4 Advantages and Applications of Molecular Imprinted Polymers

1. They are excellent separation materials and have high selectivity so they are used in sample preparations (Mosbach 1994; Turiel and Martín-Esteban 2010).
2. They are used as sensors and can replace biological molecules; they can mimic antibodies so they are used in antigen detection (Mosbach 1994).
3. MIPs are used as selective adsorbents in solid-phase extraction due to their high selectivity (Xiao et al. 2018).
4. They are used in the detection of illegal drugs and additives (Xiao et al. 2018).
5. They are used as agents for drug delivery (Ansari and Karimi 2017).

6. They are highly durable and have high thermal stability and stability against pH fluctuations, making them suitable for widespread applications. They are used in chromatographic separations as well (Zaidi 2016).

Despite all the advantages, we still need a convenient preparation method for MIPs. Problems such as template leakage are yet to be solved (Xiao et al. 2018).

57.2 Materials and Methods

Some of the suggested monomers to build an MIP are Disacrylamide, 1-vinylimidazole, AMPSA, UA (Imidazole 4-acrylic acid), etc. Microcystin - LR, 2-acrylamido-2-methyl-1-propanesulfonic acid (AMPSA), methylenesuccinic acid, cyclohexanecarbonitrile, imidazole-4-acrylic acid ethyl ester (urocanic acid ethyl ester, UAEE), ethylene glycol dimethacrylate (EGDMA), dimethyl sulfoxide (DMSO), 100 mM HCl in 50% methanol, 100 mM NaOH in 50% methanol, distilled water in 50% methanol and pure methanol (Chianella et al. 2002). All the solvents used were purified by distillation followed by filtration to remove the volatile impurities.

57.2.1 Synthesis of Polymer

- 1 mg of 1×10^{-6} M of microcystin-LR is dissolved in 300 μL of DMSO and mixed with 1×10^{-6} M of 2-acrylamide-2-methyl-1 propanesulfonic acid (AMPSA), 6×10^{-6} M of UAEE, EDGMA as cross-linker and cyclohexanecarbonitrile as an initiator in a glass bottle of 2 mL capacity.
- Nitrogen is added to the solution to remove oxygen, it is sealed and kept at 80 °C for 24 h.
- The polymer that has been synthesized is ground in a mortar, and the powder obtained is washed in 100 mM HCl in 50% methanol, 100 mM NaOH in 50% methanol, distilled water in 50% methanol, and pure methanol.
- The particles of polymer are wet sieved in acetone using two sieves of sizes 63 and 45 μm.
- The collected polymer is dried in an oven overnight at 60 °C (Chianella et al. 2002).

57.2.2 Reactivity Study

Enzyme-linked competitive assay was performed to study the reactivity between, the molecular imprinted polymer and microcystin-LR. This was done by adding the MC-LR and its analogue to the polymer suspension and testing it further with an ELISA test (Chianella et al. 2002).

57.3 Conclusion

Molecular imprinted is still an emerging technique although considerable progress has been made, these MIPs are capable of mimicking the natural and biological recognition bodies like antibodies. Because of this nature, they are used in the detection of toxic substances like Microcystin-LR. Detection and removal of microcystin-LR from water sources are necessary as they can have harmful reactions on plants, animals, and humans.

References

Abdallah MF, Van Hassel WH, Andjelkovic M, Wilmotte A, Rajkovic A (2021) Cyanotoxins and food contamination in developing countries: Review of their types, toxicity, analysis, occurrence and mitigation strategies. Toxins 13(11):786.

Ansari S, Karimi M (2017) Recent progress, challenges and trends in trace determination of drug analysis using molecularly imprinted solid-phase microextraction technology. Talanta 164:612–625

Chianella I, Lotierzo M, Piletsky SA, Tothill IE, Chen B, Karim K, Turner APF (2002) Rational design of a polymer specific for microcystin-LR using a computational approach. Anal Chem 74:1288–1293

McElhiney J, Lawton LA (2005) Detection of the cyanobacterial hepatotoxins microcystins. Toxicol Appl Pharmacol 203:219–230

Mosbach K (1994) Molecular imprinting. Trends Biochem Sci 19:9–14

Msagati TAM, Siame BA, Shushu DD (2006) Evaluation of methods for the isolation, detection and quantification of cyanobacterial hepatotoxins. Aquat Toxicol 78:382–397

PubChem (2004) PubChem Compound Summary for CID 445434, Microcystin-LR. National Library of Medicine (US), National Center for Biotechnology Information, Bethesda (MD). https://pubchem.ncbi.nlm.nih.gov/compound/Microcystin-LR

Turiel E, Martín-Esteban A (2010) Molecularly imprinted polymers for sample preparation: a review. Anal Chim Acta 668:87–99

Van Apeldoorn ME, Van Egmond HP, Speijers GJA, Bakker GJI (2007) Toxins of cyanobacteria. Mol Nutr Food Res 51:7–60

Xiao D, Jiang Y, Bi Y (2018) Molecularly imprinted polymers for the detection of illegal drugs and additives: a review. Microchim Acta 185:247. https://doi.org/10.1007/s00604-018-2735-4

Zaidi SA (2016) Molecular imprinted polymers as drug delivery vehicles. Drug Deliv 23:2262–2271

Molecular Identification of the Cyclic Peptide Hepatotoxins in Cyanobacteria

58

F. S. Idroos, P. M. Manage, and S. A. Kulasooriya

Abstract

Various cyanobacteria can produce cyclic peptide hepatotoxins such as Microcystin and Nodularin. Cyanobacterial orders, *Oscillatoriales, Chroococcales, Stigonematales*, and *Nostocales*, have been recorded as Microcystin producers. Nodularin production is a distinguishing feature of the *Nostocales* genus *Nodularia*. There are numerous ways for monitoring and quantifying cyanotoxins; nevertheless, a single rapid method is required to consistently discover cyanobacteria capable of producing these hepatotoxins. PCR is a well known as a technique to detect all potentially cyclic peptide hepatotoxin (microcystin and nodularin) generating cyanobacteria in laboratory cultures and harmful algal blooms. The aminotransferase (AMT) domain, found on the mcyE and ndaF modules of the microcystin and nodularin synthetase enzyme complexes, was chosen as the target sequence because it is required for the synthesis of all microcystins and nodularin. It is possible to produce a 472 bp PCR product from the AMT domains of hepatotoxic species using the specified PCR. Through bioinformatic analyses of the AMT in microcystin and nodularin synthetases, sequence data provide further insight into the evolution of the

F. S. Idroos
Centre for Water Quality and Algae Research, Department of Zoology, University of Sri Jayewardenepura, Nugegoda, Sri Lanka

P. M. Manage
Centre for Water Quality and Algae Research, Department of Zoology, University of Sri Jayewardenepura, Nugegoda, Sri Lanka

Faculty of Graduate Studies, University of Sri Jayewardenepura, Nugegoda, Sri Lanka

S. A. Kulasooriya (✉)
National Institute of Fundamental Studies, Kandy, Sri Lanka
e-mail: anada@kulasooriya.com

N. Thajuddin et al. (eds.), *Protocols for Cyanobacteria Sampling and Detection of Cyanotoxin*, https://doi.org/10.1007/978-981-99-4514-6_58

microcystin and nodularin synthetases, with consistency between the evolution of 16S rRNA and the AMT domain.

Keywords

Cyanobacteria · Cyanotoxins · Microcystins · Nodularin · PCR · Cyclic peptide hepatotoxin

58.1 Introduction

Cyanobacteria are autotrophic microorganisms with a long evolutionary history and intriguing metabolic properties. Cyanobacteria perform plant-like photosynthesis that produces oxygen. Cyanobacteria can be found in a variety of settings, including water (fresh and brackish water, oceans, and hot springs), land (soil, deserts, and glaciers), and symbiosis (with plants, lichens, and primitive animals). Cyanobacteria are essential primary producers and constituents of phytoplankton in aquatic environments. Biofilms and mats can also form by benthic cyanobacteria. Cyanobacteria typically develop mass occurrences, referred to as blooms in eutrophic water. Cyanobacterial blooms are capable of producing toxins. Possible putative functions include a feeding deterrent, quorum-sensing, or iron scavenging. In addition, the sporadic distribution of the ability to synthesize these compounds in several cyanobacterial lineages, and the congruent evolution between the phylogenetic relationship of 16S rDNA16S rDNA, rpoC1, and specific regions of the microcystin synthetase geneMicrocystin synthetase gene in hepatotoxicHepatotoxic cyanobacteria, has led to the hypothesis that the evolution of toxin production is an ancestral relict influenced by spontaneous horizontal gene-transfer and intragenomic recombination (Jungblut and Neilan 2006). Presence of Cyanobacterial toxin in environment can impose health issues in human and animals (Sivonen n.d.).

Cyanobacteria can produce different types of toxic compounds, which include hepatotoxins (such as microcystins), neurotoxins (such as saxitoxins), cytotoxins, dermatotoxins, and irritant toxins (Sangolkar et al. 2006). The most studied group of cyanobacterial toxins is the hepatotoxic cyclic peptides, which includes microcystins and nodularin. Cyclic heptapeptide microcystins are the most frequently found cyanobacterial toxins in freshwater blooms. *Microcystis*, *Anabaena*, *Planktothrix* (*Oscillatoria*), *Nostoc*, *Hapalosiphon*, *Anabaenopsis*, etc. are common microcystin-producing cyanobacterial genera (McElhiney and Lawton 2005). The toxin microcystin is produced non-ribosomally via a multifunctional enzyme complex consisting of peptide synthetase and polyketide synthase modules coded by the mcy gene cluster. Microcystins are considered as a cyclic heptapeptide (D-alanine-X-D-MeAsp-Z-Adda-D-glutamate-Mdha). D-MeAsp is D-b-erythro-methyl-aspartic acid, and Mdha is N-methyldehydroalanine; X and Z are variable L-amino acids (Jungblut and Neilan 2006). Currently, more than 80 microcystin variants have been identified, and they vary largely by the degree of methylation, peptide sequence, and

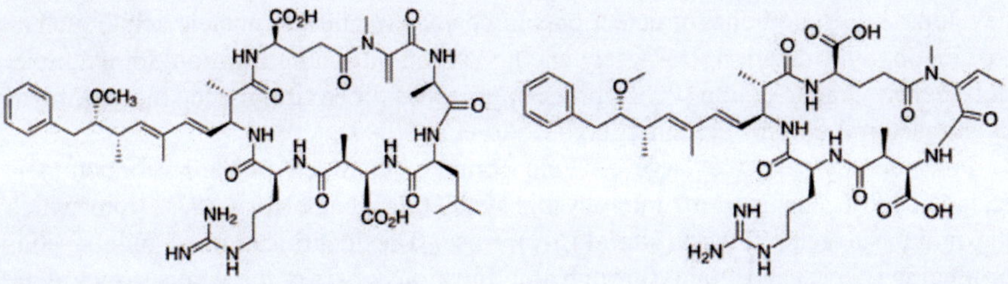

(A) (B)

Fig. 58.1 Chemical structure of Cyanobacterial hepatotoxins (**a**) Microcystin and (**b**) Nodularin

toxicity (Sivonen n.d.). The most common types are Microcystin-LR, -RR, and -YR, where the variable amino acids (X and Y) are leucine-arginine and tyrosine (Jungblut and Neilan 2006).

Nodularin, structurally related to Microcystin and has a similar mode of toxicity, has been isolated from one species of cyanobacteria, *Nodularia spumigena*. In brackish water, cyclic pentapeptide Nodularin is common (Sivonen n.d.) (Fig. 58.1).

The general structure of nodularin is cyclo-(D-MeAsp-L-arginine-Adda-D-glutamate-Mdhb), and Mdhb is 2-(methylamino)-2-dehydrobutyric-acid. Only a small number of nodularin variants have been found in nature, such as [D-Asp]-nodularin and [dmAdda]-nodularin (Jungblut and Neilan 2006).

Microcystin and nodularin are synthesized non-ribosomally by large multi-enzyme complexes consisting of different modules, including non-ribosomal peptide synthetases (NRPS) and polyketide synthases (PKS), as well as several tailoring enzymes. These modules catalyze specific amino acids' activation, modification, and condensation (Kurmayer and Christiansen 2009).

Microcystin and nodularin are considered as hepatotoxic as they can enter the hepatocytes via the bile acid carriers in their cell walls. Hence, death in exposed animals is caused by internal hemorrhage and subsequent shock. The toxic effect depends on the animal's size; thus, death may be delayed in larger animals. Toxicities of microcystins and nodularin vary (LD50 values vary from highly toxic—50 mg kg^{-1}, intravenously injected mice—to nontoxic). Oral toxicities of microcystins and nodularin are less than by intravenous injection. The nontoxic variants of these toxins include the 6Z-stereoisomer of Adda, the esterified carboxylic group in glutamate, and the linear variants of these compounds. Both microcystins and nodularins are serine/threonine-specific protein phosphatase (1 and 2A) inhibitors and tumor promoters (Sivonen n.d.).

Toxic and nontoxic populations of cyanobacteria can coexist in a single ecosystem and are indistinguishable by microscopy, rendering this technique unreliable for determining potential microcystin production. Methods commonly used for toxin monitoring include high-performance liquid chromatography (LC) with photodiode array (PDA) detection and/or mass spectrometry (MS), a protein phosphatase inhibition assay (PPIA), and enzyme-linked immunosorbent assays (ELISA). With the development of methods for detection and characterization of nucleic acids, such as extraction, hybridization, DNA sequencing, and in vitro amplification, for example, polymerase chain reaction (PCR), novel approaches for environmental monitoring of cyanobacterial blooms are emerging (Hotto et al. 2007).

For microcystins, the large enzyme complex involved in their production is encoded by a cluster of 10 microcystin synthetase genes (mcy A-J), from which the nodularin gene cluster (nda cluster) arose. The identification of microcystin-producing species and strains through amplifications of six of these genes (mcy A-E) became most commonly applied. Due to the absence of mcy genes in non-toxigenic species or strains, standard PCR with relevant mcy-specific primers provides a quick qualitative tool to discriminate between potentially toxic and nontoxic algae in water monitoring and general ecological studies. To get information about actual toxin concentrations through molecular tools, efforts were made to correlate cyanotoxin biosynthesis gene abundance with cyanotoxin occurrence and concentrations. Up to now, these tools were most successful for microcystins and nodularin and the estimation of cyanotoxin gene abundance in field populations; a quantitative PCR (qPCR, or real-time PCR) was developed (Radkova et al. 2020).

The level of toxin produced depends on the stage of blooms, species and strains involved, which limits the microscopic examination for assessing the toxic nature of a sample. However, with the application of immunological, molecular and instrumental techniques, researchers made it possible to detect, differentiate, quantify, and monitor the toxic cyanobacteria and toxins.

58.2 Materials

Cyanobacterial strains and bloom samples

XS extraction buffer (1% potassium-methyl xanthogenate; 800 mM ammonium acetate; 20 mM EDTA; 1% SDS; 100 mM Tris–HCl, pH 7.4)

Vortex

Incubator

Centrifuge

4 M ammonium acetate

Isopropanol

70% ethanol

Sterile water

Taq DNA polymerase

2.5 mM $MgCl_2$

1× Taq-polymerase buffer

0.2 mM dNTPs

0.5 pmol of forward and reverse primer

1 μL of template DNA

Thermocycler

HEP-primers

2% Agarose gels
Ethidium bromide (1 μg mL^{-1})
Reference sequences from GenBank (NCBI)

58.3 Methodology

58.3.1 DNA Extraction (Sangolkar et al. 2006; Jungblut and Neilan 2006; Köker et al. 2017)

1. Use approximately 200 mg of cyanobacterial cells (wet weight) for extraction.
2. First, combine the cells with 600 μL of XS extraction buffer.
3. Vortex-mix the mixture and incubate at 65 °C for 2 h and cool the extracts for 10 min on ice.
4. Remove the cell debris by centrifugation at 12,000 g for 10 min.
5. Precipitate the DNA by adding 1 volume of isopropanol and 1/10 volume of 4 M ammonium acetate for 15 min at 4 °C.
6. Pellet the precipitated DNA by centrifugation at 12,000 g for 10 min and wash with 70% ethanol.
7. Re-suspend the extracted DNA in 100 μL of sterile water.

58.3.2 Polymerase Chain Reaction (PCR) (Jungblut and Neilan 2006; Hotto et al. 2007; Köker et al. 2017; Moffitt and Neilan 2004)

1. Perform PCR using 0.2 U of Taq DNA polymerase in a 20 μL reaction volume containing 2.5 mM $MgCl_2$, 1× Taq-polymerase buffer, 0.2 mM dNTPs, 0.5 pmol of forward and reverse primer.
2. For PCR, use 1 μL of template DNA at a concentration of approximately 100 ng μl^{-1}.
3. Perform thermal cycling in a thermocycler.
4. Perform PCR amplification using the HEP-primers, HEPF (5'--TTTGGGGTTAACTTTTTTGGGCA-TAGTC-3′) and HEPR (5'--AATTCTTGAGGCTG-TAAATCGGGTTT-3′) primer-set.
5. An initial denaturation step at 92 °C for 2 min is followed by 35 cycles of 92 °C for 20 s, 52 °C for 30 s, and 72 °C for 1 min, with a final extension step at 72 °C for 5 min.
6. The PCR amplify a 472 bp gene fragment.
7. Visualize PCR amplification products using gel electrophoresis on 2% agarose gels with 1X TAE-buffer and stain with ethidium bromide (1 μg mL^{-1}) for 10 min.

58.3.3 Sequencing and Accession Numbers (Jungblut and Neilan 2006; Köker et al. 2017)

1. Perform sequencing using HEPF- and HEPR primers for each amplification product.
2. Perform automated sequencing using a cycle sequencing system.
3. Check the availability of observed sequences in the GeneBank (NCBI).

58.3.4 Phylogenetic Sequence Analysis (Jungblut and Neilan 2006; Hotto et al. 2007; Köker et al. 2017)

1. Analyze sequence data using the Applied Biosystem Auto-Assembler computer program.
2. Sequences from the cultures were aligned and phylogenetically analyzed according to their translated amino acid sequences.
3. Compare translated amino acid sequences in a multiple alignment.
4. Obtain the phylogenetic tree and calculate the confidence levels via bootstrapping.
5. Obtain reference sequences from GeneBank (NCBI).
6. Analyze maximum-parsimony and maximum likelihood, respectively, using software (Fig. 58.2).

58.3.5 Observation

The efficiency of the PCR-based method is much greater than other methods in cases of low abundance of toxigenic strains. Therefore, it can underline the sensitivity of molecular methods for identifying cyclic peptide hepatotoxins in cyanobacteria. However, until a single unique method is adopted, combining different approaches is more desirable for cyanotoxin monitoring practices. However, the affiliation of most of the isolated *myc*E strains from different studies to genus level due to unclassified species sequences suggested that the currently available cyanobacterial genomic sequence data are still insufficient to resolve fully the species phylogenetic identification of the isolated *mcy* pool sequences covering the whole range of *mcy*E diversity and toxicity.

Fig. 58.2 Schematic diagram of molecular identification approaches of cyanotoxins

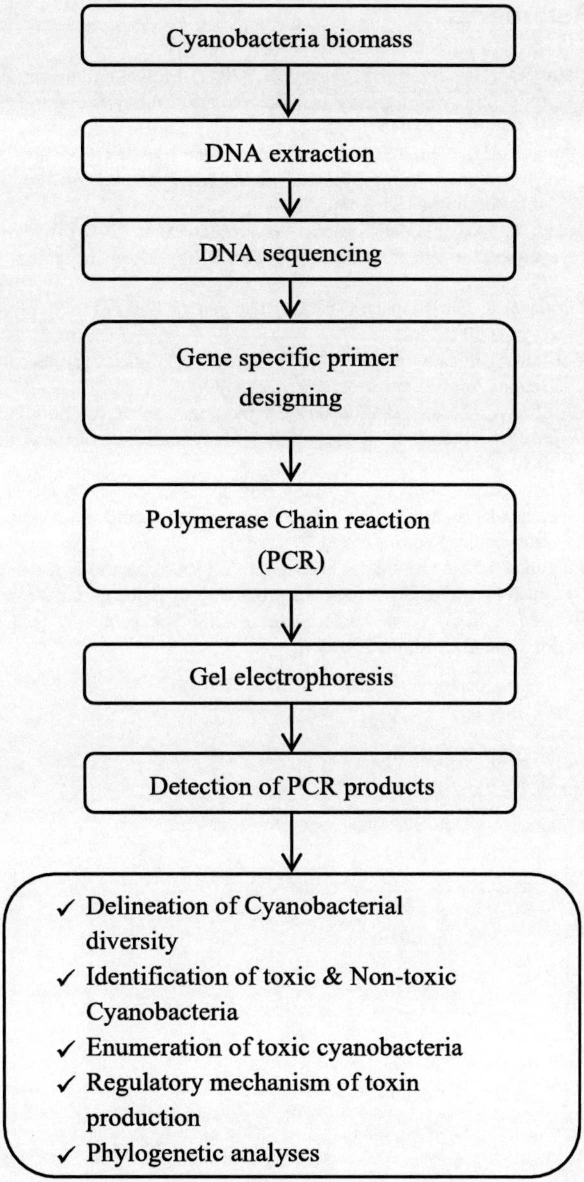

Acknowledgment Centre for Water Quality and Algae Research, Department of Zoology, University of Sri Jayewardenepura, Gangodawila, Nugegoda, 10250, Sri Lanka.

References

Hotto AM, Satchwell MF, Boyer GL (2007) Molecular characterization of potential microcystin-producing cyanobacteria in Lake Ontario embayments and nearshore waters. Appl Environ Microbiol 73(14):4570–4578

Jungblut AD, Neilan BA (2006) Molecular identification and evolution of the cyclic peptide hepatotoxins, microcystin and nodularin, synthetase genes in three orders of cyanobacteria. Arch Microbiol 185(2):107–114

Köker L, Akçaalan R, Albay M, Neilan BA (2017) Molecular detection of hepatotoxic cyanobacteria in inland water bodies of the Marmara region, Turkey. Adv Oceanogr Limnol 8(1):52–60

Kurmayer R, Christiansen G (2009) The genetic basis of toxin production in cyanobacteria. Fr Rev 2(1):31–50

McElhiney J, Lawton LA (2005) Detection of the cyanobacterial hepatotoxins microcystins. Toxicol Appl Pharmacol 203(3):219–230

Moffitt MC, Neilan BA (2004) Characterization of the nodularin synthetase gene cluster and proposed theory of the evolution of cyanobacterial hepatotoxins. Appl Environ Microbiol 70(11):6353–6362

Radkova M, Stefanova K, Uzunov B, Gärtner G, Stoyneva-Gärtner M (2020) Morphological and molecular identification of microcystin-producing cyanobacteria in nine shallow Bulgarian water bodies. Toxins 12(1):39

Sangolkar LN, Maske SS, Chakrabarti T (2006) Methods for determining microcystins (peptide hepatotoxins) and microcystin-producing cyanobacteria. Water Res 40(19):3485–3496

Sivonen K (n.d.) Cyanobacterial toxins. In: Schaechter M (ed) Encyclopedia of microbiology. Elsevier, Oxford, pp 290–307

Molecular Identification of Microcystin Synthetase Genes *mcyE* in Cyanobacteria Using PCR

59

G. Y. Liyanage, S. A. Kulasooriya, S. P. Manage, D. M. P. I. Dissanayake, and P. M. Manage

Abstract

Cyanobacteria are important primary producers that are abundant in nature, yet some create noxious chemicals known as cyanotoxins that are harmful to humans and animals. Due to their considerable toxicity, rapid and precision detection of the toxin-producing cyanobacteria is essential. However, due to the need for a morphological distinction between toxic and non-toxic strains within the same cyanobacterial species or genus, typical microscopic approaches have considerable limitations for reliable identification. Due to the intrinsic conservation of their genetic patterns, nucleic acid-based assays can provide more accurate identification. Molecular assays also offer high throughput and significantly shorten the time it takes to get test results. Such benefits make these assays a benchmark technique for quick identification and early notification of toxic

G. Y. Liyanage
Department of Zoology, Centre for Water Quality and Algae Research, University of Sri Jayewardenepura, Nugegoda, Sri Lanka

Faculty of Urban and Aquatic Resources Management, University of Sri Jayewardenepura, Nugegoda, Sri Lanka

S. A. Kulasooriya (✉)
National Institute of Fundamental Studies, Kandy, Sri Lanka
e-mail: ananda@kulasooriya.com

S. P. Manage · D. M. P. I. Dissanayake
Centre for Water Quality and Algae Research, Department of Zoology, University of Sri Jayewardenepura, Nugegoda, Sri Lanka

P. M. Manage
Department of Zoology, Centre for Water Quality and Algae Research, University of Sri Jayewardenepura, Nugegoda, Sri Lanka

Faculty of Graduate Studies, University of Sri Jayewardenepura, Nugegoda, Sri Lanka

N. Thajuddin et al. (eds.), *Protocols for Cyanobacteria Sampling and Detection of Cyanotoxin*, https://doi.org/10.1007/978-981-99-4514-6_59

strains of cyanobacteria using their toxin-producing genes (DNA) as marker genes.

For example, Microcystins, widely occurring non-ribosomal oligopeptides produced by several genera of cyanobacteria (e.g., *Anabaena, Hapalosiphon, Microcystis, Nostoc)*, are encoded by a cluster of microcystin synthetase genes (*mcy*). Therefore, numerous molecular assays targeting such cyanotoxin synthetase genes have been developed to identify toxigenic cyanobacteria at various taxonomic levels. Polymerase chain reaction (PCR)-based assays are the most prevailing. Among different versions of PCR assays, real-time quantitative PCR can quantify the genes of interest in samples, fulfilling the purpose of both taxonomic recognition and biomass estimation. Reverse transcription (RT)-PCR assays can be used to detect transcripts (i.e., mRNAs) from toxin synthetase genes, probably enhancing the predictive value of PCR detection for toxin production from observed cyanobacterial species. Nevertheless, the utility of the toxin synthetase gene- or its transcript-based PCR assays for routine cyanotoxin monitoring needs to be further evaluated on a large scale.

Keywords

Microcystin · Genes · Polymerase chain reaction (PCR) · *mcy*

59.1 Introduction

Microcystin (MC) is a cyclic heptapeptide that inhibits protein phosphatase type 1 and 2A in humans and animals by forming an irreversible covalent bond to a cysteine in the catalytic domain of these enzymes (Schreidah et al. 2020). It consists of the following amino acids: D-alanine, X, D-MeAsp (D-erythro-ß-methyl-aspartic acid), Z, Adda ((2S,3S,8S,9S)-3-amino-9-methoxy-2,6,8-trimethyl-10-phenyldeca-4,6-dienoic acid), D-glutamic acid, and Mdha (N-methyldehydroalanine) (Thenuwara et al. 2022). The X and Z represent variable L amino acids. It reportedly has over 80 variants, mainly differing in the amino acids at positions X and Z (Fig. 59.1).

Microcystin is a non-ribosomal oligopeptide, which means unlike most peptides and proteins, it is not synthesized by cellular ribosomes. The enzymes responsible for its synthesis contain the non-ribosomal peptide synthetases (NRPS) and polyketide synthases (PKS) modules and tailoring functional domains. All the enzymes are the protein products encoded by the microcystin synthetase genes (*mcy*) located as a cluster in the genome.

In *Microcystis*, ten *mcy* genes (*mcyA-J*) span 55 kb near the center of its 5.8 Mb circular chromosome and form two operons (*mcyABC* and *mcyD-J*), of which the transcription proceeds in discretely opposite directions (Yuan and Yoon 2021). The 55.4 kb *mcyA-J* gene cluster in *Anabaena* also forms two operons (*mcyABC* and *mcyGDJEFIH*), one of which the gene order differs from *Microcystis* (Wenski et al. 2022). In contrast, *Planktothrix* has a 55.6 kb mcy cluster including eight essential

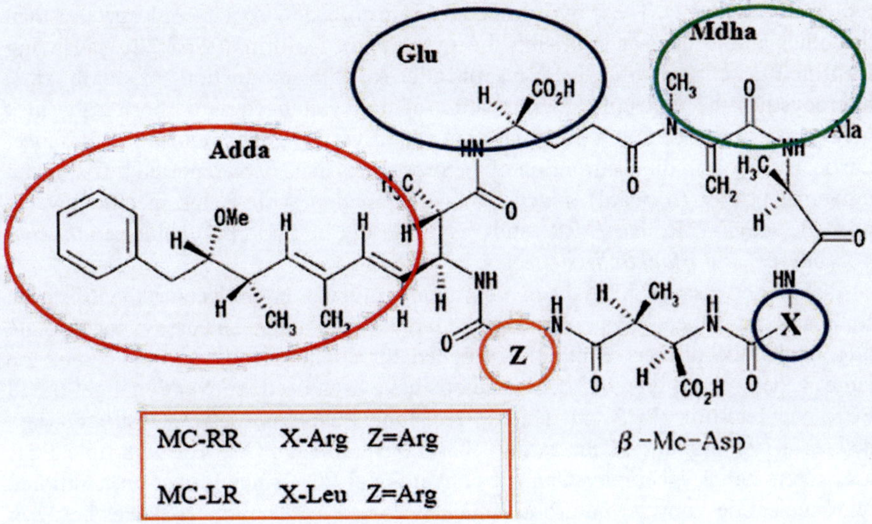

Fig. 59.1 General structure of microcystin (MC)

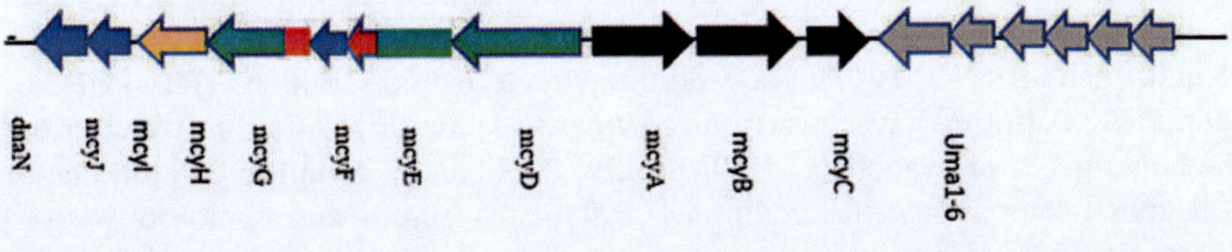

Fig. 59.2 Biosynthetic gene cluster in *Microcystis* (Blue-Tailoring enzymes; Green-Polyketide synthases; Black-non-ribosomal peptide synthetases; Grey-Non-microcystin synthetase; Yellow-ABC transporter)

genes (*mcyABCDEGHJ*) that form a single operon and one unique gene (*mcyT*), and the arrangement and sequence of specific domains in the gene products differ from those in other genera (Fig. 59.2).

Per annotation of *mcy* genes, microcystin biosynthesis is initiated by *mcyG* to covalently bind a phenylacetate precursor that is then methylated by the *mcyJ* gene. Next, *mcyD* facilitates the elongation of the growing chain by accepting a malonyl-CoA, and the *mcyE* introduces another malonyl-CoA which further extends the backbone of microcystin. As a racemase, *mcyF* is involved in either the supply of D-glutamate or D-MeAsp, or the peptidyl epimerization of L-glutamate. Then *mcyA* captures an L-serine and installs it into the growing chain, and the Mdha moiety is synthesized by *mcyI*. Following adding an amino acid into position X and a D-MeAsp by *mcyB*, *mcyC* adds the last amino acid into position Z (Yuan and Yoon 2021).

The *mcyE* gene was a logical choice as it is essential for synthesizing the bioactive side-chain Adda and activating and adding D-glutamate into the microcystin molecule (Fig. 59.2). These two amino acids are crucial to toxicity and vary less than the other amino acids comprising the many toxin isoforms (Piña 2011). Having identified a cluster of genes responsible for the production of toxin (i.e., microcystin), the molecular identification of the cyanobacteria harboring the *mcy* gene cluster has become a potential application via PCR techniques. For instance, the occurrence and the distribution of the mcy genes have been reported by using the universal primer (forward), *mcyE*-F2 in conjunction with genus-specific reverse primers, *mcyE*-12R, *mcyE*-R8, and *mcyE*-plaR3, in toxin-producing *Anabaena, Microcystis*, and *Planktothrix* species, respectively.

PCR-based assays have been commonly utilized in molecular identification studies because the assays can recognize targets accurately. The assays incorporate oligonucleotide primers explicitly designed for complementary sequences of the target gene(s). Two types of PCR methods have been used: conventional gel-based PCR and real-time PCR. In general, real-time PCR has higher sensitivity (i.e., detects a low amount of the target) than conventional PCR. The real-time PCR also offers better specificity than the conventional PCR since it uses an additional oligonucleotide known as a probe, which is complementary to sequences between primer-binding sequences (Kim et al. 2022).

Furthermore, real-time PCR allows estimating the number of the intended target in samples when performed with standards with a known copy number of the target sequences. This procedure is referred to as quantitative real-time PCR (qPCR) (Kralik and Ricchi 2017). In addition, reverse transcription (RT)-PCR or RT-qPCR platforms have been utilized to detect transcripts (i.e., mRNAs) from the target genes of cyanobacteria. Typically, PCR can be completed within one or 2 h, much shorter than the traditional analytical methods and microscopy-based approaches mentioned above.

59.2 Methods

59.2.1 DNA Extraction and Purification (Rodriguez 2016)

59.2.1.1 Materials

47 mm polycarbonate filter with a 0.22 m Millipore pore size
70% methanol
50 mM Tris-HCl
5 mM EDTA
50 mM NaCl
10 mg mL^{-1}Proteinase K
10% SDS
Phenol
Chloroform
Isoamyl alcohol
MOBIO PowerWater DNA Extraction Kit

59.2.1.2 Phenol-Chloroform Method

1. Filter 500 mL of the sample through a 47 mm polycarbonate filter with a 0.22 μm Millipore pore size and filter paper is the dip in 70% methanol and kept in a freezer for 24 h.
2. Filter the sample to remove filter paper particles and centrifuge at 12000 rpm.
3. Resuspend the purified cell pellet in 500 μL lysis buffer containing 50 mM Tris-HCl, 5 mM EDTA, and 50 mM NaCl.
4. Incubate the mixture in a water bath at 55 °C for 30 min.
5. Add 10 μL of 10 $mgmL^{-1}$Proteinase K and 20 μL of 10% SDS to the sample tube.
6. Continue the incubation at 55 °C until a clear solution appeared, followed by chilling on ice for 15 min to inhibit the lysis.
7. Extract the DNA with an equal volume of Phenol: Chloroform: Isoamyl Alcohol 25:24:1 by gentle inversion of the tube for 15 min.
8. Centrifuge solution mixture at 12,000 rpm for 5 min, and the collect clear upper aqueous layer into a new sterile 2 mL size microcentrifuge tube.
9. Add an equal volume of 4 M Ammonium Acetate and gently mix.
10. Add two volumes of ice-cold isopropanol to achieve precipitation of DNA.
11. Centrifugate the sample at 14,000 rpm for 5 min.
12. Wash the DNA pellet with 70% ethanol.
13. Air dry the resulting DNA pellet from the above step and dissolve in 12 μL of PCR water.
14. Further purified extracted DNA with NucleoTrap PCR purification (BD Biosciences) and QuickStep PCR purification (Edge BioSystems) kits according to the manufacturer's instructions.
15. Store the DNA sample at −20 °C for PCR analysis.

59.2.1.3 DNA Extraction Using a Commercial Kit (Hinlo et al. 2017)

1. Filter 500 mL of water required for the DNA extraction through a 47 mm polycarbonate filter with a 0.22 μm Millipore pore size.
2. Until required for the extraction, filter papers must be kept at −20 °C after being soaked in 10 mL of 70% methanol.
3. Extract DNA from the filer paper using the MOBIO PowerWater DNA Extraction Kit (Qiagen, USA) according to the manufacturer's instructions.
4. Store the purified DNA at −20 °C until the PCR analysis.

59.2.2 Conventional PCR to Identify Toxin-Producing Genes (Ramya et al. 2018)

59.2.2.1 Materials for the PCR

0.5 μL of Target Primer
Go Taq reaction buffer
dNTPs
25 mM MgCl2

DNA polymerase
TAE Buffer

59.2.2.2 Preparation of PCR Master Mix

1. Add 0.5 μL of target primer (10 μL), 5 μl of Go Taq reaction buffer, 0.5 μl of dNTPs, 2.0 μL of 25 mM $MgCl_2$, and 0.1 μL of GotaqDNA polymerase to the PCR mixture and adjust to a total volume of 25 μL.
2. Use purified DNA (5 L) as the template.
3. Use optimal conditions for the primers which are shown in Table 59.1.

59.2.2.3 PCR Amplification

PCR amplification cycles are given below for the analysis of *mcyE*

1. Initial denaturation at 94 °C for 5 min, followed by 35 cycles of 95 °C for the 60 s, 52 °C for 30 s, and 72 °C for 60 s.
2. Utilize the strain LEGE 00063 as a positive control in all microcystins PCR experiments.
1. Visualization of amplified DNA products using gel electrophoresis. Prepare 1% agarose-TAE gel with 4% ethidium bromide.
2. Run the PCR products on the agarose gel (in TAE buffer) at 40 V continuous voltages for 1 h.
3. Use a 100 bp DNA ladder to measure the amplicon fragment sizes of each PCR product.

59.2.3 Quantitative Real-Time PCR (QRT-PCR)(Pham et al. 2021)

59.2.3.1 Preparation of Standard Curves

1. Prepare external standards to determine *mcyE* copy numbers using genomic DNAs of *Microcystis* sp. strains GL 260735, PCC 7806, and PCC 7941, as well as those of *Anabaena* sp. strains 90, 315, and 202A1.
2. Measure the genomic DNA concentration of these DNAs with a spectrophotometer set at 260 nm.
3. Determine the purity by calculating the ratio of the absorbance measured at 260 nm (A260) to the absorbance measured at 280 nm (A280).
4. Use approximate genome sizes (4.70 Mb for *Microcystis* and 5.15 Mb for *Anabaena*) to calculate the *mcyE* copy number.
5. Estimate these genome sizes based on the genome sizes of *Microcystis* sp. strain PCC 7941, *Anabaena* sp. strain PCC 6309, and *Anabaena* sp. strain PCC 7122.

Table 59.1 Optimal condition for *mcyE* primers

Target	Primer	Primer sequence 5–3
mcyE	PKEF	CGCAAACCCGATTTACAG
	PKER	CCCCTACCATCTTCATCTTC

6. Calculate the *mcyE* copy numbers of the DNAs of the standard strains using the following equation, assuming that each genome had only one *mcyE* gene and that the molecular weight of 1 bp was 660 g mol^{-1}: number of copies per microliter = (6×10^{23})(DNA concentration)/molecular weight of one genome, where 6×10^{23} is the number of copies per mole, the DNA concentration is given in grams per microliter, and the molecular weight of one genome is given in grams per mole.
7. Prepare a series of tenfold dilutions of genomic DNAs of the standard strains, and these dilutions are amplified with *Microcystis* and *Anabaena mcyE* QRT-PCR.
8. Calculate linear regression equations for obtained cycle threshold values (Ct values, i.e., the first turning points of the fluorescence curves as a function of cycle numbers) as a function of known *mcyE* copy numbers.

59.2.3.2 Procedure

1. Perform the QRT-PCR with 1 μL of DNA from a standard strain or lake water sample, 3 mM $MgCl_2$, 0.5 μM concentrations of both primers (Sigma-Genosys, Ltd.), and 1 μL of the hot start reaction mix to a final volume of 10 μL (LightCycler FastStart DNA Master SYBR green I kit; Roche Diagnostics).
2. Amplification is performed as follows: an initial preheating step of 10 min at 95 ° C, followed by 45 cycles, with 1 cycle consisting of 2 s at 95 °C, 5 s at 58 °C, and 10 s at 72 °C.
3. Monitor the generation of the products after each extension step at 78 °C in *Microcystis* and 77 °C in *Anabaena mcyE* QRT-PCR by measuring the fluorescence of double-stranded DNA binding SYBR green 1 dye using LightCycler QRT-PCR (Roche Diagnostics).

References

Hinlo R, Gleeson D, Lintermans M, Furlan E (2017) Methods to maximise recovery of environmental DNA from water samples. PLoS One 12(6):e0179251

Kim JH, Kim H, Yoon KJ (2022) Effect of mismatch between types of viral nucleic acid and intended targets of extraction kits on polymerase chain reaction-based testing. BioTechniques 73(2):75–79

Kralik P, Ricchi M (2017) A basic guide to real time PCR in microbial diagnostics: definitions, parameters, and everything. Front Microbiol 8:108

Pham TL, Yen TTH, Thai TT, Quang NX (2021) Using quantitative real-time polymerase chain reaction (qRT-PCR) for detection microcystin producing cyanobacteria. Sci Technol Develop J 24(2):1954–1961

Piña SJ (2011) Evolutionary genetics of the AMT domain of the microcystin gene cluster in various cyanobacteria genera

Ramya M, Kayalvizhi M, Haripriya G, Rathinasabapathi P (2018) Detection of microcystin-producing cyanobacteria in water samples using loop-mediated isothermal amplification targeting mcyB gene. 3 Biotech 8(9):1–7

Rodriguez NM (2016) A paper-based point-of-care molecular diagnostic platform for the developing world. (Doctoral dissertation, Boston University)

Schreidah CM, Ratnayake K, Senarath K, Karunarathne A (2020) Microcystins: biogenesis, toxicity, analysis, and control. Chem Res Toxicol 33(9):2225–2246

Thenuwara SI, Kiridena HS, Kirchhoff JR, Isailović D (2022) Evaluation of the mechanisms of adsorption of microcystins and nodularin-R onto rice husk-based biochar. Environ Adv 9: 100314

Wenski SL, Thiengmag S, Helfrich EJ (2022) Complex peptide natural products: biosynthetic principles, challenges and opportunities for pathway engineering. Synthetic Sys Biotechnol 7(1):631–647

Yuan J, Yoon KJ (2021) Overview of PCR methods applied for the identification of freshwater toxigenic cyanobacteria. In: Cyanobacteria-Recent Advances in Taxonomy and Applications. IntechOpen, Rijeka

Molecular Identification of Cyr C Gene in Toxic Cyanobacteria in Photobiont

60

Lekhavani Ramesh, Karthiyayini Balakrishnan,
Vishnu Raja Vijayakumar, and D. Dhanasekaran

Abstract

Cyanobacteria, also known as blue-green algae, are Gram-negative, unicellular organisms found in bodies of water. They obtain their energy from photosynthesis and can grow on their own. Some cyanobacteria produce toxins that are toxic to humans and other living beings. Cylindrospermopsin is a cyanotoxin produced by *Cylindrospermopsis raciborskii*, a cyanobacterium. It is a strong cyanotoxin that harms the liver and kidneys. It has a low molecular weight of about 415 Da. Several molecular techniques have been used to identify this toxin. PCR amplification was used to identify the CyrC gene, which is responsible for toxin production. The ELISA immunoassay measures the concentration of cylindrospermopsin in a sample. This protocol describes the molecular identification of the cyrC gene, which is responsible for the production of cylindrospermopsin.

Keywords

Cyanobacteria · Toxins · Cylindrospermopsin · Cyr C gene · *Cylindrospermopsis raciborskii*

L. Ramesh · K. Balakrishnan · V. R. Vijayakumar (✉)
Department of Microbiology, School of Life Sciences, Bharathidasan University, Tiruchirappalli, Tamil Nadu, India

D. Dhanasekaran
Department of Microbiology, National Repository for Microalgae and Cyanobacteria - Freshwater (NRMC-F), Bharathidasan University, Tiruchirappalli, Tamil Nadu, India

N. Thajuddin et al. (eds.), *Protocols for Cyanobacteria Sampling and Detection of Cyanotoxin*, https://doi.org/10.1007/978-981-99-4514-6_60

60.1 Introduction

Cyanobacteria are Gram-negative bacteria that do not require a fixed carbon source and are thus widely distributed in aquatic environments. These include freshwater and marine environments, as well as certain soils (Percival & Williams 2014). Cyanobacteria yield a wide range of secondary metabolites, some of which are toxic to humans and animals, and cyanobacterial blooms are commonly linked to human or animal poisoning (Huang & Zimba 2019). So far, cyanobacterial toxins have been classified into several classes, including hepatotoxins, neurotoxins, cytotoxins, and irritants (Apeldoorn et al. 2007). Over 46 species of cyanobacteria belonging to cyanobacterial genera are known to produce cyanotoxins (Sivonen and Jones 1999).

Cylindrospermopsin is a cytotoxin formed by the cyanobacterium *Cylindrospermopsis raciborskii* and has been linked to human poisoning. (Mazmouz et al. 2010). Cylindrospermopsin is made up of a tricyclic guanidine group and a hydroxylmethyl uracil group (Ohtani et al. 1992). Cylindrospermopsin (CYN) is a potent cyanotoxin that has effects on the kidney and liver function (Liyanage et al. 2013). *Cylindrospermopsis raciborskii* is primarily found in tropical to subtropical climates, though it has recently been isolated from cooler, temperate climates (Schembri et al. 2001). Its structure is zwitter ionic and has a low molecular weight (415 Da) (Falconer & Humpage 2006).

Since then, it has been demonstrated that cyanobacteria from other genera, including *Anabaena, Aphanizomenon, Lyngbya, Raphidiopsis,* and *Umezakia*, produce this molecule. (Mazmouz et al. 2010). Some species affect livestock due to their ability to produce toxins, and high cyanotoxin concentrations have been linked to animal deaths and human health hazards through drinking and recreational waters (Köker et al. 2017). Cylindrospermopsin (CYN) is a potent cyanotoxin that affects kidney and liver function (Liyanage et al. 2013).

Several systematic techniques have been established to measure CYN concentrations in cultures or in the environment, including an enzyme-linked immunosorbent assay (ELISA) and liquid chromatography assays coupled to both a UV detector (40) or a mass spectrophotometer (Mazmouz et al. 2010).

60.1.1 Materials Required

60.1.2 Cyanobacterial Collection

- Wide mouth container/Zip lock cover
- Hand gloves
- Label
- Marker

60.1.3 Cultivation of Photobiont

- A laminar air flow cabinet
- Autoclave
- Chemicals required for different growth media
- Petri plate
- Test tube
- Conical flask
- Measuring cylinder
- Volumetric flask
- Spirit lamp
- Loop
- pH meter
- L-rod
- Micropipette
- Cotton
- Tissue paper

60.1.4 Isolation of Photobiont

- Lichen sample
- 0.25 M Sucrose
- 80% Potassium iodide
- 10 mM phosphate buffer

60.1.5 Isolation of Cyanobacterial DNA

- Eppendorf tube
- 50 mM Tris
- 5 mM EDTA
- 50 mM NaCl
- Lysozyme
- Proteinase K
- 10% SDS
- Phenol: Chloroform: Isoamyl alcohol (25:24:1)
- 4 M ammonium acetate
- Isopropanol
- 70% ethanol
- TE Buffer

60.1.6 PCR Amplification

- Double distilled water
- 10X Taq polymerase buffer
- DNA template
- 20 mM forward and reverse primers
- 10 mM dNTP's

- Taq polymerase

60.2 Collection of Cyanobacterial Strain

1. Collect cyanobacteria from the sample area using a sterile wide mouth container or a sealable plastic pouch.
2. Wear disposable hand gloves to avoid any physical contact with the sample.
3. Submerge the container in water and collect the sample. After collecting the sample, replace the lid on the container.
4. Ensure all caps are "finger tight." Overtightening caps can result in cracking.
5. After collection, label all the containers.
6. Place the bottles in an ice-filled cooler after collection.

(Note: If there is a cyanobacterial bloom present, avoid direct contact with the water because some cyanobacteria can cause skin irritation. The use of hip waders and disposable gloves will aid in the prevention of contact.)

60.3 Isolation of Photobiont

1. To remove contaminants, rinse 0.5 g of lichen thallus with distilled water. With 10 mL of distilled water, homogenize the lichen thallus.
2. Filter the homogenate through six layers of cheesecloth and centrifuge the filtrate for 5 min at 5000 rpm.
3. Re-suspend the pellet in 8 mL of 0.25 M sucrose after discarding the supernatant.
4. Add 5 mL of 80% potassium iodide, followed by 4 mL of suspension, and centrifuge at 100× for 1 min.
5. In the sucrose solution, the algal and hyphal part may settle in a wide layer above the KI, whereas the undamaged thalli sedimented in large pieces.
6. Pipette the solution's algal and hyphal pieces into 5 mL of KI solution.
7. Add 2 mL of 10 mM PBS (phosphate buffer) and centrifuge for 60 s at 1000 rpm.
8. Small thalli of fungal hyphae remain in the sucrose solution while algal spores form an interphase between phosphate buffer and sucrose.
9. Large hyphal spores form a pellet at the bottom of the centrifuge tube.
10. Eliminate the algal cells from the interphase with a micropipette and add 3 mL of phosphate buffer solution. Centrifuge the entire solution at 900× for 5 min. Repeat the previous step twice more.

11. Keep the isolated sample at 4 °C until ready to use.
12. Grow the photobiont in a specific medium after isolation to isolate the genome.

60.4 Isolation of Cyanobacterial DNA

1. Pelletize the cells after adding the appropriate amount of cyanobacterial culture. After pelleting the cells, decant the supernatant.
2. Suspend the pellet in 500 μL of 50 mM Tris, 5 mM EDTA, and 50 mM NaCl solution. To a final concentration of 1 mg/mL, add lysozyme.
3. Incubate at 55 °C for 30 min.
4. Incubate 10 μL of Proteinase K (10 mg/mL) and 20 μL of 10% SDS at 55 °C for 10 min, till the solution clears.
5. Chill the solution on ice before extracting it with an equal volume of Phenol: Chloroform: Isoamyl alcohol (25:24:1).
6. Add an equal volume of 4 M ammonium acetate to the supernatant.
7. Precipitate DNA by adding two volumes of Isopropanol. Centrifuge for 10 min.
8. After centrifugation, remove the supernatant and wash the pellet with 70% ethanol.
9. Dry the DNA pellet before dissolving it in 100 μL of TE buffer.

60.5 PCR Amplification

Amplify the DNA from the extracted Cyanobacterial DNA using a thermal cycler. Based on (Lorenz, T. C. 2012) Polymerase chain reaction protocol:

1. In an ice bucket, place a 96-well plate that can hold 0.2 mL thin-walled PCR tubes.

2. Allow PCR reagents to be introduced to cold 0.2 mL thin-walled PCR tubes to elude nuclease activity and nonspecific priming.
3. Pipette out the following PCR reagents into a 0.2 mL thin-walled PCR tube in the following order (Table 60.1):

Table 60.1 Order of PCR reagents to be added

S. No	Reagents	Volume (μL)
1.	Double distilled water	42
2.	10× Taq polymerase buffer	5
3.	DNA template	1
4.	20 mM primer (forward and reverse primer)	1
5.	10 mM dNTPs	0.5
6.	Taq DNA polymerase	0.5
	Total volume	**50**

4. After finishing the program, discard the PCR tubes. Keep the PCR tubes at 4 °C.
5. Following electrophoresis, the migrated DNA in the gel can be stained with ethidium bromide (EtBr) to identify PCR results. Using this method, aliquots of each reaction are loaded into the wells of an agarose gel.
6. In the case of a PCR product, ethidium bromide (EtBr) will be introduced between the nucleotides of the DNA strands, allowing bands to be visualized with a UV illuminator.

Forward and reverse primers for cyrC gene amplification (Kellmann et al.,2006)

SI NO	Primers	Size (BP)	cyrC primers
1.	A205PKF	558	AATGACAGAGACTTGTGCGGGG
2.	A205PKR		TTATCGGTATTGGTGGTAGCAACT

60.6 Determination of cyrC Gene Concentration by ELISA

ELISA is a quantitative and sensitive immunoassay for the detection of Cylindrospermopsin in Cyanobaceria. The ELISA assays were carried out in accordance with the manufacturer's (Attogene) instructions and calibrator concentrations. The cyanobacteria were cultured for 20–70 days as described above, and 1 mL aliquots were taken for analysis at various intervals. After centrifuging the samples (14,000 g for 5 min), the supernatant was collected, filtered (0.2 m, polyethersulfone; Nalgen), and diluted in water for analysis. Authentic CYN was used to plot standard curves, as described by the manufacturer (Attogene). Using this ELISA kit (Attogene), the comparative retorts of 7-epi-CYN and 7-deoxy-CYN to CYN were determined using pure standards.

60.7 Observation

PCR was performed on the isolated DNA. The presence of amplified DNA from the sample is indicated by band formation which confirms the DNA copying. The color formation in the wells after incubation determines the concentration of cylindrospermopsin.

References

Falconer IR, Humpage AR (2006) Cyanobacterial (blue-green algal) toxins in water supplies: Cylindrospermopsins. Environ Toxicol 21(4):299–304. https://doi.org/10.1002/tox.20194

Huang I-S, Zimba P, v. (2019) Cyanobacterial bioactive metabolites–a review of their chemistry and biology. Harmful Algae 83:42–94. https://doi.org/10.1016/j.hal.2018.11.008

Kellmann R, Mills T, Neilan BA (2006) Functional modeling and phylogenetic distribution of putative Cylindrospermopsin biosynthesis enzymes. J Mol Evol 62:267–280. https://doi.org/10.1007/s00239-005-0030-6

Köker L, Akçaalan R, Albay M, Neilan BA (2017) Molecular detection of hepatotoxic cyanobacteria in inland water bodies of the Marmara region, Turkey. Adv Oceanogr Limnol 8(1). https://doi.org/10.4081/aiol.2017.6394

Liyanage M, Magana-Arachchi DN, Chandrasekharan NV (2013) Identification of Cylindrospermopsin and *Cylindrospermopsis raciborskii* from Anuradhapura District, Sri Lanka. J Ecotechnol Res 17:23–28

Lorenz TC (2012) Polymerase chain reaction: basic protocol plus troubleshooting and optimization strategies. J Vis Exp 63:e3998. https://doi.org/10.3791/3998

Mazmouz R, Chapuis-Hugon F, Mann S, Pichon V, Méjean A, Ploux O (2010) Biosynthesis of Cylindrospermopsin and 7-Epicylindrospermopsin in Oscillatoria sp. Strain PCC 6506: identification of the cyr gene cluster and toxin analysis. Appl Environ Microbiol 76(15):4943–4949. https://doi.org/10.1128/AEM.00717-10

Ohtani I, Moore RE, Runnegar MTC (1992) Cylindrospermopsin: a potent hepatotoxin from the blue-green alga *Cylindrospermopsis raciborskii*. J Am Chem Soc 114(20):7941–7942. https://doi.org/10.1021/ja00046a067

Percival SL, Williams DW (2014) Cyanobacteria. In: Microbiology of waterborne diseases. Elsevier, Amsterdam, pp 79–88. https://doi.org/10.1016/B978-0-12-415846-7.00005-6

Schembri MA, Neilan BA, Saint CP (2001) Identification of genes implicated in toxin production in the cyanobacterium *Cylindrospermosis raciborskii*. Environ Toxicol 16(5):413–421. https://doi.org/10.1002/tox.1051

Sivonen K, Jones G (1999) Cyanobacterial toxins. In: Chorus I, Bartram J (eds) Toxic cyanobacteria in water: a guide to their public health consequences, monitoring and management. E & FN Spon, London, pp 44–111

van Apeldoorn ME, van Egmond HP, Speijers GJA, Bakker GJI (2007) Toxins of cyanobacteria. Mol Nutr Food Res 51(1):7–60. https://doi.org/10.1002/mnfr.200600185

Detection of Nodularin-Producing Cyanolichen by Polymerase Chain Reaction (PCR)

61

Vishnu Raja Vijayakumar, Lekhavani Ramesh, Karthiyayini Balakrishnan, and D. Dhanasekaran

Abstract

Cyanolichens are lichens that contain cyanobacteria (formerly known as blue-green algae) as the photosynthetic component of the entire thallus. Cyanolichens can be found in a wide range of terrestrial environments, from arctic tundra and semi-deserts to tropical montane rainforests. Cyanolichens can be found in a wide range of terrestrial environments, from arctic tundra and semi-deserts to tropical montane rainforests. Cyanolichens contribute significant amounts of nitrogen to some ecosystems because symbiotic cyanobacteria can fix atmospheric nitrogen. Nodularin is a hepatotoxin which is produced by many toxic cyanobacteria. Nodularin is mainly produced by the cyanobacteria *Nodularia spumigena*. Nodularin-producing cyanolichens are detected by mass spectra, which is widely used method for toxin detection. LC–MS confirms the presence of nodularin from the toxic cyanolichen. Extraction of genomic DNA is performed for the further detection of nodularin gene clusters. ndaF8452 and ndaF8640 primers were used in a PCR to detect all possible nodularin-producing cyanobacteria from laboratory cultures as well as harmful algal blooms.

Keywords

Lichens · Symbiosis · Cyanolichens · Nodularin · Cyanotoxin

V. R. Vijayakumar · L. Ramesh · K. Balakrishnan
Department of Microbiology, School of Life Sciences, Bharathidasan University, Tiruchirappalli, Tamil Nadu, India

D. Dhanasekaran (✉)
Department of Microbiology, National Repository for Microalgae and Cyanobacteria - Freshwater (NRMC-F), Bharathidasan University, Tiruchirappalli, Tamil Nadu, India

N. Thajuddin et al. (eds.), *Protocols for Cyanobacteria Sampling and Detection of Cyanotoxin*, https://doi.org/10.1007/978-981-99-4514-6_61

61.1 Introduction

Lichens are symbiotic unions of a fungus, or mycobiont, and a photosynthetic partner, such as a green alga or cyanobacterium, or "photobiont." The mycobiont in lichen symbiosis survives on sugars produced by the photobiont, whereas the cyanobacterial host receives nitrogen compounds in cyanobacterial symbiosis. Cyanolichens are necessary symbiotic relationships between fungi and cyanobacteria. The cyanobacterial symbiont may function as the sole partner engaged in photosynthetic processes in these partnerships, or it may function as a secondary symbiont in addition to a primary green algal photobiont. Cyanolichens can be found in a variety of terrestrial environments, including tropical highland rainforests, frozen tundra, and semi-deserts. Cyanolichens contribute a significant amount of nitrogen to various ecosystems due to the ability of symbiotic cyanobacteria to fix nitrogen from the atmosphere (Rikkinen 2015). There are numerous cyanolichens that have symbiotic genotypes of Nostoc (Nostocales, Cyanobacteria), either as primary photobionts in bipartite lichens or as auxiliary photobionts assisting green algae in tripartite lichens. Nostoc symbionts are particularly common in Peltigeralean lichens (Peltigerales, Ascomycota), with all species of *Collema*, *Leptogium*, *Nephroma*, *Peltigara*, *Pseudocyphellaria*, and *Sticta* associating with Nostoc (Rikkinen 2007). By far the most common genus of cyanobacterial symbionts found in lichens is Nostoc. Nostoc strains make up the bulk of the cyanobionts found in lichens, especially in connection with the Lecanorales fungi that generate lichens. Lichen mycobionts are highly specific about the Nostoc photobionts they use, according to recent research. (Rikkinen 2003).

Cyanobacteria produce powerful hepatotoxins and tumour promoters called microcystins and nodularins. Both of them include a distinctive beta-amino acid residue and are cyclic heptapeptides and pentapeptides, respectively (2S,3S,8S,9S) 3-amino-9-methoxy-2,6,8-trimethyl-10-phenyldeca-4(E),6(E)-dienoic acid [Adda] (Imanishi et al. 2005). Some lichen thalli contained high concentrations of nodularin, up to 60 $\mu g\ g^{-1}$. The only source of this powerful hepatotoxin and microcystin analogue until recently has been the aquatic species Nodularia. (Kaasalainen et al. 2012). Cyanobacterial planktonic species *Nodularia spumigena* and benthic species *Nodularia sphaerocarpa* produce the cyclic pentapeptide nodularin PCC7804 (Beattie et al. 2000). In Nodularia in addition Nostoc species, heterocysts, specialised cells, are responsible for dinitrogen fixation. HetR expression, which is indirectly regulated by NtcA's activity, controls heterocyst differentiation in the early stages. While nitrate was present, ammonia did not affect *N. spumigena's* ability to fix nitrogen. (Kabir and El-Shehawy 2012). According to Jodowska and Lataa (Jodłowska and Latała 2010), *Nodularia spumigena* can tolerate high light intensities between 150 and 290 μmol photons m^2s^{-1}. *Nodularia spumigena* blooms predominate in environments with high phosphorus concentrations, low N:P ratios, and moderate salinities. (Moffitt and Neilan 2004).

Drinking water sources and natural water reservoirs are thought to be at risk from cyanobacterial toxins. These poisons cause poisonings in both humans and animals when they are consumed. The main goals of water quality monitoring programmes are to predict dangerous algal blooms and to maintain the water quality of drinking water supplies. The methods used for routine monitoring cannot yet detect all of the different forms and variations of cyanotoxins. All the different forms and variations of cyanotoxins cannot yet be detected using the methods employed for routine monitoring. (Kaushik and Balasubramanian 2013). Cyanobacterial species can always be identified and counted using microscope techniques. These constraints could be overcome by the precision, dependability, and speed of molecular approaches, which would then provide data on the identification and quantification of cyanobacteria and the genes that encode toxins. Due to the high sensitivity of molecular-based approaches, hazardous cyanobacteria in water can be detected long before they manifest as cyanobacterial blooms, making them an effective monitoring strategy. Some of these molecular techniques include the Polymerase Chain Reaction (PCR), the quantitative Real-time Polymerase Chain Reaction (qPCR), and the Desoxyribonucleic Acid (DNA) Microarray (DNA chips) (Sanseverino et al. 2017). Chlorophyll A analysis can be performed to quantify pigments and to evaluate the phytoplankton abundance in water samples (Fietz and Nicklisch 2004). The analysis of microscopy is frequently used to detect cyanobacterial blooms. The two techniques that are used the most frequently are epifluorescence and inverted light microscopy with a sedimentation chamber. Identification and quantification of cyanobacteria species as well as differentiation between autotrophic and heterotrophic cells are possible through epifluorescence examination of samples stained with certain dyes (Haas 1982). The most common chemical approach used to identify cyanotoxins in waterways is High Performance Liquid Chromatography (HPLC) in conjunction with an Ultraviolet (UV), fluorescence, or mass spectrometric detector. However, because other chemicals with identical absorbance spectra exist, relying solely on UV absorbance or fluorescence for identification is insufficient (lack of specificity). Nowadays, there are currently several mass spectrometric methods with high sensitivity and selectivity for the unmistakable identification and quantification of various cyanotoxins, including Liquid Chromatography—Time-of-Flight (LC-TOF) MS, Matrix-Assisted Laser Desorption/Ionization Time-of-Flight (MALDI-TOF) MS, and Nuclear Magnetic Resonance (NMR). Microcystins (MC) and Nodularin are two cyanotoxins that have received the most attention from scientists (NOD) Sanseverino et al. 2017) (Fig. 61.1).

61.2 Materials Required

61.2.1 Sample Collection

The following characteristics are preferable for Cyanolichen sample collection.

- Shoulder/Carry bag
- About 6-inch-long chisel
- A small hammer

Fig. 61.1 (**a**) *Dirinaria* sp. and (**b**) *Lecanora* sp.

- Sharp knife Pencil Altimeter/GPS
- Paper or cloth sacs for keeping lichens
- Camera
- Field book/note book
- Others (Raincoat, water bottle, etc.)

61.2.2 Extraction of Cyanolichen DNA

- Mortar and pestle
- Cyanolichen sample
- 3% Lysis buffer
- 100 mM Tris-Hcl
- 25 mM NaCl
- 4% Mercaptoethanol
- 5% PVP
- Ethanol
- Lysozyme
- Proteinase-K
- RNase
- Phenol, chloroform, isoamyl alcohol mixture (1:1:1)
- Chloroform and isoamyl alcohol mixture (1:1)
- Isopropanol
- TE Buffer

61.3 PCR Amplification

- Double distilled water
- 10× Taq polymerase buffer
- DNA template
- 20 mM forward and reverse primers
- 10 mM dNTPs
- Taq polymerase

61.3.1 Procedure

61.3.2 Cyanolichen Sample Collection

Collect cyanolichens of genus *Leptogium, Lobaria, Nephroma, Peltigera, Pseudocyphellaria, Sticta* etc., from different habitats.

Points to be remembered during sample collection:

- A collector needs to be knowledgeable about lichen habitat.
- Lichens are found on a variety of substrates, including trees (Corticolous), dead wood (Lignicolous), rocks (Saxicolous), tombstones, mosses (Muscicolous), ground soil (Terricolous), tree twigs (Ramicolous), old monuments, and more.
- Several folicolous lichens that grow on the upper surface of living leaves can be found in tropical and subtropical evergreen forests.
- A knowledgeable collector can find several intriguing species on top of branches and under canopies.
- The optimal time to collect lichens are in the early morning, especially in mountainous areas.
- Hammer and chisel are used to remove saxicolous thalli.
- Avoid collecting bits of thalli that are too tiny.
- Avoid collecting or destroying all small or unusual lichens; instead, leave some of the thallus alone so that it can continue to grow.
- Dry paper packets or cloth bags must be used to transport collected specimens to the field.
- Wet specimens must be stored in packets of blotting paper.

61.3.3 Extraction of Cyanolichen DNA

Modified procedures of (Silva et al. 2014; Hassan et al. 2018) are as follows:

- Using a mortar and pestle, prepare a paste out of the 2 g of cyanolichen using the lysis buffer (3%(w/v) CTAB, 100 mM Tris HCl (pH 8), 2 M NaCl, 25 mM EDTA (pH 8), 4% Mercaptoethanol (v/v), and 5% PVP (w/v)).
- Transfer the powdered sample in a sterile centrifuge tube.
- Use the water bath mode of heat shock to incubate the sample at 95 °C for 1 h and the freezer method at −80 °C for 1 h.

- Thaw the sample right away by placing it back in a water bath at 95 °C.
- Add double the volume of the sample's absolute ethanol.
- Centrifuge the sample at 4000 rpm for 12 min after vortexing.
- Discard the supernatant, add 2 mL of lysis buffer, 10 µL lysozyme, 10 µL proteinase K, and 10 µL of RNase to each sample, and vortex for 10 min.
- Add buffer-saturated Phenol: Chloroform: Isoamyl alcohol (1:1:1) to the mixture in an equal volume and incubate the setup for 1.5 h at 60 °C with periodic vortexing.
- Spin the sample at 8000 rpm for 20 min at 4 °C, transfer the aqueous layer to a sterile centrifuge tube, add an equal volume of Chloroform: Isoamyl alcohol (1:1) to the aqueous layer to remove any phenol residues, and repeat the process twice.
- Transfer the supernatant to a sterile centrifuge tube with an equivalent amount of isopropanol and incubate on ice for 25 min after centrifugation.
- After centrifugation, remove the supernatant and centrifuge the particle for 10 min at 4 °C and 4000 rpm while rinsing it with 100 µL of 80% ethanol.
- Dry the DNA pellets thoroughly at room temperature by eliminating the supernatant.
- Retain the pellet at 20 °C until needed by resuspending it in 50 µL of TE buffer.

61.3.4 Amplification of gDNA

From the extracted cyanolichen DNA, amplify the DNA using thermal cycler. Based on (Lorenz 2012) polymerase chain reaction protocol:

- Place a 96-well plate that can accommodate in a thin-walled PCR tubes in an ice bucket.
- To avoid nuclease activity and nonspecific priming, allow PCR reagents to be introduced to cold thin-walled PCR tubes.
- Pipette the respective PCR solutions into a thin-walled PCR tube in the following order (Tables 61.1 and 61.2):

Table 61.1 Order of PCR reagents to be added

S. No	Reagent	Volume µL
1.	Double distilled water	42
2.	10× Taq polymerase buffer	5
3.	DNA template	1
4.	20 mM primer (forward and reverse primer)	1
5.	10 mM dNTPs	0.5
6.	Taq DNA polymerase	0.5
	Total volume	**50**

Table 61.2 Forward and reverse primers for nodularin gene amplification (Koskenniemi et al. 2007)

S. No	Primers	20 mM primers
1.	Forward primer—ndaF8452	5′-GTG ATT GAA TTT CTT GGT CG-3′
2.	Reverse primer—ndaF8640	5′-GGA AAT TTC TAT GTC TGA CTC AG-3’

Table 61.3 Temperature and time duration for PCR amplification

S. No	Steps	Temperature (°C)	Duration
1.	Initial denaturation	95	5 min
2.	Denaturation	94	40 s
3.	Annealing	55	30 s
4.	Extension	72	1 min
5.	Go to Step 2 for repetitive replication	–	30 min
6.	Final extension	72	5 min
7.	Hold	4	∞

- Mix the reagents by pipetting in and out for at least 20 times after introducing the reagents. (Note: When mixing, the micropipette should be set to approximately half the reaction volume of the master mix, and bubble formation should be avoided).
- Put caps on the 0.2 mL thin-walled PCR tubes and place them into the thermal cycler and start the programme (Table 61.3).
- Upon completion of the programme, discard the PCR tubes. Maintain the PCR tubes at 4 °C.
- After electrophoresis, DNA that has migrated into the gel can be stained with ethidium bromide to identify PCR results. This method involves loading aliquots of each reaction into the wells of an agarose gel.
- Addition of Ethidium bromide to the PCR products after electrophoresis, bands appear under UV illuminator.

61.4 Observation

From the obtained PCR product, agarose gel electrophoresis was run. Visualization of bands determine that the extracted DNA is responsible for Nodularin synthesis from the Cyanolichen (Table 61.4).

Table 61.4 List of cyanobacterial strains that produce Nodularin

S. No	Nodularin-producing cyanobacteria	Strains	References
1.	*Nodularia harveyana*	PCC7804 NROS 10 BY1 HE1 UTEX-B2093 CDAC1983/300	Rippka (1992), Bolch et al. (1999), Lehtimäki et al. (1994)
2.	*Nodularia sphaerocarpa*	PCC7804 HKVV CDAC1966/93–1	Gehringer et al. (2012), (Carmichael et al. (1988), Rippka (1992)
3.	*Macrozamia serpentina*	73.1	Gehringer et al. (2012)
4.	*M. aeruginosa*	PCC7806	Gehringer et al. (2012)
5.	*Nodularia spumigena*	NSBL05 NSBR01 NSGL02A10 NSKR07 NSLA01 NSLA02A4 NSOR04 NSOR10 NSOR12 NSPH02 NSPH05A14 L575 BY1 HEM UTEX-B2092 PCC73104	Bolch et al. (1999), Moffitt and Neilan (2001), Shi et al. (1995), Lehtimäki et al. (1994)
6.	*A. circinalis*	AWQC118a	Moffitt and Neilan (2001)
7.	*Cylindrospermopsis*	AWT205	Moffitt and Neilan (2001)
8.	*Synechocystis*	PCC6803	Lehtimäki et al. (1994)
9.	*Iningainema*	–	Berthold et al. (2021)
10.	*Nostoc sp*	–	Rinehart et al. (1988)

References

Beattie KA, Kaya K, Codd GA (2000) The cyanobacterium Nodularia PCC 7804, of freshwater origin, produces [L-Har2] nodularin. Phytochemistry 54(1):57–61. https://doi.org/10.1016/S0031-9422(00)00045-5

Berthold DE, Lefler FW, Huang I-S, Abdulla H, Zimba PV, Laughinghouse HD (2021) *Iningainema tapete* sp. nov. (Scytonemataceae, Cyanobacteria) from greenhouses in central Florida (USA) produces two types of nodularin with biosynthetic potential for microcystin-LR and anabaenopeptin production. Harmful Algae 101:101969. https://doi.org/10.1016/j.hal.2020.101969

Bolch CJS, Orr PT, Jones GJ, Blackburn SI (1999) Genetic, morphological, and toxicological variation among globally distributed strains of nodularia (cyanobacteria). J Phycol 35(2):339–355. https://doi.org/10.1046/j.1529-8817.1999.3520339.x

Carmichael WW, Eschedor JT, Patterson GM, Moore RE (1988) Toxicity and partial structure of a hepatotoxic peptide produced by the cyanobacterium Nodularia spumigena Mertens emend. L575 from New Zealand. Appl Environ Microbiol 54(9):2257–2263. https://doi.org/10.1128/aem.54.9.2257-2263.1988

Fietz S, Nicklisch A (2004) An HPLC analysis of the summer phytoplankton assemblage in Lake Baikal. Freshw Biol 49(3):332–345. https://doi.org/10.1111/j.1365-2427.2004.01183.x

Gehringer MM, Adler L, Roberts AA, Moffitt MC, Mihali TK, Mills TJT, Fieker C, Neilan BA (2012) Nodularin, a cyanobacterial toxin, is synthesized in planta by symbiotic *Nostoc* sp. ISME J 6(10):1834–1847. https://doi.org/10.1038/ismej.2012.25

Haas LW (1982) Improved epifluorescence microscopy for observing planktonic micro-organisms. Annales de Institut Oceanographique Paris 58:261–266

Hassan M, Essam T, Megahed S (2018) Illumina sequencing and assessment of new cost-efficient protocol for metagenomic-DNA extraction from environmental water samples. Braz J Microbiol 49:1–8. https://doi.org/10.1016/j.bjm.2018.03.002

Imanishi S, Kato H, Mizuno M, Tsuji K, Harada K (2005) Bacterial degradation of microcystins and nodularin. Chem Res Toxicol 18(3):591–598. https://doi.org/10.1021/tx049677g

Jodłowska S, Latała A (2010) Photoacclimation strategies in the toxic cyanobacterium Nodularia spumigena (Nostocales, cyanobacteria). Phycologia 49(3):203–211. https://doi.org/10.2216/PH08-14.1

Kaasalainen U, Fewer DP, Jokela J, Wahlsten M, Sivonen K, Rikkinen J (2012) Cyanobacteria produce a high variety of hepatotoxic peptides in lichen symbiosis. Proc Natl Acad Sci 109(15): 5886–5891. https://doi.org/10.1073/pnas.1200279109

Kabir AH, El-Shehawy R (2012) Expression of het R, and ndaF genes in Nodularia spumigena under nitrogen sources. Plant Biosystems 146(4):992–1000. https://doi.org/10.1080/11263504.2012.663811

Kaushik R, Balasubramanian R (2013) Methods and approaches used for detection of cyanotoxins in environmental samples: a review. Crit Rev Environ Sci Technol 43(13):1349–1383. https://doi.org/10.1080/10643389.2011.644224

Koskenniemi K, Lyra C, Rajaniemi-Wacklin P, Jokela J, Sivonen K (2007) Quantitative real-time PCR detection of toxic Nodularia cyanobacteria in the Baltic Sea. Appl Environ Microbiol 73(7):2173–2179. https://doi.org/10.1128/AEM.02746-06

Lehtimäki J, Sivonen K, Luukkainen R, Niemelä SI (1994) The effects of incubation time, temperature, light, salinity, and phosphorus on growth and hepatotoxin production by Nodularia strains. Arch Hydrobiol 130(3):269–282. https://doi.org/10.1127/archiv-hydrobiol/130/1994/269

Lorenz TC (2012) Polymerase chain reaction: basic protocol plus troubleshooting and optimization strategies. J Vis Exp 63:e3998. https://doi.org/10.3791/3998

Moffitt MC, Neilan BA (2001) On the presence of peptide synthetase and polyketide synthase genes in the cyanobacterial genus *Nodularia*. FEMS Microbiol Lett 196(2):207–214. https://doi.org/10.1111/j.1574-6968.2001.tb10566.x

Moffitt MC, Neilan BA (2004) Characterization of the nodularin Synthetase gene cluster and proposed theory of the evolution of cyanobacterial Hepatotoxins. Appl Environ Microbiol 70(11):6353–6362. https://doi.org/10.1128/AEM.70.11.6353-6362.2004

Rikkinen J (2003) Ordination analysis of tRNALeu (UAA) intron sequences in lichen-forming Nostoc strains and other cyanobacteria. Symbolae Botanicae Upsalienses:377–391

Rikkinen J (2007) Cyanolichens: an evolutionary overview. In: Cyanobacteria in Symbiosis. Kluwer Academic Publishers, Dordrecht, pp 31–72

Rikkinen J (2015) Cyanolichens. Biodivers Conserv 24(4):973–993. https://doi.org/10.1007/s10531-015-0906-8

Rinehart KL, Harada K, Namikoshi M, Chen C, Harvis CA, Munro MHG, Blunt JW, Mulligan PE, Beasley VR et al (1988) Nodularin, microcystin, and the configuration of Adda. J Am Chem Soc 110(25):8557–8558. https://doi.org/10.1021/ja00233a04z

Rippka R (1992) Pasteur culture collection of cyanobacterial strains in axenic culture. In: Catalogue and taxonomic handbook, catalogue of strains 1992/1993, vol 1. Institut Pasteur, Paris, pp 1–103. https://cir.nii.ac.jp/crid/1571980075695210752.bib?lang=en

Sanseverino I, António DC, Loos R, Lettieri T (2017) Cyanotoxins: methods and approaches for their analysis and detection. Publications Office of the European Union, Luxembourg

Shi L, Carmichael WW, Miller I (1995) Immuno-gold localization of hepatotoxins in cyanobacterial cells. Arch Microbiol 163:7–15. https://doi.org/10.1007/BF00262197

Silva MA, Medeiros Z, Soares CRP, Silva ED, Miranda-Filho DB, Melo FL (2014) A comparison of four DNA extraction protocols for the analysis of urine from patients with visceral leishmaniasis. Rev Soc Bras Med Trop 47(2):193–197. https://doi.org/10.1590/0037-8682-0233-2013

62 Discrimination of Nodularin Toxin-Producing Strains from Non-producing Strains Using PCR

Janet Jan-Roblero, Juan A. Cruz-Maya, and Juan C. Cancino-Diaz

Abstract

Cyanobacteria can produce blue-green algal blooms and some of them are toxin producers. Among these cyanobacterial toxins are the hepatoxins such as nodularin and microcystin. Nodularin is a cyclic pentapeptide that inhibits protein phosphates 1 and 2A in eukaryotic cells. The effect of nodularin in human is liver failure and death and is also associated with the initiation of tumor formation. Therefore, detecting the nodularin toxin in drinking water is important for human health. Nodularin is produced solely by *Nodularia spumigena* strains. The use of molecular methods for the detection of nodularin is a fast and cheap alternative. A protocol to differentiate *N. spumigena* strains that produce the nodularin toxin from those that do not produce it is described here. This protocol is based on a set of primers that detect the genes of the *ndaA-I* operon that encode the different enzymes of the nodularin synthetase gene complex in order to discriminate non-toxic *Nodularia* strains from those toxic strains that possess the machinery complete for the production of the nodularin toxin (*ndaA-I* operon).

Keywords

Nodularin · PCR · *Nodularia spumigena*

J. Jan-Roblero (✉) · J. C. Cancino-Diaz
Departamento de Microbiología, Escuela Nacional de Ciencias Biológicas, Instituto Politécnico Nacional, Mexico City, Mexico
e-mail: jjanr@ipn.mx

J. A. Cruz-Maya
Unidad Profesional Interdisciplinaria en Ingeniería y Tecnologías Avanzadas, Instituto Politécnico Nacional, Mexico City, Mexico

N. Thajuddin et al. (eds.), *Protocols for Cyanobacteria Sampling and Detection of Cyanotoxin*, https://doi.org/10.1007/978-981-99-4514-6_62

62.1 Introduction

In the blue-green algal blooms are present cyanobacteria, some of them are toxic because they produce hepatoxins such as nodularin and microcystin (Codd et al. 1999). Nodularin inhibits eukaryotic protein phosphatases 1 and 2A. The nodularin can cause a liver failure and later death in humans (Kuiper-Goodman et al. 1999). It has been observed that in subchronic concentrations nodularin can be related to the initiation and promotion of tumors (Hitzfeld et al. 2000). Therefore, it is necessary to employ cheap and fast techniques to monitor the presence of the nodularin toxin in water supplies. Nodularin is a cyclic pentapeptide (Sivonen and Jones 1999) that is synthesized non-ribosomal by nodularin synthetase genes cluster, these genes have a high variety of modules as non-ribosomal peptide synthetases (NRPS) and polyketide synthases (PKS), and tailoring enzymes. The modules can catalyze different enzymatic activities such as modification and condensation of specific amino acids of the nodularin (Moffitt and Neilan 2001). However, nodularin can have modifications in its structure as [D-Asp]-nodularin and [dmAdda]-nodularin (Beattie et al. 2000), whereas microcystins are more varied, 60 types are known. In the case of nodularin, its presence has been only confirmed in the planktonic *Nodularia spumigena* and in a strain of *N. harveyana* (Moffitt and Neilan 2001).

The complete nodularin synthetase gene (*nda*) cluster in *N. spumigena* NSOR10 is in a 48-kb fragment, this fragment consists of nine open reading frames (ORFs) which includes the *ndaA* to *ndaI* genes (Fig. 62.1). Downstream of the promoter, three NRPS modules are encoded within the two ORFs, *ndaA* and *ndaB*. Transcribed in the opposite direction, upstream of the promoter, there are three ORFs, *ndaCDF*, which encode NRPS and PKS modules, along with four other ORFs, *ndaE* and *ndaGHI*, which encode tailoring enzymes (Moffitt and Neilan 2004).

Nodularia genus comprises seven species (*N. baltica, N. armorica, N. spumigena, N. harveyana, N. sphaerocarpa, N. mediterranea, N. moravica*), of which *N. spumigena* is the only one that produces the nodularin toxin; however, some strains of *N. spumigena* do not produce it. Most of the time, synthetase genes (*nda*) are the candidate genes used to differentiate toxic from non-toxic genotypes applying a simple and cheap molecular PCR technique (Dittmann and Börner 2005; Dittmann and Wiegand 2006). The toxic *Nodularia* strains carry the peptide synthetase and polyketide synthase sequence while non-toxic strains have only some of these sequences or do not have them (Moffitt and Neilan 2001). The subunit *ndaF* gene has been used to discriminate between toxic and non-toxic strains by several

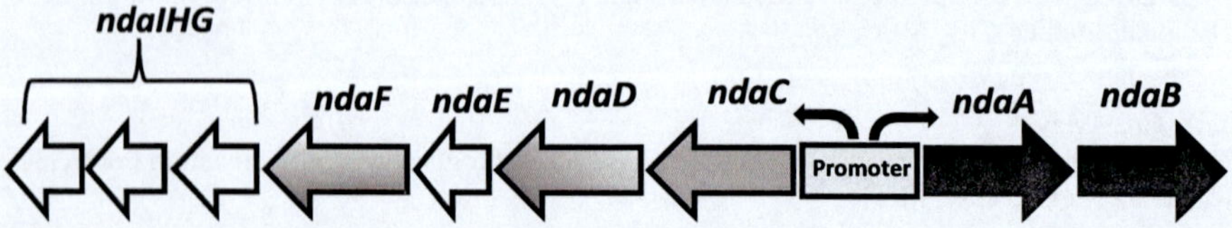

Fig. 62.1 Structure of *nda* genes of *Nodularia spumigena*

authors (Moffitt and Neilan 2001; Lyra et al. 2005; Surakka et al. 2005; Jungblut and Neilan 2006; Koskenniemi et al. 2007); however, it is not definitive that just using the *ndaF* gene means that it is a nodularin-producing strain, it is required to determine the components of the *ndaA-I* operon to verify the production of the toxin at genotype level. Krüger et al. (2009) propose a PCR method to discriminate toxic and non-toxic genotypes, which consists of detecting all *nda* cluster sequences (beyond *ndaF*, Fig. 62.1); they showed that the nodularin-producing *N. spumigena* possess all *ndaA-I* genes while the non-toxic *N. harveyana* do not. Based on this proposal, we describe the primers and the required experimental conditions for screening of different *nda* sequences between nodularin-producing and non-producing strains (Krüger et al. 2009).

62.2 Materials

- Cyanobacteria strains such as *N. spumigena*, *N. harveyana*, or others of the same genus
- XS buffer
- Isopropanol
- 4 M ammonium acetate
- 70% ethanol
- Sterile water
- Taq DNA polymerase
- $MgCl_2$
- 1× Taq-polymerase buffer
- dNTPs
- Agarose
- 1× TAE buffer
- Ethidium bromide

62.3 Methods

62.3.1 Genomic DNA Isolation Based on Jungblut and Neilan (2006)

1. Obtain a culture of cyanobacteria and recover approximately 200 mg of cell in wet weight.
2. Mix cells with 600 μL XS extraction buffer (1% potassium methyl xanthogenate; 800 mM ammonium acetate; 20 mM EDTA; 1% SDS; 100 mM Tris-HCl, pH 7.4).
3. Vortex the mixture and incubate at 65 °C for 2 h.
4. Chill the extract on ice for 10 min.
5. Separate cells debris by centrifugation at 12,000 g for 10 min.
6. Precipitate the DNA by adding 1 volume of isopropanol and 1/10 volume of 4 M ammonium acetate at 4 °C for 15 min.

7. Recover the precipitated DNA by centrifugation at 12,000 g for 10 min.
8. Wash the DNA pellet with 70% ethanol.
9. Resuspend the extracted DNA with 100 μL of sterile water.

62.4 Polymerase Chain Reaction (PCR) Based on the Method of Krüger et al. (2009)

1. For each *ndaA-I* gene, a reaction mixture with a total volume of 20 μL must be carried out with the following reagents: 2.5 mM $MgCl_2$, 1X Taq-polymerase buffer (Fischer Biotech), 0.2 mM dNTPs (Fischer Biotech), 0.5 pmol of forward and reverse primer of each *ndaA-I* gene (Table 62.1), 0.2 U of Taq DNA polymerase (Fischer Biotech, Perth, Australia) and 1 μL of DNA template at a concentration of 100 ng μL^{-1} approximately.
2. Heat the reaction mix at 94 °C for 4 min in a GeneAmp PCR System 2400 thermal cycler (Perkin Elmer, Norwalk, CT) to denature the DNA template.
3. Subsequently, carry out 40 cycles whose conditions will depend on the *nda* gene to be amplified (Table 62.1). At the end of the cycles, do a final extension step at 72 °C for 8 min.
4. Analyze the PCR product on a 2% agarose gel in 1X TAE-buffer.
5. Stain the gel with ethidium bromide (1 μg mL^{-1}) during 10 min.
6. Photodocument the gel in a Gel-DOC Bio-RAD System with Quantity One 4.1R software (BIO-RAD, USA).

Taken and adapted from Krüger et al. (2009).

62.5 Observations

This protocol detects all the genes of *the ndaA-I* operon and the detection of them genotypically indicates that it is a nodularin toxin-producing strain. Strains with the partial detection of these genes indicate that they are nodularin non-producing strains or that they produce a variant of the toxin.

On the other hand, this protocol can also be applied to total DNA samples extracted from blue-green algal blooms with different species composition, the detection of all *ndaA-I* genes indicates the presence of some cyanobacterial species, most likely *N. spumigena*, which is a nodularin toxin-producing strain. Furthermore, these primers can be used to determine the expression of the *ndaA-I* genes by means of RT-PCR. Besides, as *Nodularia* genus-specific positive PCR control, the primer pair *cpcBA*-LP and *cpc*BA-RP are included. It is important to mention that other techniques are recommended to determine the nodularin toxin content, such as protein phosphatase inhibitory assay and ELISA.

Table 62.1 Oligonucleotides and condition reaction

Gene	Primer	Sequence (5′- > 3′)	Tm (°C)	Expected fragment (bp)	Program
nda A	*nda* A-LP	GGA GTT TGG GCG ATG AGT TA	57.3	894	94 °C 20 s, 57 °C 20 s, 72 °C 1:30 min
	nda A-RP	GCC CAA CTA ACT GCA ATG GT	57.3		
nda B	*nda* B-LP	GTT ATC CAG CCG ACC GTT TA	57.3	1064	94 °C 20 s, 56 °C 20 s, 72 °C 1:30 min
	nda B-RP	TTC TCG CAG AAT CAC CAC TG	57.3		
nda C1	*nda* C1-LP	ATC TAT GCC TAG TCC TCA G	54.5	955	94 °C 20 s, 54 °C 20 s, 72 °C 1:30 min
	nda C1-RP	CTT CAG TTC TTC GGA CTG T	54.5		
nda C2	*nda* C2-LP	TAT TTG CTC AAT CTT CTC CAA G	54.7	1542	94 °C 20 s, 54 °C 20 s, 72 °C 1:55 min
	nda C2-RP	TCA ATA ATT ACC CAG AAA AGC C	54.7		
nda D	*nda* D-LP	GAA GGA GAG AAG ATG GAG A	54.5	1398	94 °C 20 s, 54 °C 20 s, 72 °C 1:55 min
	nda D-RP	TTG AGG AGG TTC GCT CAA A	54.5		
nda E	*nda* E-LP	AGC TTT GGA AAA TGT TGG C	52.2	944	94 °C 20 s, 50 °C 20 s, 72 °C 1:30 min
	nda E-RP	CAG GAA ATA TCT AAT CTG G	50.2		
nda F1	*nda* F1-LP	TTG CGG AGA GAA CAA TGG CTC	59.8	1301	94 °C 20 s, 59 °C 20 s, 72 °C 1:50 min
	nda F1-RP	CCA GGG TAT GAT TTG ACT GGG	59.8		
nda F2	*nda* F2-LP	GGG AGA AAT TCC CGT ATT TGA	55.9	858	94 °C 20 s, 55 °C 20 s, 72 °C 1:30 min
	nda F2-RP	CTG AGA GTA TTG TCC CCC ACA	59.8		
nda G	*nda* G-LP	CCA GGA GAA TTG TAT GAA TAA G	54.7	771	94 °C 20 s, 51 °C 20 s, 72 °C 1:30 min
	nda G-RP	CAC CTA GAG ACC TTA TAT ATG	54.9		
nda H	*nda* H-LP	CTA GGA GCA ATT GCA TGA CAG C	60.3	1046	94 °C 20 s, 60 °C 20 s, 72 °C 1:30 min
	nda H-RP	GTG GTT CTC ATT GCA GAT GTG G	60.3		
nda I	*nda* I-LP	GCC GCA GCT ATT TTG CTT AC	57.3	949	94 °C 20 s, 54 °C 20 s, 72 °C 1:30 min
	nda I-RP	AAG CAA GGC GTT GTT GTT CT	55.3		

(continued)

Table 62.1 (continued)

Gene	Primer	Sequence (5′- > 3′)	Tm (°C)	Expected fragment (bp)	Program
*cpc*BA	*cpc*BA-LP	AAC CTA CCA AGC TCT AGG AAC	57.9	391	94 °C 20 s, 55 °C 20 s, 72 °C 0:55 min
	*cpc*BA-RP	GTC AAA GCA CGA GCA GCT T	56.7		

Acknowledgement The authors acknowledge the financial support provided by the Instituto Politécnico Nacional (IPN) México grant SIP20230286, also the COFAA and EDI, IPN fellowships, and support from the SNI-CONACyT.

References

Beattie KA, Kaya K, Codd GA (2000) The cyanobacterium *Nodularia* PCC7804, of freshwater origin, produces [L-Har2] nodularin. Phytochemistry 54:57–61

Codd GA, Bell SG, Kaya K, Ward CJ, Beattie KA, Metcalf JS (1999) Cyanobacterial toxins, exposure routes and human health. Eur J Phycol 34:405–415

Dittmann E, Börner T (2005) Genetic contributions to the risk assessment of microcystin in the environment. Toxicol Appl Pharmacol 203:192–200

Dittmann E, Wiegand C (2006) Cyanobacterial toxins—occurrence, biosynthesis and impact on human affairs. Mol Nutr Food Res 50:7–17

Hitzfeld B, Höger S, Dietrich D (2000) Cyanobacterial toxins: removal during drinking water treatment, and risk assessment. Environ Health Perspect 108:113–122

Jungblut AD, Neilan BA (2006) Molecular identification and evolution of the cyclic peptide hepatotoxins, microcystin and nodularin, synthetase genes in three orders of cyanobacteria. Arch Microbiol 185:107–114

Koskenniemi K, Lyra C, Rajaniemi-Wacklin P, Jokela J, Sivonen K (2007) Quantitative real-time PCR detection of toxic *Nodularia* cyanobacteria in the Baltic Sea. Appl Environ Microbiol 73: 2173–2179

Krüger T, Oelmüller R, Luckas B (2009) Comparative PCR analysis of toxic *Nodularia spumigena* and non-toxic *Nodularia harveyana* (Nostocales, cyanobacteria) with respect to the nodularin synthetase gene cluster. Eur J Phycol 44:291–295

Kuiper-Goodman T, Falconer Ian R, Fitzgerald DJ (1999) Human health aspects. In: Chorus I, Bartram J (eds) Toxic cyanobacteria in water. A guide to their public health consequences, monitoring and management. E&FN Spon, London, pp 113–153

Lyra C, Laamanen M, Lehtimäki JM, Surakka A, Sivonen K (2005) Benthic cyanobacteria of the genus *Nodularia* are non-toxic, without gas vacuoles, able to glide and genetically more diverse than planktonic *Nodularia*. Int J Syst Evol Microbiol 55:555–568

Moffitt MC, Neilan BA (2001) On the presence of peptide synthetase and polyketide synthase genes in the cyanobacterial genus *Nodularia*. FEMS Microbiol Lett 196:207–214

Moffitt MC, Neilan BA (2004) Characterization of the nodularin synthetase gene cluster and proposed evolution of cyanobacterial hepatotoxins. Appl Environ Microbiol 70:6353–6362

Sivonen K, Jones G (1999) Cyanobacterial toxins. In: Chorus I, Bartram J (eds) Toxic cyanobacteria in water. A guide to their public health consequences, monitoring and management. E&FN Spon, London, pp 41–111

Surakka A, Sihvonen LM, Lehtimäki JM, Wahlsten M, Vuorela P, Sivonen K (2005) Benthic cyanobacteria from the Baltic Sea contain cytotoxic *anabaena*, *Nodularia*, and *Nostoc* strains and an apoptosis-inducing *Phormidium* strain. Environ Toxicol 20:285–292

Detection of Cyanobacterial Cell Wall Components by Cellular Signaling Biosensors

63

K. R. V. Bandara, L. D. C. Peiris, D. M. P. I. Dissanayake, P. M. Manage, and S. A. Kulasooriya

Abstract

Despite the fact that most ecotoxicological investigations on cyanobacteria have concentrated on a small number of cyanotoxins, particularly microcystins, increasing evidence suggests that additional cyanobacterial cell wall components may also have harmful effects on human and animal health. In this chapter, self-developed reporter gene-based cellular biosensors, detecting activation of the main human xenobiotic stress response pathways, PXR and NF-kB, are explained to detect novel potentially toxic bioactivities in extracts from freshwater microcystin-producing cyanobacterial blooms. Thus, the holistic approach to sample analysis with the application of cellular biosensors geared towards four separate pathways/bioactivities was validated for the identification of novel bioactivities in organisms with recognized public health significance. The ability of cellular biosensors to be activated by intact cyanobacterial cells from blooms provides proof of the concept of their direct application for environmental

K. R. V. Bandara · P. M. Manage
Centre for Water Quality and Algae Research, Department of Zoology, University of Sri Jayewardenepura, Nugegoda, Sri Lanka

Faculty of Graduate Studies, University of Sri Jayewardenepura, Nugegoda, Sri Lanka

L. D. C. Peiris
Department of Zoology, Faculty of Applied Sciences, University of Sri Jayewardenepura, Nugegoda, Sri Lanka

D. M. P. I. Dissanayake
Centre for Water Quality and Algae Research, Department of Zoology, University of Sri Jayewardenepura, Nugegoda, Sri Lanka

S. A. Kulasooriya (✉)
National Institute of Fundamental Studies, Kandy, Sri Lanka
e-mail: ananda@kulasooriya.com

N. Thajuddin et al. (eds.), *Protocols for Cyanobacteria Sampling and Detection of Cyanotoxin*, https://doi.org/10.1007/978-981-99-4514-6_63

monitoring, especially the comparison of potential threats without the need for chemical analysis and identification of toxicants.

Keywords

Cyanobacteria · Microcystins · Cyanobacterial cell wall components · Cellular biosensors · Bioactivities

63.1 Introduction

Microcystin-producing cyanobacterial blooms in freshwater constitute a significant public health threat to drinking water supplies and bathers who use the afflicted waterbodies for recreational purposes (Ibelings et al. 2014). Cyanobacterial blooms pose a substantial danger to the aquatic environment and public health due to the prevalence of main cyanotoxins, such as microcystins, cylindrospermopsin or anatoxin-a, but also other emerging bioactivities (Gągała-Borowska et al. 2022). Microcystins are determined directly from water samples using several analytical and biochemical screening methods. The most widely used analytical methods include gas chromatography (GC/FID or GC/MS), thin layer chromatography (HPTLC), matrix-assisted laser desorption/ionization-time of flight (MALDI-TOF), and different kinds of liquid chromatography (e.g., HPLC UV, PDA, FLD, LC/MS) (Lebanov et al. 2021; Welker et al. 2002; Oehrle et al. 2010). Additionally, immunoblots, ELISA, can be used as a screening method for the estimation of microcystins concentrations (Palus et al. 2007).

Moreover, screening Protein Phosphatase Inhibition Assay was proposed for the determination of the biological activity (toxicity) of microcystins (An and Carmichael 1994a). Since *Microcystis* had been identified as the predominant genus, and microcystins (a group of hepatotoxins) as toxins produced in the most significant quantities, further work entailed the development of methods for the direct assay of microcystins and other main groups of cyanotoxins (Svirčev et al. 2017). Studies have also been performed on the biological effects at the cellular and biochemical level, devoted to cytotoxicity and genotoxicity of cyanobacterial extracts containing microcystins and inhibition of protein phosphatase (Vesterkvist et al. 2012). However, the cited studies have focused only on the role of major groups of cyanobacterial toxins, never on the possibility of interactions of other compounds included in the extracts, such as cell wall components: lipopolysaccharide (LPS), lipopeptides, etc., with their respective molecular targets (Mankiewicz-Boczek et al. 2015).

Recently, the findings on the activity of unknown endogenous toxic substances contained in cyanobacterial blooms is a holistic approach to water quality monitoring based on the physiological effect of combined toxins and xenobiotics present in the cyanobacterial bloom seems to be a necessary complement for regulatory and monitoring purposes (Mankiewicz-Boczek et al. 2015). Such an approach may involve assessing toxicity mechanisms on the organism, cellular and molecular level, which are physiologically relevant to living components of the environment, especially to human health. This requires developing new tools for composite bioactivity and environmental threat detection, such as cellular biosensors.

The cellular biosensor method can identify at the molecular level the specific signaling pathway/transcription factor affected by the toxicant and the possibility of direct testing of environmental samples without complicated chemical processing in miniaturized sample sizes (Michelini et al. 2013). This chapter aims to explain the application of cellular biosensors to the analysis of the toxicity of microcystin-containing cyanobacterial extracts. The specific goal was to test the relevance of microcystin-LR as the most assayed cyanobacterial toxin to actual physiological responses of cells of human origin at the molecular signaling level within the context of two main xenobiotic stress detection pathways: PXR and NF-kB. PXR is a nuclear receptor functioning as a direct multispecific sensor of potentially toxic organic compounds, and its activation results in the upregulation of numerous detoxification and xenobiotic metabolism enzymes (Eloranta and Kullak-Ublick 2005). The NF-kB pathway, a complex set of intracellular interactions resulting in stabilization and nuclear translocation of transcription factors from the NF-kB family, plays a key role in the detection and response to several types of environmental stress (including oxidative stress) and can be activated both by specific receptor ligands and by disruption of intracellular homeostasis, leading, for example to proinflammatory response (Hoffmann and Baltimore 2006). These pathways were chosen for the present protocol as hallmarks of cellular signaling response to dangerous toxic components in cyanobacterial cell walls such as lipopolysaccharide (LPS), lipopeptides, etc.

63.2 Materials

65 μm plankton net

Plankton collecting bottles

Beakers

Ice

Whatman solid-state PTFE filters

Autoclaved plates/Petriplates

C18 columns

96-well plates

Methanol

Milli-Q water

Distilled water

DMSO

Trifluoroacetic Acid (TFA)

63.3 Methodology

63.3.1 Collection of Cyanobacterial Samples

1. Collect cyanobacterial blooms using a 65 μm plankton net.
2. Freeze dry the concentrated samples.
3. Homogenize the sample by grinding and store at −20 °C until analysis.

63.3.2 Preparation of Cyanobacterial Extracts

63.3.2.1 Crude Cyanobacterial Extracts

1. Dissolve 1000 mg of freeze-dried cyanobacterial material in 60 mL of 75% methanol.
2. Sonicate on ice for 10 min to release toxic compounds from cyanobacterial cells.
3. Centrifuge the sample for 10 min at 10,000× g at 4 °C.
4. Filter the supernatants using Whatman solid-state PTFE filters.
5. Subsequently, divide the sample into several portions and evaporate to dryness using a vacuum centrifuge.
6. Dissolve the evaporated extracts again in 75% methanol, with a subsample being taken for chromatographic microcystins analysis.
7. Dissolve the prepared crude extracts (CE) in DMSO with the concentration standardized to the initial dry weight of cyanobacterial material (Mankiewicz-Boczek et al. 2011).

63.3.2.2 Purified Cyanobacterial Extracts

1. For further purification, solid-phase extraction (SPE) on octadecyl columns (JT Baker) should be performed on evaporated crude extracts (Jurczaka et al. 2005).
2. Dissolve the crude extracts in 70% methanol and apply to methanol-preconditioned C18 columns.
3. Perform the elution with 90% methanol with 0.1% Trifluoroacetic Acid (TFA).
4. Evaporate the obtained sample and dissolve again in 75% methanol and filter using Whatman solid-state PTFE filters.
5. Finally, dissolve the prepared, purified extracts in DMSO with the concentration standardized to the initial dry weight of cyanobacterial material (Jurczaka et al. 2005).

63.3.3 Protein Phosphatase Inhibition Assay

1. For this assay, the inhibition of protein phosphatase Type 1 (PP1) is used (An and Carmichael 1994b).

2. Incubate extracted samples at 37 °C with 0.00167 U 1 – 1 of PP1 from rabbit skeleton muscle and use a 15 mM solution of the substrate nitrophenyl phosphate (pNPP).
3. Measure the quantity of p-nitrophenyl (pNP) product after the reaction of PP1 enzyme with pNPP substrate after 2 h at 405 nm in a microplate reader.
4. The product quantity should be decreased with increasing inhibition of PP1 activity.

63.3.4 Cell Viability Assay

1. Detect cell viability in all biosensor cell lines described below (O'Brian et al. 2000).
2. Seed cells in 96-well plates at 104 per well and expose to cyanobacterial extracts at varying concentrations for 24 h.
3. Subsequently, wash cells and add resazurin to a concentration of 5 M in culture medium with a further 1-h incubation to allow for resazurin reduction by viable cells
4. Determine the fluorescence of reduced resorufin in the medium using a microplate reader at 530 nm excitation and 590 nm emission wavelengths.
5. Fluorescence intensity should be directly proportional to the percentage of remaining viable cells.

63.3.5 NHRTOX Cellular Biosensor Assay

1. Use hepatocellular carcinoma-derived cells (HepG2 cell line) for stable transfection with a reporter vector containing a minimal viral promoter and nine copies of a consensus PXR binding site (5′--GATCAGACAGTTCATGAAGTTCATCTAGATC-3′).
2. Select a clonal cell line (NHRTOX-HepG2) for the highest and most consistent rifampicin-induced (PXR-mediated) luciferase expression.
3. Use the PXR-responsive reporter cell line as a cellular biosensor of toxicity of cyanobacterial extracts.
4. For a biosensor experiment, seed cells in 96-well plates at 1.5 × 104 cells per well and incubated with various concentrations of cyanobacterial extracts or environmental toxins in a cell culture medium for 24 h.
5. Determine firefly luciferase reporter gene expression by luminometric assay of luciferase activity in cell lysate on a microplate reader (Ratajewski et al. 2011).
6. Recalculate the results as a percentage of control (cells exposed only to cell culture medium).
7. As control of biosensor functionality, each experiment should also include cells exposed to 20 M rifampicin (Sigma) and verify the induction ratio to be stable.

63.3.6 OXIBIOS Cellular Biosensor Assay

1. Prepare an NF-kB-responsive reporter plasmid containing an artificial promoter sequence with six copies of a consensus NF-kB-binding site (5′-GGGAATTTCC-3′) and a minimal core promoter in a firefly luciferase-encoding vector.
2. Subsequently, use this vector for stable transfection into two cancer cell lines of different tissues of origin (ovarian epithelial—Sk-Ov-3 and hepatocellular—Hep3B), with stable clonal cell lines derived by antibiotic selection and selected to yield the highest luciferase expression enhancement ratio upon treatment with NF-kB activating stimuli.
3. Use the N-FB responsive reporter cell lines as cellular biosensors of the bioactivity of cyanobacterial extracts.
4. Obtained the results using the Hep3B-originated OXIBIOS-Hep3B reporter cell line, which allows for specific detection of NF-kB activation in a cellular background typical for liver cells, and also in an ovarian cell background (OXIBIOS-Sk-Ov-3) which shows greater sensitivity and signal enhancement due to higher endogenous expression of NF-kB pathway components.
5. For a biosensor experiment, seed cells in 96-well plates at 104 cells per well and incubate with bacterial cell wall-derived compounds in a cell culture medium for 24 h.
6. Determine firefly luciferase reporter gene expression by luminometric assay of luciferase activity in cell lysate on a microplate reader (PerkinElmer).
7. As control of biosensor functionality, include each experiment cell exposed to 1 ng ml^{-1} human recombinant TNF alpha (Sigma) and verify the induction ratio to be stable.
8. Examples of bacterial cell wall component controls are *Escherichia coli* LPS, Pam3CSK4 and FSL-1 (Wagner et al. 2011).

References

An J, Carmichael WW (1994a) Use of a colorimetric protein phosphatase inhibition assay and enzyme linked immunosorbent assay for the study of microcystins and nodularins. Toxicon 32(12):1495–1507

An J, Carmichael WW (1994b) Use of a colorimetric protein phosphatase inhibition assay and enzyme linked immunosorbent assay for the study of microcystins and nodularins. Toxicon 32: 1495–1507

Eloranta JJ, Kullak-Ublick GA (2005) Coordinate transcriptional regulation of bile acid homeostasis and drug metabolism. Arch Biochem Biophys 433(2):397–412

Gągała-Borowska I, Karwaciak I, Jaros D, Ratajewski M, Kokociński M, Jurczak T, Remlein B, Rudnicka K, Pułaski Ł, Mankiewicz-Boczek J (2022) Cyanobacterial cell-wall components as emerging environmental toxicants-detection and holistic monitoring by cellular signalling biosensors. Sci Total Environ 807:150645

Hoffmann A, Baltimore D (2006) Circuitry of nuclear factor κB signaling. Immunol Rev 210(1): 171–186

Ibelings BW, Backer LC, Kardinaal WEA, Chorus I (2014) Current approaches to cyanotoxin risk assessment and risk management around the globe. Harmful Algae 40:63–74

Jurczaka T, Tarczynska M, Izydorczyk K, Mankiewicz J, Zalewski M, Meriluoto J (2005) Elimination of microcystins by water treatment processes—examples from Sulejów reservoir, Poland. Water Res 39:2394–2406

Lebanov L, Lam SC, Tedone L, Sostaric T, Smith JA, Ghiasvand A, Paull B (2021) Radical scavenging activity and metabolomic profiling study of ylang-ylang essential oils based on high-performance thin-layer chromatography and multivariate statistical analysis. J Chromatogr B 1179:122861

Mankiewicz-Boczek J, Palus J, Gagała I, Izydorczyk K, Jurczak T, Dziubałtowska E, Stepnik M, Arkusz J, Komorowska M, Skowron A, Zalewski M (2011) Effects of microcystins-containing cyanobacteria from a temperate ecosystem on human lymphocytes culture and their potential for adverse human health effects. Harmful Algae 10:356–365

Mankiewicz-Boczek J, Karwaciak I, Ratajewski M, Gągała I, Jurczak T, Zalewski M, Pułaski Ł (2015) Application of cellular biosensors for detection of atypical toxic bioactivity in microcystin-containing cyanobacterial extracts. Aquat Toxicol 168:1–10

Michelini E, Cevenini L, Calabretta MM, Spinozzi S, Camborata C, Roda A (2013) Field-deployable whole-cell bioluminescent biosensors: so near and yet so far. Anal Bioanal Chem 405(19):6155–6163

O'Brian J, Wilson I, Orton T, Pognan F (2000) Investigation of the Alamar blue (resazurin) fluorescent dye for the assessment of mammalian cell cytotoxicity. Eur J Biochem 267:5421–5426

Oehrle SA, Southwell B, Westrick J (2010) Detection of various freshwater cyanobacterial toxins using ultra-performance liquid chromatography tandem mass spectrometry. Toxicon 55(5): 965–972

Palus J, Dziubaltowska E, Stanczyk M, Lewinska D, Mankiewicz-Boczek J, Izydorczyk K, Bonislawska A, Jurczak T, Zalewski M, Wasowicz W (2007) Biomonitoring of cyanobacterial blooms in polish water reservoir and the cytotoxicity and genotoxicity of selected cyanobacterial extracts. Int J Occup Med Environ Health 20(1):48–65

Ratajewski M, Walczak-Drzewiecka A, Sałkowska A, Dastych J (2011) Aflatoxins upregulate CYP3A4 mRNA expression in a process that involves the PXR transcription factor. Toxicol Lett 205:146–153

Svirčev Z, Drobac D, Tokodi N, Mijović B, Codd GA, Meriluoto J (2017) Toxicology of microcystins with reference to cases of human intoxications and epidemiological investigations of exposures to cyanobacteria and cyanotoxins. Arch Toxicol 91(2):621–650

Vesterkvist PS, Misiorek JO, Spoof LE, Toivola DM, Meriluoto JA (2012) Comparative cellular toxicity of hydrophilic and hydrophobic microcystins on Caco-2 cells. Toxins 4(11):1008–1023

Wagner W, Sachrajda I, Pułaski L, Hałatek T, Dastych J (2011) Application of cellular biosensors for analysis of bioactivity associated with airborne particulate matter. Toxicol In Vitro 25:1132–1142

Welker M, Fastner J, Erhard M, von Döhren H (2002) Applications of MALDI-TOF MS analysis in cyanotoxin research. Environ Toxicol 17(4):367–374

Nanosensor Devices on the Detection of Cyanotoxin

64

Charu Gupta, Mahendra K. Gupta, Mir Sajad Rabani, and Madhu Gupta

Abstract

Cyanotoxins are toxic substances produced by harmful cyanobacterial algal bloom. These harmful algal blooms pose detrimental effects to animals and human beings. There are four common types of cyanotoxins—cylindrospermopsin, saxitoxin, anatoxin-a, and microcystin-LR. Microcystin-LR (MC-LR) is one of the most hazardous cyanotoxins in this group. These toxic compounds enter into the human body and animals through contaminated water and causes problems like kidney failure, liver failure, hepatocirrhosis, and disruption with hormones and eventually resulting into death. Hence, it becomes important to develop methods to detect cyanotoxins in water samples. Although various technologies such as liquid chromatography-mass spectroscopy (LC-MS), high-performance liquid chromatography (HPLC), and enzyme-linked immunosorbent assay (ELISA) have been developed for the detection of MC-LR and provide accurate results. But these technologies require massive assembly of instrumentation and complex operation system which could be the obstacle for field-readiness and portability of the detection system. Hence, in order to achieve the goal, biosensors have been designed to detect cyanotoxins in the water system. Meanwhile, the introduction of nanomaterials to the biosensors is a better option to construct an improved quality biosensor which provides improved sensitivity and high performance. The current literature provides a brief detail of detection of harmful cyanotoxins with the advancement of nanobiosensors.

C. Gupta
School of Studies in Microbiology, Jiwaji University, Gwalior, Madhya Pradesh, India

M. K. Gupta (✉) · M. S. Rabani · M. Gupta
Microbiology Research Lab, School of Studies in Botany, Jiwaji University, Gwalior, Madhya Pradesh, India

N. Thajuddin et al. (eds.), *Protocols for Cyanobacteria Sampling and Detection of Cyanotoxin*, https://doi.org/10.1007/978-981-99-4514-6_64

Keywords

Cyanotoxins · Cyanobacteria · Nanobiosensors · Cyanobacterial algal bloom · Microcystin-LR

64.1 Introduction

Cyanobacteriaare known to produce harmful algal blooms. These cyanobacterial algal blooms are noxious and are found in different aquatic environs (Heisler et al. 2008; Anderson 2009). The enrichment of minerals in an aquatic ecosystem results into eutrophication and due to which the algal blooms are formed. These blooming cyanobacteria are responsible to produce cyanotoxins. Cyanotoxins are the secondary metabolites and distinct class of chemical compounds such as lipopolysaccharides, peptides, and alkaloids. Cyanotoxins have been classified on the basis of affected target organs in animals.

(a) **Hepatotoxins**— Cylindrospermopsin, microcystins, nodularin
(b) **Neurotoxins**—Saxitoxins, homoanatoxin-a, anatoxin-a, anatoxin-a (S), and neosaxitoxins
(c) **Cytotoxin** s—Lingbyatoxin, aplysiatoxin, lipopolysaccharide endotoxin, and debromoaplysiatoxin
(d) Skin and some gastrointestinal toxins (Kaebernick and Neilan 2001; Briand et al. 2003; Falconer 2008).

Humans are particularly vulnerable to the risks posed by cyanotoxins after ingesting them orally, through the consumption of food or drinking food (Svircev et al. 2007). The most frequent disorders brought on by ingesting these blooms are hepatotoxicosis and neurotoxicosis (Haider et al. 2003). Following direct contact with water from sources that include cyanobacterial blooms, numerous health issues could occur: abdominal pain, headache, vomiting, diarrhoea, nausea, drowsiness, exhaustion, asthma attacks, irritation of the skin and mucous membranes of the eyes, nose, and throat, trembling in the muscles of the hands and feet, dizziness, fever, paralysis, hypoxia, blurred vision, cyanosis, and cardiac arrest that results in death (Chorus and Bartram 1999; Svircev et al. 2011). Therefore, there is an urgent need to detect these noxious algal blooms in the water system to protect human health. Although a number of techniques for cyanobacteria identification and quantification have been developed include high-performance liquid chromatography, gas chromatography, mass spectroscopy, etc. However, all of these methods make it simple to monitor and measure cyanotoxins in water samples. Contrarily, the sensitivity, specificity, and repeatability of such measurements are undeniable; nonetheless, the deployment of these approaches is primarily constrained by the time and money they demand, as well as the expensive equipment and specialized personnel they require (Chawla et al. 2018). Therefore, it is imperative that systems for detecting such water contaminants are simple, rapid, and practical from an economic standpoint.

Nanobiosensors could be a promising device for fast and field-based analysis (Li et al. 2011; Kumar et al. 2016; Bala et al. 2017).

Nanosensors are the emerging nanoscale sensing device to recognize a biological component, environmental circumstance, or specific molecule. These sensors are far more specialized, portable, affordable, and sensitive than their macroscale counterparts. The three fundamental parts of a typical nanosensor device operation are as follows:

1. **Preparation of sample:** It should be a homogeneous or intricate mixture of solid, liquid, or gas phases. Sensors can target the sample that include specific functional groups of compounds, chemicals, or organisms. These targeted molecules are referred as analytes which includes molecules, biomolecules, ions, gases, organisms, and environment.
2. **Recognition:** Specific molecules or substances can identify the analytes present in the sample. These recognition molecules include enzymes, antibodies, chemical legends, and aptamers have strong affinities, specificities, and selective properties to their analytes that enable their quantification to acceptable levels.
3. **Signal transduction:** Using specific signal transduction techniques, these simple devices have been divided into several different types, including electrochemical, optical, pyroelectric, gravimetric electronic, and piezoelectric biosensors. Recognition events are transformed into computable signals, which are then further processed to provide the data.

The current literature discusses methods for developing and evaluate nanobiosensors to identify MC-LR (Microcystin-LR) in cyanobacteria culture-*Microcystis aeruginosa* (Sharma et al. 2021).

64.2 Materials Required

- In 0.01 M phosphate buffer saline, solutions of microcystin and nanomaterial-based aptamer (5' GGC GCC AAA CAG GAC CAC CAT GAC AAT TAC CCA TAC CAC ATT ATG CCC CAT CTC CGC 3′) were prepared.
- In 0.5 M carbonate buffer solution, monoclonal MC-LR antibody was prepared. At 4 °C, the solutions were kept.
- Epichlorohydrin and Coomassie brilliant blue R-250.
- Sephadex G-50 medium granules. All the chemicals used were of analytical grade.
- Buffer solutions were prepared using Milli-Q.

64.3 Procedure

64.3.1 Preparation of Activated Microgranules

The Sephadex G-50 medium granules activates using epichlorohydrin. The antibody is covalently linked to the epoxy carbon of the epichlorohydrin via the amino group. Firstly, 47 mg of granules in 1 mL of water is allowed to swell for an entire night at 4 °C. Following swelling, 400 μL of 0.1 M–1 M NaOH solution and 100 μL of epichlorohydrin is added to activate the granules. The mixture is then shaken for 3 h at room temperature. Following that, the beads are rinsed two times with water and once with a 0.5 M carbonate buffer (pH 9.5). After each washing procedure, the mixture is centrifuged at 2450 × g for 5 min to separate the beads. The amount of MC-LR antibody injects, ranged from 0.5 to 2 mL, with the antibody concentration ranging from 10–350 μg mL^{-1} in a 0.5 M carbonate buffer (pH 9.5). The mixture is incubated for 24 h at room temperature on a shaker. The suspension is incubated, centrifuged (5 min at 2450 × g), and then rinsed with 0.5 M carbonate buffer once (pH 9.5). The addition of an ethanolamine solution (85 μL/mL in 0.5 M carbonate buffer) blocks free binding sites on the surface of the granules. At room temperature, the mixture is shaken for 2 h. The suspension is centrifuged for 5 min at 2450 × g, washed two times with water and multiple times in PBS buffer and then stored in PBS buffer at 4°C.

With the help of two distinct techniques, the yield of the associated antibody on the granules is assessed. Prior to adding ethanolamine, it is visually checked by adding 0.1% Coomassie brilliant blue R-250 (99%) to 30 μL of granules. It is determined if the granules turned blue, showing that bound protein is present on the granules. Additionally, both before and after the antibody attachment process, the protein content of the antibody solution is measured spectrophotometrically. By measuring the sample protein content at 280 nm and using the absorption coefficient of 1.37 for IgG ($\varepsilon^{1\%}$), the concentration is computed.

64.3.2 Carrying out Measurements with Nanobiosensor

A sliding cap is used to partially seal the BIA system's outflow channel before 20 μL of bioactivated microgranules are injected into the measuring cell at a flow rate of 1 μL s^{-1} to create a microcolumn. For the column to be packed, 30 μL of PBS buffer are supplied at a flow rate of 2 μL/s. At a flow rate of 1 μL s^{-1}, a sample containing 150 μL of MC-LR is introduced. The flow is then halted, and the system is incubated for 30 min. To get rid of the unbound toxin, 150 μL of PBS are poured into the measurement cell at a flow rate of 2 μL s^{-1}. At a flow rate of 1 μL s^{-1}, 30 μL of MC-LR nanomaterial-based aptamer that has been labelled with a fluorescent marker is then added, and the mixture is incubated for 30 min. The marker's concentration ranges from 0.5 to 5 g mL^{-1}. 350 μL of PBS are added to the microcolumn at a flow rate of 1 μL s^{-1} in order to eliminate the unbound aptamer. The cap is opened and the apparatus rinses with PBS buffer at least four times after each measurement. The system's background signal is experimentally determined using PBS buffer without the addition of MC-LR.

A triplicate of each measurement is performed. At room temperature, all measurements are performed. Fluorescence intensity is measured at 670 nm. The signal after removing the unbound MC-LR is subtracted from the final signal to determine the signal change (signal after washing off unbound aptamer, recording started 5 min after completion of aptamer wash). The signal is captured every 1 s. To lower experimental noise after stabilization, the mean signal is determined as the average of 100 points.

64.3.3 Cultivation and Preparation of Cyanobacteria Sample

Microcystis aeruginosa (Norwegian Culture Collection of Algae, K-0540) cells are cultivated for 14 days at 16 °C under artificial light after being inoculated into 50 mL of liquid-sterilized BG11+ media [47]. The light source is an LED bulb (16 h white/ 8 h dark, 6 W 3000 K) that is kept 20 cm away from the culture jar. At −20 °C, the cells are kept safe.

The samples are concentrated after thawing. Five minutes at 10,000 × g of centrifugation is repeated to minimize the sample volume. In order to disrupt the cyanobacterial cells, an ultrasonic probe sonicator is employed for 1 min at a cycle intensity of 7/10 and a power of 75%.

64.3.4 The Characterization of the Formation of MC-LR Complexes with Size-Exclusion Chromatography (SEC)

The formation of MC-LR complexes with antibody-aptamer is investigated with liquid chromatography. Sephacryl S-200 HR gel is put into the column, then pre-expanded in PBS buffer (pH 7.2) overnight to ensure better quality packing. For estimation, the flow rate (0.18–0.39 mL min^{-1}), sample concentration (0.05–1 mg mL^{-1}) and sample volume (50 and 100 μL) are all optimized. Different proteins with molar weights ranging from 20 to 240 kDa are used to calibrate the column. The total volume of the column and the void volume are calculated using dextran blue (2000 kDa) and potassium dichromate (294 Da). At 8 °C, all calibrations, measurements, and optimizations are carried out.

The studies are conducted with a sample volume of 50 μL, an optimal flow rate of 0.18 mL/min, and a flushing solution of 70 mL of PBS buffer on the column (pH 7.2). After preparing the nanomaterial-based aptamer-microcystin mixture, nanomaterial-based aptamer is incubated with MC-LR for 30 min (1:1 molar ratio). The antibody is incubated with microcystin for 30 min before being re-incubated with the aptamer for an additional 30 min to create the nanomaterial-based aptamer-microcystin-antibody mixture (1:2:43 molar ratio).

64.4 Observation and Results

The cyanobacterial toxin MC-LR is intended to be detected using a nanomaterial-based aptasensor. It takes hardly 75 min to analyse MC-LR. The WHO permitted MC-LR limit of 1 g L^{-1} in drinking water. A cyanobacterial culture's MC-LR concentration is measured using the MC-LR nanomaterial-based aptasensor. The aptasensor utilizing nanomaterials is considered suitable for use in natural waterbodies for water quality monitoring since its sensitivity allows it to detect MC-LR in samples that contain algae (Rohtla et al. 2022).

References

Anderson DM (2009) Approaches to monitoring, control and management of harmful algal blooms (HABs). Ocean Coast Manag 52:342–347

Bala R, Dhingra S, Kumar M et al (2017) Detection of organophosphorus pesticide—malathion in environmental samples using peptide and aptamer based nanoprobes. Chem Eng J 311:111–116. https://doi.org/10.1016/j.cej.2016.11.070

Briand JF, Jacquet S, Bernard C, Humbert JF (2003) Health hazards for terrestrial vertebrates from toxic cyanobacteria in surface water ecosystems. Vet Res 34:361–377

Chawla P, Kaushik R, Shiva Swaraj VJ, Kumar N (2018) Organophosphorus pesticides residues in food and their colorimetric detection. Environ Nanotechnol Monit Manag 10:292–307

Chorus I, Bartram J (1999) Toxic cyanobacteria in water: a guide to their public health consequences, monitoring, and management. E & FN Spon, London

Falconer IR (2008) Health effects associated with controlled exposures to cyanobacterial toxins. In: Hudnell KH (ed) Cyanobacterial harmful algal blooms: state of the science and research needs. Springer, New York, pp 607–612

Haider S, Naithani V, Viswanathan PN, Kakkar P (2003) Cyanobacterial toxins: a growing environmental concern. Chemosphere 52:1–21

Heisler J, Glibert PM, Burkholder JM, Anderson DM, Cochlan W, Dennison WC, Dortch Q, Gobler CJ, Heil CA, Humphries E (2008) Eutrophication and harmful algal blooms: a scientific consensus. Harmful Algae 8:3–13

Kaebernick M, Neilan BA (2001) Ecological and molecular investigations of cyanotoxin production. FEMS Microbiol Ecol 35:1–9

Kumar N, Kumar H, Mann B, Seth R (2016) Colorimetric determination of melamine in milk using unmodified silver nanoparticles. Spectrochim Acta Part A Mol Biomol Spectrosc 156:89–97

Li H, Guo J, Ping H et al (2011) Visual detection of organophosphorus pesticides represented by methamidophos using Au nanoparticles as colorimetric probe. Talanta 87:93–99

Rohtla R, Kivirand K, Jogi E, Rinken T (2022) Biosensor for the detection of cyanobacterial toxin microcystin-LR. In: Biotechnology-Biosensors, Biomaterials and Tissue Engineering. IntechOpen, Rijeka

Sharma P, Pandey V, Sharma MMM, Patra A, Singh B, Mehta S, Husen A (2021) A review on biosensors and nanosensors application in agroecosystems. Nanoscale Res Lett 16:1–24

Svircev Z, Simeunovic J, Subakov-Simic G, Krstic S, Vidovic M (2007) Freshwater cyanobacterial blooms and cyanotoxin production in Serbia in the past 25 years. Geogr Pannonica 11:32–38

Svircev Z, Baltic V, Miladinov-Mikov M, Sajenkovic D (2011) Cyanobacterial blooms in Serbia—epidemiological studies and health risk assessment. In: 16th Academy of Studenica–International Scientific Conference "Cyanobacteria and human health". Novi Sad. Abstract Book, p 34

In Silico Insights into the Cyanobacterial Genomes to Reveal Their Metabolic Interaction

65

Gargi Sen, Sandipan Ghosh, Indrani Sarkar, and Arnab Sen

Abstract

Microbial interactions are very complex and dynamic in nature. For instance, there are inter-species as well as intra-species interactions along with the interaction between the microbes and nature. Traditional methods have several limitations in deciphering these interactions. Here we have presented an in silico-based protocol, namely "Reverse Ecology" which can depict the microbial interactions in terms of competition and complementation. The level of complexity in microbial communication is directly dependent on the metabolic convolution of the organisms. Thus, cyanobacteria, a highly versatile organism with vast distribution and assorted metabolic complications may serve as a target organism to study the inter- as well as intra-species competition and complementation. In the present study, we considered the intra-species interactions. Nevertheless, this protocol may be applied in larger populations to study their interacting mechanisms in diverse habitats.

Keywords

Cyanobacteria · Metabolic network · Metagenomics · Microbiome · Reverse ecology

G. Sen · I. Sarkar
Bioinformatics Facility, University of North Bengal, Siliguri, West Bengal, India

S. Ghosh
Department of Botany, University of North Bengal, Siliguri, West Bengal, India

A. Sen (✉)
Bioinformatics Facility, University of North Bengal, Siliguri, West Bengal, India

Department of Botany, University of North Bengal, Siliguri, West Bengal, India

Biswa Bangla Genome Centre, University of North Bengal, Siliguri, West Bengal, India

N. Thajuddin et al. (eds.), *Protocols for Cyanobacteria Sampling and Detection of Cyanotoxin*, https://doi.org/10.1007/978-981-99-4514-6_65

65.1 Introduction

Ecological interactions among organisms are highly complex and difficult to decrypt. However, unveiling these interactions is crucial for understanding the role of the organisms in the environment. Among all, microbial interactions are prominent of all interactions. Classical ecological methods have limitations in studying the complex interactions of the microbiome. With the advent of genomics and next-generation sequencing, it is possible to collect information about the microbial genomes present in the ecosystem where the traditional approach give access on a small scale (Egan et al. 2012). Metagenomics aids in understanding the species diversity of the ecosystem, but it is inadequate in analyzing the microbial interactions among themselves and their interaction with the host. This lacuna was addressed with a novel approach that is efficient in determining the metabolic interactions among the species directly from genomic information. The current approach is now widely implemented to study microbial interactions, ecological inquest and to predict the traits among the microbial communities. Various network-based reverse ecology tools, such as Net Seed (Carr and Borenstein 2012) and Net Cooperate (Levy et al. 2015), are available for studying the species' interactions with other species and the environment. However, these tools have very less applications as they could not execute the metabolic reconstruction of the species. Another free software RevEcoR has become popular as it is user-friendly, and aids in predicting the role of the microbial community, their interactions, etc. (Levy and Borenstein 2014; Borenstein et al. 2008) from high-throughput genomic data. Here we present a protocol for studying the interactions of cyanobacteria among them which is also applicable in studying their interactions with the habitat and host using this graph-theory-based algorithm called the reverse ecology approach. RevEcoR is open-source software for the study of microbial community ecology. The RevEcoR package is freely available under the GNU General Public License v. 2.0 (http://cran.r-project.org/web/packages/RevEcoR/) with the vignette and typical usage examples and the interactive Shiny web application is available at http://yiluheihei.shinyapps.io/shiny-RevEcoR, or can be installed locally with the source code accessed from https://github.com/yiluheihei/shiny-RevEcoR.

Cyanobacteria are a photosynthetic, Gram-negative, morphological, and ecologically diverse group of ancient bacteria. They are the primary producers in most of the phototrophic communities and play a crucial role in the ecosystem (Abed et al. 2009). The insights into their diversity and metabolic networks are vital for harnessing their potential for the benefit of mankind. It is nearly impossible to characterize the species in laboratory conditions. So a genomic approach is more feasible to elucidate their interactions in the microbial community and the environment.

65.2 Materials

The KO ID of the selected genomes can be downloaded from the following databases:

JGI IMG Integrated Microbial Genomes & Microbiomes (https://img.jgi.doe.gov)
NCBI (https://www.ncbi.nlm.nih.gov)
KEGG: Kyoto Encyclopedia of Genes and Genomes (https://www.genome.jp›kegg)

65.3 Methods

65.3.1 Downloading the KO Id

1. When downloading from the KEGG database, the organism code has to be entered from the KEGG organism directory. Forexample, *Microcoleus vaginatus* PCC 9802 the code is mvag.
2. Click on the genome Id (e.g., mvag—T08026) and select the KEGG genes.
3. The selected genes along with the KO Id are copied to the notepad and then to excel. The downloaded data is modified by removing the genes with no KO Id.

65.3.2 Reverse Ecology Program

1. Upload the KO Id of all genomes to be studied to the reverse ecology server—RevEcoR/shiny-RevEcoR.
2. Click on species interactions and compute interactions for competitive and complementary indexes.
3. Download the results that are available in excel format and analyze them.

The flow chart for the methodology is shown in the schematic diagram.

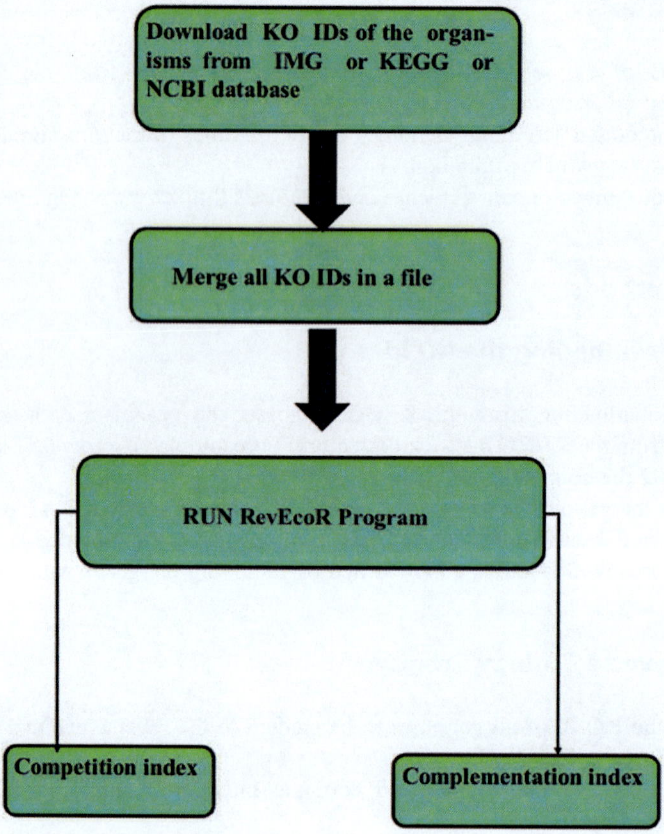

The reverse ecology studies of ten cyanobacterial genomes have been done. For this purpose, the genomes of the following cyanobacteria are downloaded from the KEGG database.

S.NO	Name of the organism	KEGG organism code
1	*Microcoleus vaginatus* PCC 9802	mvag
2	*Microcystis aeruginosa* NIES-843	mar
3	*Chondrocystis* sp. NIES-4102	chon
4	*Anabaena cylindrica* PCC 7122	acy
5	*Nostoc punctiforme* PCC 73102	npu
6	*Trichormus variabilis* ATCC 29413	ava
7	*Phormidium* sp. PBR-2020	phor
8	*Oscillatoria acuminata* PCC 6304	oac
9	*Leptolyngbya* sp. PCC 7376	lep
10	*Geminocystis* sp. NIES-3709	gen

The KEGG Id are downloaded and uploaded as per the protocol mentioned above.

65.4 Observations

The microbial interactions in terms of competition and complementation have been shown in Tables 65.1 and 65.2. A heat map has been generated for the same.

Table 65.1 Complementary index

	mvag	acy	chon	gen	lep	mar	npu	oac	phor	ava
mvag	0	0.0454545	0	0.0454545	0.0454545	0	0	0	0.0454545	0
acy	0.0384615	0	0	0.0384615	0	0.0384615	0	0	0.1538462	0
chon	0.0384615	0.0384615	0	0	0	0	0	0	0.0769231	0
gen	0.1	0	0.1	0	0.1	0.1	0	0.1	0.3	0.1
lep	0.25	0	0	0	0	0	0	0	0	0
mar	0	0	0	0	0	0	0	0	0.2	0
npu	0.0666667	0.0666667	0	0	0	0	0	0	0	0
oac	0	0	0	0	0	0.1111111	0	0	0.2222222	0
phor	0.1111111	0.4444444	0.1111111	0.1111111	0	0.2222222	0	0.1111111	0	0.2222222
ava	0	0	0	0	0	0	0	0	0	0

Table 65.2 Competition index

	mvag	acy	chon	gen	lep	mar	npu	oac	phor	ava
mvag	1	0.6818182	0.8636364	0	0	0	0.6363636	0	0	0
acy	0.5769231	1	0.6153846	0.1153846	0.0769231	0.0384615	0.5384615	0.0769231	0.0384615	0
chon	0.7307692	0.6153846	1	0.1153846	0.0384615	0.0769231	0.5769231	0.1153846	0.0769231	0.0384615
gen	0	0.3	0.3	1	0.3	0.2	0	0.3	0	0
lep	0	0.5	0.25	0.75	1	0.25	0	0.25	0.25	0
mar	0	0.1	0.2	0.2	0.1	1	0	0.2	0	0
npu	0.9333333	0.9333333	1	0	0	0	1	0	0.0666667	0
oac	0	0.2222222	0.3333333	0.3333333	0.1111111	0.2222222	0	1	0	0
phor	0	0.1111111	0.2222222	0	0.1111111	0	0.1111111	0	1	0.1111111
ava	0	0	0.25	0	0	0	0	0	0.25	1

	mvag	acy	chon	gen	lep	mar	npu	oac	phor	ava
mvag	0.00	0.05	0.00	0.05	0.05	0.00	0.00	0.00	0.05	0.00
acy	0.04	0.00	0.00	0.04	0.00	0.04	0.00	0.00	0.15	0.00
chon	0.04	0.04	0.00	0.00	0.00	0.00	0.00	0.00	0.08	0.00
gen	0.10	0.00	0.10	0.00	0.10	0.10	0.00	0.10	0.30	0.10
lep	0.25	0.00	0.00	0.00	0.00	0.00	0.00	0.00	0.00	0.00
mar	0.00	0.00	0.00	0.00	0.00	0.00	0.00	0.00	0.20	0.00
npu	0.07	0.07	0.00	0.00	0.00	0.00	0.00	0.00	0.00	0.00
oac	0.00	0.00	0.00	0.00	0.00	0.11	0.00	0.00	0.22	0.00
phor	0.11	0.44	0.11	0.11	0.00	0.22	0.00	0.11	0.00	0.22
ava	0.00	0.00	0.00	0.00	0.00	0.00	0.00	0.00	0.00	0.00

0	0.1	0.5

Heat map for complementary interaction

	mvag	acy	chon	gen	lep	mar	npu	oac	phor	ava
mvag	1.00	0.68	0.86	0.00	0.00	0.00	0.64	0.00	0.00	0.00
acy	0.58	1.00	0.62	0.12	0.08	0.04	0.54	0.08	0.04	0.00
chon	0.73	0.62	1.00	0.12	0.04	0.08	0.58	0.12	0.08	0.04
gen	0.00	0.30	0.30	1.00	0.30	0.20	0.00	0.30	0.00	0.00
lep	0.00	0.50	0.25	0.75	1.00	0.25	0.00	0.25	0.25	0.00
mar	0.00	0.10	0.20	0.20	0.10	1.00	0.00	0.20	0.00	0.00
npu	0.93	0.93	1.00	0.00	0.00	0.00	1.00	0.00	0.07	0.00
oac	0.00	0.22	0.33	0.33	0.11	0.22	0.00	1.00	0.00	0.00
phor	0.00	0.11	0.22	0.00	0.11	0.00	0.11	0.00	1.00	0.11
ava	0.00	0.00	0.25	0.00	0.00	0.00	0.00	0.00	0.25	1.00

0.00	0.50	1.00

Heat map for competition interaction

The analysis of the obtained data showed that competition ranges from 0 to 1 and complementation ranges from 0 to 0.4. The competition index within the genome is high and is denoted as one. The competition between the *Nostoc* and *Microcoleus, Anabena* is very high (0.933), and this may be attributed due to their same ecological and metabolic requirements. Similarly, the complementation within the genomes is mostly nil which is evident from the index value of zero. The highest complementation value is obtained for the interaction between *Phormidium* and *Anabaena* in the present study indicating high cooperation among them. Reverse ecology studies give an insight into the interactions among the genomes and this may be utilized for studying complex microbial interactions.

References

Abed RM, Dobretsov S, Sudesh K (2009) Applications of cyanobacteria in biotechnology. J Appl Microbiol 106(1):1–12

Borenstein E, Kupiec M, Feldman MW, Ruppin E (2008) Large-scale reconstruction and phylogenetic analysis of metabolic environments. Proc Natl Acad Sci 105(38):14482–14487

Carr R, Borenstein E (2012) NetSeed: a network-based reverse-ecology tool for calculating the metabolic interface of an organism with its environment. Bioinformatics 28(5):734–735

Egan AN, Schlueter J, Spooner DM (2012) Applications of next-generation sequencing in plant biology. Am J Bot 99(2):175–185

Levy R, Borenstein E (2014) Metagenomic systems biology and metabolic modeling of the human microbiome: from species composition to community assembly rules. Gut Microbes 5(2): 265–270

Levy R, Carr R, Kreimer A, Freilich S, Borenstein E (2015) NetCooperate: a network-based tool for inferring host-microbe and microbe-microbe cooperation. BMC bioinformatics 16(1):1–6

CRISPR Gene Finding in the Genome of *Oscillatoria* sp. and *Lyngbya* sp.

66

Fayaazuddin Thajuddin, Peer Muhamed Noorani, Upaasna Sunilkumar, and N. Thajuddin

Abstract

In order to create higher quality strains, genetic engineering of cyanobacteria and microalgae is a crucial method. It enhances biomass productivity, yield, polyunsaturated fatty acid, pigment, etc. required for commercial production of biomass. A variety of techniques are available for genome editing, which involves adding, removing, or knocking out certain targeted DNA regions in the chromosome. For most genome modifications, the CRISPR/Cas system has been used. We present a method used to analyze the presence of CRISPR genes in the whole-genome sequence of *Oscillatoria* and *Lyngbya* using CRISPRcasFinder.

Keywords

Cyanobacterial chromosome · Whole-genome sequences · Oscillatoria acuminata · CRISPRcasFinder · CRISPR genes

---

F. Thajuddin · U. Sunilkumar
School of Life Sciences, Bharathidasan University, Tiruchirappalli, Tamil Nadu, India

P. M. Noorani (✉)
Department of Microbiology, School of Life Sciences, Bharathidasan University, Tiruchirappalli, Tamil Nadu, India

N. Thajuddin
B.S. Abdur Rahman Crescent Institute of Science & Technology (Deemed to be University), Chennai, Tamil Nadu, India

Department of Microbiology, National Repository for Microalgae and Cyanobacteria - Freshwater (NRMC-F), School of Life Sciences, Bharathidasan University, Tiruchirappalli, Tamil Nadu, India

N. Thajuddin et al. (eds.), *Protocols for Cyanobacteria Sampling and Detection of Cyanotoxin*, https://doi.org/10.1007/978-981-99-4514-6_66

66.1 Introduction

It is essential to develop an alternative energy source in order to address the current problems of the world's population explosion, obstructive water pollution, energy shortages, and food scarcity. Microalgal biofuels are excellent renewable energy sources that have the ability to meet all of the world's energy needs. Microalgal cell especially green algae *Chlorella* sp. contains remarkable amount of lipid content, which facilitates the biofuel production (Praveenkumar et al. 2012; Arutselvan et al. 2021, 2022; Baldev et al. 2021). In addition to bioenergy, the bioactive compounds extracted from microalgae play a vital role in production of high value-added products in nutraceutical and pharmaceutical industries (Davoodbasha et al. 2018; Lewis Oscar et al. 2021; Marappa et al. 2020). Understanding their genetic and molecular-biological information in depth is crucial for full utilization of microalgae and the development of more affordable manufacturing procedures. The development of higher quality strains with increased biomass productivity, yield, polyunsaturated fatty acid, pigment, etc., necessary for commercial biomass production, is facilitated by genetic engineering of microalgae.

A variety of techniques are available for genome editing, which involves adding, removing, or knocking out certain targeted DNA regions in the chromosome (Gaj et al. 2013). Current technologies for genome editing includes clustered regularly interspaced short palindromic repeats (CRISPR)/Cas nucleases, Zinc finger nucleases (ZFNs), Meganucleases (MNs) and Transcriptional activator-like effector nucleases (TALENs) (Zhang et al. 2019). CRISPR/Cas system has been generally employed for modification of the genome. It has developed into a key tool for editing target genes to increase the production of biochemicals in microalgae. For example, this system has been used with the following organisms: *Chlamydomonas reinhardtii*, *Synechocystis* sp., *Synechococcus* sp., etc. (Jiang et al. 2014; Yao et al. 2016; Markley et al. 2015). The substantial rise in specific CRISPR arrays also provides their effective application in bacterial isolates (Grissa et al. 2009). Various systems have been designed to recognize CRISPR arrays in genomic sequences. The most repeatedly cited are CRISPRFinder, PILER-CR, and CRT (Grissa et al. 2007; Bland et al. 2007; Edgar 2007). The CRISPRCasFinder is an advanced and combined version of CasFinder and CRISPRFinder with publicly accessible third-party software dependencies. CRISPRCasFinder currently incorporates a standalone version and presents improved CRISPR detection performance.

66.2 Materials Required

1. CRISPRcasFinder.
2. Whole-genome sequences of *Oscillatoria* sp. and *Lyngbya* sp.
3. National Center for Biotechnology Information (NCBI) database.

66.3 Procedure

1. To access NCBI use the following link—https://www.ncbi.nlm.nih.gov/
2. Search genome sequence for *Oscillatoria* sp. in NCBI and download FASTA sequence.
3. Open CRISPRcasFinder [Online] using following link https://crisprcas.i2bc.paris-saclay.fr/CrisprCasFinder/Index.
4. Upload the FASTA sequence of *Oscillatoria* sp. in Choose file option.
5. Click Run CRISPRcasFinder at the bottom of webpage.
6. It will take 2–5 min to complete the process.
7. Observe and analyze the presence of CRISPR genes in the whole-genome sequence of *Oscillatoria* sp.
8. Same protocol is followed for *Lyngbya* sp. using CRISPRcasFinder.

66.4 Conclusion

Among the available CRISPR finding tools, the CRISPRcasFinder not only provides a unique addition, but also earns reputation for very high accuracy in its individual CRISPR annotations. *Oscillatoria* sp. and *Lyngbya* sp. were selected to study the presence of CRISPR genes in their whole-genome sequences. Out of these two cyanobacterial species, the presence of CRISPR genes is substantially higher in *Oscillatoria* sp. linked with 80 CRISPR genes, while *Lyngbya* sp. exhibited 1 CRISPR gene in its whole-genome sequence. The heritable defense offered by the adaptive CRISPR-Cas immune system takes the form of spacers, which are short nucleic acid sequences (28–36 bp) acquired through prior interactions with mobile genetic elements (MGEs). These are stored in the cyanobacterial chromosome in CRISPR arrays (Jackson et al. 2017). Presence of spacers differ on each cyanobacterial whole-genome sequence and the CRISPR locus with higher number of spacers in *Oscillatoria* sp. and *Lyngbya* sp. are depicted in Table 66.1.

Table 66.1 CRISPR locus with higher number of spacers in *Oscillatoria* sp. and *Lyngbya* sp.

S. No	CRISPR ID	Starting base pair of CRISPR gene	Ending base pair of CRISPR gene	Spacers count
Oscillatoria sp.				
1	NC0196931 Oscillatoria acuminata PCC6304 complete sequence 9	475,830	476,937	14
2	NC0196931 Oscillatoria acuminata PCC6304 complete sequence 28	3,497,270	3,497,448	2
3	NC0196931 Oscillatoria acuminata PCC6304 complete sequence 39	4,394,995	4,398,974	54
4	NC0196931 Oscillatoria acuminata PCC6304 complete sequence 54	5,530,351	5,530,620	4
5	NC0196931Oscillatoria acuminata PCC6304 complete sequence 65	6,251,489	6,253,282	24
6	NC0196931 Oscillatoria acuminata PCC6304 complete sequence 61	5,789,186	5,790,985	23
7	NC0196931 Oscillatoria acuminata PCC6304 complete sequence 73	7,038,938	7,043,138	57
8	NC0196931 Oscillatoria acuminata PCC6304 complete sequence 78	7,504,261	7,509,286	69
9	NC0196931 Oscillatoria acuminata PCC6304 complete sequence 53	5,519,080	5,519,338	3
Lyngbya sp.				
10	NZ_JTHE03000027.1	63,892	65,600	23

References

Arutselvan C, Narchonai G, Pugazhendhi A, Lewis Oscar F, Thajuddin N (2021) Evaluation of microalgal strains and microalgal consortium for higher lipid productivity and rich fatty acid profile towards sustainable biodiesel production. Bioresour Technol 339:125524

Arutselvan C, Narchonai G, Pugazhendhi A, Kumar Seenivasan H, Lewis Oscar F, Thajuddin N (2022) Phycoremediation of textile and tannery industrial effluents using microalgae and their consortium for biodiesel production. J Clean Prod 367:133100

Baldev E, MubarakAli D, Sivasubramanian V, Pugazhendhi A, Thajuddin N (2021) Unveiling the induced lipid production in Chlorella vulgaris under pulsed magnetic field treatment. Chemosphere 279:130673

Bland C, Ramsey TL, Sabree F, Lowe M, Brown K, Kyrpides NC, Hugenholtz P (2007) CRISPR recognition tool (CRT): a tool for automatic detection of clustered regularly interspaced palindromic repeats. BMC Bioinformatics 8(1):1–8

Davoodbasha M, Edachery B, Nooruddin T, Lee SY, Kim JW (2018) An evidence of C16 fatty acid methylesters extracted from microalga for effective antimicrobial and antioxidant property. Microb Pathog 115:233–238

Edgar RC (2007) PILER-CR: fast and accurate identification of CRISPR repeats. BMC Bioinformatics 8(1):1–6

Gaj T, Gersbach CA, Barbas CF III (2013) ZFN, TALEN, and CRISPR/Cas-based methods for genome engineering. Trends Biotechnol 31(7):397–405

Grissa I, Vergnaud G, Pourcel C (2007) CRISPRFinder: a web tool to identify clustered regularly interspaced short palindromic repeats. Nucleic Acids Res 35(suppl_2):W52–W57

Grissa I, Vergnaud G, Pourcel C (2009) Clustered regularly interspaced short palindromic repeats (CRISPRs) for the genotyping of bacterial pathogens. In: Molecular epidemiology of microorganisms. Humana Press, Totowa, NJ, pp 105–116

Jackson SA, McKenzie RE, Fagerlund RD, Kieper SN, Fineran PC, Brouns SJ (2017) CRISPR-Cas: adapting to change. Science 356(6333):eaal5056

Jiang W, Brueggeman AJ, Horken KM, Plucinak TM, Weeks DP (2014) Successful transient expression of Cas9 and single guide RNA genes in Chlamydomonas reinhardtii. Eukaryot Cell 13(11):1465–1469

LewisOscar F, Nithya C, Vismaya S, Arunkumar M, Pugazhendhi A, Nguyen-Tri P, Alharbi SA, Alharbi NS, Thajuddin N (2021) In vitro analysis of green fabricated silver nanoparticles (AgNPs) against Pseudomonas aeruginosa PA14 biofilm formation, their application on urinary catheter. Prog Org Coat 151:106058

Marappa N, Ramachandran L, Dharumadurai D, Nooruddin T (2020) Plant growth-promoting active metabolites from Frankia spp. of actinorhizal Casuarina spp. Appl Biochem Biotechnol 191(1):74–91

Markley AL, Begemann MB, Clarke RE, Gordon GC, Pfleger BF (2015) Synthetic biology toolbox for controlling gene expression in the cyanobacterium Synechococcus sp. strain PCC 7002. ACS Synth Biol 4(5):595–603

Praveenkumar R, Shameera K, Mahalakshmi G, Akbarsha MA, Thajuddin N (2012) Influence of nutrient deprivations on lipid accumulation in a dominant indigenous microalga Chlorella sp., BUM11008: Evaluation for biodiesel production. Biomass Bioenergy 37:60–66

Yao L, Cengic I, Anfelt J, Hudson EP (2016) Multiple gene repression in cyanobacteria using CRISPRi. ACS Synth Biol 5(3):207–212

Zhang YT, Jiang JY, Shi TQ, Sun XM, Zhao QY, Huang H, Ren LJ (2019) Application of the CRISPR/Cas system for genome editing in microalgae. Appl Microbiol Biotechnol 103(8): 3239–3248

Photosynthetic organisms, 18, 48, 86, 132, 162, 206, 274
Phycocyanin (PC), 8, 9, 13, 86, 128, 144, 230, 398
Phycotoxins, 162, 262
PKS genes, 376, 382–385
Planktothrix, 8, 39, 72, 73, 78, 127–130, 156, 206, 212, 286, 304, 340, 388, 391–393, 422–425, 436, 444, 446
Plant bioassays, 30, 236
Polyketide synthase (PKS) gene, 437
Polyketide synthases (PKSs), 218, 375–379, 382–385, 404, 437, 444, 470
Polymerase chain reaction (PCR), 19, 29, 54, 65, 73, 101, 134, 152, 163, 190, 219, 244, 377, 384, 388, 398, 404, 423, 438, 446, 455, 461, 470
Polymerization, 430, 431
Postmortem, 189–194
Potassium permanganate, 304
PrestoBlue assay, 255–256
Prochlorococcus, 18
Protein kinase C activators, 41, 108
Protein phosphatase inhibition assay (PPIA), 31, 150, 153, 157, 164, 213, 364–367, 437, 476
Protein phosphatase inhibitor assay, 31
Protein phosphatases (PPs), 124, 215, 254, 260, 296, 404, 422, 476
Protein phosphatases PP1/PP2A, 296
Proteinase K, 57, 190, 192–194, 221, 384, 406, 408, 423, 446, 447, 453, 455, 462, 464

Q
QPCR assay, 73, 399, 404, 407, 409–413, 416, 417

R
Raphidiopsis, 119–121, 262, 316, 341, 452
Reactive oxygen species (ROS), 254, 260
Real-time PCR, 22, 23, 134, 244, 417, 418, 438, 446, 448–449
Research Collaboratory for Structural Bioinformatics Protein Data Bank (RCSB PDB), 276
Restriction fragment length polymorphism (RFLP), 377–379
RevEcoR, 490, 491
Reverse ecology, 490–492, 495
RPMI medium, 260
RT-PCR, 472

S
Salt pan, 48, 50
Saxitoxins (STXs), 19, 28, 72–74, 79, 88, 100, 114, 206, 212, 234, 262, 267, 274, 330–333, 340, 349, 354, 355, 359, 398, 399, 429, 484
Schizothrix, 108–110
Scums, 5, 8, 14, 39, 72, 132, 304
Scytonemin, 212
SDS PAGE, 66–67
Selected Reaction Monitoring (SRM), 357, 358
Semi-permeable membrane (SPM), 289, 290
Serological methods, 228, 244
Shellfish, 4, 62, 73, 178, 206, 212, 228, 359, 360
Sinapis test, 236
Solid-phase extraction (SPE), 32, 51, 159, 245, 287, 290–292, 298, 300, 304, 305, 307, 311, 327, 337, 344, 346, 431, 478
Solid-phase extraction cartridges, 245, 304, 311
16S rDNA, 65, 68, 100
16S rRNA, 73, 103, 153, 398, 401, 408, 424
STX immunoaffinity column, 331, 332
STX Standard solutions, 333
Surface Plasmon Resonance (SPR), 365–368
sxtA, 73, 397–401
SYBR green, 73, 398, 399, 407, 412, 449
SYBR-green dye, 405
Symbiosis, 100, 218, 460

T
Tectonic estuary, 36
Therapeutic, 69, 382, 389
Tolerable daily intake (TDI), 169–171, 359
Toxic blooms, 7–15, 18–23, 28–32, 35–44, 96, 133–135, 137–139, 422, 429
Toxic cyanobacteria, 4, 5, 8, 19–21, 54–59, 100, 142–147, 153, 164, 166, 198, 388–392, 398, 438, 452–456
Toxic cyanobacterial blooms, 37, 100, 316
Toxic water blooms, 162
Toxins, 8, 28, 37, 48, 63, 72, 79, 86, 100, 109, 114, 120, 128, 150, 156, 162, 167, 183, 198, 206, 212, 219, 228, 234, 244, 260, 274, 289, 304, 309, 316, 326, 330, 344, 356, 364, 376, 382, 388, 404, 429, 436, 446, 452, 460, 470, 476, 484
Transparency, 8, 12–13, 108, 128
Trichomes, 43, 86, 97, 98, 108, 115, 119, 121, 123, 127, 128, 132, 137–139

Trifluoroacetic acid (TFA), 157, 159, 305, 311, 327, 341, 342, 345, 346, 477, 478
Tumour promotion, 29, 166, 260

U
Ultra-high-performance liquid chromatographic-tandem mass spectrometry (UHPLC-MS/MS), 354–360
Urine analysis, 201

V
Vibrio fischeri, 228, 229
Visual inspection, 19, 21–22

W
Waterbodies, 4, 5, 31, 37, 39, 62, 73, 78, 86, 166, 175, 198, 206, 296, 344
Water bloom, 4, 5, 8, 109
WC medium, 128, 129
Whole genome sequencing (WGS), 392

X
X-ray crystallography, 68, 69

Z
Z8 medium, 125
Zooplankton, 4, 39, 54, 62, 183, 228